International Conference on Ship and Offshore Technology 2009

ICSOT India 2009: Developments in Ship Design and Construction

Khararagpur, India
10-11 December 2009

ISBN: 978-1-61567-850-1

Printed from e-media with permission by:

Curran Associates, Inc.
57 Morehouse Lane
Red Hook, NY 12571

Printed by Curran Associates, Inc. (2010)

The Royal Institution of Naval Architects
10 Upper Belgrave Street
London SW1X 8BQ UK

Phone: 44 207 235 4622
Fax: 44 207 259 5912

publications@rina.org.uk

Additional copies of this publication are available from:

Curran Associates, Inc.
57 Morehouse Lane
Red Hook, NY 12571 USA
Phone: 845-758-0400
Fax: 845-758-2634
Email: curran@proceedings.com
Web: www.proceedings.com

TABLE OF CONTENTS

PREFACE

The global maritime industry today is experiencing a downturn in activity that is in marked contrast to the historically high level of activity which it had enjoyed for a number of years. Whilst a reduction in such a level of activity might well have been expected, the global financial crisis which has impacted on the maritime industry as much as any, has led to the dramatic decline from which few sectors of the industry have been immune. Even for those who have experienced the highs and lows which are a regular feature of the maritime industry, the speed and depth of the current decline has perhaps been beyond their experience.

At such a time of uncertainty it is understandable that for many companies, their priority has been survival today, rather than thinking about tomorrow. But even at such a difficult time, it is important and necessary that companies in the maritime industry should look ahead if they are to be in a position to respond to the upturn in activity which surely will come, and to meet the challenges which they will face. Such challenges will come from the increasing demands of operators, regulators and society for greater efficiency, safety and the protection of the environment. Meeting these challenges will require innovative thinking from those in all sectors of the industry, and particularly those involved in ship design and construction.

When the Royal Institution of Naval Architects was founded in 1860, its mission statement was to promote and facilitate the exchange and discussion of scientific and technical information. That mission statement remains equally true 150 years later in 2010, when it is achieved through the Institution's international publications and conferences such as the first *International Conference on Ship & Offshore Technology – India* which takes "*Developments in Ship Design & Construction*" as its theme

The Royal Institution of Naval Architects is an international organisation with members in over 90 countries, and with links to over 60 universities. The Institution is proud to join with IIT Kharagpur in organizing its first conference in India, bringing together members of the international maritime industry to present and discuss the latest developments in ship design and construction which will provide the technological innovation which will be the key to the industry's future success.

Nisith R. Mandal
Chairman
Organizing Committee

Trevor Blakeley
Chairman
Organizing Committee

INTERNATIONAL CONFERENCE
ON
SHIP AND OFFSHORE TECHNOLOGY
(ICSOT India - 2009)

Patron: Prof. Damodar Acharya, Director, IIT Kharagpur, Kharagpur

ORGANIZING COMMITTEE

Chairmen

Prof. Nisith R. Mandal, IIT Kharagpur
Mr. Trevor Blakeley, Royal Institution of Naval Architects

Members

1. Prof. Debabrata Sen, IIT Kharagpur
2. Prof. Trilochan Sahoo, IIT Kharagpur
3. Prof. Hari V. Warrior, IIT Kharagpur
4. Prof. Ashoke Bhar, IIT Kharagpur
5. Prof. Prasad K. Bhaskaran, IIT Kharagpur
6. Prof. Nisith R. Mandal, IIT Kharagpur
7. Prof. S. K. Bhattacharjee, IIT Kharagpur
8. Prof. P.K. Das, IIT Kharagpur
9. Prof. A. Chakraborty, IIT Kharagpur
10. Prof. Suresh C.Misra, NSDRC, Visakhapatnam
11. Capt. K.J.H. Christie, GRSE, Kolkata
12. Mr. P. Barman Roy, Bharati Shipyard
13. Dr. P. Misra, MMD, Chennai

INTERNATIONAL TECHNICAL COMMITTEE

1. Prof. Yasumi Kawamura, Yokohama National University, Japan
2. Prof. Purnendu K. Das, Universities of Glasgow & Strathclyde, UK
3. Prof. Richard Lee Storch, Professor and Director, Industrial Engineering, USA
4. Prof. R. Ajit Shenoi, School of Engineering Sciences, University of Southampton, U.K.
5. Capt.K.J.H.Christie IN (Retd), General Manager (S)(D), GRSE Ltd. India
6. Prof. R.P Gokarn, India
7. Prof. Anantha Subramanian, Ocean Engineering Centre, IIT Madras, India
8. Prof. Howard Bunch, University of Michigan, USA
9. Prof. Suresh C.Misra, IIT Kharagpur
10. Prof. O.P.Sha, IIT Kharagpur, India

SPONSORS

Naval Research Board

Garden Reach Shipbuilders & Engineers Limited

Goa Shipyard Limited

Shipping Corporation of India

Indian Resister of Shipping

Napa Software Services India Pvt. Limited

Smart Engineering and Design Solutions Limited

PERFORMANCE OF INDIAN SHIPYARDS

SR. NO.	FOR THE PERIOD	ORDER BOOK (RS. IN CRORES)	TURN OVER (RS. IN CRORES)	INVESTMENT (RS. IN CRORES)
1.	BETWEEN 1997-2002 (IE. BEFORE SUBSIDY)	816	1,017	43
2.	BETWEEN 2002-2007 (IE. AFTER SUBSIDY)	14,877	3,657	843
3.	PERCENTAGE DIFF.(%)	1,723%	259%	1,860%

India China Facts

India China facts

	India	China
Area (million sq km)	3.3	9.6
Population (billion)	1.15	1.33
GDP (ppp 2007) trillion USD	3	7
GDP (per capita 2007) USD	2700	5300
Ships (>1000 GRT)	490	1820
No of Shipyards	26	1200 700 > 15000 dwt

China Market Share

Market Share

Shipbuilding Index for 2007 (China)		Global Market share
Delivery	18.93 million dwt	23%
Contracts	98.45 million dwt	42%
Order book	158.89 million dwt	33%

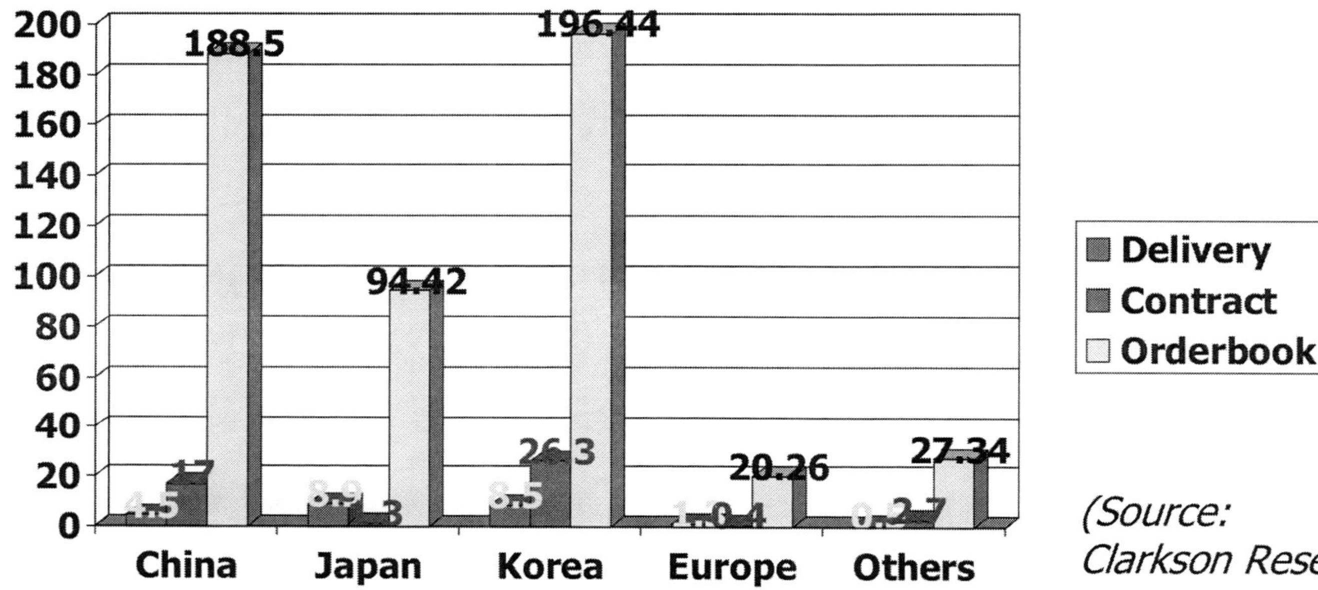

(Source: Clarkson Research)

China is a rapidly developing giant in the global shipbuilding market

EXTRACTS FROM THE NATIONAL MANUFACTURING COMPETITIVENESS CONCIL REPORTS

- MANUFACTURING HAS BEEN RECOGNISED AS THE MAIN ENGINE FOR GROWTH AND ACCORDINGLY EMPHASIS WAS PLACED ON GROWTH OF INDUSTRY IN MOST OF OUR FIVE YEAR PLANS.

- MANUFACTURING WILL REMAIN ONE OF THE PRINCIPAL BY WHICIH WEALTH IS CREATED.

- IT IS A MATTER OF CONCERN THAT THE CONTRIBUTION THAT THE CONTRIBUTION OF MANUFACTURING TO GDP HAS REMAINED STAGNANT FOR OVER TWO DECADES SINCE 1985.

- THE LOW LEVELS OF MANUFACTURING GROWTH HAD ITS ADVERSE IMPACT ON EMPLOYMENT GENERATION.

AS PER THE KMPG REPORT SUBMITTED TO THE MINISTRY IN 2007

- THE MULTIPLIER EFFECT OF GROWTH IN SHIPBUILDING CAN CREATE A TOTAL INVESTMENT OF RS. 37,000 CRORES IN THE INDIAN ECONOMY BY 2012 AND AROUND RS. 222,000 CRORES BY 2017.
- CAN INCREASE OVERALL ECONOMIC OUTPUT TO AROUND INR. 342,000 CRORES BY 2017

GLOBAL SHIPBUILDING SUBSIDIES

The growth of shipbuilding industry in almost all the shipbuilding countries has demonstrated a strong correlation with Government Support to the sector comprising Direct Subsidy alongside a number of indirect incentives. These measures have enabled new entrants like Japan in 1960s, Korea in 1970 and now China, to establish their presence and slowly dominate the global shipbuilding activity. These measures have also enabled players like Germany to maintain their share in global market. In the following section, we provide a brief outline of the direct subsidy provided by various countries to their shipbuilding sector. In addition, an overview of the indirect support made available to shipbuilding sector is also provided.

SOUTH KOREA:

Since 1960s, South Korea has promoted the growth of shipbuilding sector through a host of measures directed at creating an enabling environment including legal and policy measures, as well as a host of targeted fiscal and non-fiscal incentives. The following section captures in brief, the details on the above aspects:

•Enabling Environment:

•Shipbuilding Encouragement Act, promulgated on March 1958

•Shipbuilding Promotion Act, promulgated on March 1967

•Machinery Promotion Act, established in July 1967

•Long-term Shipbuilding Industry Promotion Plan, introduced in March 1973

•National Investment Fund (INF), created in late 1970s

•Rationalization Plan of the Maritime Industry of July 1985

•Enforcement of Rationalization of Management of Shipbuilding Industry, 1989

•Direct Subsidy

•Under the Shipbuilding Encouragement Act and Shipbuilding Promotion Act, in 1950s and 1960s, the government provided direct subsidies upto 40% of contract value till 1968, and thereafter 30%.

•Other Fiscal and Non-Fiscal Incentives

•Besides providing direct subsidy under the Shipbuilding Encouragement Act and Shipbuilding Promotion Act, the government also guaranteed loans upto 50% of the vessel construction costs. These loans carried a nominal interest rate of 5% and maturity terms of 5 to 10 years. The beneficiary was required to finance only 10% of the total construction cost with its own capital.

In 1960s, a debt guarantee scheme was introduced for lenders to shipbuilding sector. The intention was to address the lack of the main beneficiary's collateral as well as to reduce the main beneficiary's burden.

Vessel construction was funded by government-guaranteed loans during the construction period, and after completion of construction the ship was pledged as collateral for the loan. The government also reduced the beneficiary's annual burden by extending the maturity term of loans to 15 years with an annual interest rate of 5%.

In 1960's, Society of Naval Architects of Korea, a R&D institution, developed and propagated designs of ships to standardize and simplify the shipbuilding raw materials and construction process which resulted in enhancing the quality of the ships, shortening manufacturing period and reducing construction costs. In addition, shipbuilding companies with poor technologies were able to construct ships using these standard designs and thereby improved their technologies within a very short period. In addition, they were able to enhance operations with more planned organized shipbuilding. A Ship laboratory was also founded in November 1976 to introduce foreign advanced technique in shipbuilding, and thereafter, to develop such technique independently

The Machinery Promotion Act provided a legal basis for government to extend support to the machinery industry, including machinery for shipbuilding sector. The machinery industry was identified as one of the six infrastructure industry.

In 1960s, Government provide exemption on import duties for main raw materials, such as imported steel and timber that are used for shipbuilding which reduced construction costs

As part of the Long-term Shipbuilding Industry Promotion Plan, The Korea export-import bank (KEXIM), provided deferred payment loans to ship owners at a nominal 7% interest rate to promote export of ships. KEXIM also provided Korean shipyards with refund guarantees for advance payment

Presently, KEXIM provides export loan which covers up to 100% of the export contract value less the cash payment at a floating interest rate linked to LIBOR, with a repayment spread over 12 years

Also as part of the "The Planned Shipbuilding Program" the government introduced the policy of "our cargoes in our own ships". It allowed the movement of government cargo only through domestically built ships.

Also, in the late 1970's, owners of new bulk-carriers, which were built through the Planed Shipbuilding Program (PSP), were given benefits such as a right of transporting imported fertilizer for 5 years and a higher freight tariff.

In the late 1970s', the government reduced the price of steel plate and thus manufacturing cost of shipbuilding decreased as a consequence.

In 1978, the government fostered specialty companies in the area of shipping equipments for ensuring stable supply of ship equipments to shipbuilding sector.

In the late 1970s', the government created the National Investment Fund which provided funds to shipbuilding sector at 11% interest rate1980's, in order to reserve foreign currency and localize equipments used in shipbuilding, the Korean Government gave financing and tax benefits on domestic suppliers of equipments

As part of the 'The Rationalization Plan of the Maritime Industry of July 1985, the government gave tax incentives and financial support to participating companies as follows:

The maturities of loan was to deferred for 3 years.

Banks provided roll-forward of the loans

To facilitate restructuring of companies, banks provides guarantee on loans.

CHINA:

The central government of China has an aggressive policy to support the development of China's shipbuilding sector. In addition, the local governments also provide incentives and subsidies to shipbuilders.

China's shipbuilding industry is still in the developing stage. The production of new ships was 1.75 million tons in 1995 which increased to 3.46 million tons in 2000 and then jumped to 14.52 million tons in 2006 recording an annual growth of 27% over the 2000-2006 period.

Enabling Environment:

The mid-long term shipbuilding industry development plan of 2006

The nine incentives to encourage shipbuilding industry development at Liaoning Shipbuilding Industry Zone, 2003

The circular on financing of shipbuilders in Wenzhou city, 2006

The national tenth five-year development plan for shipbuilding industry, 2005

The circular on accelerating shipbuilding industry development in Zhejiang province, 2004

Direct Subsidy

•Chinese government provides subsidy to Chinese ship buyers equivalent to 17% of contract ship price if they purchase domestically produced ships.

•The central government and local government provides fixed subsidy for inland sailing ships irrespective of value of ship.

•Other Fiscal and Non – Fiscal Incentives

•Shipbuilders importing technologically advanced equipment are entitled to zero customs duty and an import VAT refund.

•China provides 17% tax refund if ships are exported.

•The exchange rate of Yuan remained unchanged in the 14 years prior to 2006, providing an effective subsidy to its shipbuilders. China's central bank started to broaden the range of the foreign exchange rate only since 2005. As a result the RMB exchange rate has climbed from RMB 8.27 yuan per US dollar in July 2005 to RMB 7.59 per US dollar in July 2007, and is still considered substantially undervalued.

•The Export-Import Bank of China provides project financing to shipbuilders and overseas buyers. The interest rate for buyers is 2.7%.

•China provides scientific research subsidy to maritime universities, and provides R&D and living allowance to postgraduate students

•Financial institutions provide working capital loans at nominal interest rates using ships under-construction as security

•Poorly performing state-owned shipbuilders can apply for conversion of debt to equity

•Chinese government provides loss reimbursement to domestic shipbuilders. This incentive only applies to major state-owned shipbuilders.

•The government encourages steel producers and shipbuilders to establish strategic alliances in order to better position themselves on the market. Some steel companies have become shareholders of shipbuilders, for instance Bao Steel invested over RMB 2 billion to purchase 18.18% of Shanghai Waigaoqiao Shipyard's shares

•Provincial Government Support

•Zhejiang province:

- Since 2004, if the R&D expenditure increased by over 10% year on year, then the shipbuilders are entitled for 50% of R&D expenditure to be tax deductible.
- Since 2006, in Wenzhou city of Zhejiang province, local government provides additional RMB 3 million to support technology advancement projects of shipbuilders
- Shipbuilders are eligible to use accelerated depreciation method in their accounting practice
- If shipbuilders use domestically produced equipment to improve their technical capability, this amount can be deducted from the company's income tax

- Jiangsu Province
    - Provincial Government provides debt guarantee schemes to financiers and shipbuilders from the province
- Liaoning Province
    - Since 2003, Newly set-up shipbuilders can enjoy 50% of VAT refund in the first two years, and 25% of VAT refund in the following three years.
    - In addition, the companies enjoy income tax holiday in the first two profitable years and pay half the income tax in the following three years.

EUROPEAN UNION, UK, GERMANY AND ITALY

Direct Subsidy

Subsidies governed by European regulation

European Commission has set the maximum level of aid to shipbuilding, as a percentage of the value of the contract. The aid ceiling is fixed by the Commission with reference to the prevailing price difference between cost structures of the most competitive yards in the Union and the prices charged by their main international competitors with particular regard to the market segments in which the EU yards remain relatively most competitive. Over the years the aid levels allowed have been lowered.

EU aid ceiling 1987-1994 as a percentage of contract value		
Year	**Value > € 6 million**	**Value < € 6 million**
1987	*28*	*20*
1988	28	20
1989	26	16
1990	20	14
	Value > € 10 million	**Value < € 10 million**
1991	13	9
1992	9	4.5
1993	9	4.5
1994	9	4.5

•The EU Council Regulation in 1998 allowed shipbuilding companies to partly benefit from state aid until 31 December 2000, for each shipbuilding contract up to an amount equal to 9% of "contract value before aid" for orders of over 10 million euros in value, and 4.5% for orders of less than 10 million euros

•20% direct shipbuilding subsidy for innovation is applicable at present in Germany, France, Italy, Spain, Norway and the Netherlands. Since the European Union are engaged in construction of high technology oriented ships, most of their work fall under the shipbuilding subsidy scheme. Moreover the innovative subsidy include both cost of the capital expenditure in the shipyard, revenue expenditure for the design, testing etc and the cost pertaining to it in construct ion of the vessel .

Direct Subsidies in UK

•The Shipbuilding Intervention Fund (SIF) subsidised many shipbuilding companies from the 1960s until 2000. The fund paid production aid to UK yards under the terms of various UK directives and between 1960 and 2000, as much as 36% of the cost of each vessel being produced was subsidised through the fund. In 1987, the sixth directive (an EU directive) placed a ceiling on both contract and overall support to each European ship builder which limited the total subsidy to not more than 28% of contract value.

Direct Subsidies in Germany

•From 1994-96, in Germany, 2.9% of total state aid went to the shipbuilding industry which reduced to 2.0% from 1997-99.

•36% subsidy was given to the erstwhile East German Shipyards after the unification of Germany.

•Germany pumped in more than $ 1 billion in the East German yards so that productivity in those yards could be brought to the level of West German Yards

Direct Subsidies in Italy

Since 1991, for the new construction of ships by national firms, a subsidy was granted on the value of the vessel, not exceeding 13% of contract value for 1991 and 9% of contract value for 1992 to 1993. This was reduced to 9% and 4.5% for contracts with a value less than 10 million euros

In 1994 there were a number of government assisted programs to aid merchant ships with gross tonnage of over 50 tons

The grant was paid after work had commenced, in instalments on the basis of the work undertaken to date

It included a maximum of 70% of the cost of the construction, transformation or modification; 80% of the cost of large-scale repair to merchant ships; and 40% of the investments made by the shipyard to improve working conditions

<u>Other Fiscal and Non – Fiscal Incentives</u>

EU aid to research and development (R&D) increased as a proportion of EU industrial support, from 7.3% in 1994-96 to 8.8% in 1997-99. This increase was particularly evident in Austria, Germany, the Netherlands and Sweden.

Aid granted for innovation in existing shipbuilding, ship repair or ship conversion yards was further enhanced in December 2005 to a maximum aid intensity of 20% gross of innovation expenditure.

- Research and development aid, allowable under the Framework on State aid to Shipbuilding (OJ 2003/C 317/06) is as follows:
    - 100% for fundamental research;
    - 50% (75% maximum if particular conditions are met) for industrial research;
    - 25% (50% maximum if particular conditions are met) for pre –competitive development
- EU framework provides for regional investment aid for upgrading and modernising existing yards not exceeding 22.5% or 12.5% of the total costs incurred, depending on the region's compliance with certain criteria in the EC Treaty to State Aid.
- In UK, the Home Shipbuilding Credit Guarantee Scheme (HSCGS) offered long-term loans at a fixed rate of interest of 7½%. This scheme was run in parallel with the SIF until 2000.
- In UK, The aid to innovation was further enhanced on 31st December 2005 to 20 % through EC document (2003/C 317/06) and the above provision has been extended till 2008.
- Germany provides soft loans with below market interest rates and extended grace periods in their industrial support schemes.

In Italy, despite a Europe-wide decline in state aid to shipbuilding (from 3.1% of EU State aid in 1994-96 to 2.8% in 1997-99), there was an overall increase in the share going to shipbuilding in this period in Italy

- Between 1996 and 1998, Italy allocated around 1.5% of GDP, equivalent to 3% of total public spending, for state aids across all industries
- From 1994-96, in Italy, 2.1% of total state aid went to the shipbuilding industry
- From 1997-99, this figure was increased to 2.2%

➢Italy, in 2001, allowed incentives to investment, by granting a tax credit for the construction of ships (which did not enter into the computation of taxable income) and also granted a reduction on the commercial interest rate up to a maximum of 3.8% per year having a maturity of 12 years.

➢Italy has also set up a budget of 75 million euros for 2007 to 2009 for naval shipbuilding industries innovation which is in line with the EU policy on the naval industry.

REQUIREMENTS FOR THE WAY FORWARD

- COUNTRY'S VISION FOR SHIPBUILDING
- A PRACTICABLE AND WORKABLE LONG TERM POLICY
- EFFECTIVE EXECUTION OF THE POLICY
- MONITORING OF THE POLICY ON A CONTINOUS BASIS.

SHIPBUILDING IS DIFFERENT FROM OTHER MANUFACTURING INDUSTRIES

- IT IS A CUSTOM BUILT PRODUCT
- IN SHIPBUILDING WE FIRST SELL AND MANUFACTURE ONLY THEREAFTER
- THE MANUFACTURING PROCESS ITSELF IS LONG – IE. AFTER SIGNING A SHIPBUILDING CONTRACT IT TAKES ANYWHERE BETWEEN 2 TO 3 YEARS TO BUILD THE SHIP.

MR. MANMOHAN SINGH, HON'BLE PRIME MINISTER'S MESSAGE DATED 17-02-2006 TO THE NMCC (REFERENCE: NMCC REPORT "THE NATIOINAL STRATEGY FOR MANUFACTURING")

- ..I AM CONCERNED THAT THE SHARE OF MANUFACTURING IN NATIONAL INCOME HAS SHOWN ONLY A MARIGINAL IMPROVEMENT FROM 15.8% IN 1991 TO 17% IN 2003. THIS SHOULD BE SOMEWHERE IN THE RANGE OF 25% TO 35%. THIS REQUIRES MANUFACTURING TO KEEP GROWING AT 12%-14% IN THE NEXT DECADE.
- INADEQUATE GROWTH IN MANUFACTURING HAS HAD ITS ADVERSE IMPACT ON EMPLOYMENT GENERATION.
- I DONOT ACCEPT THE PROPOSITION THAT INDIA CAN SKIP THE MANUFACTURING STAGE OF DEVELOPMENT AND GO FROM BEING AN AGRARIAN SOCIETY DIRECTLY TO BECOMING A SERVICES AND KNOWLEDGE BASED SOCIETY.
- THIS IS A MISTAKEN VIEW. A SUBSTANTIAL MANUFACTURING BASE IS ESSENTIAL TO ABSORB THE WORKFORCE AND ENSURE SUBSTANTIAL GROWTH OF THE ECONOMY

EXTRACTS FROM THE NATIONAL MANUFACTURING COMPETITIVENESS COUNCIL REPORT ENTITLED "THE NATIONAL STRATEGY FOR MANUFACTURING

- MANUFACTURING IS A FORCE MULTIPLIER.
- IT CREASES PRODUCTIVE EMPLOYMENT, PROMOTES AGRICULTURE AND SERIVCE SECTOR AND SPINS A CYCLE OF WEALTH CREATION.
- INVESTMENTS IN MANUFACTURING PRODUCE A KEYNESIAN MULTIPLIER EFFECT ON GDP GROWTH TO THE TUNE OF FOUR TIMES IN INDIA AS EVERY RUPEE INVESTED ADDS FOUR RUPEES TO GDP.

EXTRACTS FROM THE NMCC REPORT ENTITLED "THE NATIONAL STRATEGY FOR MANUFACTURING"

- THE NATIONAL SAMPLE SURVEY (NSS) HELD IN 1999-2000 POINTS TO A SLOWDOWN IN THE ANNUAL AVERAGE GROWTH IN EMPLOYMENT. WHILE THE AVERAGE ANNUAL GROWTH RATE IN EMPLOYMENT WAS 2.89% BETWEEN 1983 AND 1988 AND 2.50% BETWEEN 1987 AND 1994, IT DECLINED DRASTICALLY TO 1.07% BETWEEN 1993 AND 2000. IT IS ESTIMATED THAT NEARLY 10 MILLION PEOPLE WILL JOIN THE WORKFORCE ANNUALLY BETWEEN 2005-2015 AND WOULD REQUIRE TO BE PROVIDED WITH EMPLOYMENT. IN THIS CONTEXT, IT NEEDS TO BE NOTED THAT THE PRESENT GROWTH PATTERN LED BY HIGH SERVICE SECTOR GROWTH AND A STAGNANT MANUFACTURING SECTOR IS LEADING TO A RURAL-URBAN DIVIDE.

Proposed Commercial New Building Yard Locations

Approx 7,000 km Coast line

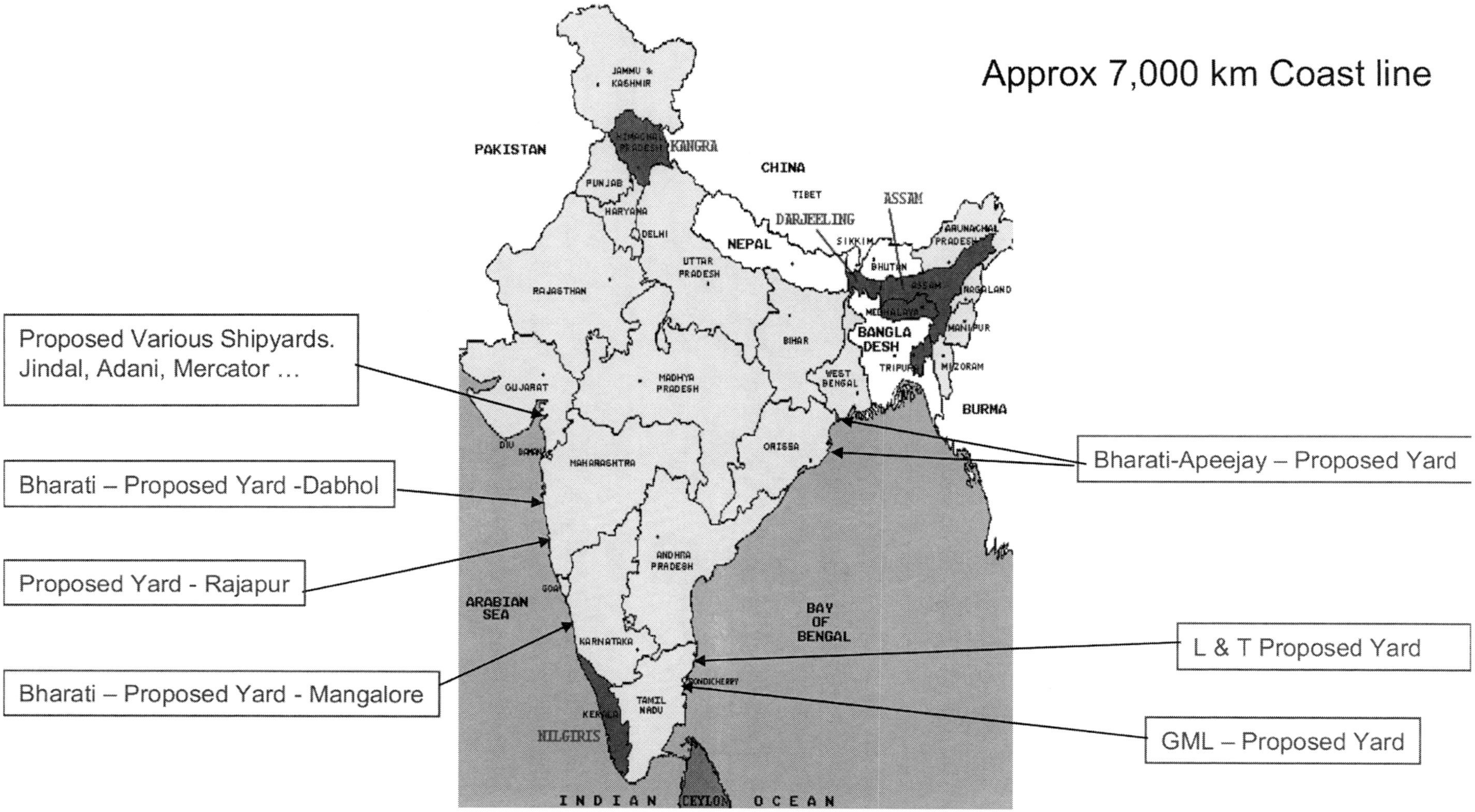

Some more shipyards are not listed

Existing Commercial New Building Yard Locations

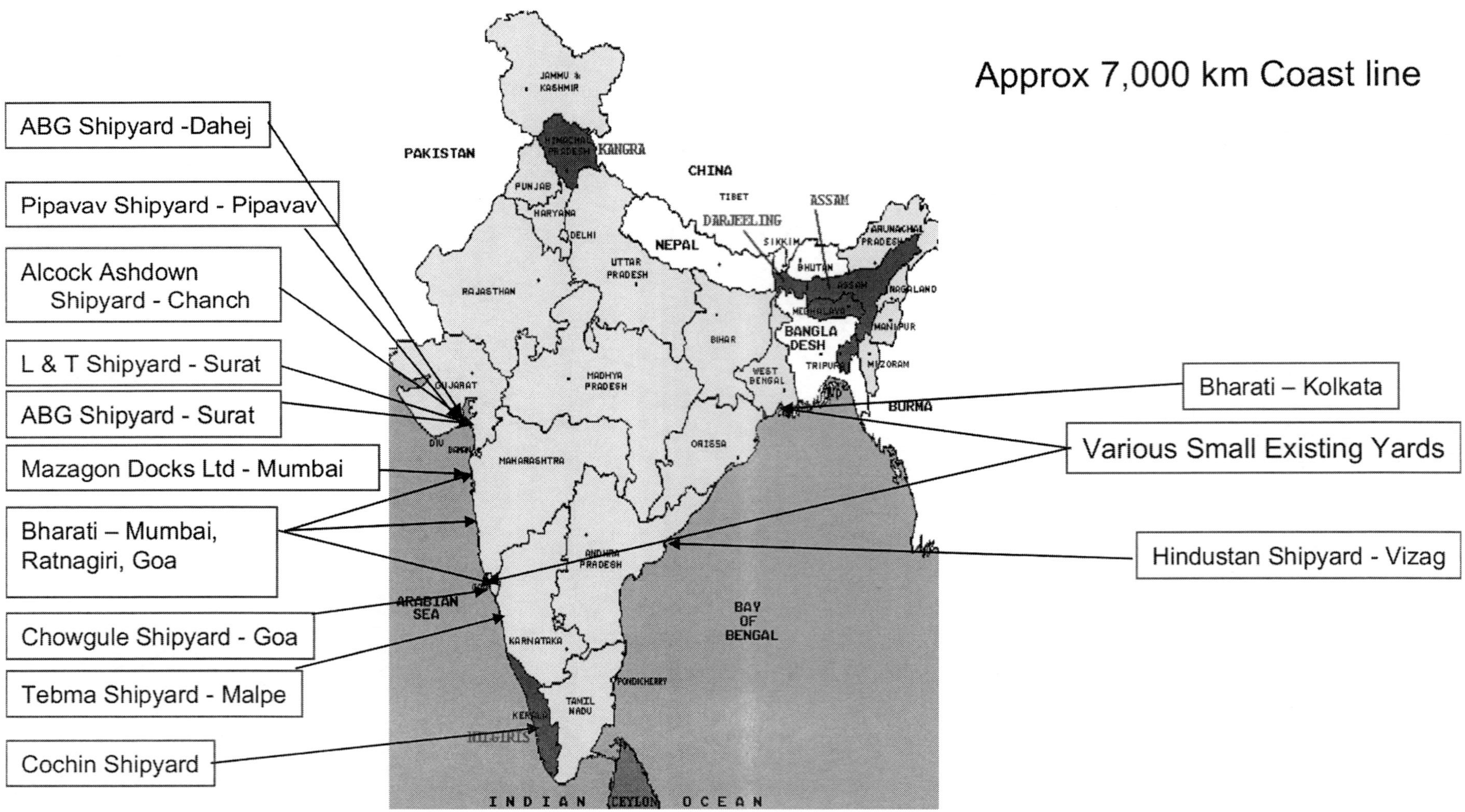

Some more shipyards are not listed

PRESS REPORT

- FRENCH PRESIDENT NICOLAS SARKOZY SAID HE HAS CALLED HIS 'FRIEND' AND INDIAN-BORN BILLIONAIRE LAKSHMI N MITTAL TO EXPLAIN JOB LOSSES FROM A PROPOSED SHUTDOWN OF ONE OF ARCELOR MITTAL'S PLANTS IN THAT COUNTRY.

- ARCELOR MITTAL, THE WORLD'S LARGEST STEEL MAKER, HAD ON JANUARY 16 ANNOUNCED CLOSURE OF ITS GANDRANGE PLANT IN EASTERN FRANCE BY EARLY NEXT YEAR, POTENTIALLY AFFECTING 600 JOBS.

- THE COMPANY HAD SAID THAT RELOCATION OF AFFECTED EMPLOYEES WOULD BE ITS PRIORITY AND EVERY EMPLOYEE WOULD BE OFFERED A "SOLUTION".

AS PER THE KMPG REPORT SUBMITTED TO THE MINISTRY IN 2007

- MULTIPLE EFFECT OF GROWTH IN SHIPBUILDING OVER NEXT 10 YEARS CAN GENERATE ADDITIONAL OVER ALL EMPLOYMENT OF APPROXIMATELY 480,000 EMPLOYEES BY 2012 AND AROUND 2,450,000 EMPLOYEES BY 2017.

PRESS REPORT

- EMPLOYMENT GENERATED IN THE MANUFACTURING SECTOR FROM 1998 TO 2005 OF ALL THE STATES PUT TOGETHER WAS ONLY 2.4 MILLION JOBS.
- ON THE OTHER HAND ONE INDUSTRY – VIZ. THE SHIPBUILDING INDUSTRY ITSELF CAN CREATE ABOUT 2.5 MILLION JOBS.

LEMAN & CO. LTD. | ESTABLISHED 1838 | NEW DELHI | MONDAY, JULY 7, 2008 PAGES 58 | CAPITAL | PR

HE TIMES OF

INCLUSIVE OF DELHI TIMES AND EDUCATION TIMES | TIMESOFINDIA.COM | EPAPER.TIMESOFIN

TER EXPENSIVE HOMES, MITTALS EYE
GH-END HERITAGE HOTELS IN UK | 23

MENDIS MES
LANKA RUN

Rich Maharashtra, Gujara fare poorly in job creatio

ndhra Tops Non-Farm obs Chart

Rema Nagarajan
TIMES INSIGHT GROUP

Delhi: For long, Maha-a and Gujarat have been as the commercial and acturing hubs of the y. Figures on foreign di-vestment also suggest ey are among the states ing the biggest invest- Yet, they have been the laggards when it to creating non-agricul-r manufacturing jobs. states like Haryana,

ADDING HANDS

Major States '98-'05	Manufacturing Jobs Added	% Growth
Rajasthan	2,85,735	37.3
Kerala	3,12,004	37.0
Haryana	1,88,585	36.7
Karnataka	3,98,706	29.8
AP	5,28,808	27.7
Tamil Nadu	5,79,887	24.9
Punjab	1,34,341	21.4
Orissa	1,67,088	21.3
MP*	2,08,037	16.4
Bihar*	85,525	13.1

Major States '98-'05	Manufacturing Jobs Added	% Growth
West Bengal	2,27,522	7.9
Gujarat	37,004	2.6
UP*	49,506	2.1
Maharashtra	25,485	1.1
Assam	–57,647	–12.6
Delhi	–3,43,849	–23.9

* Bihar includes Jharkhand, UP Uttarakhand & MP Chhattisgarh

Punjab and Rajasthan in the north and the four southern states that have registered the highest growth both in the number of non-agricultural business establishments and in the number of employees working in these. This emerges clearly from a comparison of the data from the economic censuses of 1998 and 2005.

In terms of growth in the number of such establishments, Assam, Tamil Nadu, Haryana and Kerala top the list among the bigger states.

However, in terms growth in employm such establishments, tops, followed by Ha Kerala, Tamil Nad Punjab in that order.

There are smaller like Jammu and Kash some in the north-eas have even higher rates, but given thei the comparison is obv meaningless.

When it comes to th ufacturing sector, whil Nadu now has the h number of units, Keral in growth, with the nu more than doubling. K followed by Assam, Nadu, Punjab, Karnata jasthan and Haryana, istering over 60% incr

► Guj behind Orissa, Bih

otesters go erserk at residence

elhi: About 100 pro- from various Kash-andit groups in the allegedly damaged nce around Union r Saifuddin Soz's res- and broke flower er he reportedly re-

ist vote today, Cong

Guj behind Orissa, Bihar in manufacturing sector

Rema Nagarajan | TIMES INSIGHT GROUP

TIME TO BUCK UP

New Delhi: Rajasthan, Kerala and Haryana top the list among states that have registered growth in manufacturing jobs according to comparison of the data from the economic censuses of 1998 and 2005. The other southern states too have recorded rather high growth in employment in the manufacturing sector.

While Maharashtra registered a growth of over 56% in manufacturing establishments, the growth in employment in manufacturing is a meagre 1.1% in the state. And in Gujarat, where the number of manufacturing establishments increased by just over a quarter, employment in the sector has grown by just over 2.5%.

Even states like Orissa and Madhya Pradesh have done better than Gujarat in terms of growth in the number of manufacturing establishments and in terms of employment in this sector.

In the last census, Uttarakhand, Jharkhand and Chhattisgarh did not exist. This makes the performance of Uttar Pradesh, Bihar and Madhya Pradesh seem poor at first glance, since establishments in the newly carved out states are now counted separately. However, if we add their numbers, it is obvious that even undivided Bihar, Madhya Pradesh and UP have done better than Gujarat.

West Bengal which figures among the top five on most counts in absolute numbers, whether it is in number of establishments or persons employed in the sector, shows a much slowed rate of growth in comparison to most states, which explains why the Left front is so eager to push industrialization in the state.

However, even Bengal's performance is far better than Maharashtra and Gujarat on most counts. In the case of Assam, while the number of manufacturing establishments has more than doubled, employment in the sector has actually fallen, one of the few states where this has happened.

Delhi is another instance of a territory that has seen a decline in manufacturing units as well as jobs. This may seem strange, but is probably explained by the Supreme Court mandated shift of a number of units from the Capital.

IAS officer, kin booked

PRESS REPORT

STRATEGIC IMPORTANCE OF SHIPBUILDING

- CHIEF OF NAVAL STAFF HAS BEEN QUOTED:

BECAUSE OF SHORTAGE OF SHIPYARDS CAPACITY IN INDIA, THE INDIAN NAVY WILL HAVE TO SOURCE A LARGE NUMBER OF CONSTRUCTION OF NAVAL SHIPS FROM OUTSIDE INDIA.

ANCILLIARISATION

- ONE OF THE COST DISADVANTAGE SUFFERED BY AN INDIAN SHIPYAD IS THAT PRESENTLY A SUBSTANTIAL PART OF THE BOUGHT OUT ITEMS HAVE TO BE IMPORTED. HOWEVER, ONCE THE "MOTHER" SHIPBUILDING INDUSTRY REACHES IT'S CRITICAL MASS, ANCILLIARIES KICK IN.

ANCILLIARISATION

- JAPAN - IN MID 50'S – ABOUT 30% TO 40% : PRESENTLY ALMOST 100%
- KOREA - IN MID 70'S - ABOUT 20% TO 25% : PREENTLY ABOUT 88- 90%
- CHINA - IN EARLY 90'S – ABOUT 25% : PRESENTLY ABOUT 60-65%

Location of Large shipyard in Korea
ClassNK
HYUNDAI
HEAVY INDUSTRIES CO.,LTD.
Gunsan (New Yard)
HYUNDAI SAMHO
SAMSUNG
SAMSUNG HEAVY INDUSTRIES CO., LTD.
HYUNDAI
HEAVY INDUSTRIES CO.,LTD.
HYUNDAI
MIPO DOCKYARD CO.,LTD.
HHIC
HANJIN HEAVY INDUSTRIES & CONSTRUCTION CO., LTD.
DSME
stx
STX SHIPBUILDING CO., LTD.
NORTH
KOREA
Demarcation
Line
SEOUL
Inch'on
Suwŏn
Taejŏn
P'ohang
Taegu
Chŏnju
Ulsan
Kwangju
Masan
Pusan
Mokp'o
Yosu
Kunsan
Yellow
Sea
Sea
of
Japan
JAPAN

RATIONALE FOR SUPPORT – MACRO FACTORS

- STRATEGIIC INDUSTRY – VITAL FOR INDUSTRIAL DEVELOPMENT OF A NATION
- LARGEST GENERATION OF EMPLOYMENT PER CAPITA OF INVESTMENT.
- LARGEST GENERATION OF INDIRECT EMPLOYMENT. FOR EVERY 1 PERSON EMPLOYED IN A SHIPYARD, INDIRECT EMPLOYMENT GENERATION FOR ABOUT 7 PERSONS IN ANCILLIARY AND SUPPORT INDUSTRY.
- SINCE MOST OF THE SHIPYARD WOULD BE IN RURAL AREAS SHIPBUILDING AND CLUSTER OF ANCILLIARIES THAT WOULD GROW AROUND IT, WOULD DRAMATICALLY ENHANCED THE RURAL DEVELOPMENT.
- HUGE POTENTIAL FOR INVESTMENTS IN ENGINEERING INDUSTRIES AS THE SUPPLY CHAIN CAN BE DEVELOPED WITHIN THE COUNTRY.
- REQUIRES HUGE SKILLED, SEMI- SKILLED AND UNSKILLED WORK FORCE. IMPROVES THE TECHNICAL QUALIFICATIONS OF THE WORK FORCE.
- CREATES INFRASTRUCTURE.

RATIONALE FOR SUPPORT – IMPACT ON TRADE DEFICIT

SHIPBUILDING INDUSTRY HAS THE CAPABILITY TO SUBSTANTIALLY REDUCE THE TRADE DEFICIT OF INDIA

FOREIGN TRADE IN MILLION USD IN 2007

	INDIA	KOREA
EXPORT	124,629	325,465
IMPORT	181,368	309,383
TRADE BALANCE	-56,738	16,082

SHIPBUILDING EXPORTS OF KOREA: 30,000 MILLION USD

- IN CASE OF KOREA, THE SHIPBUILDIING EXPORT ACCOUNT FOR 9.2% OF THE TOTAL MERCHANDISE EXPORT, WHEREAS IN CASE OF INDIA, IT IS NEGLIGIBLE.
- A FULL –FLEDGED SHIPBUILDING INDUSTRY LIKE IN THE CASE OF KOREA CAN HELP INDIA DECREASE THE NEGATIVE BALANCE OF TRADE

Rationale for support – Employment Potential

The shipbuilding industry is labour intensive and hence has a huge employment potential

Employment generated by investment of INR 1 crore in different sectors	
Automobile	6
Shipbuilding	16
Auto with ancillary	28
Shipbuilding with ancillary	80
Rural roads	109

- Inspite of growth in automotive sector, the number of people employed is decreasing year on year
- The core shipbuilding industry in India has more potential than the automobile sector in providing employment
- According to a research in University of Bremen, the ratio of sub-contractors to direct employment in shipbuilding industry is 4:1 which explains its the huge potential in the future

NATIONAL RURAL EMPLOYMENT GUARANTEE SCHEME (NREG)

- FOR EVERY RS.1.0 CRORES OF INVESTMENT NREG'S ACTUALLY EMPLOYS 1000 PEOPLE. HOWEVER IT ASSURES EMPLOYMENT FOR 100 DAYS ONLY. THEREFORE CONSIDERING THE MANDAYS THE NUMBER REDUCES TO **275** PERSONS EMPLOYED.
- AS SEEN ON AN EARLIER SLIDE SHIPBUILDING (DIRECT + INDIRECT) EMPLOYEES **80** PERSONS (MOSTLY IN THE RURAL AREAS), FOR EVERY Rs.1.0 CRORES INVESTED.
- HOWEVER, SHIPBUILDING INDUSTRY EMPLOYS THESE 80 PERSONS THE WHOLE YEAR AND YEAR ON YEAR AND FOREVER. THE NREG SCHEME EMPLOYS 275 PERSONS FOR ONLY 1 YEAR !

POLICY SUPPORT REQUIRED

- ➢TO MONITOR SUPPORT AND DEVELOPMENT OF THE SHIPYARDS.
- ➢NATIONAL CO-ORDINATION COMMITTEE ON SHIPBUILDING (TO BE ESTABLISHED UNDER NSB) TO ACCELERATE AND MONITOR DEVELOPMENT OF SHIPYARDS.
- ➢CENTRAL AGENCY FOR INDUSTRY REPRESENTATION BEFORE GOVERNMENT AUTHORITIES FOR VARIOUS CONCESSIONS/ SUBSIDIES/ TAX HOLIDAY, ETC.
- ➢CO-OPERATION BETWEEN VARIOUS SHIPYARDS AND TRAIINING INSTITUTES FOR DEVELOPMENT & TRANSFER OF TECHNOLOGY AND R&D WORK.
- ➢DEVELOPMENT OF RESOURCES (HUMAN, TECHNICAL, MANUFACTURING, SUPPLY CHAIN, ETC.) REQUIRED FOR LARGE SCALE SHIPBUILDING).
- ➢TO DRAW UP A MASTER PLAN TO
    - INCRREASE MARKET SHARE FROM PRESENT 1.13% TO 3-4% IN NEXT 3 YEARS, 7% IN NEXT 7 YEARS AND 15% BY 2020.
    - CONSTRUCTION OF EVERY SOPHISTICATED VESSELS LIKE RIGS, FPSO, PASSENGER VESSELS, OFFSHORE STRUCTURES, LNG CARRIERS, ETC.

POLICY SUPPORT REQUIRED TO TRIGGER GROWTH OF THE INDUSTRY

- SINGLE WINDOW CLEARANCE - FAST TRACK CLEARANCE
- SUBSIDY – TO NEGATE THE IMPACT OF TAXES AND OTHER COST AND PRICE DISADVANTAGES
- SETTING UP OF TRAINIING INSTITUTES FOR DEVELOPING SHIPBUILDING SKILLS.
- DEVELOPMENT OF ANCILIARY INDUSTRIES
- PRODUCTION OF SHIPBUILDING QUALITY STEEL INDIGEOUSLY AT COMPETITIVE INTERNATIONAL PRICES.
- R&D WORK IN THE FIELD OF SHIPBUILDING, RELATED HRD, PRODUCTIVITY OPTIMIZATION AND THE USE OF TECHNOLOGY LEVELS AT VARIOUS STAGES OF SHIPBUILDING TAKING INTO CONSIDERATION IMPORTANT FACTORS LIKE PRODUCT QUALITY, COST OF BUILD AND EMPLOYMENT GENERATION.

EXTRACTS FROM THE MINUTES OF 110TH MEETEING OF THE NATIONAL SHIPPING BOARD HELD ON 12TH AUGUST 2008 AT PARADIP,.

- THE ISSUE (OF SUBSIDY) WAS DELIBERATED AT LENGTH AND IT WAS DECIDED TO RECOMMEND TO THE GOVERNMENT TO EXTEND THE SCHEME OF SUBSIDY TO THE SHIPBUILDING INDUSTRY FOR ATLEAST ANOTHER FIVE YEARS SO THAT THE INDUSTRY COULD TAKE ADVANTAGE OF THE CURRENT BOOM IN DEMAND.

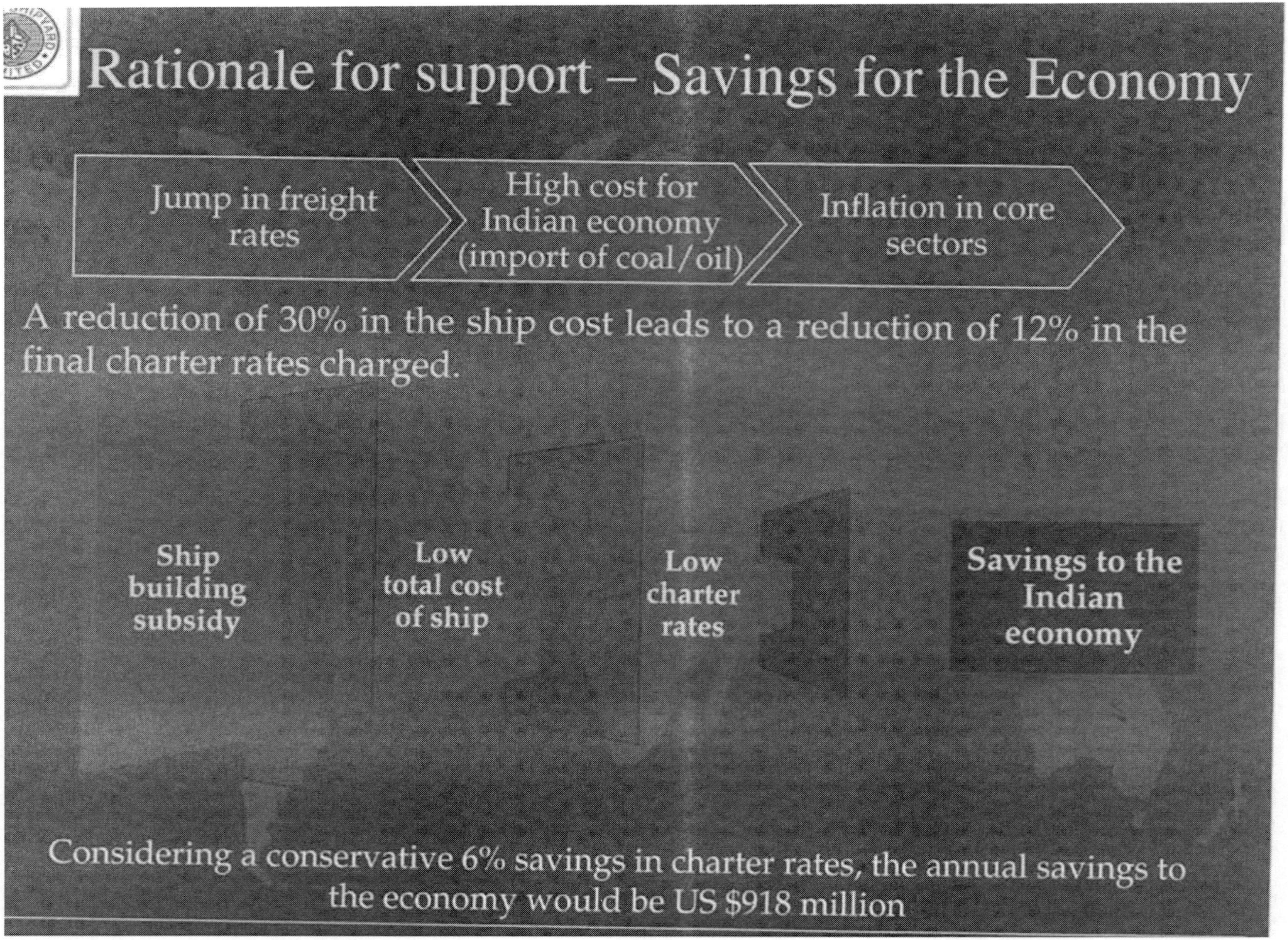

Rationale for support – Savings for the Economy
Jump in freight rates
High cost for Indian economy (import of coal/oil)
Inflation in core sectors
A reduction of 30% in the ship cost leads to a reduction of 12% in the final charter rates charged.
Ship building subsidy
Low total cost of ship
Low charter rates
Savings to the Indian economy
Considering a conservative 6% savings in charter rates, the annual savings to the economy would be US $918 million

Rationale for support – Defense Requirements

The shipbuilding industry has to flourish in India to meet the strategic requirements of Indian navy.

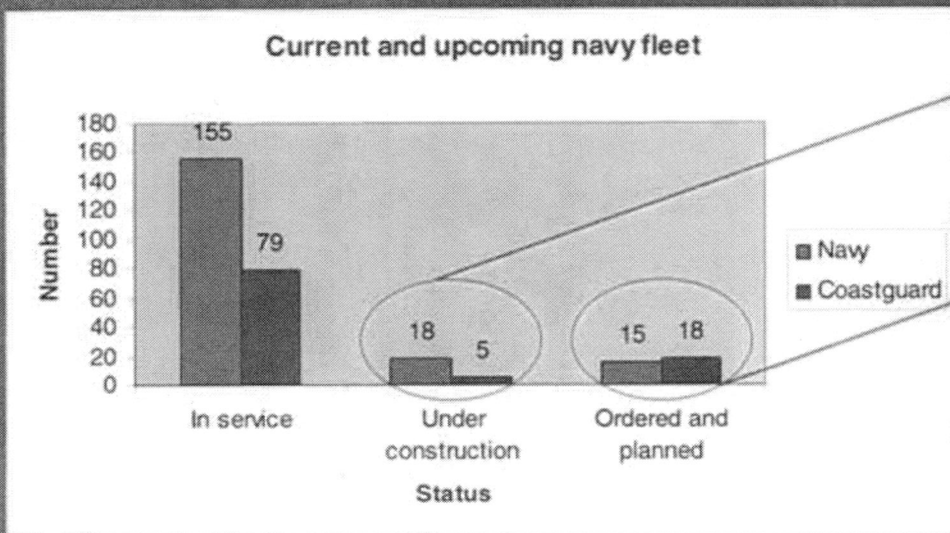

- The fleet size under construction is 23 and would come to use within the next five years
- According to the former Vice Chief of the Naval Staff, Vice Admiral P.J. Jacob, India should encourage the setting up of 'Greenfield' shipyards if the Navy is to, as planned, acquire around 35 ships over the next decade
- Navy gets stealth frigates from foreign shipyards – inability of the indian shipyards to deliver on time and unavailability of modern equipment

- The Defense Ministry planning to acquire Vishakhapatnam-based Hindustan Shipyard Ltd. completely by 2010 shows that it needs shipyards with good facilities for its own use

Indian shipyards should have their own servicing capabilities which will boost Indian Navy's ship building capacity and would cater to its repair requirements

Rationale for support – Rural Development

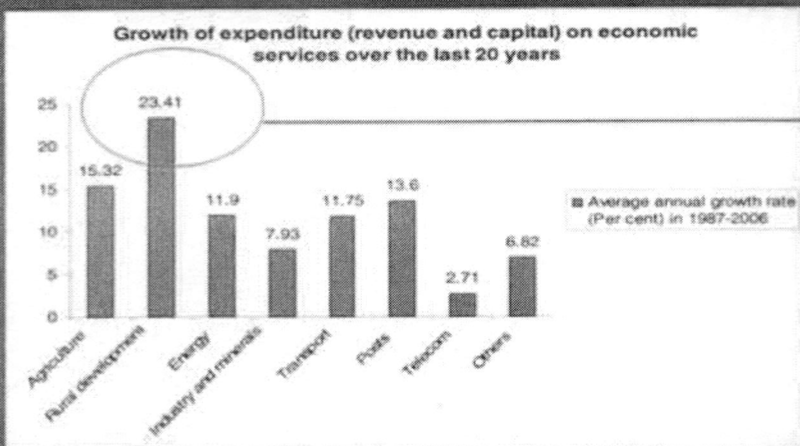

A significant positive shift in rural development reflects the Government's overwhelming priorities for the sector particularly in the context of generating employment opportunities in rural areas

Location of new shipyards:

1. Pipavav
2. Mundra
3. Kakinada
4. Dhamra
5. Other non-urban areas

Flourishing shipbuilding industry would not only employ rural people but also help uplifting the rural economy as a whole.

THE DELIBERATION OF THE NATIONAL MANUFACTURING COMPETITIVENESS COUNCIL ESTABLISHES--------

- "THE NEED FOR HAVING A STRONG AND VIBRANT SHIPBUILDING SECTOR AS A STRATEGIC INPUT FOR THE COUNTRY'S LONG TERM VISION"

FUTURE CHALLENGES IN THE AREAS OF COMPUTER AIDED SHIP DESIGN AND PRODUCTION

Tae-wan Kim, Seoul National University, Republic of Korea

SUMMARY

Ships are important and strategic engineering structures. The modern engineering designs are becoming increasingly more complex because of the pressing demands of economies of scale, clean environment, and advancements in material science and technologies. Ships are one of the most complex engineering structures. The design and analysis time for a ship runs in months. The advent of the computer in the early fifties changed the scope, style and methodology of mathematical form definition in a dramatic way. Today, modern curve and surface design around the world can no longer be imagined without intensive use of computers and computational methods. It is worth noting here that shipbuilding industries have been one of first users of computers and computational methods for geometric modeling, design and analysis for ship geometries. The aim of this paper is to report on the key challenges in the areas of computer aided ship design and production.

1. INTRODUCTION

Ships are important and strategic engineering structures. The modern engineering designs are becoming increasingly more complex because of the pressing demands of economies of scale, higher speed, lower motions, operational ability in adverse weather conditions, energy efficiency, clean environment, and advancements in material science and technologies. Ships are one of the most complex engineering structures whose design and analysis time run in months and production may take anywhere between months to years. This complexity in various engineering structure is shown in Fig. 1, [2]. The advent of the computer in the early fifties changed the scope, style and methodology of mathematical design and analysis approach in a dramatic way. Today, modern engineering design and analysis around the world can no longer be imagined without intensive use of computers and computational methods. It is worth noting here that shipbuilding industries have been one of first users of computers and computational methods for modeling, design and analysis for ship structures.

Computer-aided design (CAD) is the process of creation and manipulation of design model on a computer to assist the engineer in the design and analysis process. In modern environment, CAD can be used to design, simulate, develop and optimize engineering structures. The main advantage of a CAD model is its ability to be used for detailed engineering of 3D models and/or 2D drawings of physical components, and it can be used throughout the complete engineering process from conceptual design and layout of structure, through strength and dynamic analysis of assemblies to definition of manufacturing methods of components. Apart from the theoretical interest, CAD models are used as computational tools in science, engineering, computer aided geometric design, and animation, etc. So the subject offers a rich mixture of problems in pure mathematics, applications of mathematics, and computational mathematics.

Today in ship and floating structure design and analysis, it is possible to use the CAD model not only for visualization purposes but also for other analytical purposes (e.g. basic structural form design, configurational design, design of final dimensions, simulation of motion characteristics, structural form design with outfittings and machinery items, propulsion system design with/without rudder, bulkhead arrangement design, and detailed layout designs, design of required plate sizes and thickness for production, design of heating pattern, material requirement planning and scheduling, and work assignment for technical workers, etc.). It is therefore important to use industrial standardized CAD representations for geometries and structures; so that the data can be exchanged between various design and analytical systems and tools in practice without any loss of information. Fundamentally, ships and floating structures are defined by 'offset points', geometric arrangements and basic dimensions. The basic CAD model is generated with these 'offset points', geometric arrangements, basic dimensions and satisfying constraints of volume, space, and derivational/geometric continuities, etc. And, the basic CAD model is input to further analysis. Hence, in our opinion accurate and efficient CAD model with efficient computational methods is the key to ship and floating structural design and analysis.

Computers allow easy creation and manipulation of design model to assist the engineer in the design, analysis and production process. In modern integrated environment, it is possible to use computer aided design model and subsequently the computational methods related to CAD model for design and development. Further downstream applications of basic CAD model are in product layout design, life cycle management systems, manufacturing analysis, etc. The schematic diagram of computer applications in the area of ship and production is shown in Fig. 2.

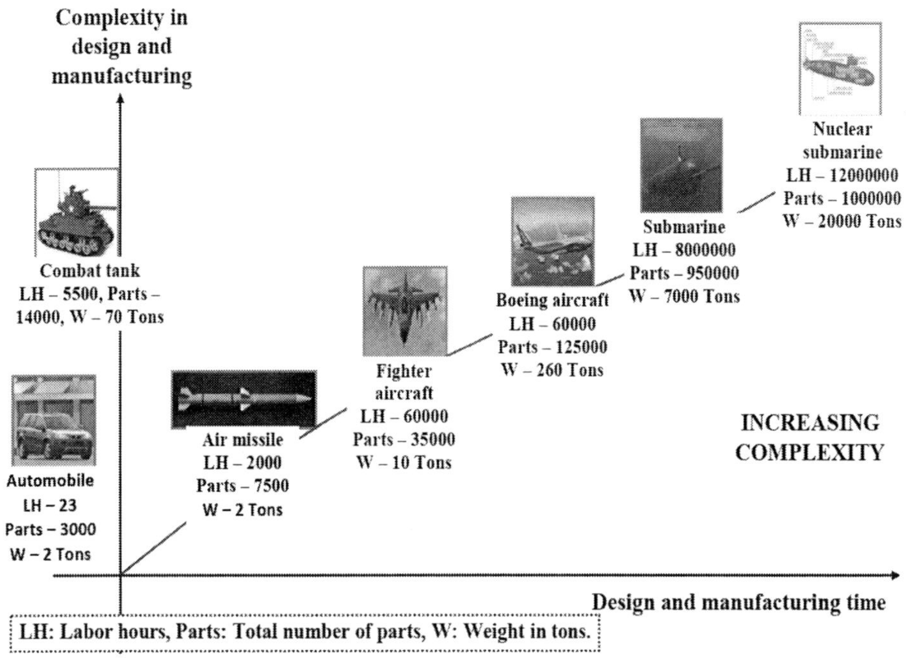

Figure 1: The complexity in design and manufacturing of modern engineering structures.

Since, ships are complex and complicated products which are mainly manufactured with steel plates and assembly of outfit items and machinery. Thus a healthy shipbuilding industry also pushes the demand of materials and items, and hence creating a network of ancillary industries revolving around it, [2]. Utilizing this many countries have used shipbuilding to build their industrial infrastructures. Initially UK used the shipbuilding to develop and maintain not only its territorial but industrial empire also; and later Japan had used shipbuilding in the 1950's and 1960's to rebuild its industrial infrastructure of mainly manufacturing based industries. South Korea had used shipbuilding as a strategic industry to develop its economy around manufacturing based large industrial groups; and currently China is in the process to repeat these successful models with large governmental investments to develop its manufacturing base. Slowly other developing nations like Vietnam are trying to build their shipbuilding capabilities. Similarly, other European nations like Poland, Denmark, and Spain are either trying to re-enter into the world of shipbuilding or are slowly thinking to support and develop their existing shipbuilding industry.

As is clear from Fig. 1, ships are the one of the most complex engineering structures. The design and analysis time for a ship runs in months. We believe that to bring down the design and analysis time and to remain competitive, the resources in engineering design and analysis should be utilized efficiently and optimally and that can only be achieved with an automatic decision support system.

However, the shipbuilding industry still remains labor intensive with little or no automation of production and decision processes. Since shipbuilding is an important industry for nations which are trying to develop their industrial base, nations show a very strong tendency towards directly or indirectly supporting their shipbuilding industries; and because of this shipbuilding industry is fiercely competitive and utilization of resources becomes very critical.

Therefore, to remain competitive and cost efficient it is essential for a shipyard to design a new ship at a shorter lead time and manufacture it with low cost and time. Ships are huge structures with ever increasing dimensions and that adds to its complexity. The ships are capital intensive. So, a decision to order a new ship is taken only after a rigorous techno-economic study. The ship design process is highly interdisciplinary and draws its fundamentals from Computational Fluid Dynamics (Fluid Mechanics), Computer-Aided Design (CAD)/Computer Aided Manufacturing (CAM), Communication Systems and Networking (CSN), Finite Element Analysis (FEA), etc. This is shown in Figure 3.

Because of its inherent complexities the ship design process is time consuming. A ship is either designed for the specific requirements placed by the owner or order placing agency or designed by the shipyard and later marketed amongst various owners/shipping companies/order placing agencies.

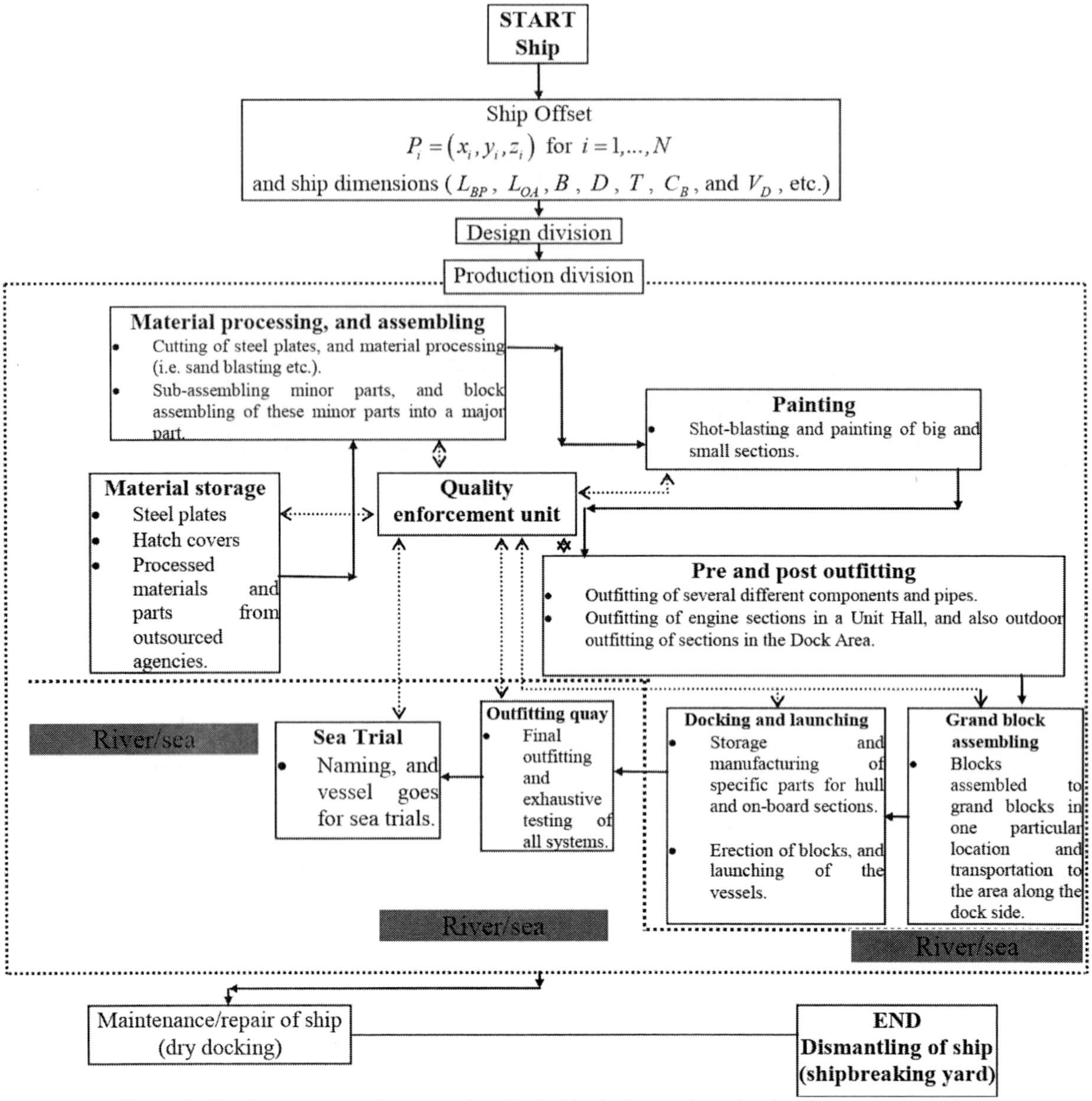

Figure 2: The key processes in computer aided ship design and production from start to end.

In both the cases, it starts with the preliminary design which takes about 2 - 6 months, e.g. for a competitive shipyard: bulk cargo ship = 2 - 3 months; small to medium sized container ships = 2 – 4 months; very large container ships, and cruise ships = 4 – 6 months; chemical carrier ships = 4 – 6 months; and LNG/LPG carrier ships = 4 – 6 months. This preliminary design is essential for the contract signing. This existing process of ship design and production is shown in Fig. 3.

Once the contract is signed then a detailed design is completed. The detailed design process takes about 6-12 months depending upon the type of ship, e.g. for a competitive shipyard: bulk cargo ship = 3 – 4 months; small to medium sized container ships = 4 – 8 months; very large container ships, and cruise ships = 6 – 10 months; chemical carrier ships = 6 – 8 months; and LNG/LPG carrier ships = 8 – 12 months. The detailed design consists of several different designs:

Engineering design: This is technical design which consists of FEA, hydrodynamic analysis, structural configurations, etc.

Classification design: This is design for classification societies. Each ship is classified by an international classification society depending upon its type, configuration, capacities etc. The classification design is proposed to the classification society and classification society reviews it. They may demand some changes depending upon their review. This process of review and modification is important in ship design.

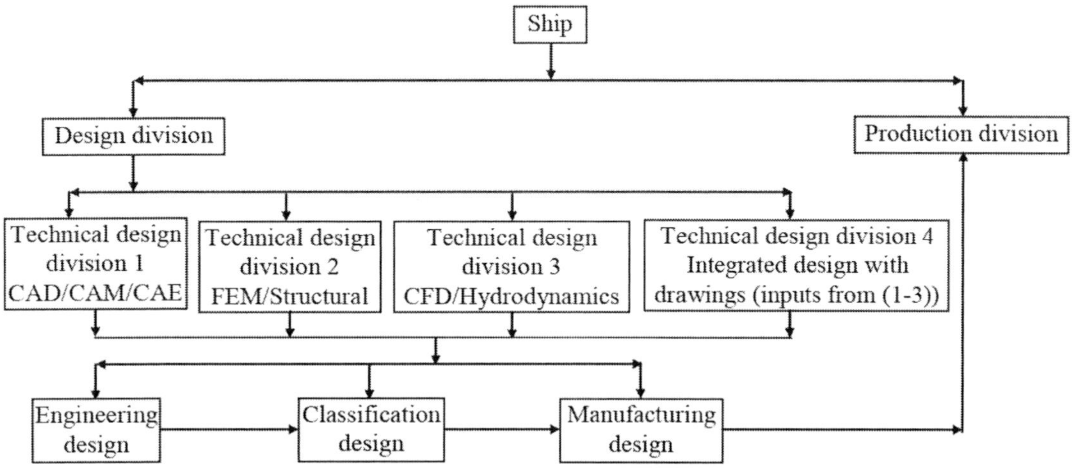

Figure 3: The schematic of existing process of ship design and production.

It takes time and in the design team it is important to have the technical workers who are familiar with codes and practices of classification societies.

Manufacturing design: From a classified design a manufacturing design is derived. The manufacturing design consists of all the design sectional details, material requirements, production planning, etc.

In a fiercely competitive market it is important to develop new products at a shorter lead time. And, in our opinion computers can play an important to achieve the design of new, novel, and greener ships at a shorter lead time in future.

The remaining of this paper is organized as follows: Section 2 presents some of our research at SNU, Republic of Korea in the areas of computer aided ship design and production (CASDP); Section 3 documents of the pressing challenges in the areas of CASDP and Section 4 conclude the paper with identifications on current status and future evolving directions.

2. OUR RESEARCH IN THE AREAS OF COMPUTER AIDED SHIP DESIGN AND PRODUCTION (CASDP)

At the 'Computer Aided Design and Information Technology Laboratory (CADITL)', Department of Naval Architecture and Ocean Engineering, Seoul National University, Republic of Korea; we focus on Engineering Design and Production, Deepwater Floating Structure Design, Marine Robotics, Geometric and Solid Modeling, Advanced Manufacturing, Distributed Information Systems, Computational Geometry, Shipbuilding, Producibility, and Life Cycle issues, etc.

Some of our recent research problems that are being explored by us are as:

- **Surface generation over T-junction with G^1 continuity:** Ships normally have a 'T-junction', in the forward region (e.g. bulbous bow). Currently, modelling practices do not allow modelling/design over T-junctions. In this research our aim is to design/implement/test geometric modelling algorithms for surfaces defined over T-junction. This research is briefly explained in Fig. 4.
- **Development of 'Bill of Materials (BOM)' kernel of product lifecycle management model for shipbuilding industries:** The BOM is an important information model that can be extracted to create efficient and accurate models for material estimation and requirement (i.e. material and labor) planning. However, so far BOM has not been investigated to extract all the relevant information from it. In this research we are addressing this. This research is briefly explained in Fig. 5.
- **Development of a logic based product lifecycle management (PLM) System for shipbuilding industry:** PLM (Product Lifecycle Management) has become something like 'a magic wand' for various industries because of its capability to integrate different product modules via on-line network through the product's complete lifecycle, and hence providing one window access; thereby making the whole processes of product conception, design, manufacturing, delivery, maintenance and disposal integrated with a reduction in product development time and cost. However, heavy industries (i.e. shipbuilding and shipbreaking) are different from consumer product industries because of high customization in design process and engineering software, widely varying scales of operations and less compatibility between different design and production processes, e.g. ship production is planned in activity driven network scheduling system in-general, and is assumed more as a construction process or assembling process rather than a production process.

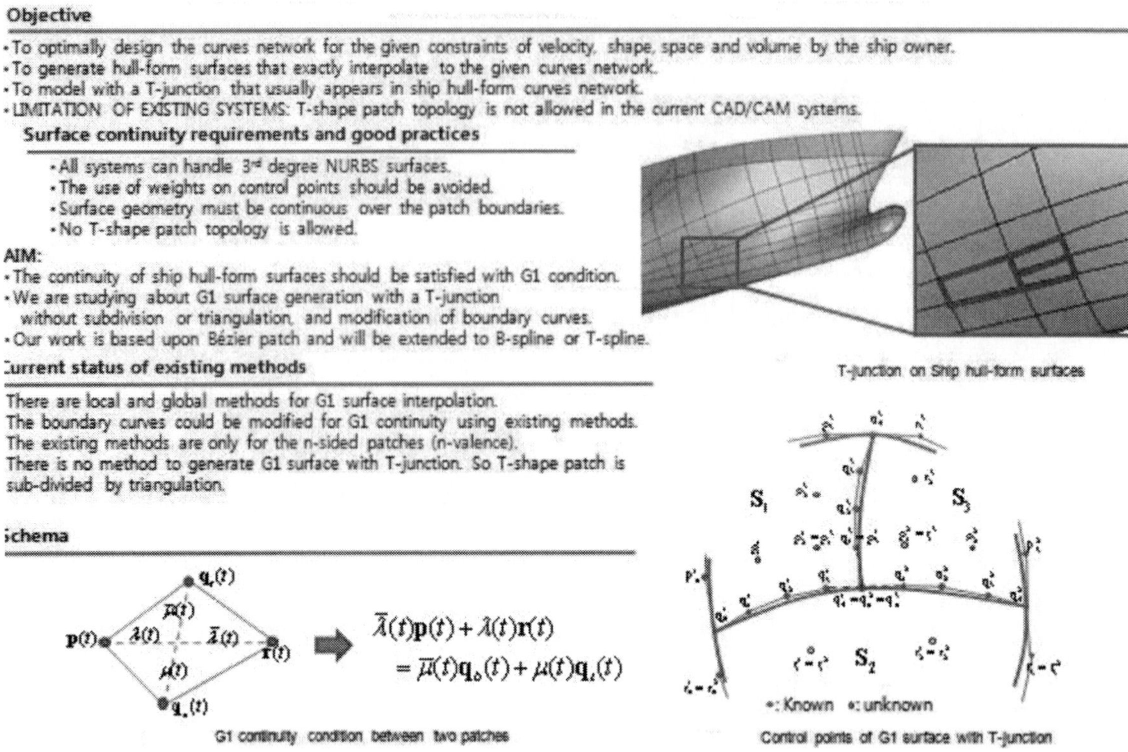

Figure 4: A brief description about surface generation over T-junction with G^1 continuity.

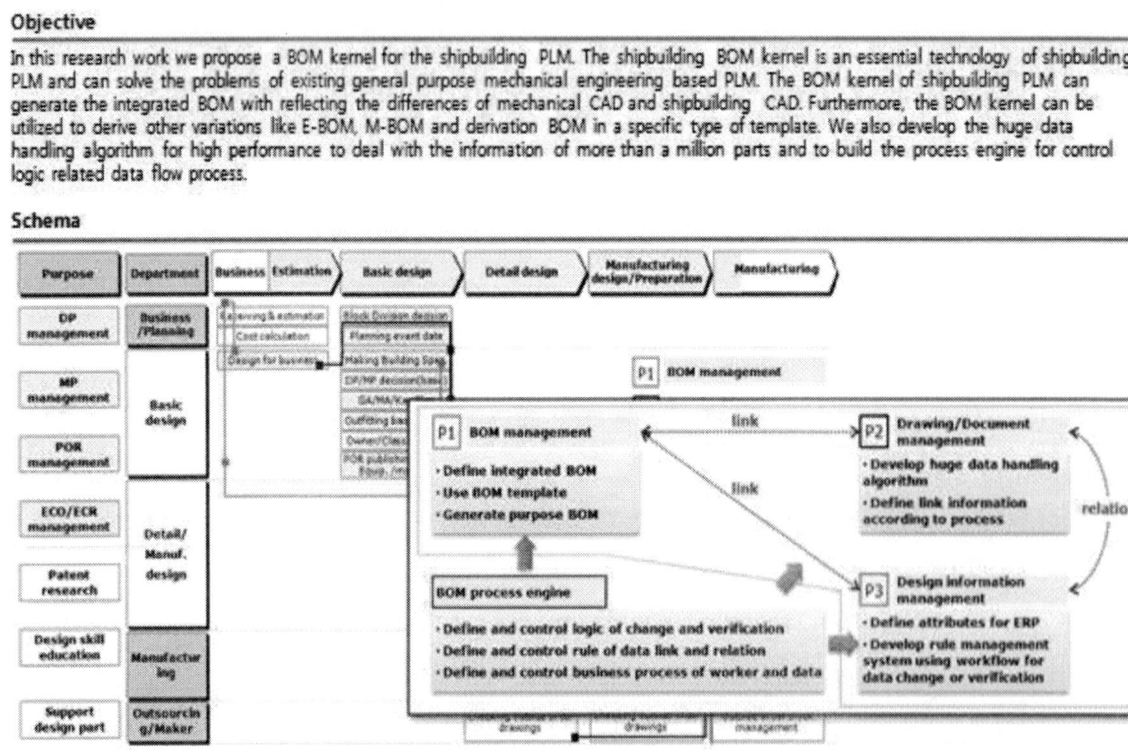

Figure 5: A brief description about development of 'Bill of Materials (BOM)' kernel of product lifecycle management model for shipbuilding industries.

In this research we aim to develop a logic based PLM system for shipbuilding industry. This research is briefly explained in Fig. 6.

- **Optimum design of an ultra-low motion semi-submersible floating oil and gas production system:** Economic and efficient energy resources are keys to a nation's development. Because of their low cost and advancement in drilling and exploration technologies oil and gas based energy systems are most widely used in practice throughout the world. The inexpensive oil and gas based energy systems are used for everything from transportation of goods and people to the harvesting of crops for food to being able see at night. As the energy demand continues to rise, this is pushing further the demands for inexpensive energy solutions. Since, available reservoirs of energy rich fossil fuels at shallow and moderately deep waters have been exploited already, the fossil fuel exploration and production is being forced to move into extremely deep waters. At deep water the expenses associated with fixed production platforms are no longer feasible, and that makes a floating production platform design a far more economical choice. This research investigates a critical parameter driven optimum conceptual design of an ultra-low motion semi-submersible floating oil and gas production system. This research is briefly explained in Fig. 7.

3. CHALLENGES IN THE AREAS OF CASDP

The main advantage of a CAD model and computational methods is their ability to be used for detailed engineering of 3D models and/or 2D drawings of physical components throughout the complete engineering process from conceptual design to manufacturing analysis to production.

Ships are produced mainly with material (i.e. steel or aluminum) plates of varying thicknesses (e.g. 4 mm to 80 mm). The plates are bended and/or heated with plasma arc/laser torches to provide the desired shapes. The shaped plated are welded with structural elements, fitted with other items, and painted. It is not possible to build a ship from aft to forward end because of size and complexity in the scale of various manufacturing and fitting/painting operations. All these processes make ship production as significantly complicated manufacturing process. An efficient ship design and production system requires an efficient CAD model definition, computational methods for manufacturing analysis, efficient product information model, efficient production planning and scheduling model, and efficient and lean production management system, etc.

It is better to use a standardized computer model (e.g. Standard for the Exchange of Product Model Data (STEP) compatible) right from the initial stage to seamless flow of information from one design and analysis process to another. At present, shipbuilding design drawings are prepared with insufficient and untimely information of material/equipment availability that result in drawing changes and errors, increased labor and material costs, and increased ship production time. An efficient CAD model definition integrated with computational methods will allow easy changes in design drawings, estimation of material/equipment quantity/availability, easy and efficient communication among shipyard design engineers, vendors, regulatory authority personnel, and shipyard construction department managers, and end-to-end connectivity between various processes. Hence, in our opinion accurate and efficient ship CAD model integrated with computational methods is the key to ship design and production.

The major research challenges in area of computer applications in ship design and analysis for economies of scale, higher speed, lower motions, operational ability in adverse weather conditions, energy efficiency, clean environment, and advancements in material science and technologies can be listed as:

- Efficient CAD for the ship design for economies of scale.
- CAD for finer design of ship for higher speed.
- CAD for motion simulation of ship.
- CAD for compact and finer design of ship for lower motion.
- CAD for design and development of 'hybrid' ship.
- Computer models for the operational simulation of ship in adverse weather conditions.
- Computer models for energy efficiency analysis of ship.
- Geometric and structural form optimization with CAD and CFD for ship.
- Optimum design of propeller to improve propeller/hull interaction and propeller-rudder combinations with CAD and CFD.
- Computer models for efficient weather routing, speed or trim optimization.
- Computer aided environmental impact analysis of ship from design to construction to dismantling.
- Computer aided conversion and recycling model for ship.
- CAD for propulsion system design with/without rudder, mid-ship section, bulkhead arrangement, and detailed layout.
- CAD for material requirement planning and scheduling, and work assignment for technical workers, etc.
- CAD for the design and analysis of 'green (for clean environment)' ship.
- CAD for the design and analysis of ship with advanced materials and control systems.

Objective

Logic simplifies and discreteness the information that needs to be exchanged and shared amongst different systems for efficient communication and co-ordination.

- **ADVANTAGE:** This facilitates development and integration of 'point-to-point' connections.
- **AIM:** To develop a logic based PLM (LBPLM) system for the entire lifecycle.

Limitations of existing PLM solutions

- Customers are not convinced about the offered capabilities and added values of the current solution systems.
- Early support for early conceptual design, detailed design and fully functional system design lack in the current PLM systems.
- Efficient communication and co-ordination among users is not properly supported by data exchange between application Systems.

Schema

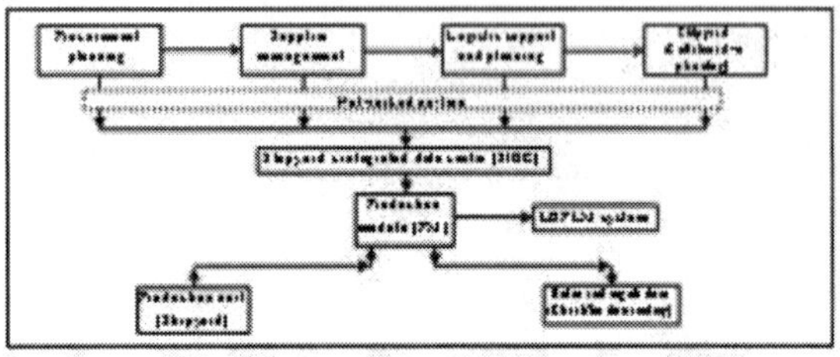

The architecture of integrated backward supply chain.

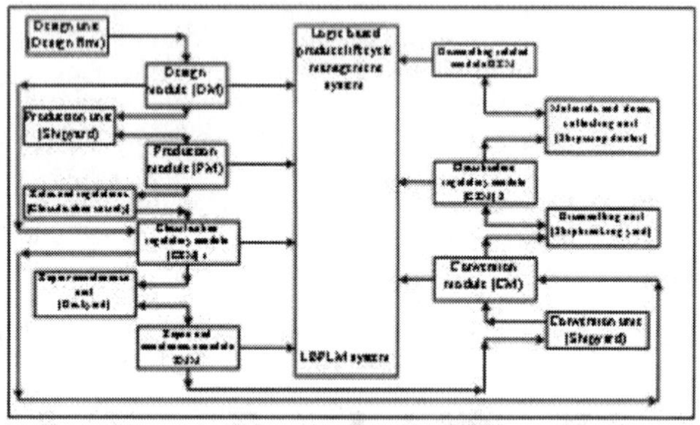

The system architectural configuration of different modules in our proposed LBPLM system.

Figure 6: A brief description about development of a logic based product lifecycle management (PLM) System for shipbuilding industry.

Objective

- A critical parameter driven optimum design of an ultra-low motion semi-submersible floating oil and gas production system.
- The design is carried to satisfy the given parameter of weights and dimensions of the platform topsides required for production and drilling, water depth location, and low motion requirements.
- The proposed design process is highly iterative process of altering the key dimensions of a 'n-column ring' pontoon, and it meets applicable regulations and requests of the order placing agency while minimizing costs and satisfying the chosen parameters.

Current status of existing designs

- There is a growing need for deepwater solutions.
- The exploration and production requirement in deep sea demands that the motion characteristics of the structure are low.
- The ultra low motion can be achieved with novel use of column sizes and spacing, floaters spacing and arrangement, and application of dynamic position systems (DPS).

Schema

The graphical design of ultra-low motion semi-submersible floating oil and gas production system.

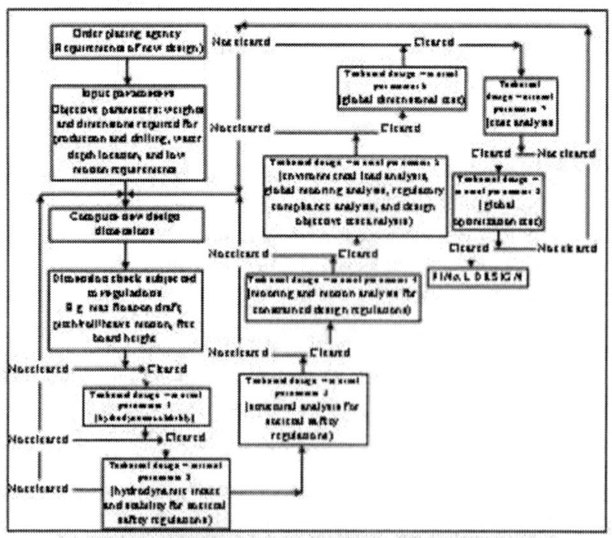

The schematic diagram of our proposed critical parameter driven iterative optimum design process model.

Figure 7: A brief description about optimum design of an ultra-low motion semi-submersible floating oil and gas production system.

The major research challenges in area of computer applications in ship production are listed as:

- Production friendly geometric design of curves and surfaces for ship geometries.
- Design of nearly developable, developable and low curvature surfaces.
- Geometric design of curves and surfaces under constraints of volume and space.
- CAD for the architectural/production layout design of ships.
- Computational models for efficient design and analysis of heating patterns.
- Computational models for efficient analysis and prediction for deformations in thin and large stiffened panels.
- Computational models for efficient material estimation and production scheduling.
- Computational models for strength analysis and structure design of lifting systems.
- Computational models for efficient pipe routing design.
- Computer information management systems for efficient ship production.
- CAD for ship product design, product life cycle management, lean and integrated manufacturing, and environmental impact analysis for ships from design to production to dismantling.
- Computational models for strength analysis and structure design of conversion and recycling for ships.
- CAD/CAM/CAE, FEM, PLM integration.

4. CONCLUSIONS

In this work, we have presented current status and challenges in the areas of computer applications in ship design and production. We believe that there is an urgent need to develop new analytical tools with computer support for design and production to ensure competitive utilization of various technical parameters in the design of ships and offshore structures. Since, a ship and offshore structure use many independent systems (e.g. CAD/CAM/CAE system, CFD system, and FEM systems) for design; a dedicated interface to exchange data will facilitate the integration process. For this purpose expandable (flexible) middle data format that synchronizes CAD/CAM/CAE DB, CFD DB, and FEM DB, might hold the key in future research. Our future work shall go in this direction, and currently this is under investigation.

(A) Geometry is the important in analysis but mainly thought as representative of shape. Because, of the belief that geometry is just related to shape, modern methods of computational geometry, do not utilize geometric forms from CAGD, CG, Computational Geometry, etc. The reason is that the Finite Element Method (FEM), was developed in the 1950-1960, before the advent and widespread use of Computer Aided Design (CAD) programs, which occurred in the 1970-1980.
Many technical difficulties encountered with FEM originate from its approximate, polynomial based geometry, such as, for example, mesh generation, mesh refinement, sliding contact, flows about aerodynamic shapes, buckling of thin shells, etc. However, since an integrated approach can potential develop new shapes with less time, it is desired that the powerful descriptions of geometry are used in computational mechanics.

Main points in future research:

- Use exact and popular geometric representations from CAGD, CG, Computational Geometry in computational mechanics. For more details on this see [2].

(B) It is well known that in FEA the processes of mesh construction, mesh refinement, and formation of boundary conditions are computationally expensive and constitute a significant part in the overall analysis time. This is further compounded in the analysis of complex engineering shapes like ships, automobiles etc. because in them approximation of geometry discretization, and material composition needs to be finer and that is computationally more complex and complicated. Additionally, the separation of design and analysis model increases the time and complexity of the overall process of design and analysis. It is in this regard that 'Iso-geometric Analysis', offers the possibility of development of an integrated design and analysis approach. The integrated model will allow efficient geometric modeling, improve the CAD/CAM related data exchange between design and analysis, and improve the analysis efficiency. The structure and the exchange of internal information in an integrated design and analysis system which is based upon a unified representation of NURBS for the geometry and computational analysis, FE mesh and shape function in FEA is shown in Fig. 8. For more details on this see [1].

Main points in future research:

- The basic objective is integration of computational mechanics with CAD/CAM/CAE and CAGD.
- The 'NURBS' too have their problems. A better approach will be to look and borrow concepts from physics.
- NURBS are 'ONLY' suitable to describe a '2D descriptive geometric object' like an open 2D/3D surface. Geometric represenattion of closed surfaces or surfaces defined over solid is computationally too expensive with NURBS. Hence, if one is willing to explore FEM for volumetric objects then computational geometric resresentstion will be better.

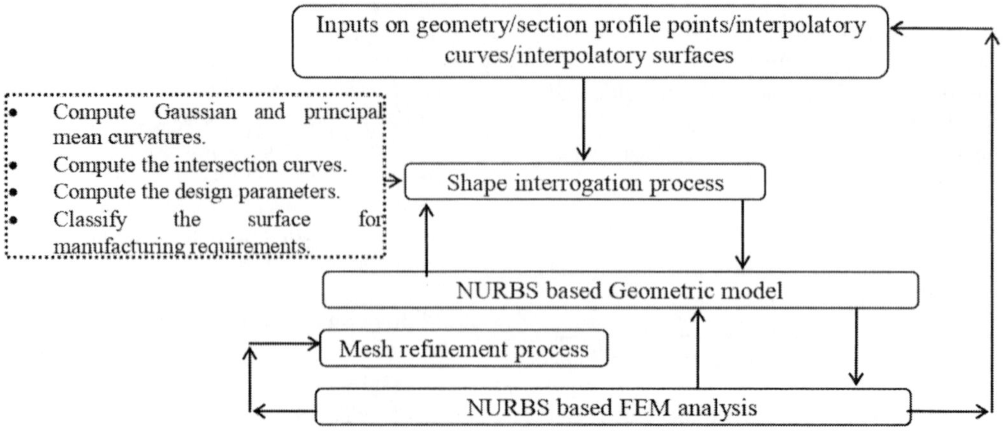

Figure 8: The integrated design and analysis system with 'Iso-geometric analysis'.

5. ACKNOWLEDGEMENTS

This research was supported by the Korea Science and Engineering Foundation (KOSEF), Ministry of Education, Science and Technology (MEST), Government of Korea via grant number: 2009-0081219.

6. REFERENCES

1. COTTRELL, J. A., HUGHES, T. J. R., and BAZILEVS, Y., 'Isogeometric Analysis: Toward Integration of CAD and FEA', 1st edition, Wiley International, USA, 2009.
2. GORDON, S., 'Complexity of Engineering Structures', Electric Boat Division, General Dynamics, USA, 2009.

7. AUTHOR BIOGRAPHY

Tae-wan Kim holds the current positions of Associate Professor in the Department of Naval Architecture and Ocean Engineering, College of Engineering; Director, Planning and Coordination Office, Advanced Institutes of Convergence Technology, Associate Head, Future IT-Convergence Program, and Vice President, Korean Society of CAD/CAM Engineers, Seoul National University, Seoul, Republic of Korea. He is responsible for the teaching and research in the areas of computer-aided design (CAD), ship production management, system modeling and simulation, information technology application for shipbuilding industry, and PDM/PLM (Product Development/Lifecycle Management).

BACK TO CONTENTS

Keynote address: ICSOT-2009,

IIT Kharagpur, India, December 10-11, 2009

Future Challenges in Computer Aided Ship Design and Production

2009. 12. 10 (Thursday)

Seoul National University
Department of Naval Architecture
and Ocean Engineering

Tae-wan Kim

taewan@snu.ac.kr
http://caditlab.snu.ac.kr/

Contents

1. Introduction
   Current status of ship design and production
2. **Our researches:** computer aided ship design and production
   Surface generation over T-junction with continuity
3. **Our researches:** computer aided ship design and production
   Development of 'Bill of Materials (BOM)' kernel of product lifecycle management model for shipbuilding industries
4. **Our researches:** computer aided ship design and production
   Development of a logic based product lifecycle management System for shipbuilding industry
5. **Our researches:** computer aided ship design and production
   Optimum design of an ultra-low motion semi-submersible floating oil and gas production system
6. **Listing of some of the challenges of computer aided ship design and production.**
7. **Conclusions and future directions of research**

World shipbuilding company rankings

NO	Nation	Company	Max dock Capacity				Amount of order		
			Length	Beam	DWT	CGT	EA	DWT	CGT
1	K	Hyundai H.I.	642	92	9,032	3,861	253	33,509	9,754
2	K	Daewoo SB	530	131	5,559	2,396	186	27,900	8,861
3	K	Samsung H.I.	640	98	5,618	2,531	188	23,411	8,792
4	K	Hyundai Mipo	380	76	2,898	1,543	227	8,726	4,811
5	K	STX Shipbuild.	385	74	2,643	1,149	178	19,593	4,811
6	K	Hyundai Samho	500	100	3,535	1,086	117	14,303	4,363
7	C	Dalian Shipbld. Ind.	674	96	3,262	820	111	18,458	3,419
8	C	Jiangnan Changxing	580	120	1,455	290	119	13,684	3,253
9	C	Jiangsu Rongsheng	2,900	140	701	141	85	16,884	2,826
10	K	Sungdong S.B.	-	-	1,291	328	93	13,144	2,629
11	J	Oshima S.B. Co.	299	80	2,007	552	124	8,444	2,265
12	C	Waigaoqiao S/Y	540	106	3,389	657	63	12,608	2,077
13	J	Tsuneishi Zosen	330	55	2,784	670	96	8,206	1,967
14	C	Jiangsu New YZJ	1,000	90	674	212	79	5,235	1,910
15	C	Zhoushan Jinhaiwan	1,096	120	217	47	68	10,264	1,865
16	K	SLS Shipbuilding	-	-	522	247	79	3,691	1,721
17	C	Hudong Zhonghua	360	92	1,120	452	76	5,304	1,714
18	C	STX Dalian	680	135	115	33	78	4,697	1,546
29	C	New Times S.B.	588	106	1,163	275	53	8,417	1,533
20	P	HHIC-Phil. Inc.	-	-	257	129	44	5,562	1,518
21	C	Nantong Cosco KHI	500	80	1,300	226	44	7,135	1,509
22	C	Jinling Shipyard	500	60	403	227	79	4,595	1,425
23	K	Hanjin H.I.	302	50	1,160	611	47	4,261	1,374
24	C	Bohai Shipbld.	480	107	754	173	58	7,119	1,369
25	K	SPP Shipbuilding	230	45	451	219	91	3,951	1,362
26	C	Dayang S.B.	360	80	484	205	123	3,953	1,351
27	J	Namura Zosensho	450	70	1,664	334	37	8,500	1,314
28	J	Imabari S.B.	420	89	2,347	387	38	7,889	1,264
29	C	Guangzhou S.Y. Int.	200	36	793	394	56	2,495	1,259
30	K	SPP Plant & Shipbldg	310	86	703	341	58	3,697	1,208
31	T	CSBC	950	92	1,145	331	37	2,563	1,185
32	J	Mitsubishi H.I.	990	100	1,671	688	31	2,105	1,183
33	J	Universal S.B.	600	75	1,840	357	33	7,480	1,153
34	C	Shanghai S.Y.	860	44	488	246	51	3,366	1,108

K	Korea
J	Japan
C	China
P	Philippines
T	Taiwan

*DWT, CGT unit: million
*Length, Beam unit: meter

* Clarkson, World Shipyard Monitor, 2009. 7

Introduction (1/5)

- Ships are important and strategic engineering structures (because they are important for industrial growth and defense).

- The modern engineering designs are becoming increasingly more complex because of the pressing demands of economies of scale, clean environment, and advancements in material science and technologies.

- Ships are one of the most complex engineering structures. The design and analysis time for a ship runs in months.

- Therefore, to remain competitive and cost efficient it is essential for a shipyard to design a new ship at a shorter lead time and manufacture it with low cost and time.

Introduction (2/5)

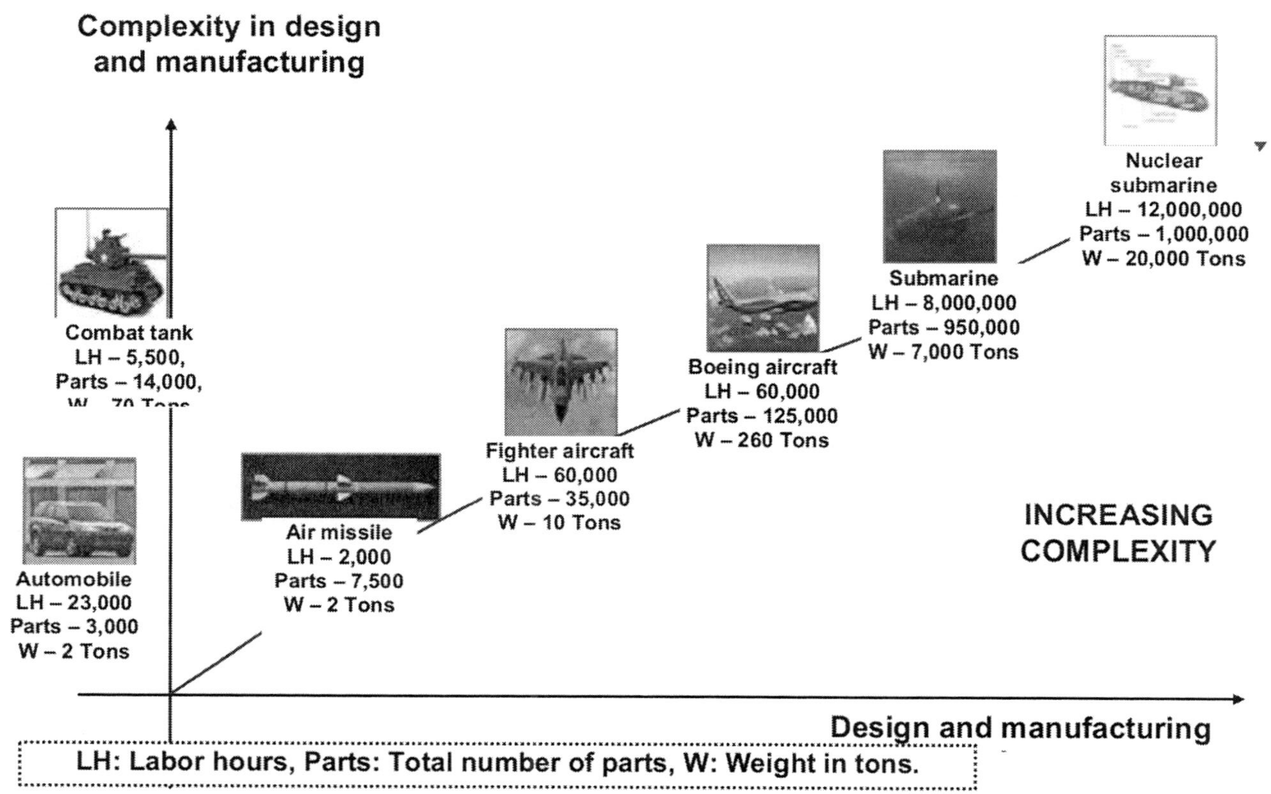

Figure 1: The complexity in design and manufacturing of modern engineering structures.

Introduction (3/5)

Table 1: The various times stages of detailed design for different types of ships.

(This table shows total time taken in various design stages for ships)

S. No.	Technical design division 1 CAD/CAM/CAE	Technical design division 2 FEM/Structural	Technical design division 3 CFD/Hydrodynamics	Technical design division 4 Integrated design with drawings (inputs from (1-3))
1. General/multi purpose cargo ships Deadweight = 200 K tons	0 – 0.74 months Number of d_i = 1 to 128	0.6 – 1.38 months Number of d_i = 1 to 152	1.00 – 2.25 months Number of d_i = 1 to 96	2.15 – 3.45 months Number of d_i = 1 to 250
2. Small to medium sized container ships Capacity ≤ 10 K	0 – 1.34 months Number of d_i = 1 to 173	1.0 – 2.87 months Number of d_i = 1 to 216	2.50 - 4.65 months Number of d_i = 1 to 130	2.25 – 6.56 months Number of d_i = 1 to 285
3. Very large container ships, and passenger carrier ships Capacity > 10 K	0 – 1.67 months Number of d_i = 1 to 293	1.25 – 2.84 months Number of d_i = 1 to 350	2.00 – 4.92 months Number of d_i = 1 to 210	4.50 – 8.27 months Number of d_i = 1 to 444
4. Chemical carrier ships Deadweight = 20 – 30 K tons	0 – 1.36 months Number of d_i = 1 to 300	1.20 - 2.65 months Number of d_i = 1 to 447	2.50 – 4.15 months Number of d_i = 1 to 214	3.75 – 6.78 months Number of d_i = 1 to 468
5. LNG/LPG carrier ships LNG MOSS	0 – 2.15 months Number of d_i = 1 to 324	2.00 – 4.22 months Number of d_i = 1 to 447	4.00 – 7.16 months Number of d_i = 1 to 211	6.50 – 9.79 months Number of d_i = 1 to 491
Nomenclature: start time – finish time, 0th is the beginning of the design process.				

Introduction (4/5)

Table 2: The key event times of various manufacturing stages for different types of ships.

(This table shows total time required in design and production for various types of ships)

S. No.	Technical design division 4 Integrated design with drawings (inputs from (1-3))	Start of manufacturing process			
		WC – KL	KL – LC	LC – DL	Total time from design to production
1. General/multi purpose cargo ships Deadweight = 200 K tons	2.15 – 3.45 months Number of d_i = 1 to 250	2.15 – 5.11 months Number of d_i = 1 to 48	5.11 – 7.11 months Number of d_i = 1 to 32	7.11 – 8.15 months Number of d_i = 1 to 17	8.15 months
2. Small to medium sized container ships Capacity ≤ 10 K	2.25 – 6.56 months Number of d_i = 1 to 285	2.25 – 5.21 months Number of d_i = 1 to 48	5.21 – 7.21 months Number of d_i = 1 to 32	7.21 – 8.41 months Number of d_i = 1 to 20	8.41 months
3. Very large container ships, and passenger carrier ships Capacity > 10 K	4.50 – 8.27 months Number of d_i = 1 to 444	4.50 – 7.49 months Number of d_i = 1 to 50	7.49 – 9.44 months Number of d_i = 1 to 32	9.44 – 11.00 months Number of d_i = 1 to 25	11.00 months
4. Chemical carrier ships Deadweight = 20 – 30 K tons	3.75 – 6.78 months Number of d_i = 1 to 468	3.75 - 7.50 months Number of d_i = 1 to 60	7.50 – 9.375 months Number of d_i = 1 to 30	9.375 – 11.03 months Number of d_i = 1 to 27	11.025 months
5. LNG/LPG carrier ships LNG MOSS	6.50 – 9.79 months Number of d_i = 1 to 491	6.50 – 11.40 months Number of d_i = 1 to 79	11.40 – 14.20 months Number of d_i = 1 to 45	14.20 – 20.15 months Number of d_i = 1 to 96	20.15 months

Nomenclature: start time – finish time, 0th is the beginning of the design process, WC = work commencement and start of plate cutting operations, KL = keel laying, LC - launching commencement, and DL – delivery line.

Key aims and focuses of research

AIMS - In a fiercely competitive market it is important to develop new products at a shorter product development time.

FOCUSES: And, in our opinion computers can play an important role in achieving the designs of new, novel, and greener (with less energy consumption and less polluting to the marine/global environment) ships at a shorter product development time in future.

Contents

Curve Network Interpolation Problems in Ship Hull-form Design (1/6)

Given :
curve network of hullform

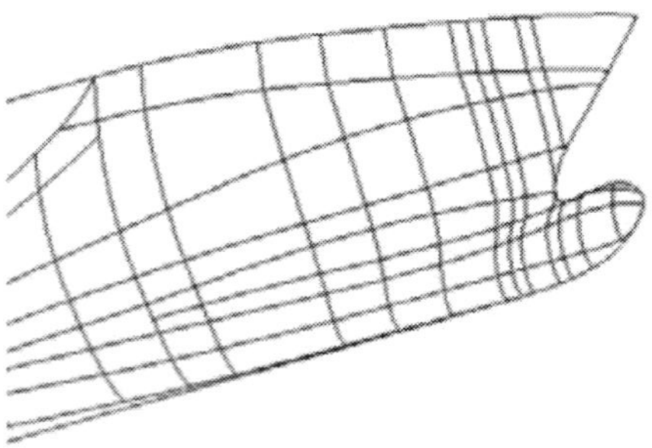

- **Irregular topology**
- **Knuckle edges**

Find :
smooth hull-form surfaces

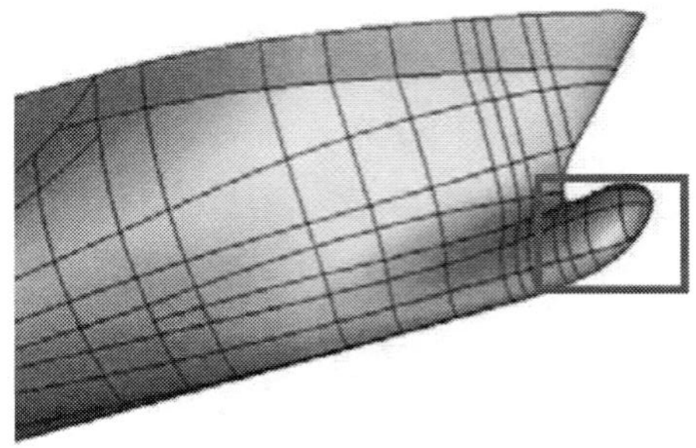

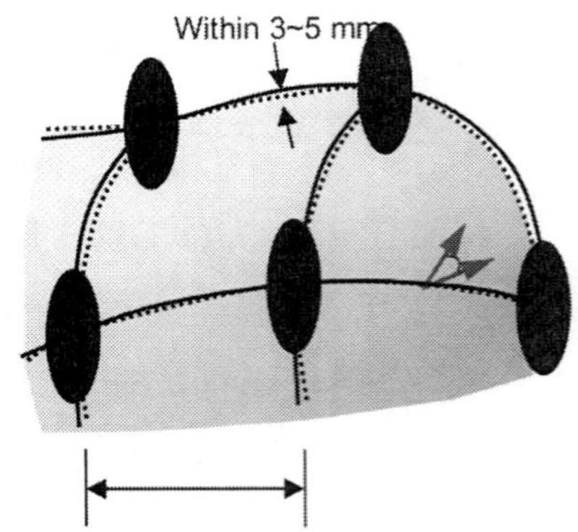

Requirements

- **Exact Interpolation** of given curve network *or* **Max. distance error** between given curve network and generated surface **< tolerance***
- Smoothness: **exact G^1**

→ Generation of G^1 surface that interpolates over the given irregular curves network without any major changes in the given curves network.

* Acceptable tolerance in shipbuilding industry is about 3~5 mm for ships of 200~300m length

Why keep the curve network unchanged ? (2/6)

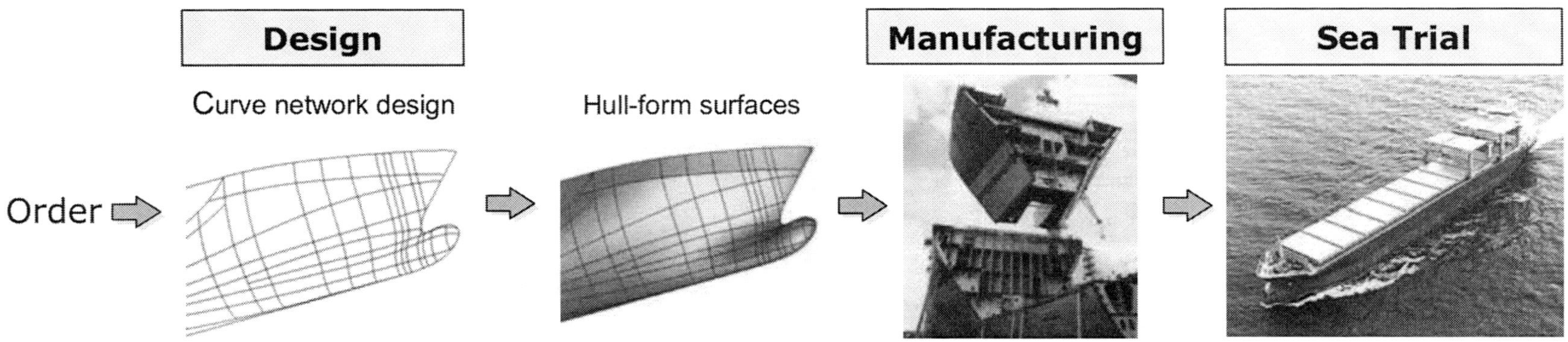

- Ship hull-form is deeply related to water resistance and propulsion efficiency which affect ship velocity

- Constraints from hull-form designers
  - The curve network is almost optimal design result by hull form designers for velocity constraints given by the ship owner (i.e. service speed $\geq$ 20 knots)
  - If finally produced ship does not satisfy the service speed, ship builder can be fined with a penalty (i.e. its amount can be millions of dollars)
  - Therefore, it is important to generate hull-form surfaces that exactly interpolate given curve network*

* Actually, up to errors of about 3~5mm for 200~300m length ship is allowed In ship industry

T-junction Problem in Ship Hull-form Design (3/6)

- T-junction usually appears in ship hull-form curves network.
- Usually T-shaped patches for irregular topology are not allowed in current CAD/CAM systems.
- Regions with a T-junction are subdivided into 3-sided sub-regions in previous works*.
- We want to generate G^1 surface with a T-junction without subdivision or triangulation, and modification of boundary curves. This is one of the most challenging problems for the last 30 years in CAD/CAM area. If this is resolved, then many problems can be resolved accordingly, e.g. analysis related problems because the topology will become easy to handle for mesh generation.

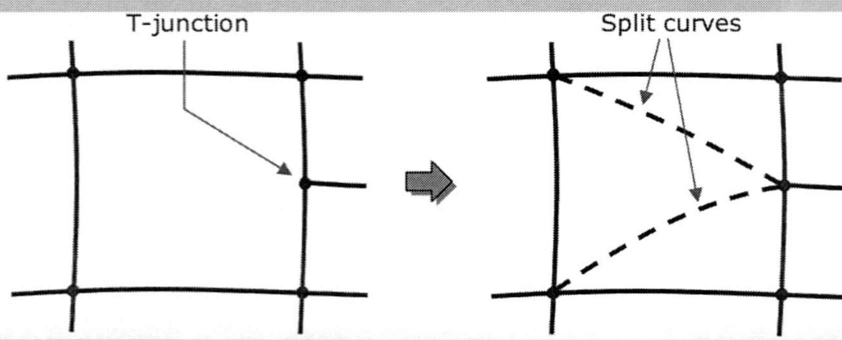

* Doo-Yeoun Cho, Kyu-Yeul Lee, and Tae-wan Kim, "Interpolating G^1 Bézier surfaces over irregular curve networks for ship hull design," Computer-Aided Design, Vol. 38, pp. 641-660, 2006.

G^1 Condition of Two Adjacent Surfaces (4/6)

- Two surfaces share a common boundary curve.
- Common tangent plane at any point on the boundary curve.

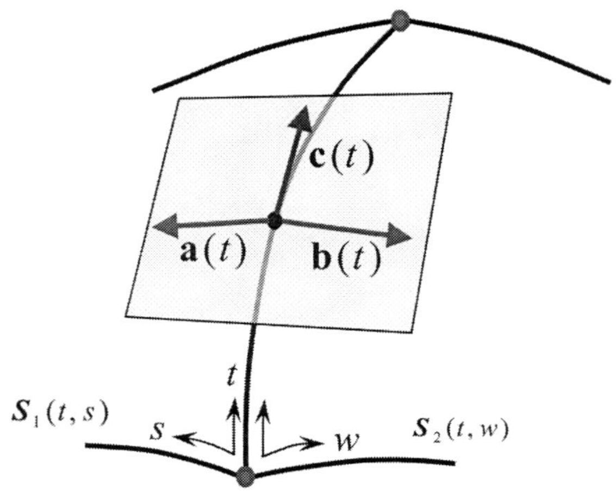

3 vectors in a plane

$$\alpha(t)\cdot\mathbf{a}(t)+\beta(t)\cdot\mathbf{b}(t)+\gamma(t)\cdot\mathbf{c}(t)=\vec{0}$$

Farin's G^1 condition*

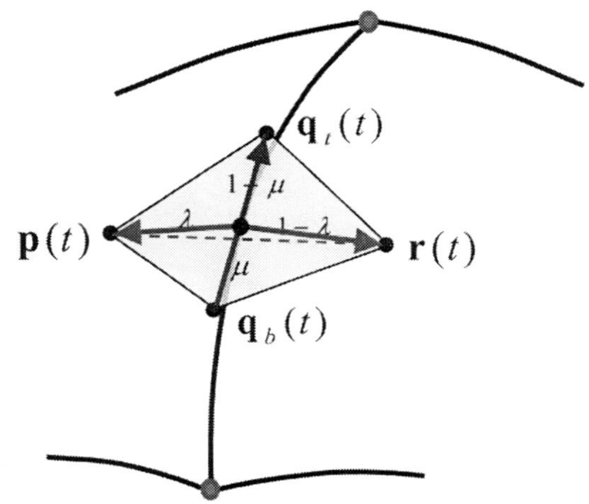

4 points in a plane

$$[1-\lambda(t)]\cdot\mathbf{p}(t)+\lambda(t)\cdot\mathbf{r}(t) = [1-\mu(t)]\cdot\mathbf{q}_b(t)+\mu(t)\cdot\mathbf{q}_t(t)$$

→ Scalar weight functions were assumed to be lienar

* G. Farin, "A construction for the visual C^1 continuity of polynomial surface patches," Computer Graphics and Image Processing, Vol. 20, 1982.

G^1 Condition of Three Adjacent Surfaces with a T-junction (5/6)

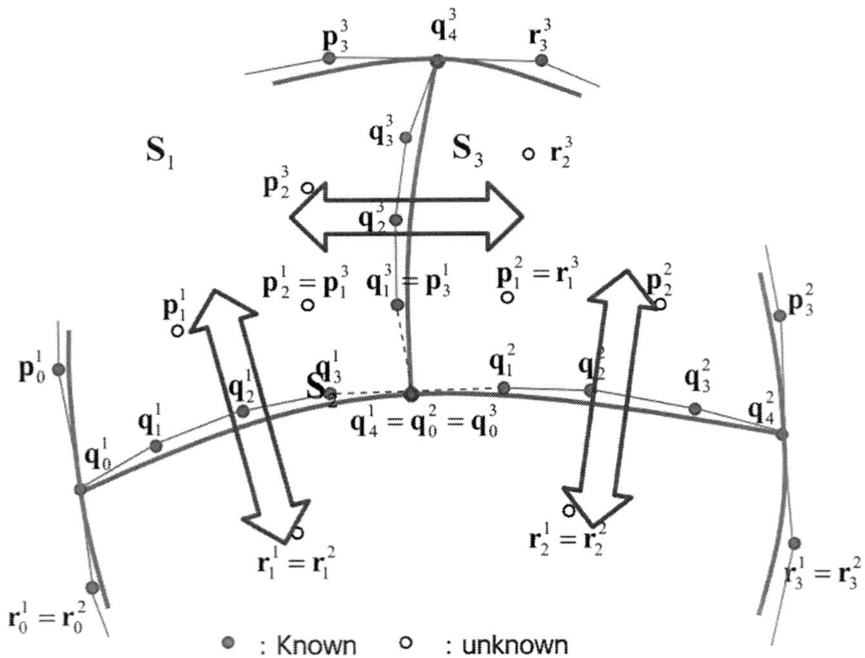

- G^1 continuity condition

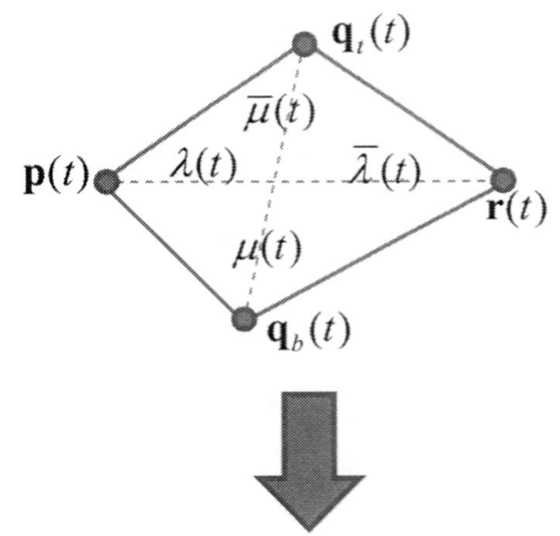

$$\overline{\lambda}(t)\mathbf{p}(t) + \lambda(t)\mathbf{r}(t)$$
$$= \overline{\mu}(t)\mathbf{q}_b(t) + \mu(t)\mathbf{q}_t(t)$$

Example of G^1 Surface with a T-junction (6/6)

- Initial surface with T-junction using Coons' Boolean sum method

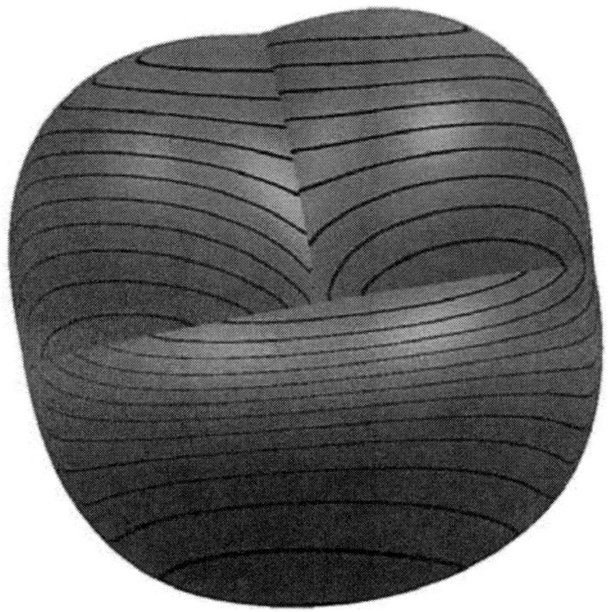

- Generated G^1 surface from the given initial curves network or surface (implicit).
  - On-going research by us, SNU, Korea.

Contents

1. **Introduction**
2. **Our researches:** computer aided ship design and production
   Surface generation over T-junction with continuity
3. **Our researches:** computer aided ship design and production
   Development of 'Bill of Materials (BOM)' kernel of product lifecycle management model for shipbuilding industries
4. **Our researches:** computer aided ship design and production
   Development of a logic based product lifecycle management System for shipbuilding industry
5. **Our researches:** computer aided ship design and production
   Optimum design of an ultra-low motion semi-submersible floating oil and gas production system
6. **Listing of some of the challenges of computer aided ship design and production.**
7. **Conclusions and future directions of research**

The range of BOM kernel (1/14)

BOM kernel ⇨ **Base platform to customize and develop for constructing shipbuilding PLM**

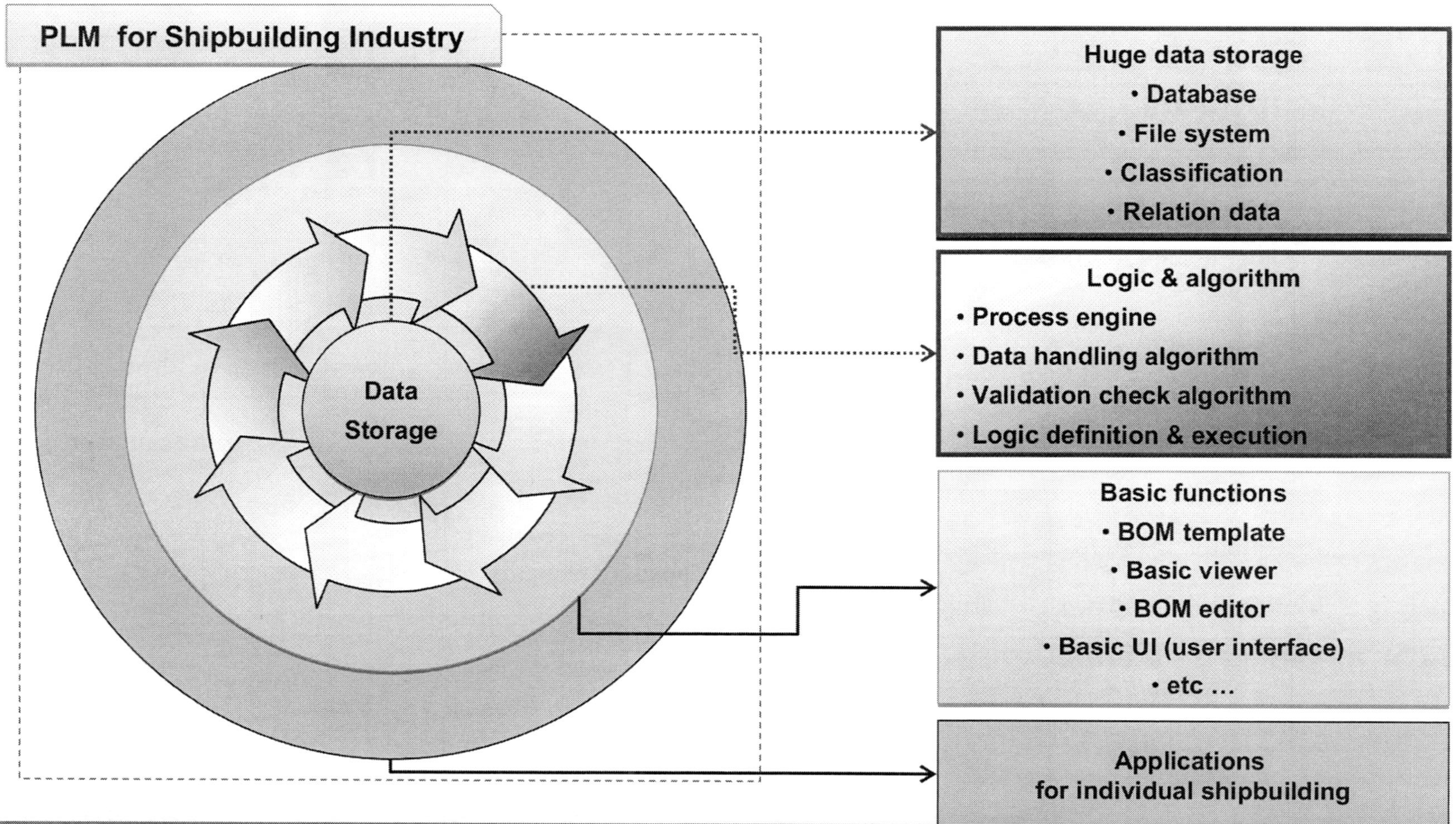

Requirements & needs (2/14)

- ✓ regular process control
- ✓ irregular process control

- ✓ dynamic data structure
- ✓ rich API set

- ✓ functional modularization
- ✓ reconfiguration of module

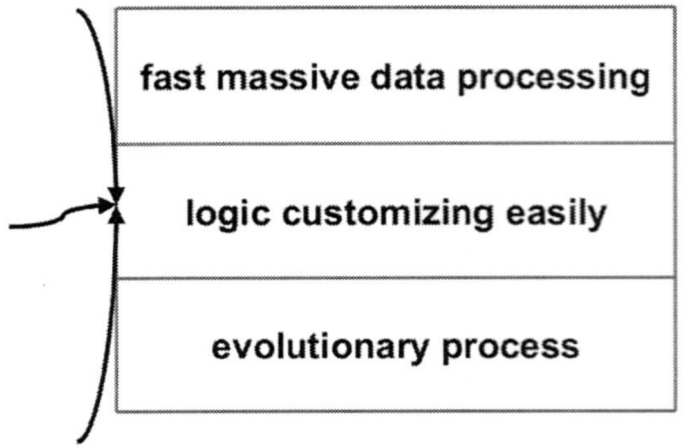

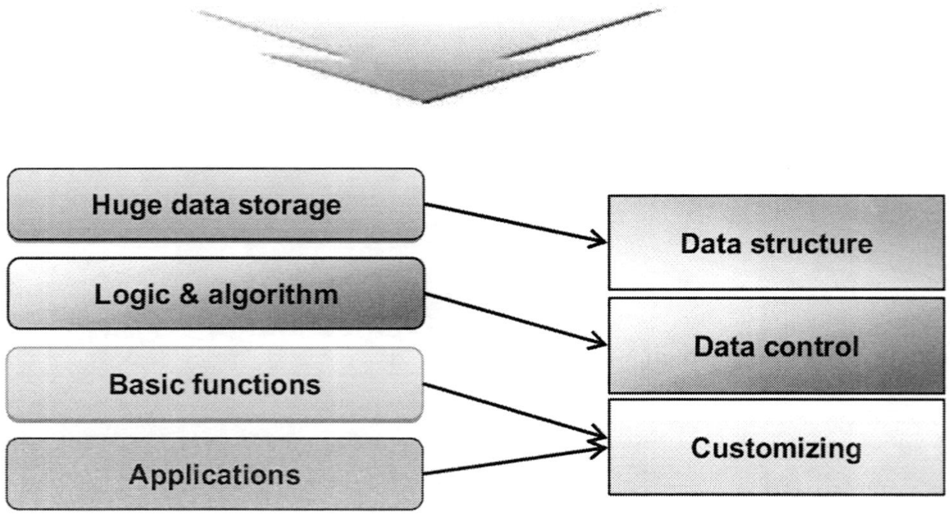

Data structure

Knowledge map based structure

- Acquiring of data independence
- Flexibility of data reconfiguration

Data control

Process based rule management

- Data control
- Building a process from logic
- Applying BPM based evolutionary process

Customizing

Module based development environment

- Modularization of components
- Kernel level API

Concept of shipbuilding BOM structure (3/14)

Contract & estimation → Detail design → Manufacturing design

Base	Building Specification
Structure	Structure of Building Specification
Mapping	Structure of components and equipments - Sentence (attributes – variable or parameter)
Example	

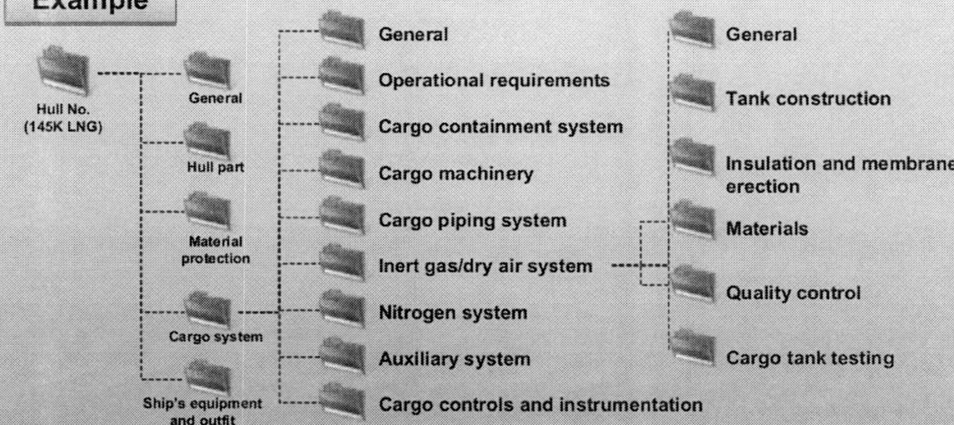

Concept	Step by step growing data, purposed exchanging structure
Purpose	Generation of the BOM by design part or task Generation of related documents Calculation of amount of material and equipments
Use	Improvement of design task (data search & reference, document auto generation)
Solution	Making the BOM structure and data by output of design stage

Concept of shipbuilding BOM structure (4/14)

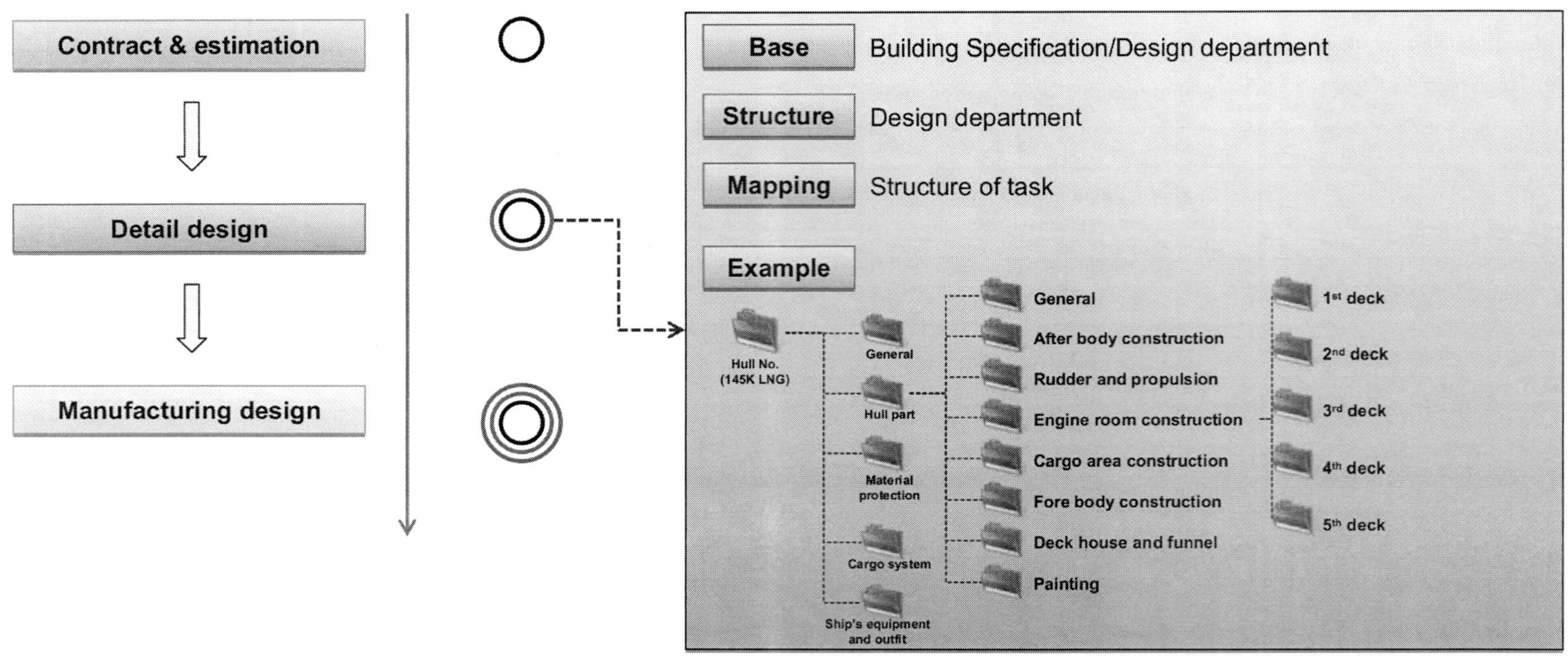

Concept	Step by step growing data, purposed exchanging structure
Purpose	Generation of the BOM by design part or task Generation of related documents Calculation of amount of material and equipments
Use	Improvement of design task (data search & reference, document auto generation)
Solution	Making the BOM structure and data by output of design stage

Concept of shipbuilding BOM structure (5/14)

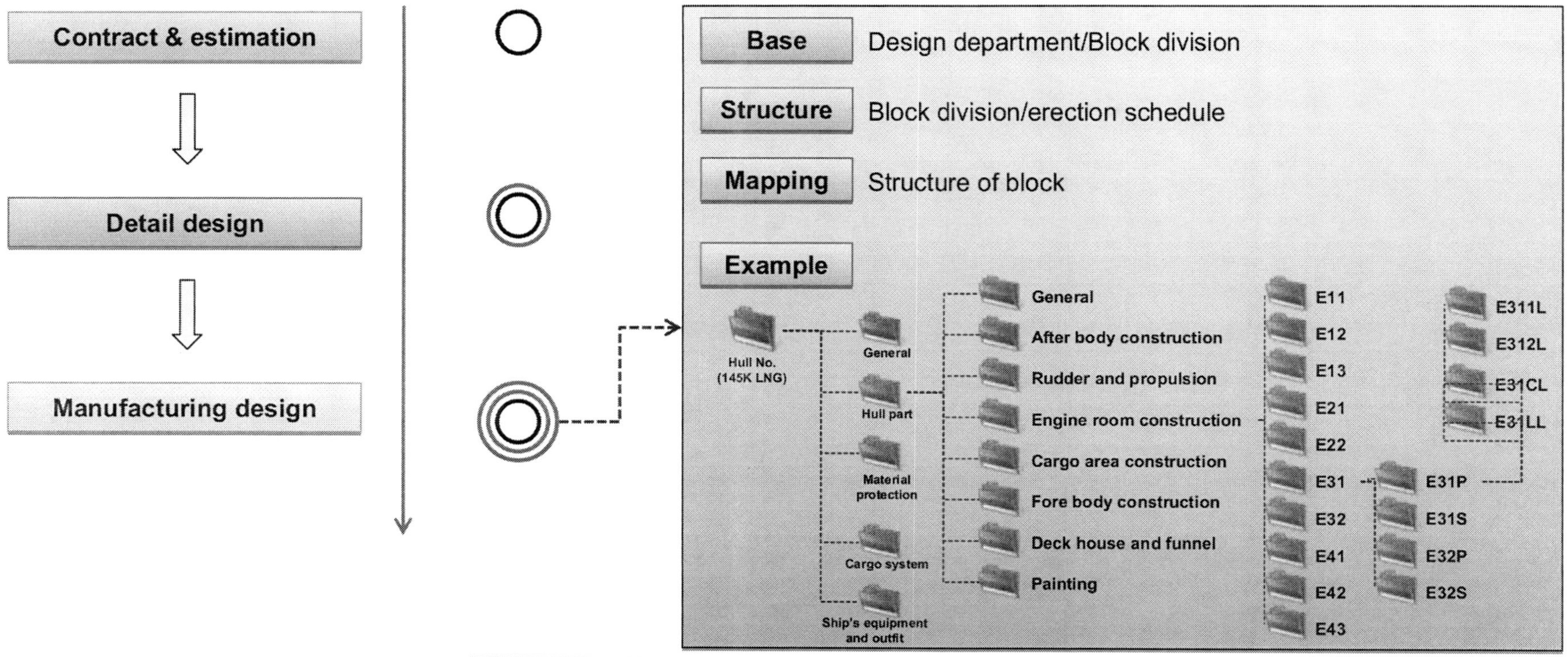

Concept	Step by step growing data, purposed exchanging structure
Purpose	Generation of the BOM by design part or task Generation of related documents Calculation of amount of material and equipments
Use	Improvement of design task (data search & reference, document auto generation)
Solution	Making the BOM structure and data by output of design stage

Data control (6/14)

Knowledge map based database relation

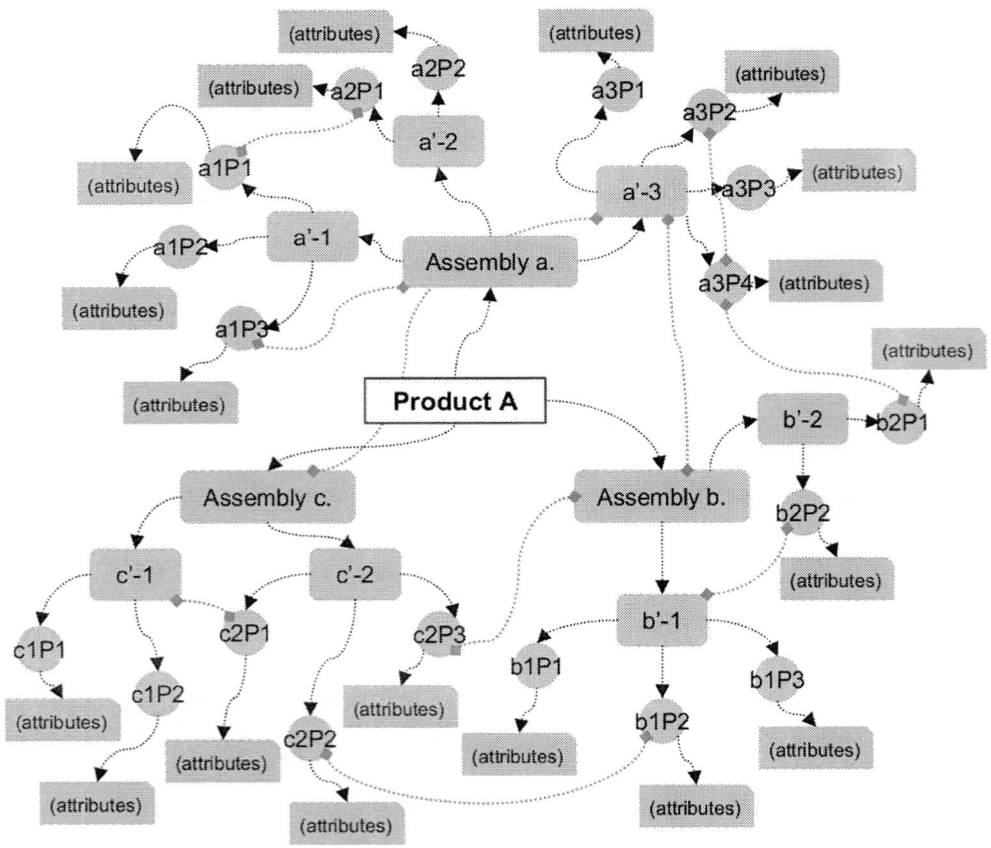

Data flow control using process engine

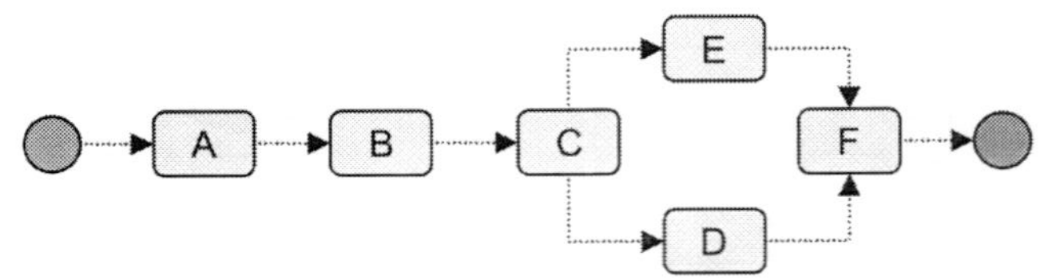

Data control (7/14)

Knowledge map based database relation

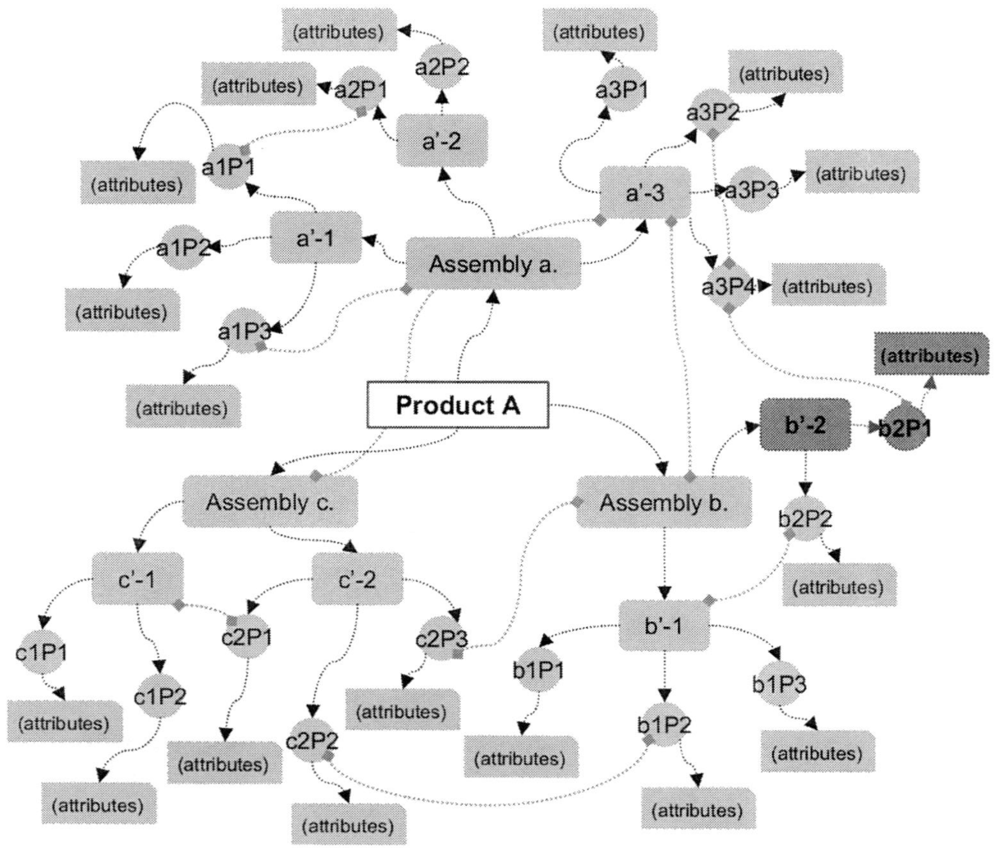

Data flow control using process engine

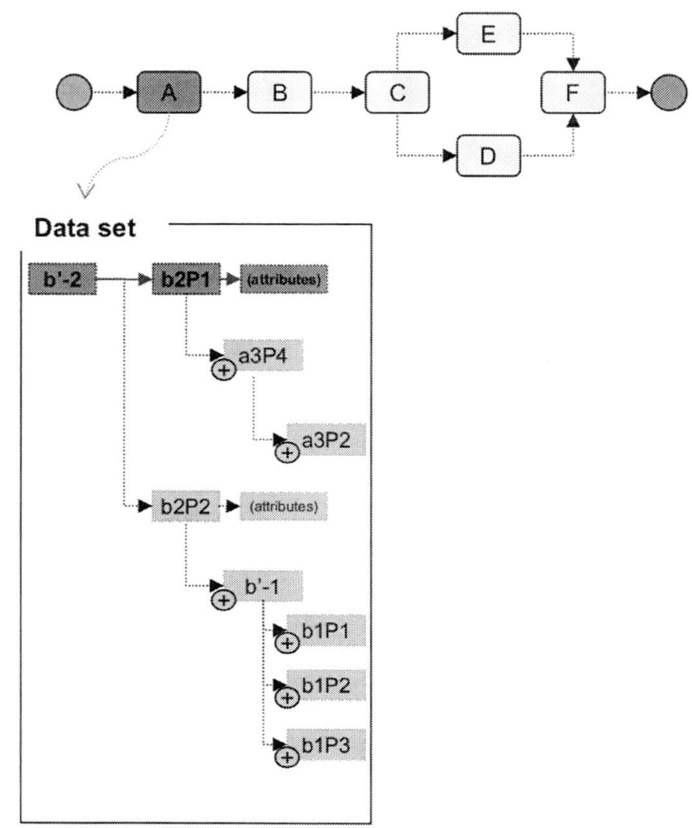

Data control (8/14)

Knowledge map based database relation

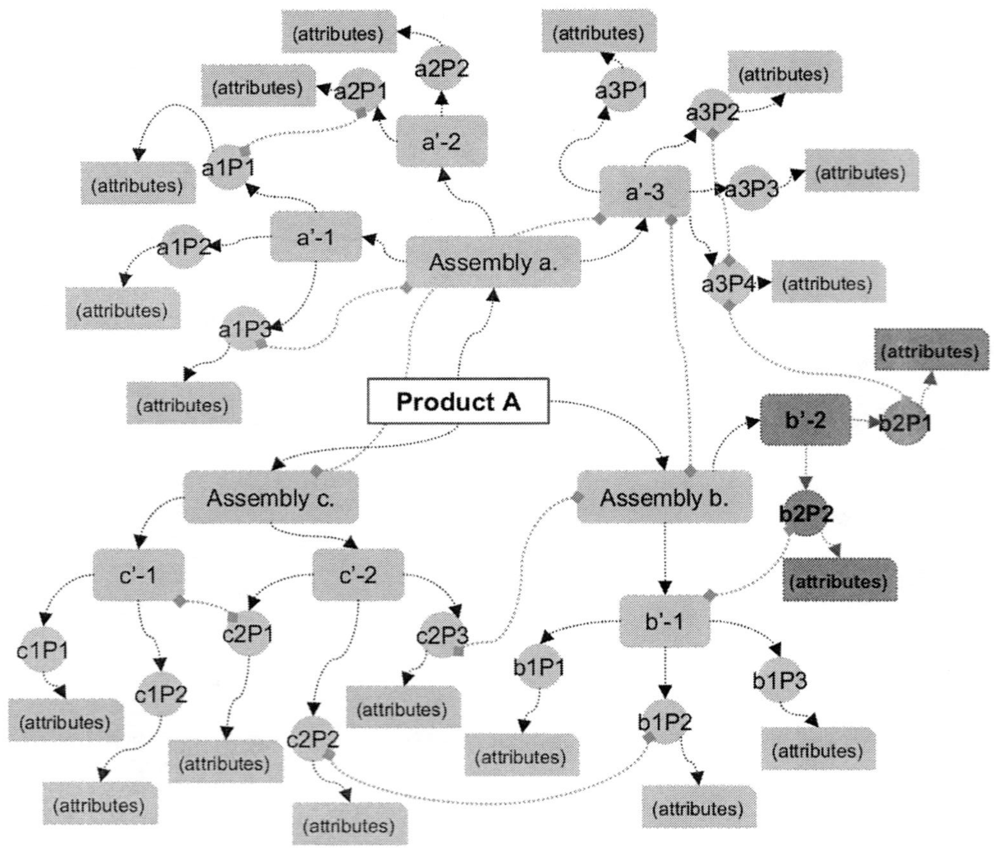

Data flow control using process engine

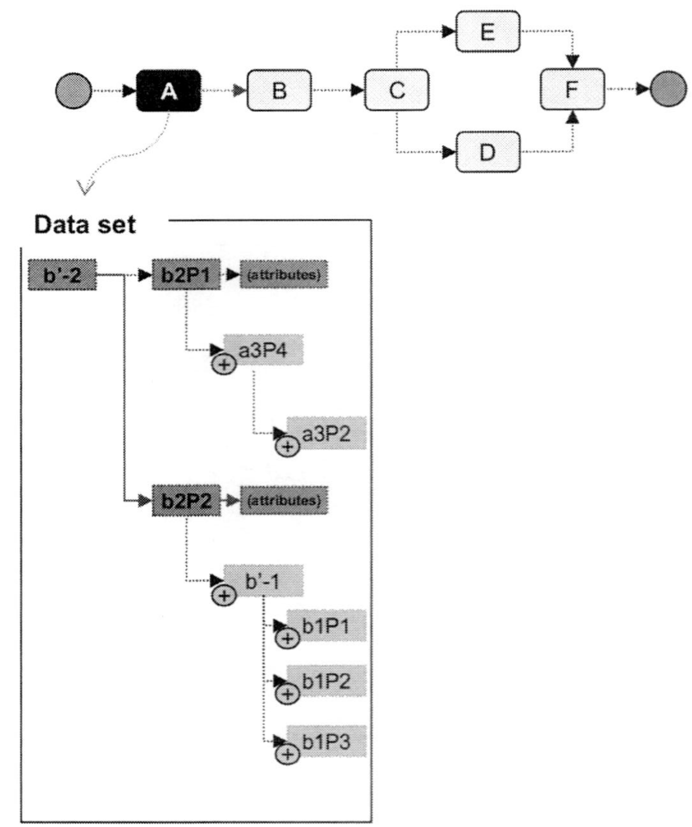

Data control (9/14)

Knowledge map based database relation

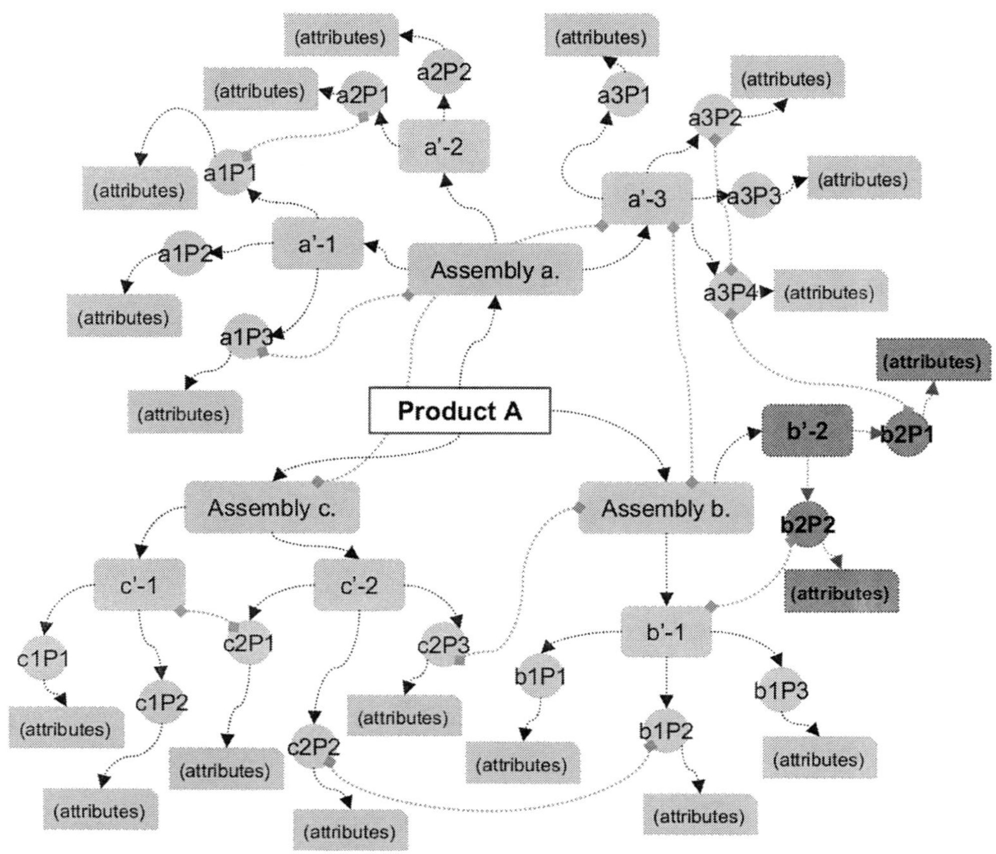

Data flow control using process engine

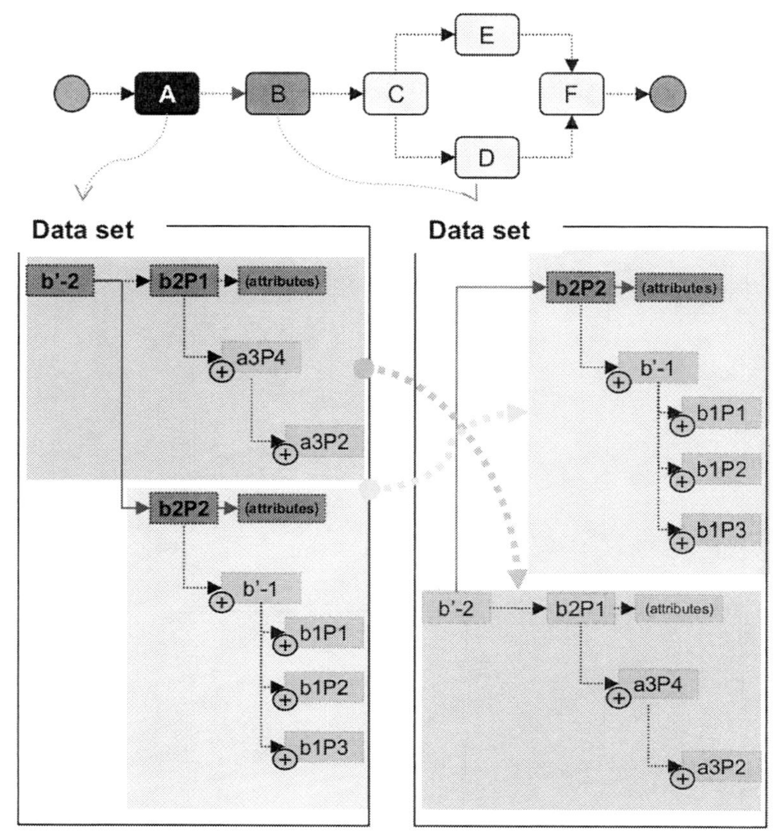

Data control (10/14)

Knowledge map based database relation

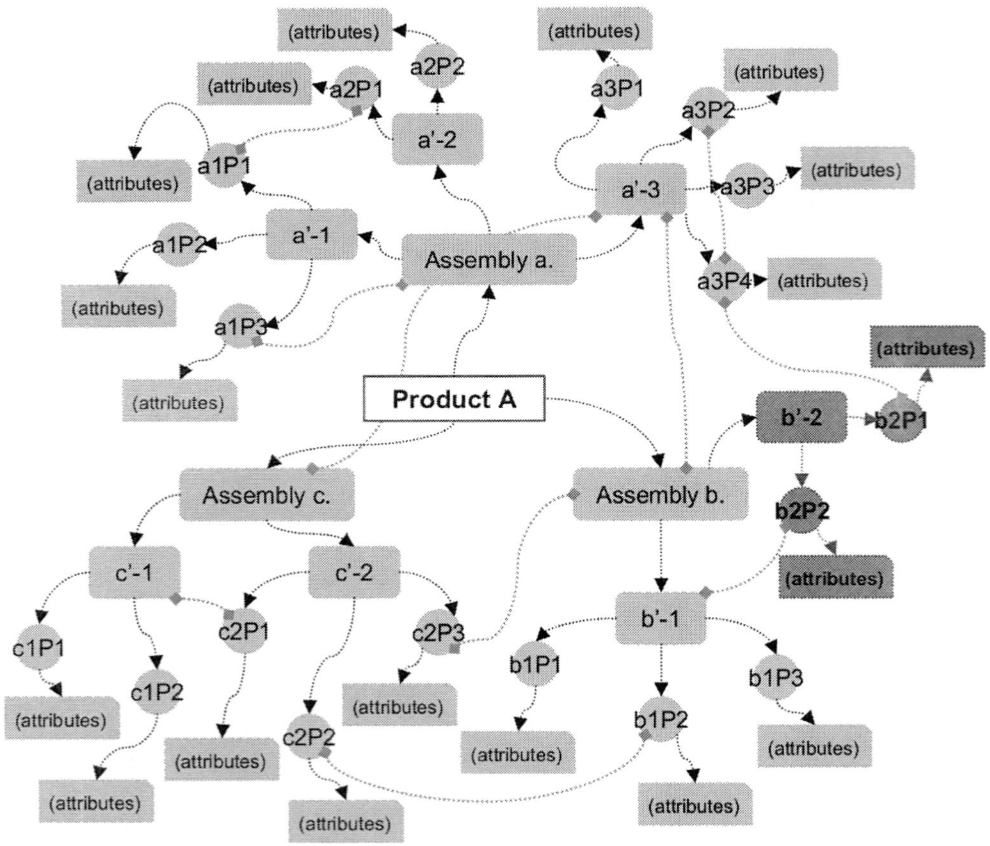

Data flow control using process engine

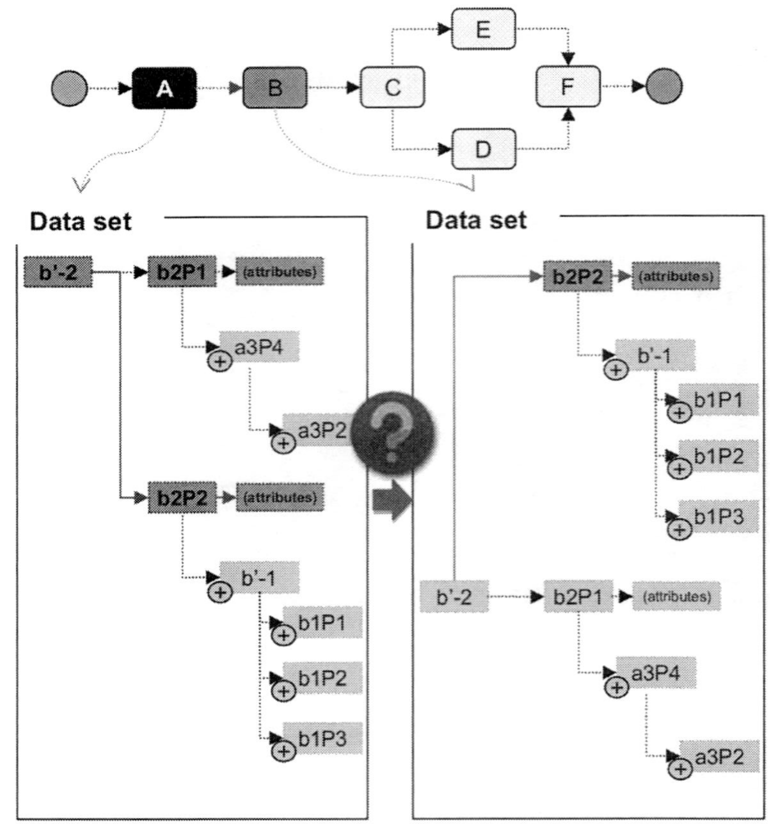

Data control (11/14)

Knowledge map based database relation

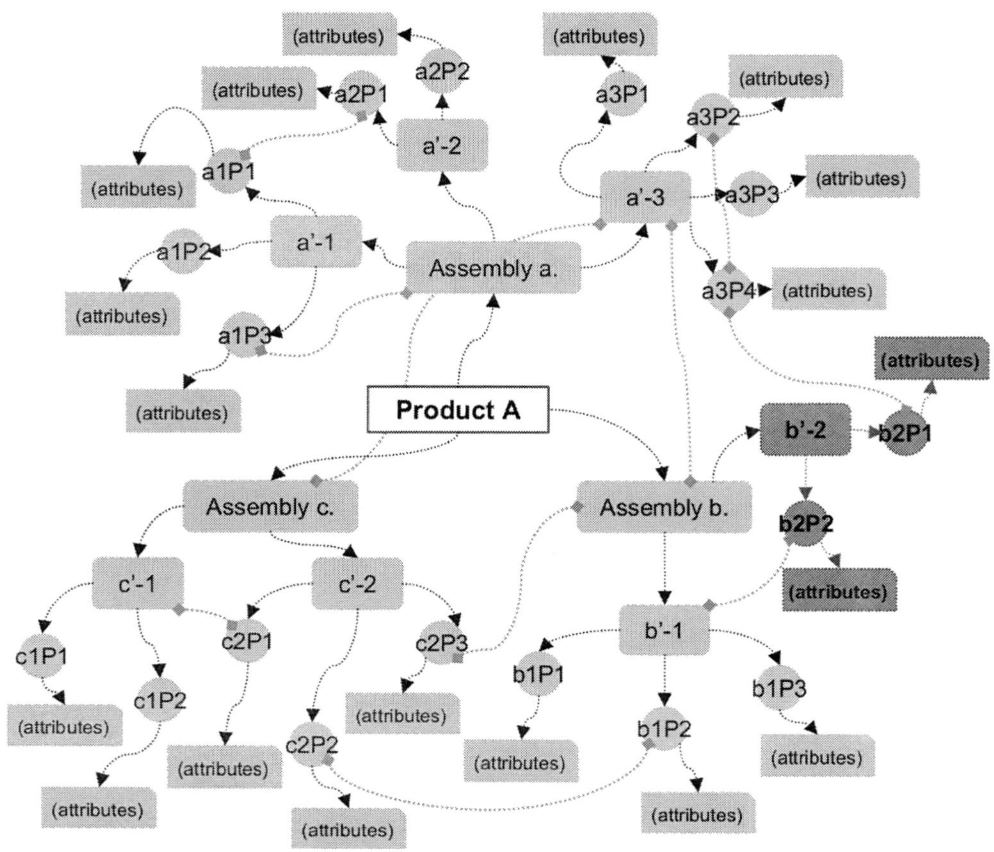

Data flow control using process engine

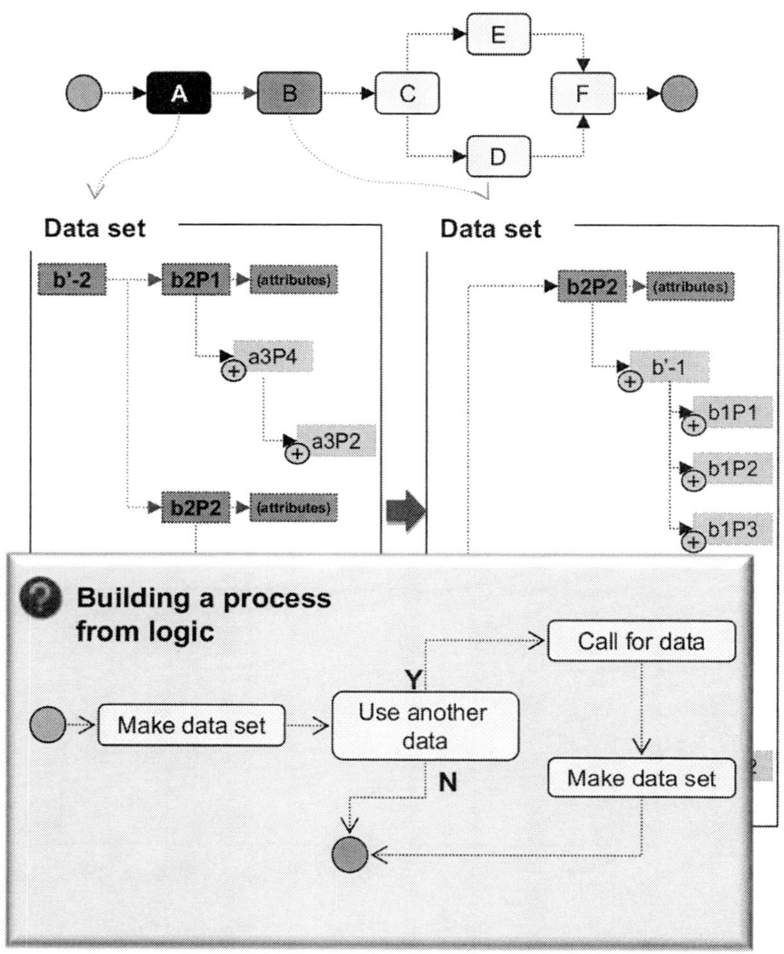

Data control: evolutionary process (12/14)

Defined process of all field

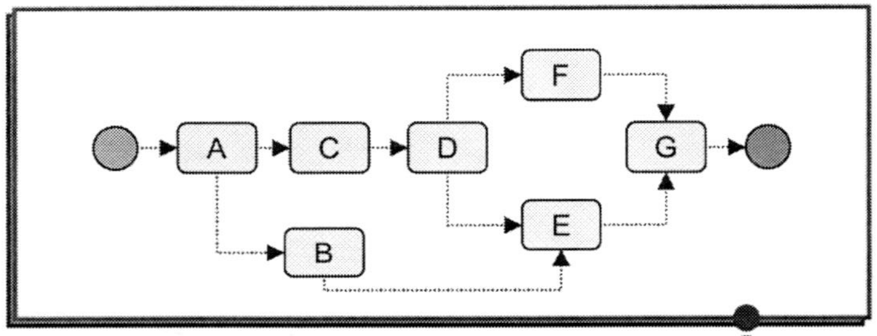

Changed process of all field

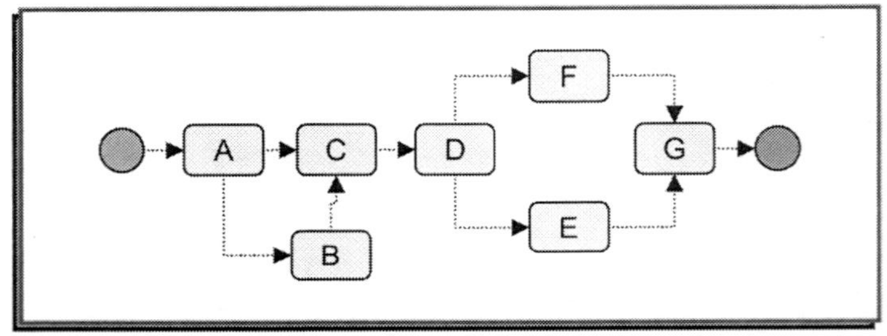

Process instance 1 : 70 times

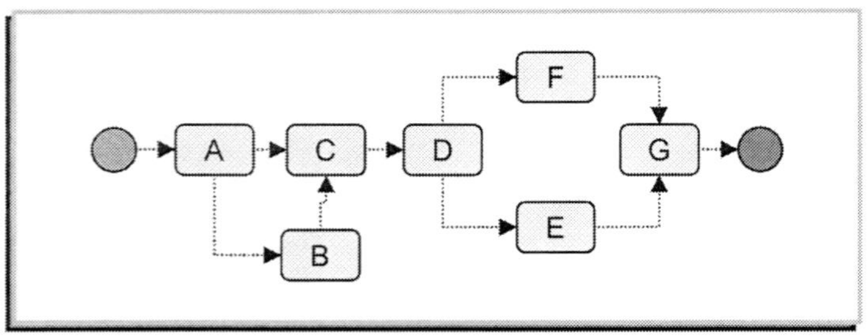

Process instance 3 : 4 times

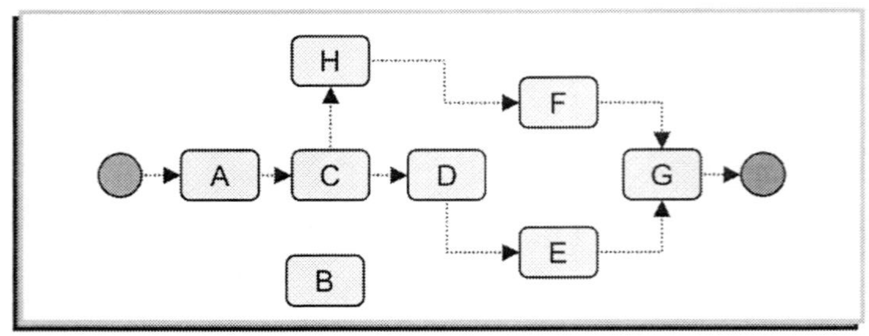

Process instance 2 : 6 times

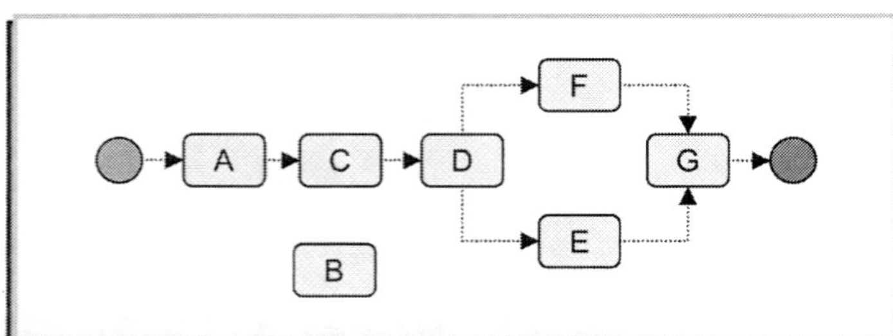

Process instance 4 : 10 times

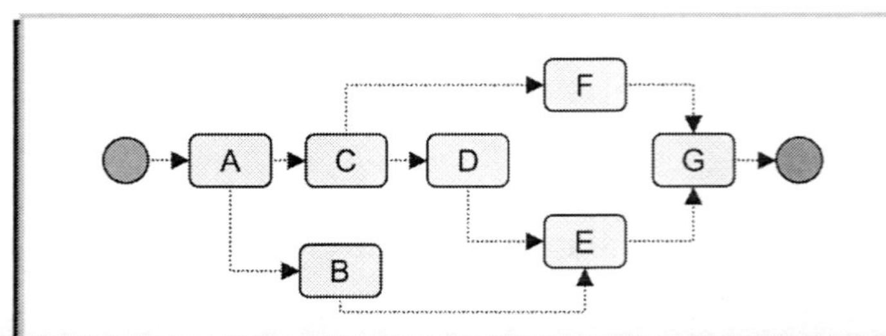

Master plan (13/14)

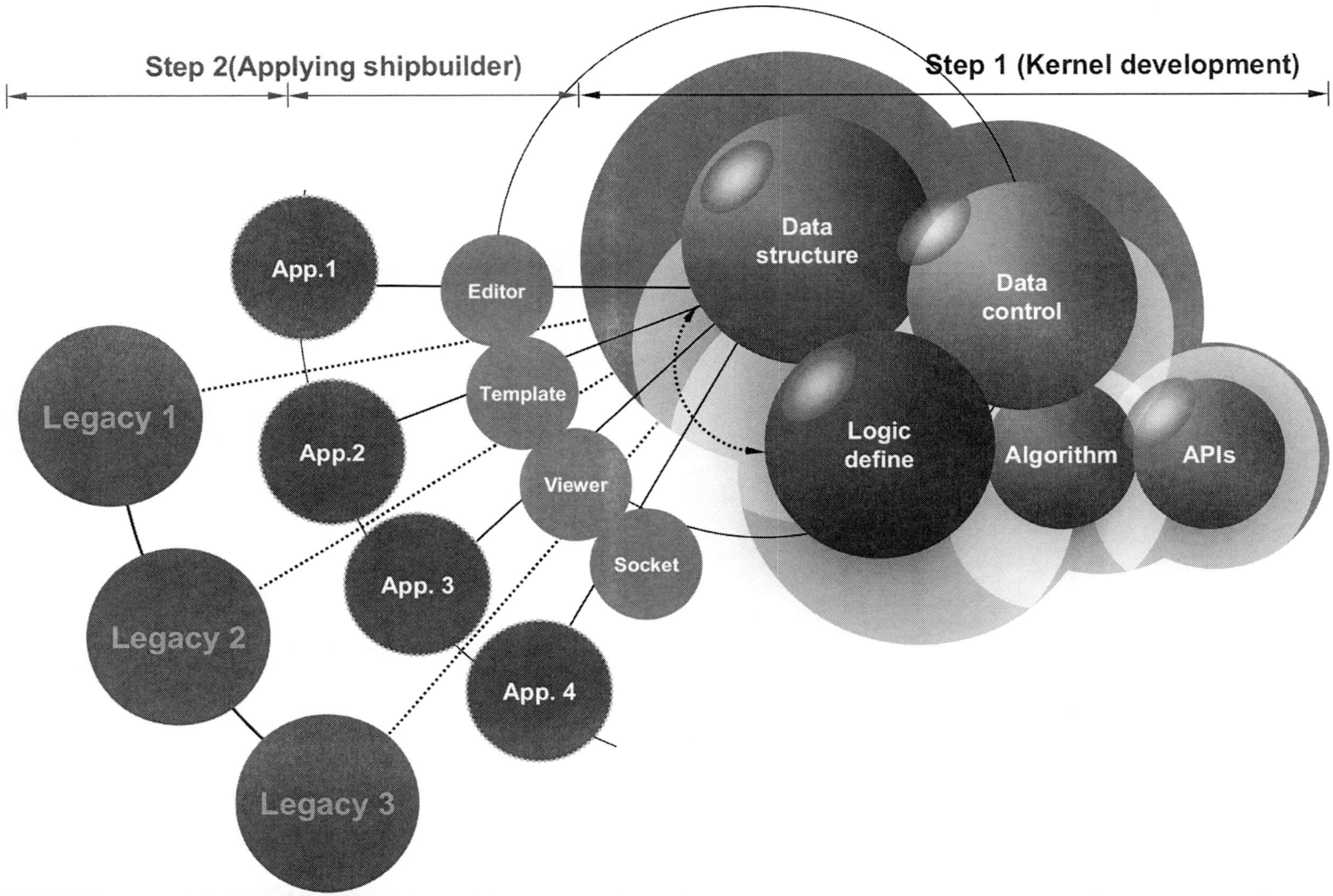

Direction of development (14/14)

Huge data storage
- Database
- File system
- Classification
- Relation data

Logic & algorithm
- Process engine
- Data handling algorithm
- Validation check algorithm
- Logic definition & execution

Basic functions
- BOM template
- Basic viewer
- BOM editor
- Basic UI (user interface)
- etc …

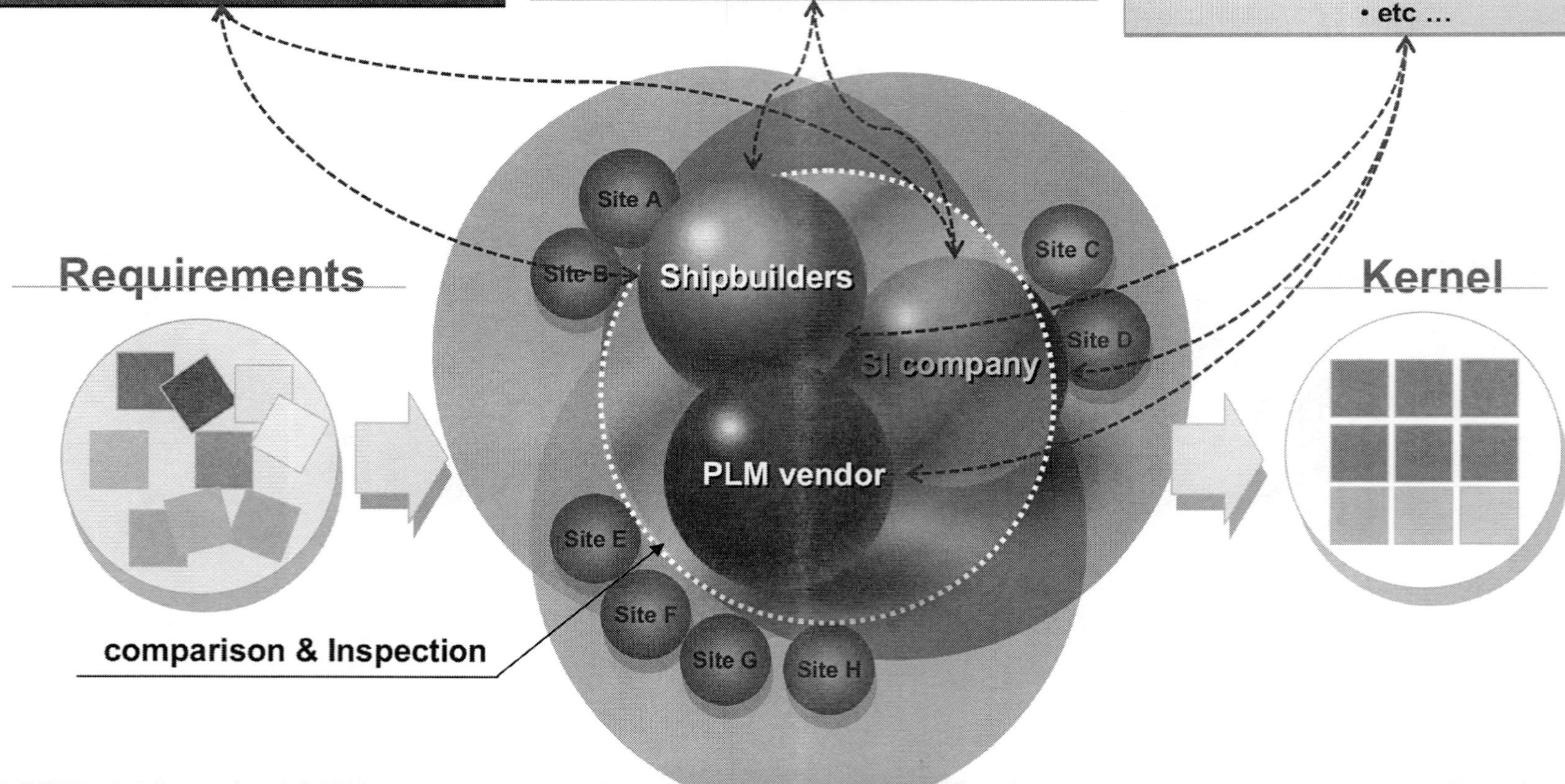

Contents

1. **Introduction**
2. **Our researches:** computer aided ship design and production
   Surface generation over T-junction with continuity
3. **Our researches:** computer aided ship design and production
   Development of 'Bill of Materials (BOM)' kernel of product lifecycle management model for shipbuilding industries
4. **Our researches:** computer aided ship design and production
   Development of a logic based product lifecycle management System for shipbuilding industry
5. **Our researches:** computer aided ship design and production
   Optimum design of an ultra-low motion semi-submersible floating oil and gas production system
6. **Listing of some of the challenges of computer aided ship design and production.**
7. **Conclusions and future directions of research**

Development of a Logic Based Product Lifecycle Management (LBPLM) System for Shipbuilding Industry* (1/5)

- PLM - Process of managing the entire lifecycle of a product.

- PLM systems have been developed and implemented since the early 1990s.

- **Limitations:**
  - Customers are not convinced about the offered capabilities and added values of the current solution systems.

  - Early support for early conceptual design, detailed design and fully functional system design lack in the current PLM systems.

  - Efficient communication and co-ordination among users is not properly supported by data exchange between application systems.

*This work is to be published in a series of papers in the "Journal of Ship Production".

Development of a Logic Based Product Lifecycle Management (LBPLM) System for Shipbuilding Industry (2/5)

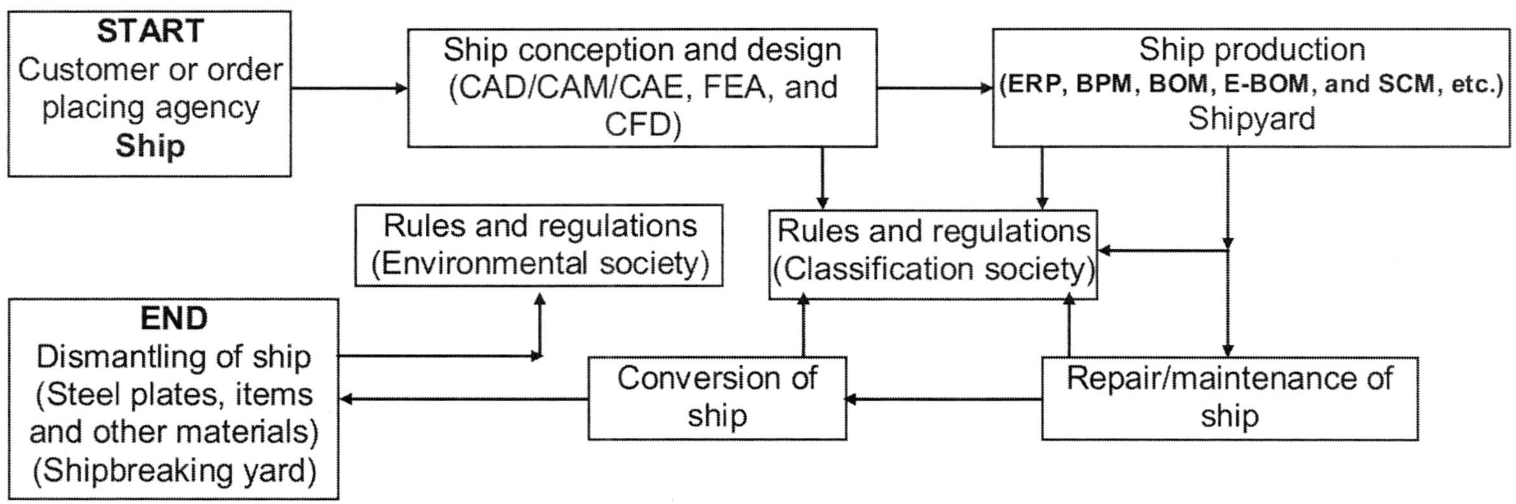

Figure 2: The complete lifecycle of ship/boat.

Development of a Logic Based Product Lifecycle Management (LBPLM) System for Shipbuilding Industry (3/5)

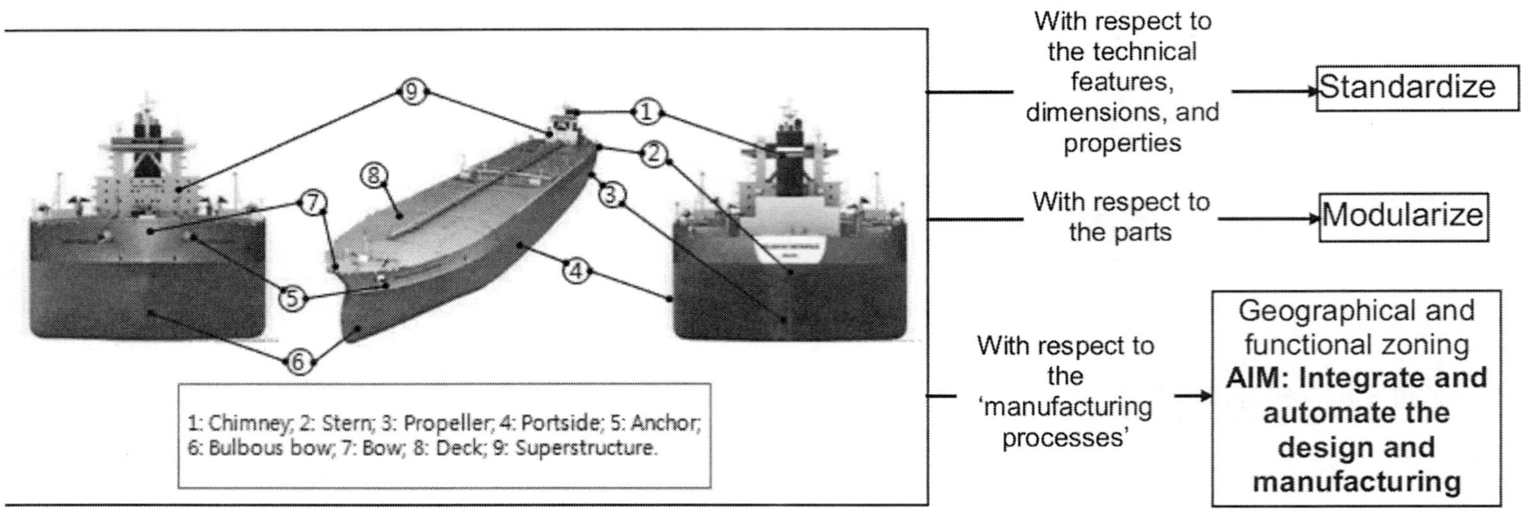

Figure 3: The key logic bases of the LBPLM system for shipbuilding industry.

OUR PROPOSED APPROACH

- Logic simplifies and discretizes the information that needs to be exchanged and shared amongst different systems for efficient communication and co-ordination.

- **ADVANTAGE:** This facilitates development and integration of 'point-to-point' connections.

- **AIM:** To develop a logic based PLM (LBPLM) system for the entire lifecycle.

Development of a Logic Based Product Lifecycle Management (LBPLM) System for Shipbuilding Industry (5/5)

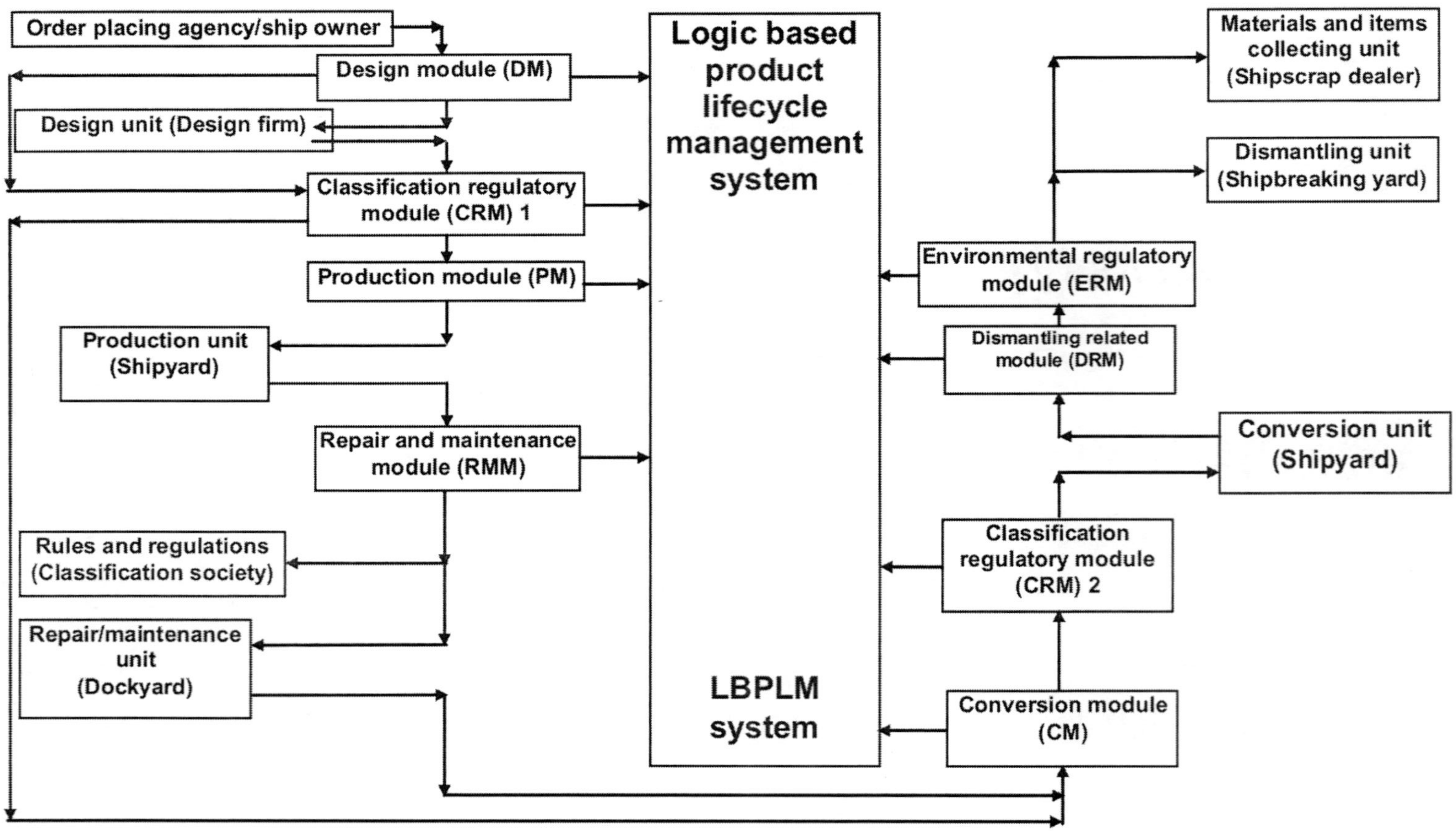

Figure 4: The system architectural configuration of different modules in our proposed LBPLM system.

* **Module means a part in the overall software.**
* **Unit means the specific production shop in the shipyard.**

Contents

1. **Introduction**
2. **Our researches:** computer aided ship design and production
   Surface generation over T-junction with continuity
3. **Our researches:** computer aided ship design and production
   Development of 'Bill of Materials (BOM)' kernel of product lifecycle management model for shipbuilding industries
4. **Our researches:** computer aided ship design and production
   Development of a logic based product lifecycle management System for shipbuilding industry
5. **Our researches:** computer aided ship design and production
   Optimum design of an ultra-low motion semi-submersible floating oil and gas production system
6. **Listing of some of the challenges of computer aided ship design and production.**
7. **Conclusions and future directions of research**

Optimum conceptual design of an ultra-low motion semi-submersible floating oil and gas production system ** (1/4)

- Since, available reservoirs of energy rich fossil fuels at shallow and moderately deep waters have been exploited already, the fossil fuel (petroleum oil and gas) exploration and production is being forced to move into extremely deep waters (water depth > 2500 m).
- At deep water the expenses associated with fixed (like jacket and concrete platforms) production platforms are no longer feasible (economic), and that makes a floating production platform design a far more economical choice.
- The exploration and production requirement in deep sea demands that the motion characteristics (e.g. heave/pitch/roll etc.) of the structure are further lower as compared to the inherently low motion characteristics of the semi-submersible.

**To be presented at ICSOT – 2009, to be held at IIT, Kharagpur, India by Dr. Rajiv Sharma.

Optimum conceptual design of an ultra-low motion semi-submersible floating oil and gas production system ** (2/4)

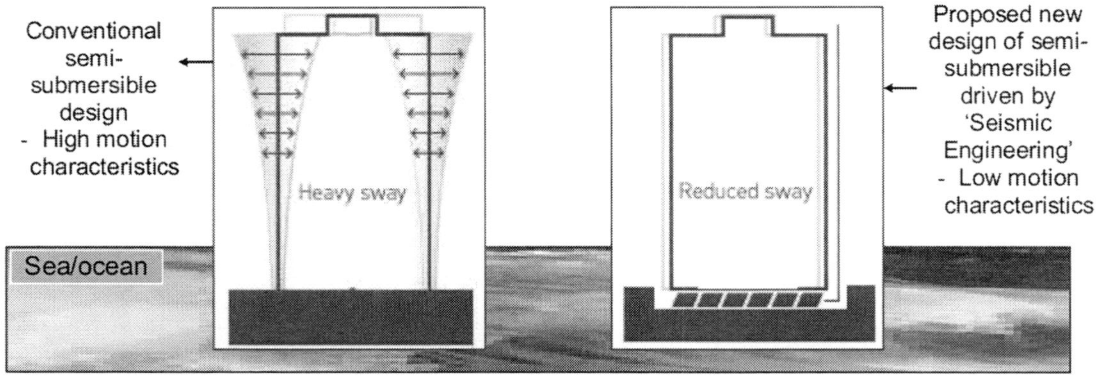

Figure 5: The conceptual description of our proposed ultra-low motion design idea for semi-submersibles.

Figure 6: The additional pendulum based control of centralized trusses of semi-submersible.

Optimum conceptual design of an ultra-low motion semi-submersible floating oil and gas production system ** (3/4)

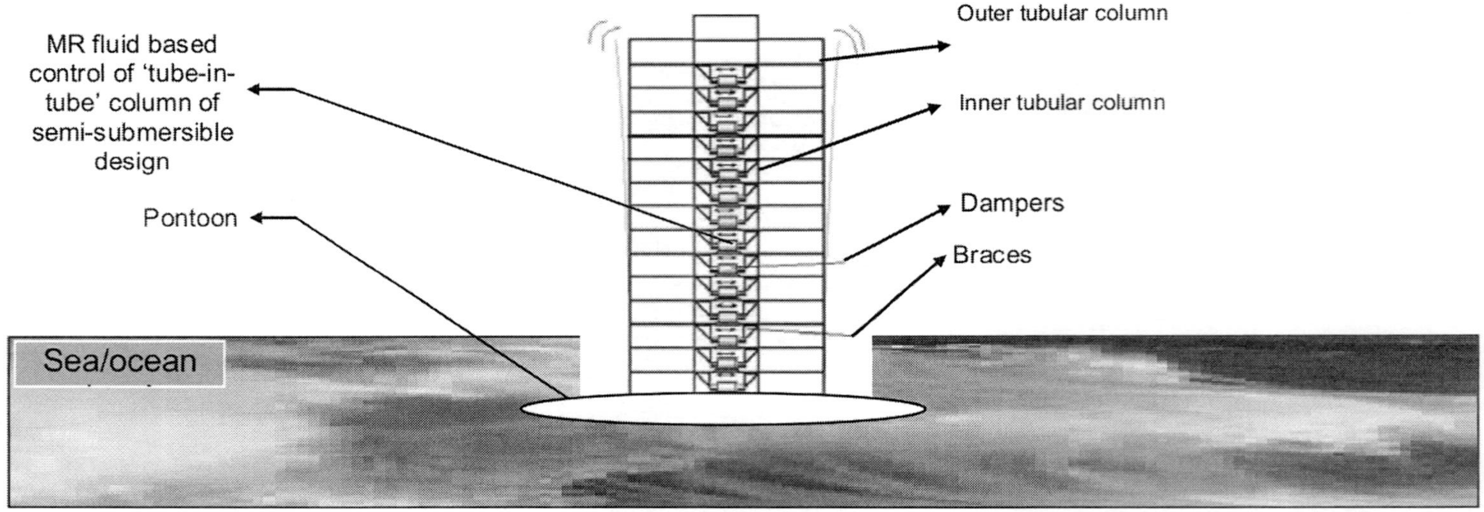

Figure 7: The control design concept for the column of the semi-submersible.

Optimum conceptual design of an ultra-low motion semi-submersible floating oil and gas production system ** (4/4)

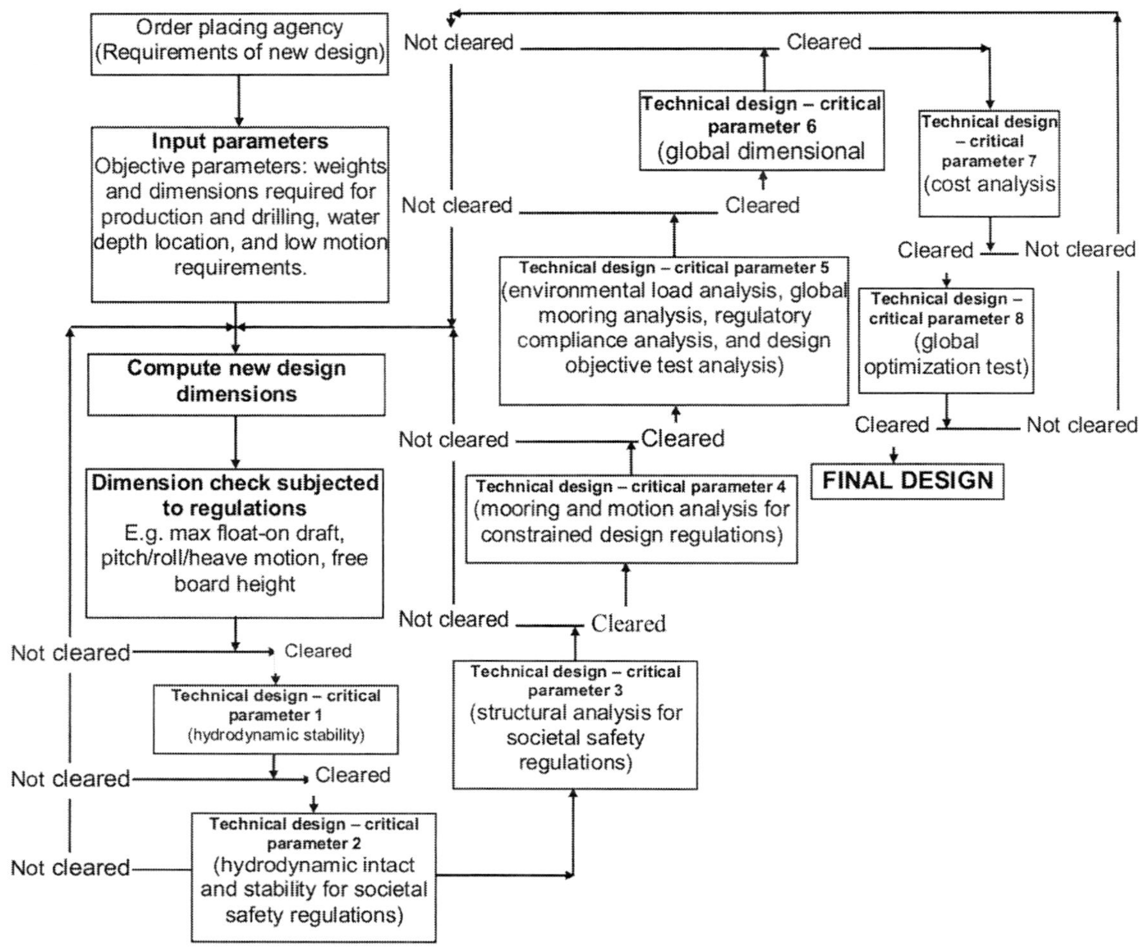

Figure 8: The schematic diagram of our proposed critical parameter driven iterative optimum conceptual design process model.

Contents

Listing of some of the challenges of computer aided ship design and production (1/5)

- Efficient CAD for the ship design for economies of scale.
- CAD for finer design of ship for higher speed.
- CAD for motion simulation of ship.
- CAD for compact and finer design of ship for lower motion.
- CAD for design and development of 'hybrid' ship.
- Computer models for the operational simulation of ship in adverse weather conditions.
- Computer models for energy efficiency analysis of ship.
- Geometric and structural form optimization with CAD and CFD for ship.

Listing of some of the challenges of computer aided ship design and production (2/5)

- Optimum design of propeller to improve propeller/hull interaction and propeller-rudder combinations with CAD and CFD.
- Computer models for efficient weather routing, speed or trim optimization.
- Computer aided environmental impact analysis of ship from design to construction to dismantling.
- Computer aided conversion and recycling model for ship.
- CAD for propulsion system design with/without rudder, mid-ship section, bulkhead arrangement, and detailed layout.
- CAD for material requirement planning and scheduling, and work assignment for technical workers, etc.
- CAD for the design and analysis of ‘green (for clean environment)’ ship.

Listing of some of the challenges of computer aided ship design and production (3/5)

- CAD for the design and analysis of ship with advanced materials and control systems.
- Production friendly geometric design of curves and surfaces for ship geometries.
- Design of nearly developable, developable and low curvature surfaces.
- Geometric design of curves and surfaces under constraints of volume and space.
- CAD for the architectural/production layout design of ships.
- Computational models for efficient design and analysis of heating patterns.
- Computational models for efficient analysis and prediction for deformations in thin and large stiffened panels.

Listing of some of the challenges of computer aided ship design and production (4/5)

- Computational models for efficient material estimation and production scheduling.
- Computational models for strength analysis and structure design of lifting systems.
- Computational models for efficient pipe routing design.
- Computer information management systems for efficient ship production.
- CAD for ship product design, product life cycle management, lean and integrated manufacturing, and environmental impact analysis for ships from design to production to dismantling.
- Computational models for strength analysis and structure design of conversion and recycling for ships.
- CAD/CAM/CAE, FEM, PLM integration.

Listing of some of the challenges of computer aided ship design and production*** (5/5)

Journal of Computer-Aided Design special issue*** - WE are editing a special issue for the journal of 'Computer-Aided Design' on 'applications in ship and floating structure design and analysis', Guest editors - Prof. Tae-wan Kim, REPUBLIC OF KOREA; Prof. Richard Lee Storch, USA; Prof. J. J. Hopman, The Netherlands; and Prof. Stein Ove Erikstad, NORWAY.

Journal of Ship Production special issue*** - WE are editing a special issue for the 'Journal of Ship Production' on 'computer applications in ship design and production', Guest editors - Prof. Tae-wan Kim, REPUBLIC OF KOREA; Prof. Richard Lee Storch, USA; and Prof. George Bruce, UK.

***All of you who are interested are invited to contribute an article in the above mentioned special issues.

Contents

1. **Introduction**
2. **Our researches:** computer aided ship design and production
   Surface generation over T-junction with continuity
3. **Our researches:** computer aided ship design and production
   Development of 'Bill of Materials (BOM)' kernel of product lifecycle management model for shipbuilding industries
4. **Our researches:** computer aided ship design and production
   Development of a logic based product lifecycle management System for shipbuilding industry
5. **Our researches:** computer aided ship design and production
   Optimum design of an ultra-low motion semi-submersible floating oil and gas production system
6. **Listing of some of the challenges of computer aided ship design and production.**
7. **Conclusions and future directions of research**

Conclusions and future directions of research (1/4)

A. Geometry is the important in analysis but mainly thought as representative of shape. Because, of the belief that geometry is just related to shape, modern methods of computational geometry, do not utilize geometric forms from CAGD, CG, Computational Geometry, etc.

Main points in future research:

- Use exact and popular geometric representations from CAGD, CG, Computational Geometry in computational mechanics.

Conclusions and future directions of research (2/4)

B. It is well known that in FEA the processes of mesh construction, mesh refinement, and formation of boundary conditions are computationally expensive and constitute a significant part in the overall analysis time.

- This is further compounded in the analysis of complex engineering shapes like ships, automobiles etc. because in them approximation of geometry discretization, and material composition needs to be finer and that is computationally more complex and complicated.

- Additionally, the separation of design and analysis model increases the time and complexity of the overall process of design and analysis.

- It is in this regard that 'Iso-geometric Analysis', offers the possibility of development of an integrated design and analysis approach.

Conclusions and future directions of research (3/4)

Main points in future research:

•The basic objective is integration of computational mechanics with CAD/CAM/CAE and CAGD.

•The 'NURBS' too have their problems. A better approach will be to look and borrow concepts from physics.

•NURBS are 'ONLY' suitable to describe a '2D descriptive geometric object' like an open 2D/3D surface. Geometric representation of closed surfaces or surfaces defined over solid is computationally too expensive with NURBS. Hence, if one is willing to explore FEM for volumetric objects then computational geometric representation will be better.

Conclusions and future directions of research (4/4)

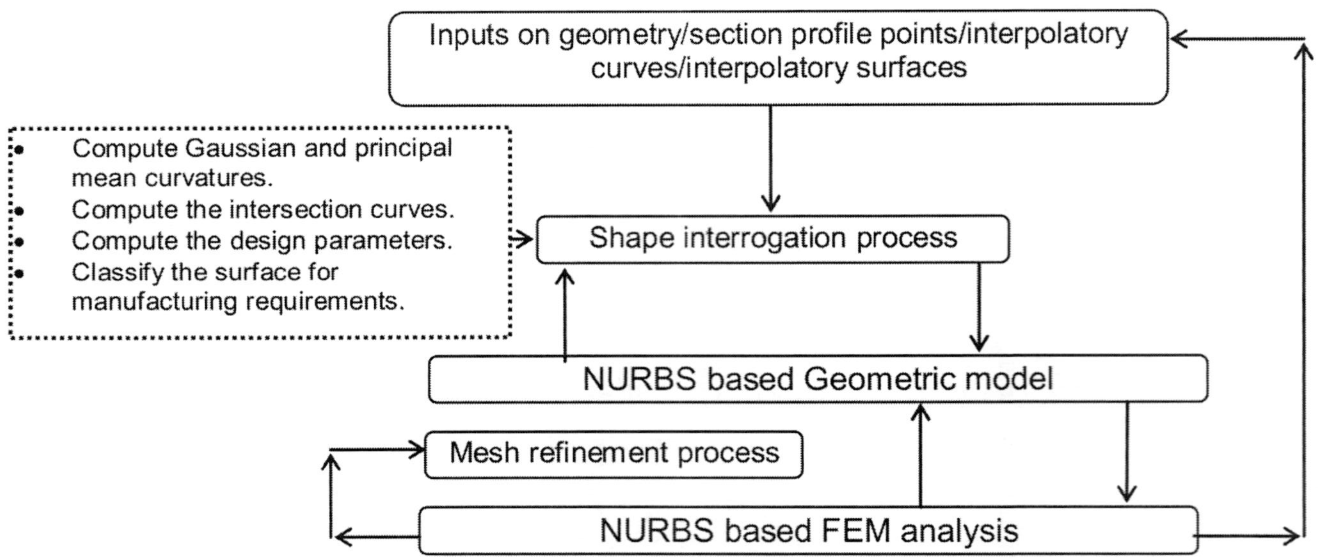

Figure 9: The integrated design and analysis system with 'Iso-geometric analysis'.

Thank you very much for your kind attention.

We are looking for graduate students and post-doctors to work for the challenges mentioned in this talk. Those interested are welcome to contact us at:

taewan@snu.ac.kr

http://caditlab.snu.ac.kr/

FLOW OF PRODUCTION INFORMATION IN A SHIPYARD IN THE ERA OF IT ENABLED SYSTEMS

K J H Christie, Garden Reach Shipbuilders & Engineers Ltd, India
S P Ghosh, Garden Reach Shipbuilders & Engineers Ltd, India
A Bhattacharya, Garden Reach Shipbuilders & Engineers Ltd, India

SUMMARY

The IT revolution has completely changed the design process and the manner in which the design is converted into the product for most of the industries. Although the shipbuilding industry in India has been trying to adopt this efficiency multiplier and have made considerable investments towards this, the progress has been slow with limited success so far. One of the reasons for the limited success is the manner in which IT is being applied by the shipbuilding industry. While applying IT to an existing process can surely speed it up, the full benefit of IT can be realized only when the process is re-engineered to take full advantage of IT. At GRSE, we already use design software Tribon for generating design information and ERP software SAP in a limited way for controlling production. We are currently in the process of upgrading SAP to bring the complete production activities under the purview of ERP. This paper makes an attempt to bring out the challenges being faced by the shipyard while doing the process re-engineering to effectively adopt ERP in shipbuilding.

1. INTRODUCTION

Shipbuilding is viewed as a huge fabrication industry with a lot of steel cutting and welding. Most of our shipyards have been developed as ones with extensive facilities for steel fabrication. While these facilities are essential for shipbuilding, focusing only on these facilities have kept our shipbuilding industry virtually primitive as compared to modern shipyards elsewhere in the world. It is true that hull work remains a major constituent of the shipbuilding activities, especially in merchant shipbuilding. However with more advanced and complex ships such as warships and special purpose vessels with extensive and sophisticated equipment fit, outfitting has assumed predominance over steel fabrication and it is this capability of outfitting which decides how successful the shipbuilder is going to be.

2. TRADITIONAL SHIPBUILDING

Traditional methods of outfitting call for the fabrication of the complete hull before outfitting begins. This is a pretty simple affair which requires the minimum planning and is practiced by shipyards with the bare minimum infrastructure. In ships where the majority of the shipyard efforts are related to hull fabrication this is a cost effective method where the overheads can be reduced to the minimum. Most of the small private shipyards which have come up in the recent past follow this method to give them the competitive edge in terms of the cost without the need to invest heavily in infrastructure.

3. CHALLENGES OF COMPLEX SHIPS

When we need to build complex ships, we need to look at things from a different perspective. A ship becomes complex mostly from the point of view of the mission it needs to accomplish and this has more to do with the equipment / systems than the hull which houses them. The outfitting effort is directly proportional to the complexity of the systems and the only way to reduce the build period of the complex ships is by adopting methods which reduce the outfitting effort as well as time. It is an established fact that outfitting effort in a shipyard is greatly reduced by carrying out outfitting in parallel to hull fabrication at the panel or block level where the structure is easily accessible and work can be carried out under the convenience of a shop floor. Such simultaneous progress of hull fabrication and outfitting work forms the very basis of *Integrated Construction* adopted successfully by many world renowned shipyards in construction of warships.

4. FLOW OF INFORMATION

Flow of information plays a big role in reducing the outfitting effort. Typically the production information generated in a shipyard can be classified into the following broad types of information depending on who generates them.

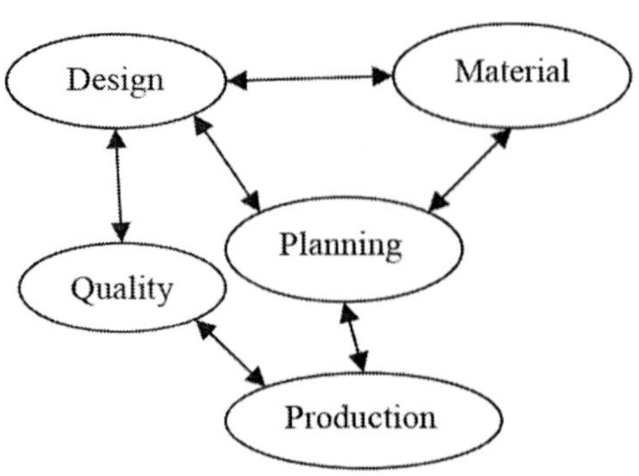

Departments involved in Production Information Flow

4.1 DESIGN

- Drawings which give a visual appreciation of space and location various components of the ship and which facilitates the measurement of distance. Examples: GA drawings, Lines plan, Compartment layout etc.
- Drawings which provides the functional insight into the working of various systems. Examples: Ship System Layouts.
- The bill of material (BOM) which lists the name of the parts and a material code and also the group to which the material belongs.
- Process information on how to carryout various production tasks. Examples: Welding procedures etc
- Material Indents / Requisitions which are carved out of BOM but has the added information with regards to Procurement specifications and sources of supply.
- Quality / Performance parameters of all equipment and systems.

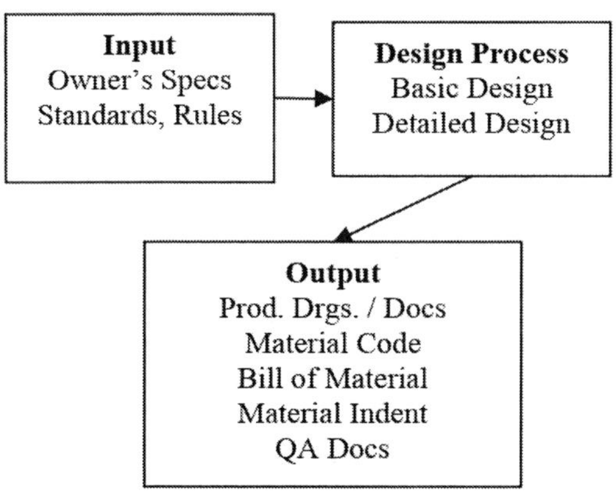

Information Flow - Design

4.2 PLANNING

- Work Breakdown Structure (WBS) which divides the shipbuilding activities into work elements of easily manageable tasks to which specific BOM, workforce and information can be attached. This is usually carried out with the assistance of Design Dept.
- The schedule and duration of the WBS tasks and their interdependence.
- Manpower and material resources required to complete each task.
- Budgeted cost of each activities

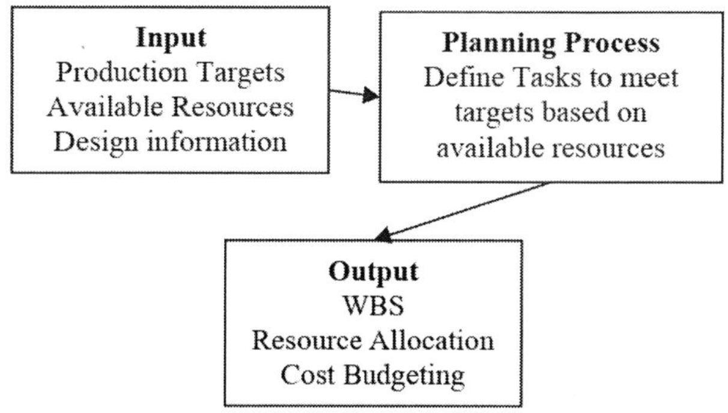

Information Flow - Planning

4.3 MATERIAL PROCUREMENT

- Data base of Sources and cost of material
- Lead times for material
- Cost and expected delivery date of material ordered.

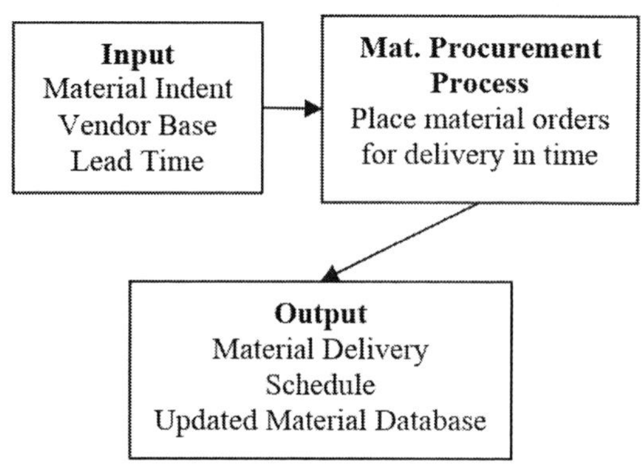

Information Flow - Material

4.4 QUALITY CONTROL

- Quality Assurance Procedures
- Inspection Schedules and Inspection Records

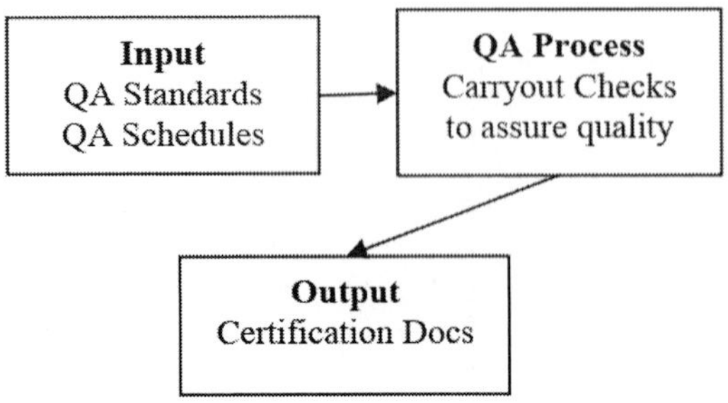

Information Flow - QA

4.5 PRODUCTION

- Manpower employed for production of various tasks
- Progress of various tasks

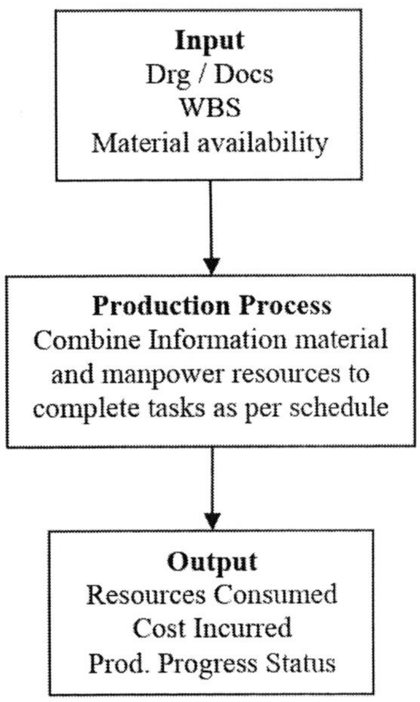

Information Flow - Production

5. IT MANAGEMENT OF INFORMATION

One of the fundamental principles of information management is that, the agency within the organization which generates the information has full control over the information and takes full responsibility for the accuracy and updating the information. The information should be stored in a central place as a master data and when required, it should be directly accessed from master data or copies which are constantly synchronized with the master data. This ensures that authentic and valid information is made available to everyone across the organization at all times. Modern IT enabled systems are specifically engineered to ensure this.

5.1 ERP AS AN IT TOOL IN SHIPBUILDING

For any ERP system, to remain a solution and not become a problem by itself, it has to meet certain conditions as given below:

- The users should be able to quickly adapt to the new methods of working brought about by the ERP.
- The users should clearly perceive a significant advantage while working in the new system in terms of shortened timeframe to carry out the same task and general reduction in effort to complete the same task. The users should eventually demand the availability of the system rather than finding excuses not to use the system.
- The reports generated by the system should provide intelligent information on processes that is not going well which can be leveraged to find solutions to the problem.
- It should be possible to integrate the old legacy systems to the new system and migration should be made possible with minimum effort.
- The advantage should clearly justify the additional and often exorbitant cost of implementation.

5.2 CHALLENGES OF ERP AS AN EFFICIENCY MULTIPLIER IN SHIPBUILDING

ERP Software was originally intended to be a financial software which would bring in certain amount of financial discipline to an organization. However it would have become apparent pretty soon that without an organized way of working where each task is linked to an individual or a group of individuals in the organization with associated accountability, no real financial discipline can be achieved. Typically an individual would get his/her wage or salary simply by marking his presence in an organization. However for the organization to benefit by his presence he would have to be attached to a specific job which needs to fetch a higher return than the salary he is paid. However in order to assess his specific contribution to the job, the job has to be well defined and the progress need to be tracked accurately. ERP systems in an organization should be able to achieve this. It is neither the intention nor the scope of this paper to cover the entire range of processes involved in a typical ERP system. The focus of this paper is on those information which, if channeled properly, would make the shipbuilding activities more efficient and cost effective.

5.3 ERP IN GRSE

Garden Reach Shipbuilders and Engineers, (GRSE) Kolkata where the authors are employed are currently in the process of implementing a SAP based ERP solution to the shipbuilding processes. To ensure a successful implementation, there would be a need for extensive re-engineering in the processes to make them efficient while being handled through an ERP system. Many of the issues discussed in this paper are related to this need for re-engineering.

5.4 STRUCTURED HANDLING OF INFORMATION

There are many processes in the shipyard which would derive great benefits from the implementation of a structured handling of information through ERP. Most of these benefits would be related to conversion of information from Design Model to Production Model as discussed below:

- In the traditional shipbuilding methods, the production model closely follows the design model. A typical design process would produce the structural drawings early in design followed by the GA and design of various systems leading to layout and BOM for each system. The production would follow almost in an identical fashion with the hull being fabricated first followed by filling the hull with various systems.

- In an *Integrated Construction* which is the need of the day, production would be following a different sequence than the design process. There would be a need to divide the ship into various modules which would be constructed as separate entities and then joined together in a convenient manner. Each module would be a hull block with a high level of outfitting containing, pipes, cables, compartments and equipment. There would be need to create self contained drawings which would facilitate the fabrication and outfitting of these blocks. Also a BOM and a material procurement / handling plan would have to be put into place to ensure that material flows into the production floor in a segregated and timely fashion to facilitate the construction of these integrated blocks.

- The activities for the construction of an integrated block needs to be broken down into appropriate WBS elements which are handled by cohesive work groups needing specific resources and material package. These WBS elements eventually form the building blocks of the production process having definite production information and material package attached to the WBS element.

6. BENEFITS OF ERP

Some of the specific shipbuilding processes which would directly benefit from an organized and manageable information flow are discussed below:

6.1 MATERIAL PROCUREMENT AND SEGREGATION

- Material procurement process starts with the process of indenting the material by the design dept. It is procured by the material dept. based on the description and quantity given by the design dept and the material requirement schedule given by the planning dept. The indenting closely follows the design process where the material gets defined as each system is designed. The material procurement schedule should closely follow the production plan, which in a *Integrated Construction* would try to group material from different systems to form part of a particular block outfit. The procurement process would like to club together similar material so that they can be tendered to specific groups of suppliers to get the full cost benefit procurement by large volumes.

- Just in time delivery of material is important for overall efficiency in terms of reduced inventory levels. The task elements in the WBS to which different materials are attached would give a good indication of the desired schedule of material procurement based on the actual production plan. It should be possible to appropriately schedule the ordering of the material from this information by allowing for procurement lead times and any anticipated delays in the material delivery.

- On receipt of the material in the yard, they are normally kept in groups based on their functionality. When they are issued for production, they need to be segregated as per the production plan. Typically the production floor would like to demand the material as per the WBS to which groups of material are attached.

- Grouping the material and segregating them to take the full advantage of various processes is a tedious task and is an exercise in database management. Material code plays an important part in this segregation. The coding process should not be too complicated for the designer to code the material but at the same time should provide sufficient information to segregate the material as demanded by the production.

- Each material should have a unique identifier based on its functionality to group them for easy procurement. Thereafter it should have tag to identify its belonging to a particular ship system, and then it should also have a tag to identify its belonging to a particular block for the convenience of outfitting. The coding system should take care of all this needs in a convenient fashion.

6.2 AVAILABILITY OF UPDATED DRAWINGS & DOCUMENT FOR PRODUCTION

- The drawings and documentation required for the production are generated by Design dept. and these are not necessarily generated in the manner in which the production would find it convenient to use. As mentioned earlier, the design would produce them as per various systems and production would want them as per systems as well as zone wise. Breaking them zone wise and attaching them to specific tasks would be essential to manage them efficiently.

- To execute a WBS one would need to refer to a set of drawings and documents. These drawings and documents would be stored in a central database which would be controlled by the design dept but would be accessed by the WBS concerned. In a project still under development, the changes to these drawings and documents are inevitable and it is essential that production dept access the current version of the drawings and documents to minimize the errors. The document management facilities in the ERP would be able to conveniently handle this requirement.

6.3 MONITORING OF PRODUCTION AND APPLYING NECESSARY CORRECTION

Frequent and accurate monitoring of production is the most important step for timely completion of any project. While monitoring is not so difficult in general terms, if one has to identify the root cause of delays each task would have to be monitored separately. For this, the tasks (WBS) would have to be broken down to a level to which specific resources (men, material, information) can be associated. By doing so it would be possible to identify which of the resources or the lack of it is the root cause for delays in that particular task. Without clear identification of the cause it would not be possible apply corrective measures to arrest the delay.

6.4 COMPLETION OF A TASK WITHIN THE ALLOCATED BUDGET

An enterprise can sustain only when it makes more money than it spends. Unlike a manufacturing industry like automobiles where a large number of products go through a relatively lesser number of processes, in a shipbuilding industry the same product goes through a large number of processes and the build period is relatively large. In order to have a tight control over the cost of production it is essential that the overall production cost is broken down to elements which can be controlled timely and conveniently at every stage. Though estimating the cost of production for every task would be ideal, it is not so simple to achieve in practice. A judicious selection of WBS with a realistic attachment of resources and completion time would allow such an estimation of cost with great benefits.

7. SOFTWARE

7.1 Any discussion on information flow in a shipyard would be incomplete without a mention of the software tools which would facilitate this. Here the authors would like to limit the discussions to the software currently in use at GRSE. Like most of the shipyards in India, GRSE has been using Tribon for the design to a large extent and for the transfer of information from design to production in a limited sense. SAP is also being used in a limited manner for indenting, procurement and issue of material. As part of modernization, GRSE is going for the full implementation of SAP which is currently under process. Some of the issues related to this implementation of SAP and the manner in which it is expected to co-exist with other software is discussed below:

7.2 In a typical scenario, the design would be created completely in Tribon and the production information would be transferred to SAP which would make it available to other processes. However the way things stand Tribon is still being used as a design tool (mostly for 3-D layout) than a product information model. With respect to hull structure the complete information with respect to nesting and structure BOM is being transferred to production for the plate cutting and planning for fabrication of structure blocks. However information with respect to outfitting is limited to extraction of 2-D layout drawings from the 3-D model and preparation of pipe sheets for the fabrication and installation of pipe systems onboard. One major constraint has been the relative inaccessibility of Tribon database by other software.

7.3 In a warship construction, design and construction are almost concurrent with an essential and minimum telescoping between various activities. Often various equipment and system components are required to be indented even before the Tribon model is finalized. Therefore mostly a material package for a ship is defined first in a SAP database even before it is created in Tribon. Even after it is created in Tribon, it is difficult to have a one to one correspondence between material in the Tribon database and in the SAP database because of the lack of a two way communication between both software.

7.4 In addition to Tribon and SAP, there are other software already in use or under installation at GRSE such as MS Exchange, MS Project and MS SharePoint. Most of the users are already familiar with these software and it is expected that they would provide a cost effective solution while co-existing with SAP.

7.5 Even though SAP is a reputed ERP package widely used by many industries, there are not many examples of successful implementation of SAP in shipbuilding industry in India. While there are many experts with good implementation skills in general industry, so far none could be made available with a good domain knowledge of shipbuilding Industry. A solution to the problem is likely to come from shipbuilding experts who would be willing to learn and leverage the capabilities of SAP than the other way around.

7.6 While certain re-engineering of the processes would be essential the dilemma is to decide how much of it is actually essential and not just desirable. One has to strike a good balance between the efficiency brought about by process re-engineering and difficulty faced by the workforce to adapt to new methods of working. The bottom line is not how efficient the system theoretically is, but how efficient the system becomes after the uses adapt to the system.

8. CONCLUSION

This paper has mostly focused on production information, whereas for a shipyard to work efficiently, it is essential to handle the information related to financial and personnel matters. When an employee of a shipyard marks his presence in the shipyard on a typical day, he has made sure that he gets is pay. Now it is up to the management of the shipyard to put him through a system where he would produce an output which would give adequate return to the organisation. This would need the system to attach him to a well defined task, provide him with material and facilities to efficiently carryout the task, estimate the cost of such a task and define the responsibility and accountability of each employee with respect to specific tasks. Such a system can function only when the information related each and every aspect of various shipyard activities flows accurately and seamlessly within the organization.

9. AUTHORS' BIOGRAPHY

Capt (IN Retd) K J H Christie, a graduate in Naval Architecture from IIT Kharagpur and a post graduate in Naval Arch, Marine and Mechanical Engineering from MIT, USA holds the position of Director(Shipbuilding) in GRSE. He is responsible for the entire shipbuilding operations of the company. His previous experience includes 26 years in the Indian Navy in different capabilities including the Head of Design of Indigenous Aircraft Carrier.

Shyama Pada Ghosh, a graduate in Naval Architecture from IIT Kharagpur holds the position of Addl. General Manager(Design)(Hull Outfit) in GRSE. He is in charge of Outfit Design in the Central Design Office of GRSE

Aparajita Bhattacharya, a graduate in Naval Architecture from I I T Kharagpur holds the position of Addl. General Manager(Design)(CAD) in GRSE. He is in charge of CAD implementation in the Central Design Office of GRSE

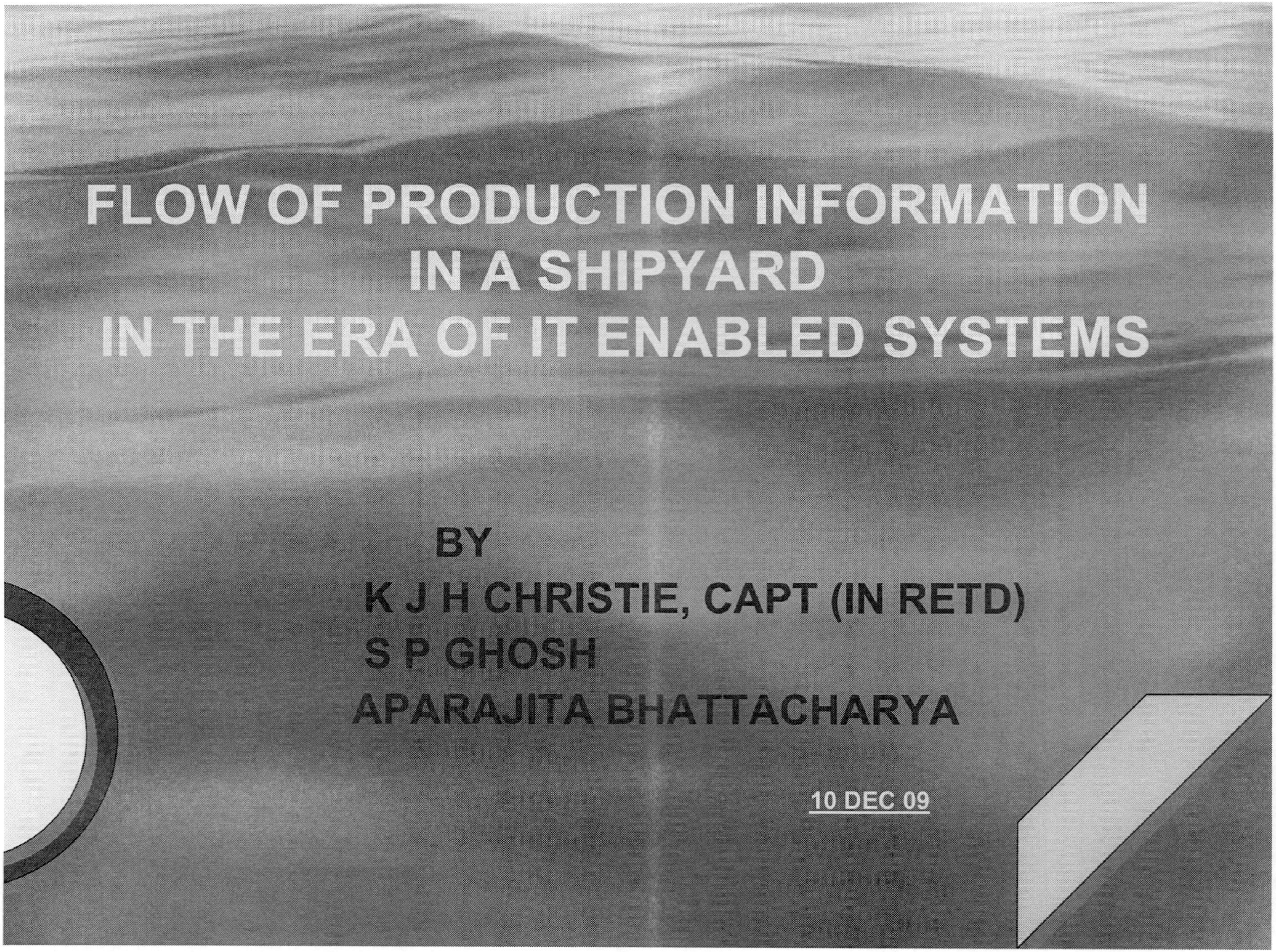
FLOW OF PRODUCTION INFORMATION
IN A SHIPYARD
IN THE ERA OF IT ENABLED SYSTEMS
BY
K J H CHRISTIE, CAPT (IN RETD)
S P GHOSH
APARAJITA BHATTACHARYA
10 DEC 09

SCENARIO OF COMMERCIAL SHIPBUILDING – *TRADITIONAL WAY*

In conventional shipbuilding for **commercial vessels:**

- ❑Hull fabrication - main manufacturing acitivity accounting for 65-70% of total efforts

- ❑Balance 30-35% of efforts spent for outfitting (machinery & piping, accommodation and electrical)

- ❑Outfitting commences after hull fabrication is complete following a build strategy with acitivities in series

SCENARIO OF COMMERCIAL SHIPBUILDING – *TRADITIONAL WAY*

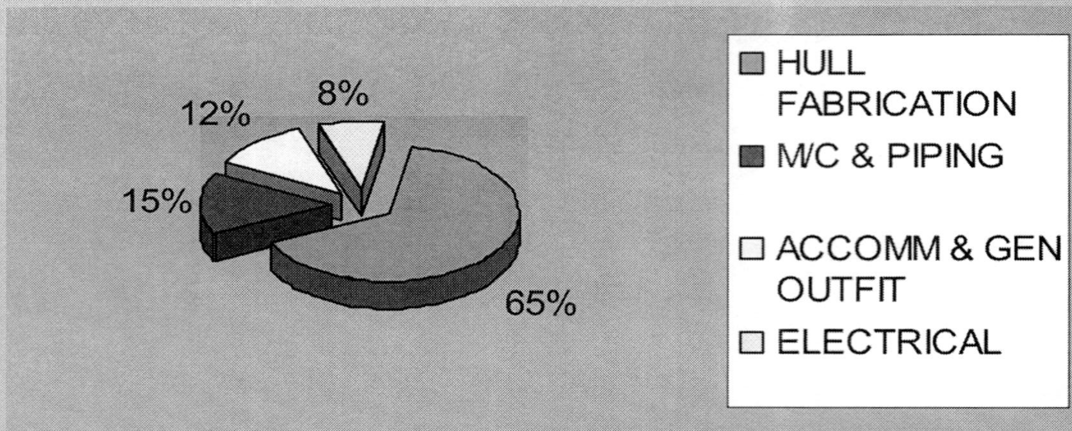

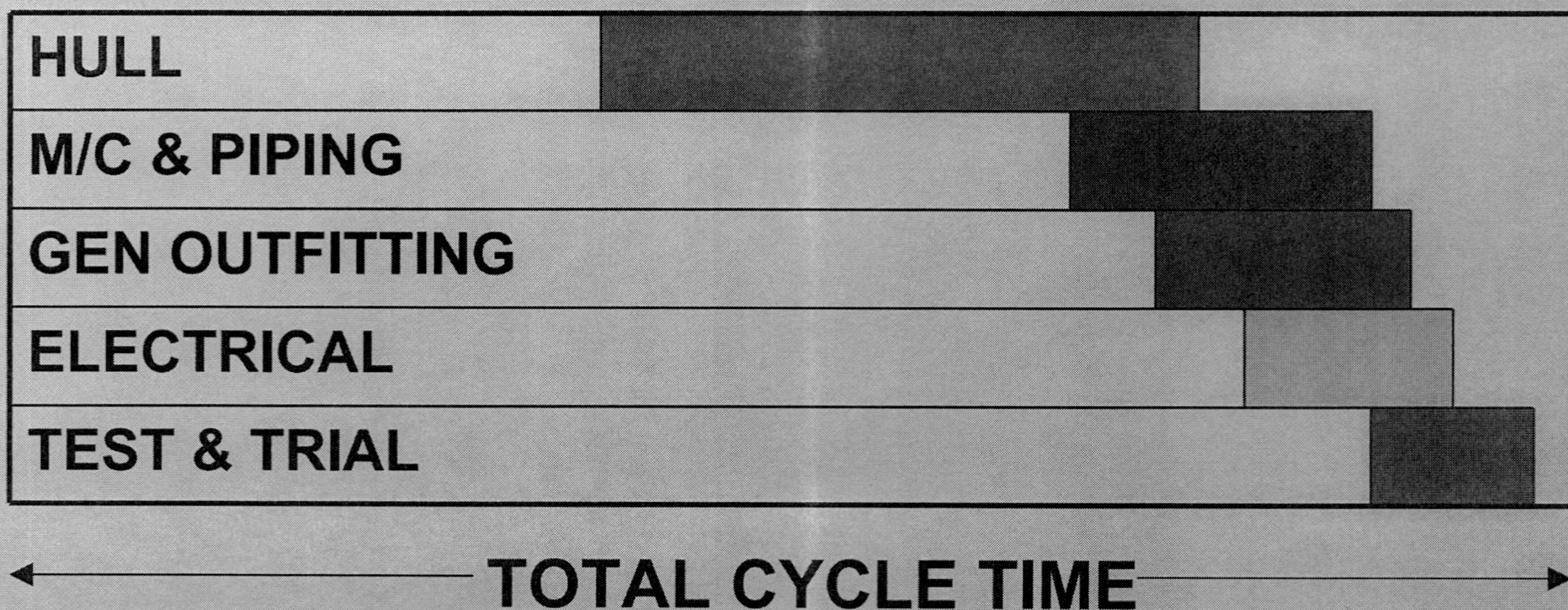

SCENARIO WITH *MORE COMPLEX SHIPS / WARSHIPS*

Challenges of a modern complex ship:

- ❑**Hull fabrication accounts for only 25-30% of total effort**
- ❑**Mission requirement demands more sophisticated systems & equipment-fit**
- ❑**Complexity of systems / equipment far higher increasing total outfitting effort to 70-75%**
- ❑**Production cycle with conventional build strategy far too long**
- ❑**To reduce build period, outfitting need to be progressed concurrently with hull fabrication - *Integrated Construction***

REQUIREMENTS OF *INTEGRATED CONSTRUCTION*

- **Outfitting to progress concurrently with hull fabrication:-**
  - **Availability of a large gamut of different types of outfitting material at panel/sub-block/block stage**
    - **Meticuluous micro level planning ensuring finalization of design and timely procurement of material**
      - **Efficient flow of correct information among different sections of shipyard carrying out different functions**

FLOW OF INFORMATION

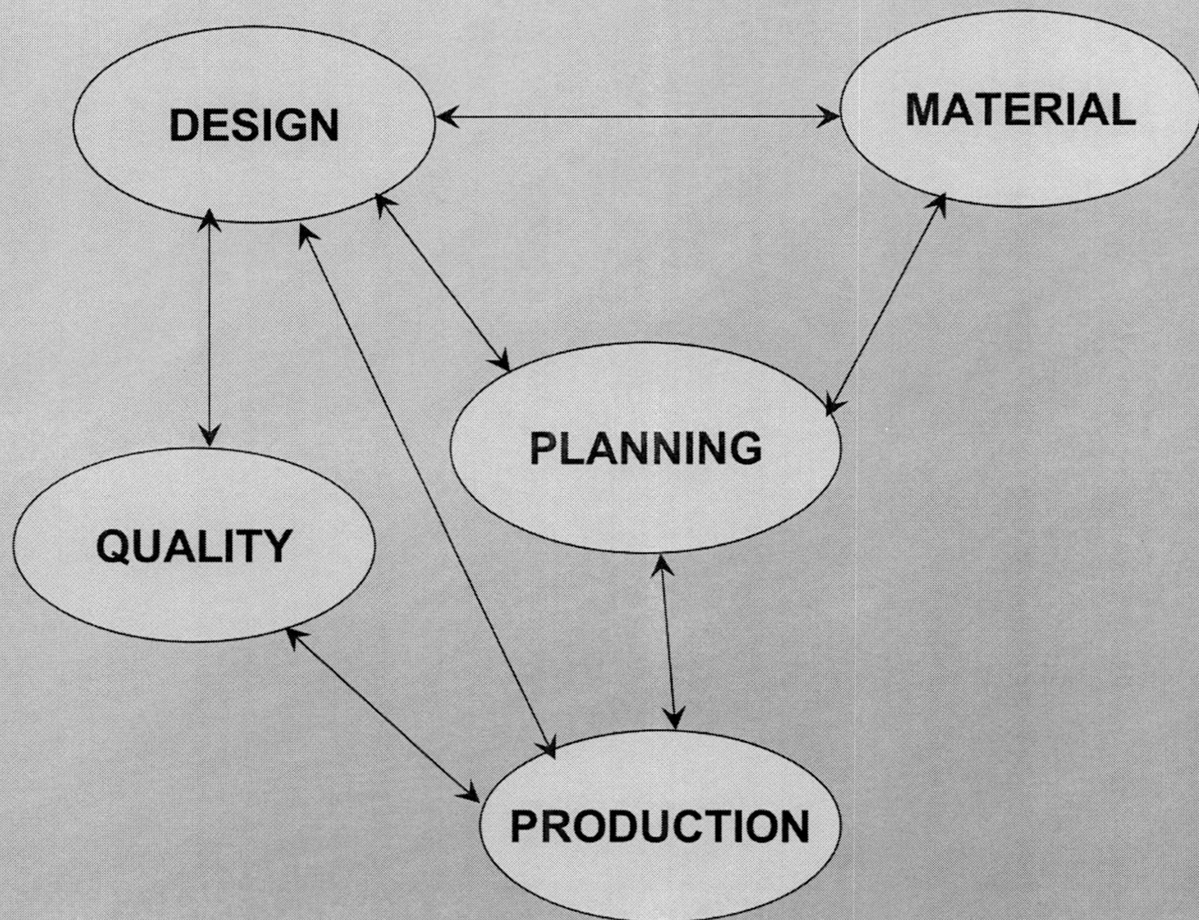

FLOW OF INFORMATION- DESIGN

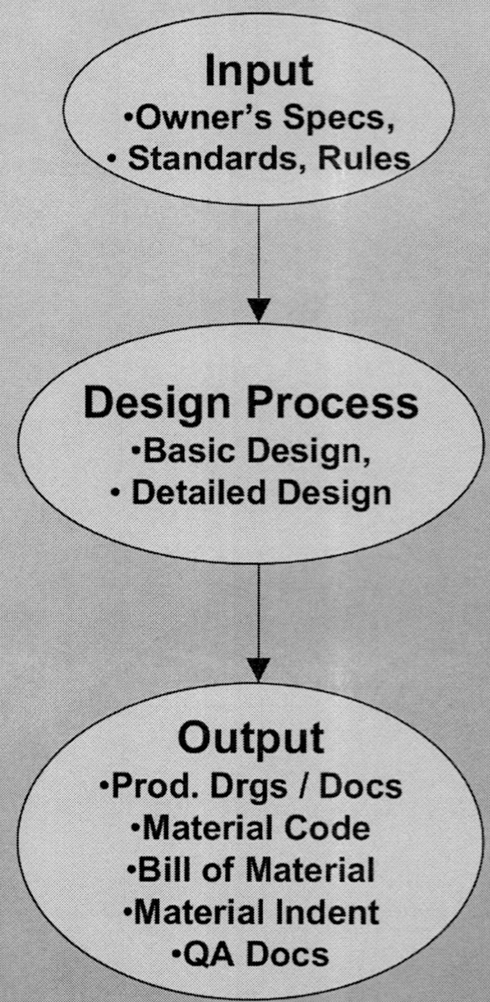

FLOW OF INFORMATION- PLANNING

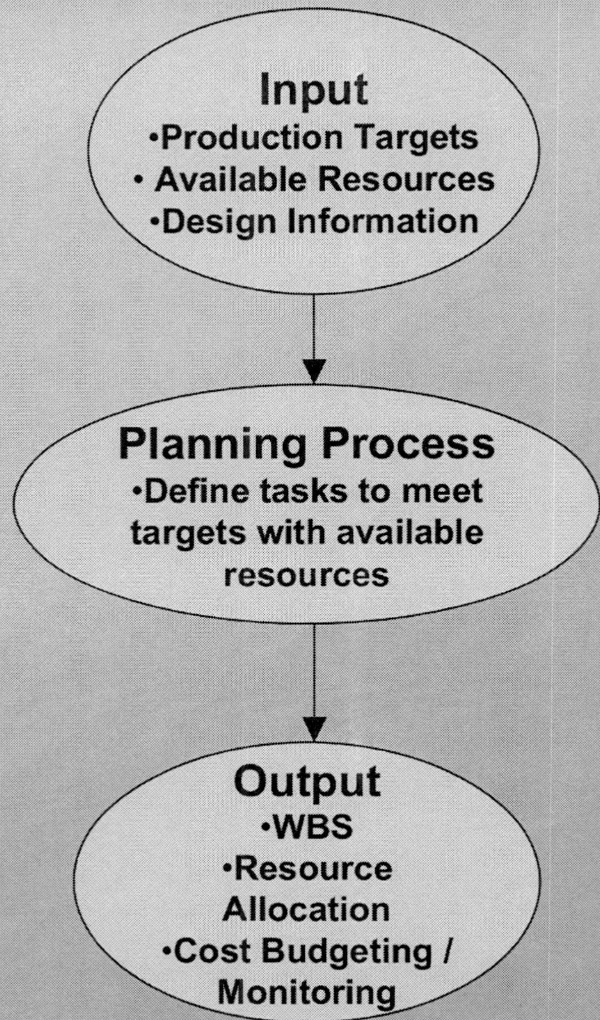

FLOW OF INFORMATION- MATERIAL

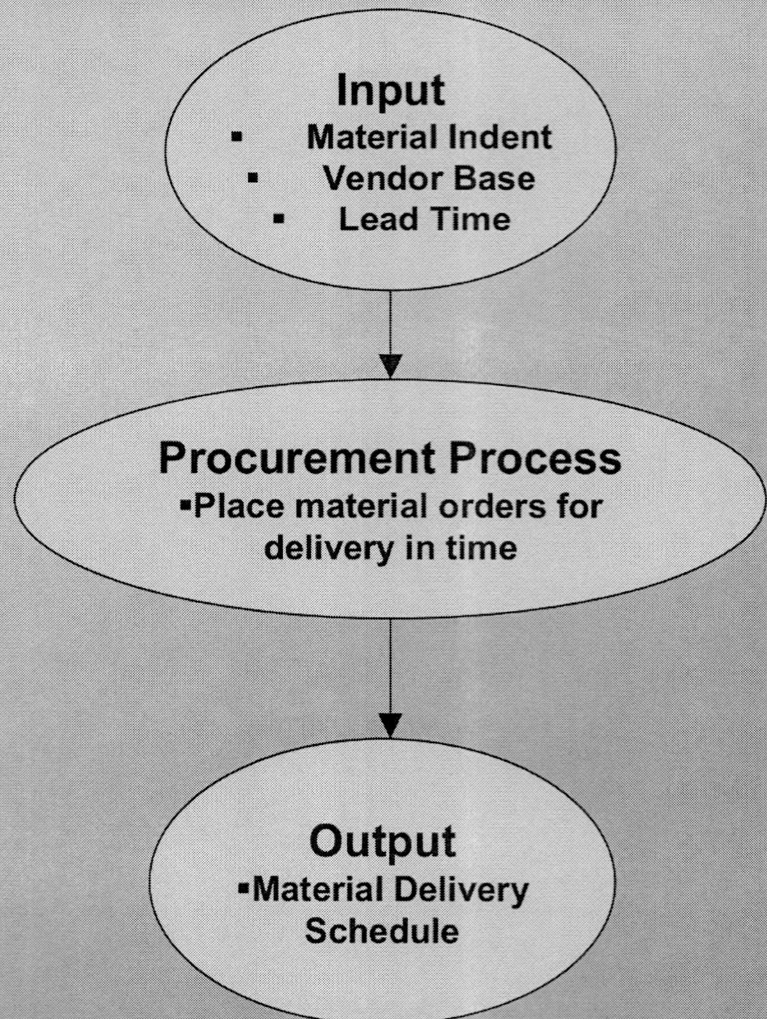

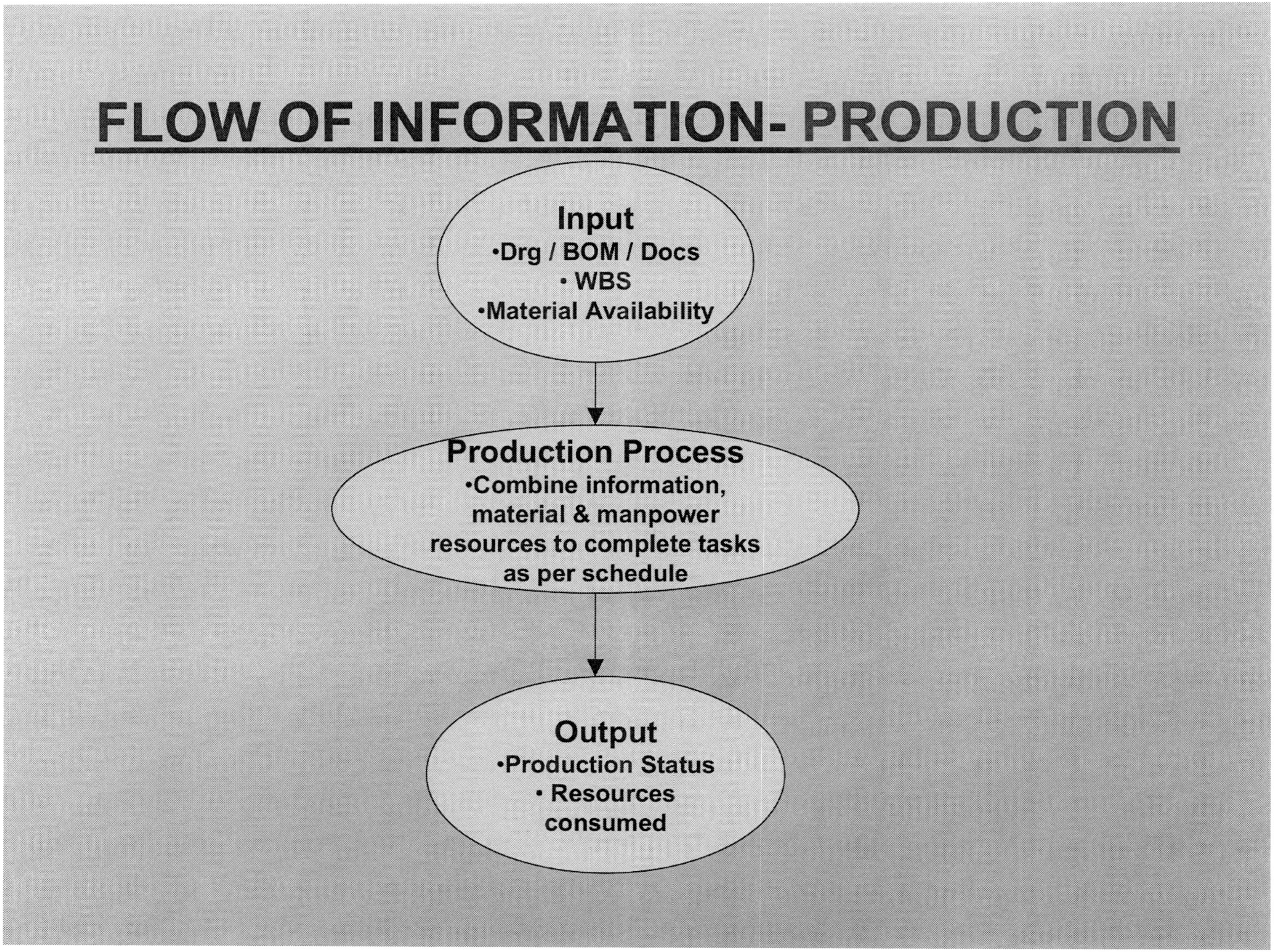
FLOW OF INFORMATION- PRODUCTION
Input
•Drg / BOM / Docs
• WBS
•Material Availability
Production Process
•Combine information, material & manpower resources to complete tasks as per schedule
Output
•Production Status
• Resources consumed

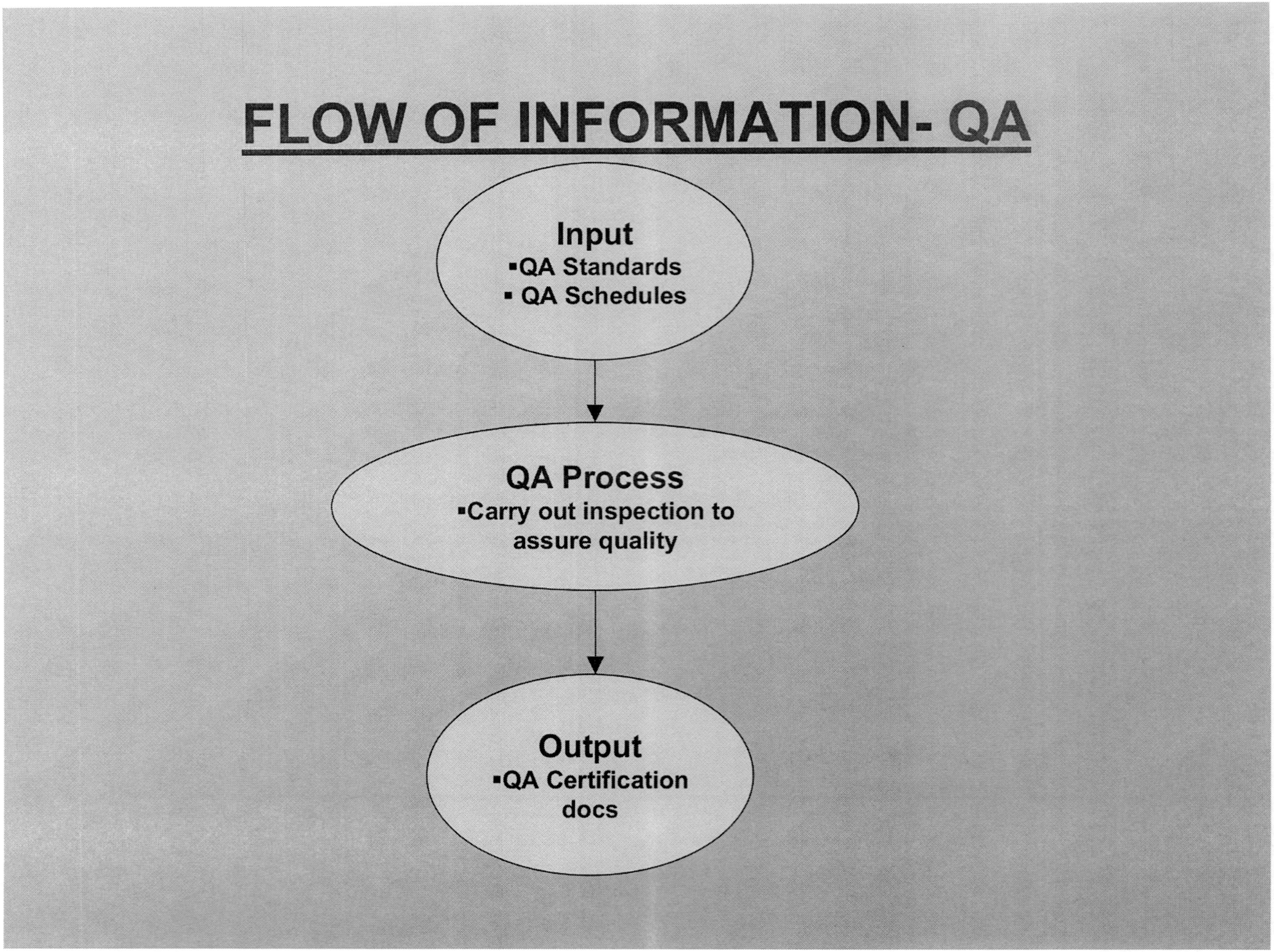
FLOW OF INFORMATION- QA
Input
▪QA Standards
▪ QA Schedules
QA Process
▪Carry out inspection to assure quality
Output
▪QA Certification docs

IT MANAGEMENT OF INFORMATION - ERP

- **Any such tool must meet following criteria:**
  - User friendly
  - Reduction in time & effort
  - Intelligent & articulate reporting
  - Easy migration
  - Cost Effective

IMPLEMENTATION OF ERP IN *GRSE*

- **ERP Phase-I implemented in 2002:**

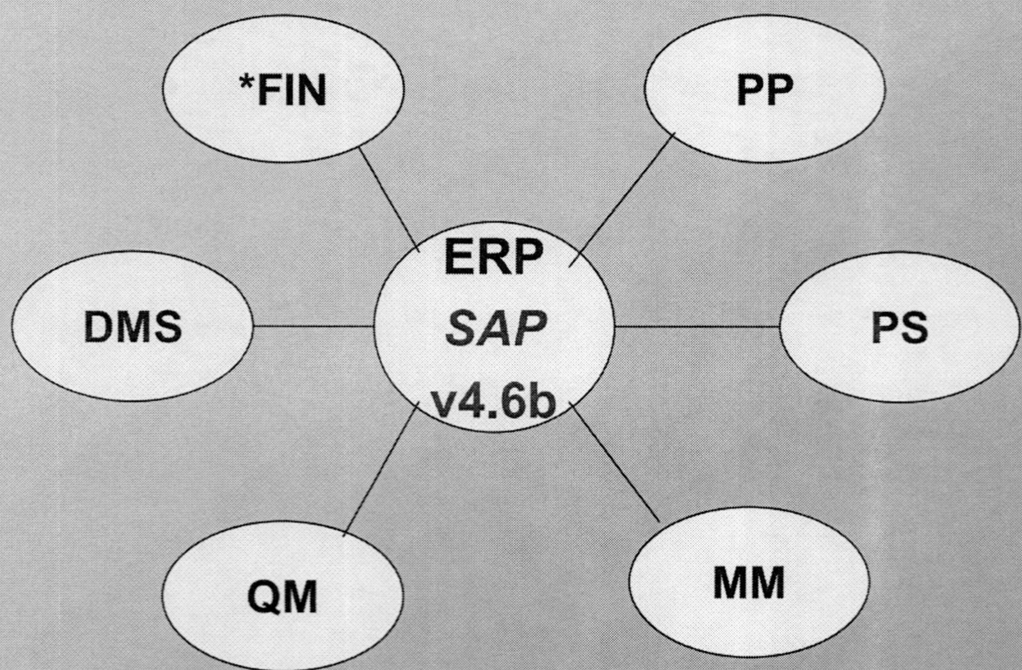

PP – Prod. Planning

PS – Project Structure

MM – Matl Module

QM – Quality Module

DMS – Doc Mgmt Module

***FIN – Finance (truncated)**

PRE-REQUISITES FOR ERP

- Material Codification for every item
- Work Break-Down Structure (WBS)
- Project Planning Tool
- Documents Management for Drgs, BOM etc.

SCENARIO - POST ERP IMPLEMENTATION

- **Automatic slippage alerts**
- **Pinpointing cause of delay & necessary remedial course**
- **Efficient tracking of material**
- **Possibility of *Just-In-Time* material – reduced inventory level**
- **Possibility of efficient grouping /segregation/ storage of material for easy retrieval**
- **Project cost monitoring & control**

UPGRADATION OF ERP PHASE-II FOR *INTEGRATED CONSTRUCTION*

- Block / Zone wise WBS instead of discipline-wise
- WBS broken down to identify each activity
- Attachment of appropriate material at each level
- Deployment of identified manpower at each level to bring in accountability
- Time schedule for each WBS element
- Feed back of progress on each WBS element

UPGRADATION OF ERP PHASE-II

- ERP Phase-II implementation presently under way

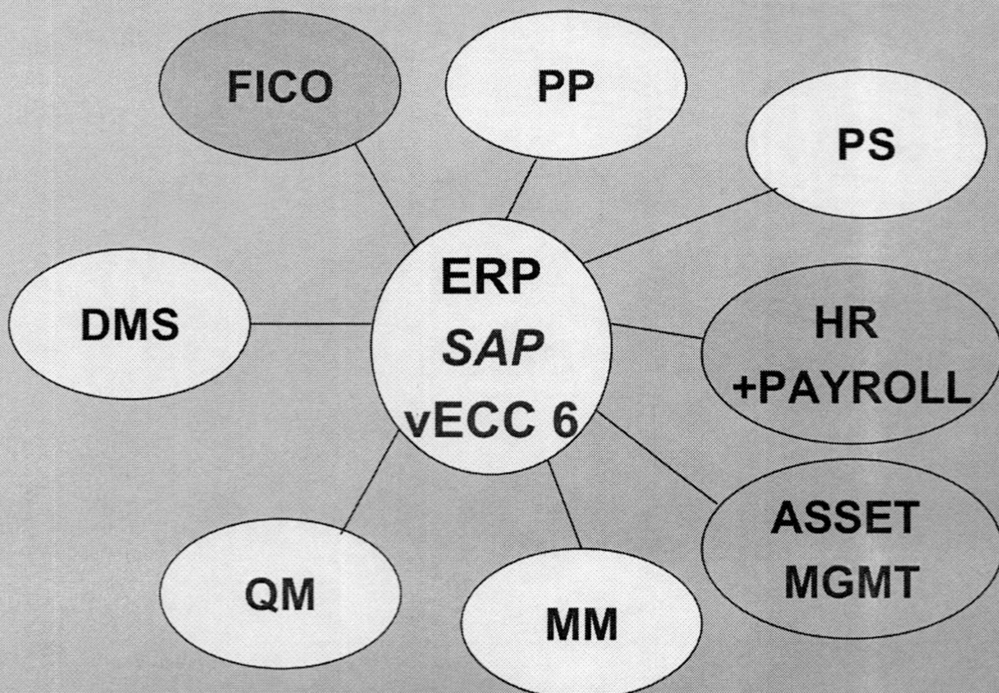

PP – Prod. Planning

PS – Project Structure

MM – Matl Module

QM – Quality Module

DMS – Doc Mgmt Module

FICO – Finance & Control

INTEGRATED CONSTRUCTION METHODOLOGY

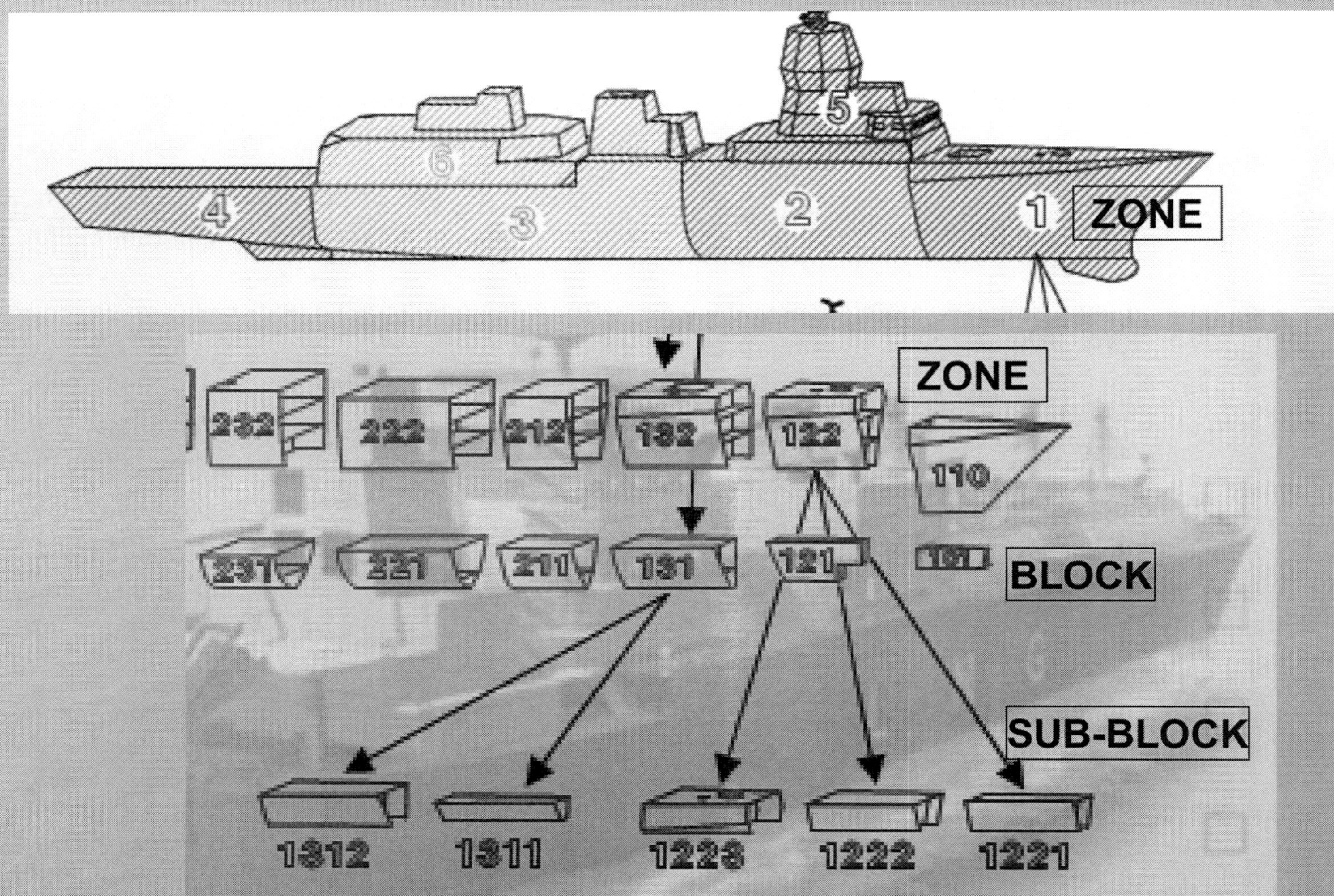

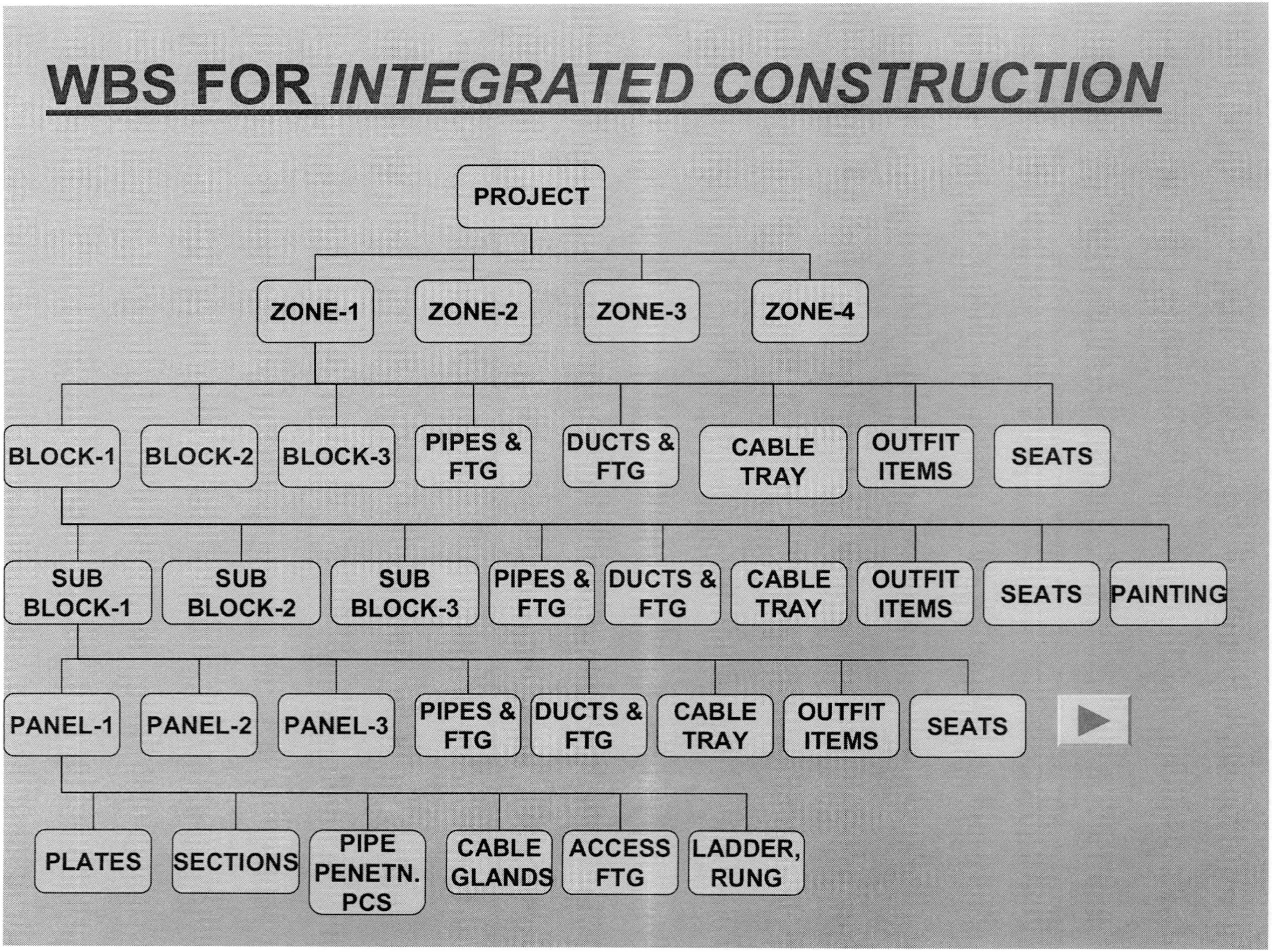
WBS FOR INTEGRATED CONSTRUCTION
PROJECT
ZONE-1
ZONE-2
ZONE-3
ZONE-4
BLOCK-1
BLOCK-2
BLOCK-3
PIPES & FTG
DUCTS & FTG
CABLE TRAY
OUTFIT ITEMS
SEATS
SUB BLOCK-1
SUB BLOCK-2
SUB BLOCK-3
PIPES & FTG
DUCTS & FTG
CABLE TRAY
OUTFIT ITEMS
SEATS
PAINTING
PANEL-1
PANEL-2
PANEL-3
PIPES & FTG
DUCTS & FTG
CABLE TRAY
OUTFIT ITEMS
SEATS
PLATES
SECTIONS
PIPE PENETN. PCS
CABLE GLANDS
ACCESS FTG
LADDER, RUNG

OUTPUT FROM ERP

- Demand schedule of drawing & other prod. info.
- Requirement schedule of material
- Histogram of required manpower
- Cash outflow required

SOFTWARE USED

- **Core Design Software - TRIBON (v. M3 SP5)**
  - » Excellent 3D Modeller with capability of attaching all necessary attributes
  - » Effective Output of Production Information
  - » Efficient customized Report Generation

  BUT

  - » Lack of 2-way information flow to / from other s/w
- **Other Design Software**
  - » AutoCAD
  - » PIPENET
- **Other Software**
  - » MS Office
  - » MS Project
  - » MS Exchange
- **ERP Software - *SAP* version 4.6b / ECC 6**

CONCLUSION

- Implementation of ERP, as a tool for efficient flow of information in shipbuilding yet to reach optimum level
- Non-availability of professionals with ERP implementation skill as well as thorough domain knowledge in shipbuilding
- Only possible if, shipbuilding experts master & leverage capabilities of ERP.

THANK YOU

DOCUMENT SCHEDULE

List Edit Goto System Help

Display Char. Data of Doc. Type : (Drw / Sor / Graq / D787 / PRS)

Project Number: 3017
Document Type: DRW
Department: HOF

Sl.No	Document Number	Docket No	Rev No	Document Description	Completion Date Scheduled	Completion Date Actual	NSM Approval Scheduled	Recv. Dat Actual
1	3017-18/093002			M. S. STOOL FOR SIDE BLOCKS				
2	3017/093001		C	ARRGT_KEEL & SIDE BLOCKS IN BUILD BERTH	11-12-07	15-12-07		
3	3017/HO/171101			CO'S DAY CABIN DK 01 FR 23 - 27 (S)	10-02-10			
4	3017/HO/171102			CO'S NIGHT CABIN DK 01 FR 23 - 26½ (S)	10-02-10			
5	3017/HO/171201			D. CABIN -1(N.O) DK 01 FR 25½ - 28 (P)	31-01-10			
6	3017/HO/171202			JOM (4 MEN) DK 01 FR 30½ - 33 (P)	31-12-09			

MODEL - TRIBON

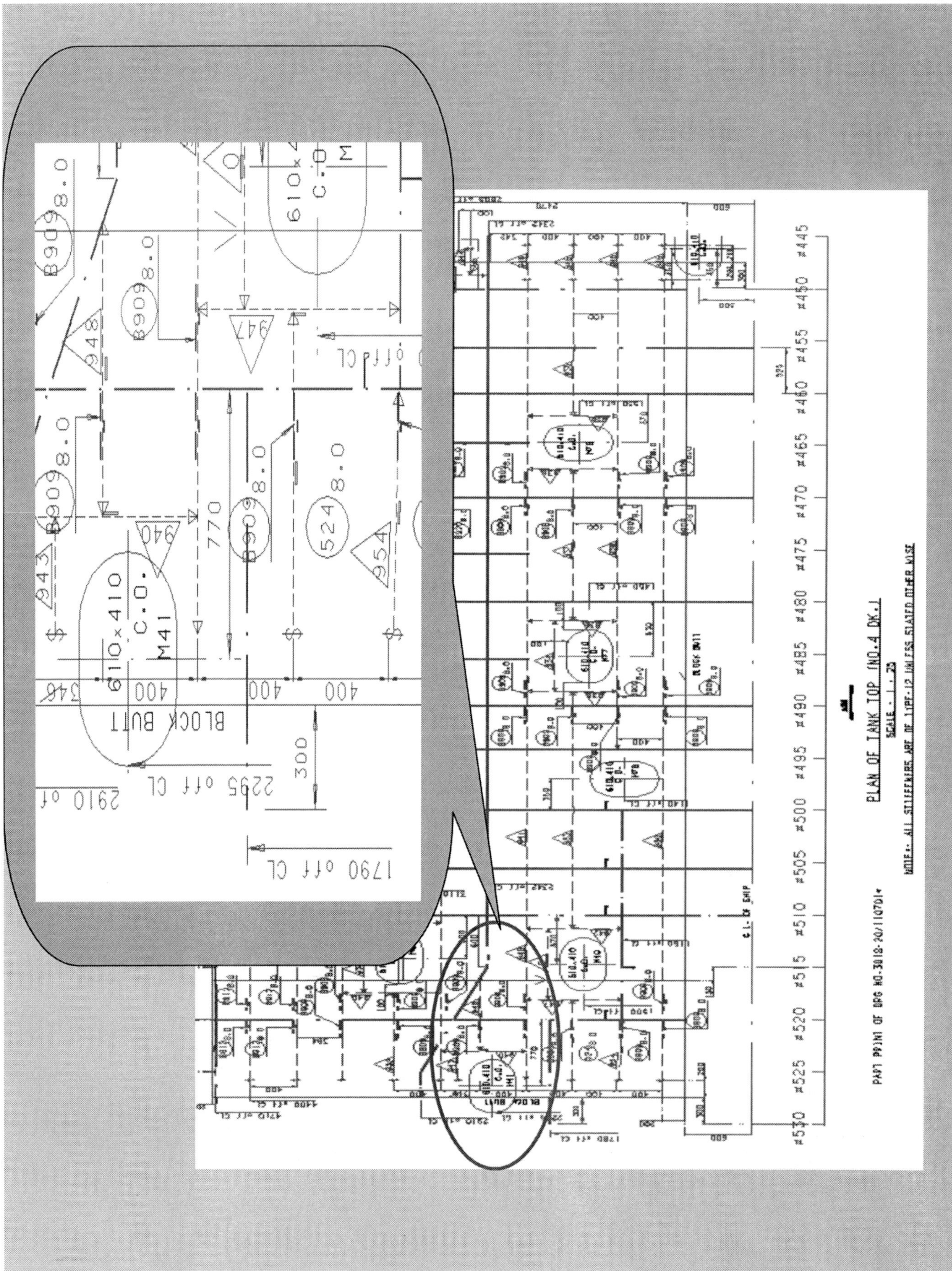
PLAN OF TANK TOP (NO.4 DK.)
BLOCK BUTT
2295 off CL
1790 off CL
610×410 C.O. M41

BOM

	B	C	D	E	F	G	H	I	J	K	L
1	TB NAME	ASS.	SB	PS	NESTING	QUALITY	PT_LEN	PT_WIDTH	THK	WT_KG	CHARGE
236	BP5-STGR-DK3-23BP	A20	1	0	20BP50851	ABA	194	110	8	0.964607	C015114114447
237	BP5-STGR-DK3-25BP	A20	1	0	20BP50851	ABA	194	110	8	0.964044	C015114114447
238	BP5-STGR-DK3-26BP	A20	1	0	20BP50851	ABA	195	110	8	0.967259	C015114114447
239	BP5-STGR-DK3-24BP	A20	1	0	20BP50853	ABA	130	125	8	0.798877	C015114114434
240	BP5-STGR-DK3-28BP	A20	1	0	20BP50853	ABA	130	125	8	0.792971	C015114114434
241	BP5-STGR-DK3-27BP	A20	1	0	20BP50853	ABA	130	125	8	0.794839	C015114114434
242	BP5-DEC40-3P	A30	1	0	20BP50801	ABA	4135	1347	8	350.635683	015114117747.
243	BP5-DEC40-2P	A30	1	0	20BP51553	ABA	8301	1660	15	1615.193289	C015114257547
244	BP5-DEC40-1P	A30	1	0	20BP51551	ABA	8301	2468	15	2402.507250	C015114257547
245	BP5-FLRFR47A-1P	A301	1	0	20BP51002	ABA	2351	887	10	128.425203	015114157747.
246	BP5-FLRFR47-1P	A301	1	0	20BP51002	ABA	1844	1152	10	134.954455	015114157747.
247	BP5-FLRFR48A-1P	A302	1	0	20BP51001	ABA	2351	884	10	113.737706	015114157747.
248	BP5-FLRFR48-1P	A302	1	0	20BP51001	ABA	1844	1151	10	119.577623	015114157747.

	A	B	C	D	E	F	G	H	I
1	Part Name	Tribon Name	Assembly	SB	PS	Quality	Profile Type	Prof_param1	Prof_param
163	S3020BP5-0	BP5-DEC40/S39P	A30	0	1	ABA	20	120	6.5
164	S3020BP5-934	BP5-DEC40/S36P	A30	0	1	ABA	20	120	6.5
165	S3020BP5-934	BP5-DEC40/S38P	A30	0	1	ABA	20	120	6.5
166	S3020BP5-935	BP5-DEC40/S35P	A30	0	1	ABA	20	120	6.5
167	S3020BP5-936	BP5-DEC40/S30P	A30	0	1	ABA	20	120	6.5
168	S3020BP5-936	BP5-DEC40/S31P	A30	0	1	ABA	20	120	6.5
169	S3020BP5-936	BP5-DEC40/S33P	A30	0	1	ABA	20	120	6.5
170	S3020BP5-936	BP5-DEC40/S34P	A30	0	1	ABA	20	120	6.5
171	S3020BP5-937	BP5-DEC40/S32P	A30	0	1	ABA	20	120	6.5
172	S3020BP5-938	BP5-DEC40/S29P	A30	0	1	ABA	20	120	6.5
173	S3020BP5-939	BP5-DEC40/S28P	A30	0	1	ABA	20	120	6.5
174	S3020BP5-940	BP5-DEC40/S22P	A30	0	1	ABA	20	120	6.5
175	S3020BP5-940	BP5-DEC40/S26P	A30	0	1	ABA	20	120	6.5
176	S3020BP5-940	BP5-DEC40/S27P	A30	0	1	ABA	20	120	6.5
177	S3020BP5-941	BP5-DEC40/S23P	A30	0	1	ABA	20	120	6.5
178	S3020BP5-942	BP5-DEC40/S24P	A30	0	1	ABA	20	120	6.5
179	S3020BP5-944	BP5-DEC40/S21P	A30	0	1	ABA	20	120	6.5
180	S3020BP5-945	BP5-DEC40/S20P	A30	0	1	ABA	20	120	6.5
181	S3020BP5-946	BP5-DEC40/S12P	A30	0	1	ABA	20	120	6.5
182	S3020BP5-946	BP5-DEC40/S13P	A30	0	1	ABA	20	120	6.5
183	S3020BP5-946	BP5-DEC40/S14P	A30	0	1	ABA	20	120	6.5
184	S3020BP5-946	BP5-DEC40/S15P	A30	0	1	ABA	20	120	6.5
185	S3020BP5-946	BP5-DEC40/S16P	A30	0	1	ABA	20	120	6.5
186	S3020BP5-946	BP5-DEC40/S17P	A30	0	1	ABA	20	120	6.5
187	S3020BP5-946	BP5-DEC40/S18P	A30	0	1	ABA	20	120	6.5
188	S3020BP5-946	BP5-DEC40/S19P	A30	0	1	ABA	20	120	6.5
189	S3020BP5-947	BP5-DEC40/S11P	A30	0	1	ABA	20	120	6.5

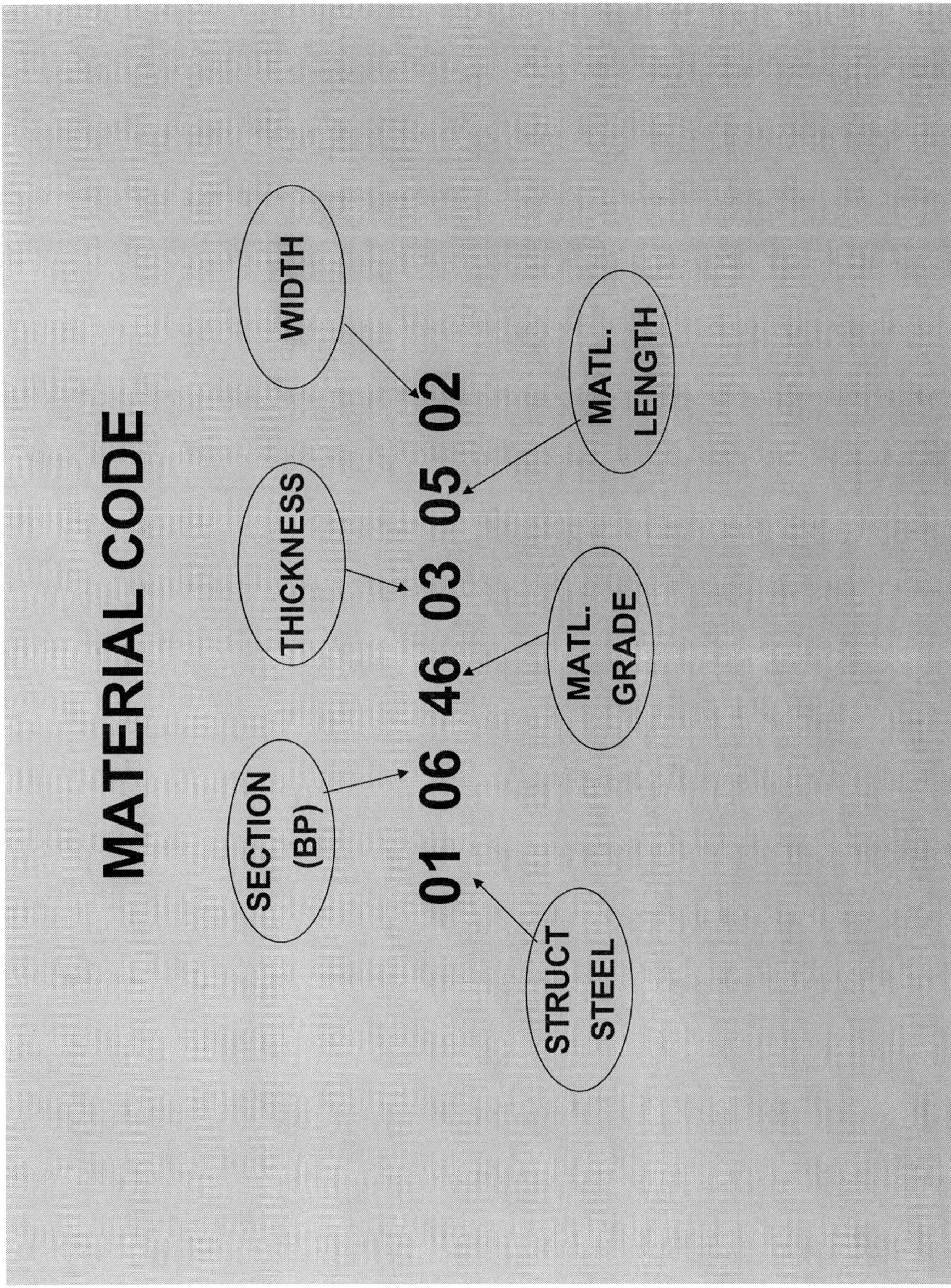
MATERIAL CODE
SECTION (BP)
WIDTH
THICKNESS
01 06 46 03 05 02
STRUCT STEEL
MATL. GRADE
MATL. LENGTH

INDENT

Garden Reach Shipbuilders & Engineers Ltd.
(Ministry of Defence, India)
Design and Planning Office-1,
43/46, Garden Reach Road, Kolkata - 700024.

INDENT

YARD NO: 2072 IPV

010646030502

Sl No	Material Code		…antity Required	UOM	Req Dlv Date	Pur Grp	Fitting Number	WBS Element
00010	010646030502.	ABS GR AH36 STEEL BULB FLAT 60X5mm	5.270	TO	30.09.2009	IMP	QTY REQD. 260 PCS@6M(CUTTING LG)=5.27 TO	I1.01.01
00020	010646030506.	ABS GR AH36 STEEL BULB FLAT 80X5mm	11.060	TO	30.09.2009	IMP	QTY REQD. 400 PCS@6M(CUTTING LG)=11.06TO	I1.01.01
00030	010646050513.	ABS GR AH36 STEEL BULB FLAT 120X7mm	0.210	TO	30.09.2009	IMP	QTY REQD. 4 PCS@6M(CUTTING LG)=0.210TO	I1.01.01

Components of Service

Inspection Authority / Scope Of Inspection:
INSPECTION AUTHORITY: ABS& 3 COPIES OF INSPECTION & MATERIAL TEST CERTIFICATE TO BE SUPPLIED.

ENCLOSURES:

(DRAWING)
(SOR)

Signature of Planning officer:

Signature of Dealing Officer:
Signature of Mgr./ DM.:
Signature of DGM/CM:

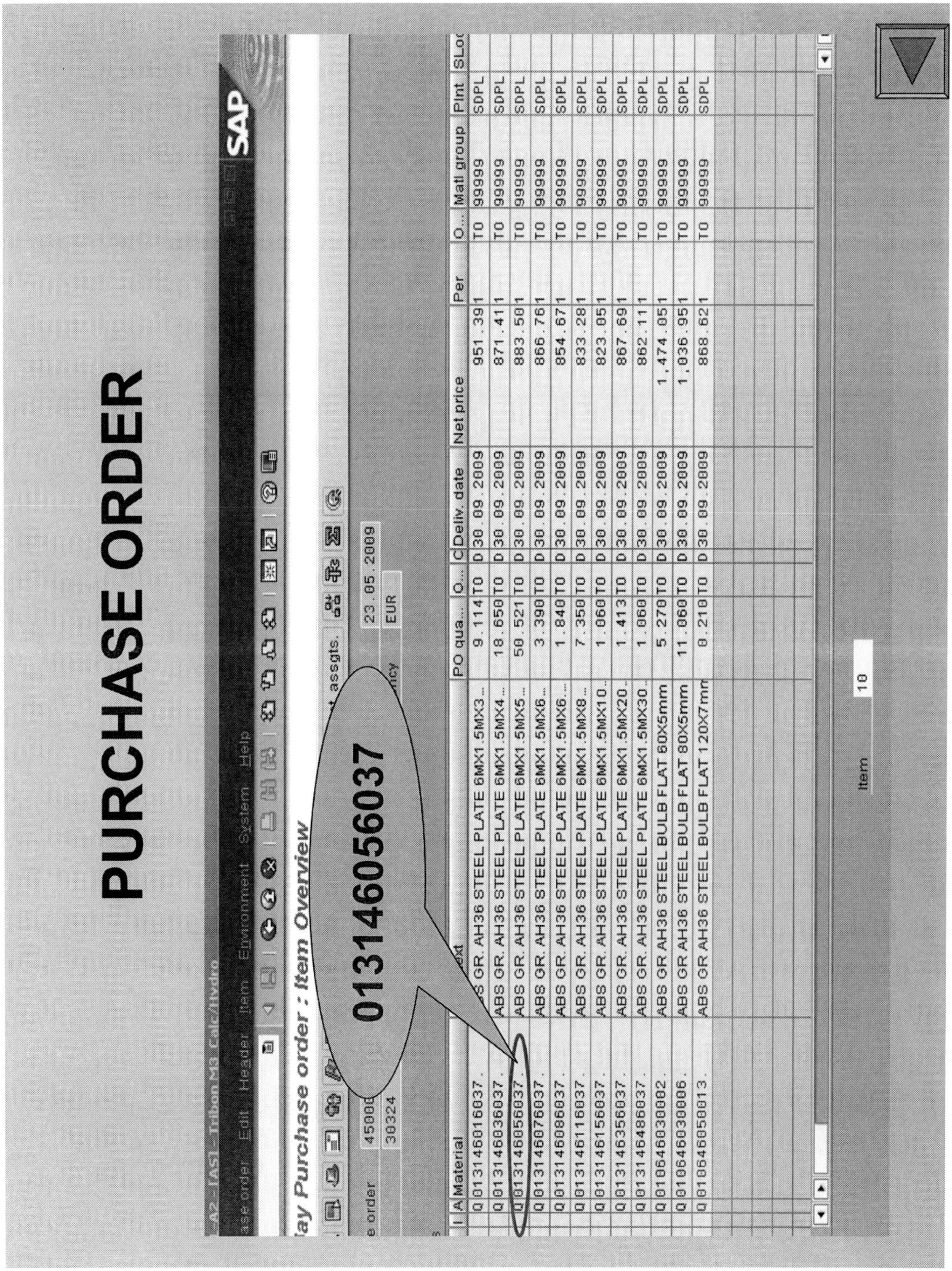
PURCHASE ORDER
Purchase order : Item Overview
0131460 56037
Material
01314601 6037.
01314603 6037.
01314605 6037.
01314607 6037.
01314608 6037.
01314611 6037.
01314615 6037.
01314635 6037.
01314648 6037.
01064603 0002.
01064603 0006.
01064605 0013.
ABS GR. AH36 STEEL PLATE 6MX1.5MX3 ...
ABS GR. AH36 STEEL PLATE 6MX1.5MX4 ...
ABS GR. AH36 STEEL PLATE 6MX1.5MX5 ...
ABS GR. AH36 STEEL PLATE 6MX1.5MX6 ...
ABS GR. AH36 STEEL PLATE 6MX1.5MX6 ...
ABS GR. AH36 STEEL PLATE 6MX1.5MX8 ...
ABS GR. AH36 STEEL PLATE 6MX1.5MX10 ...
ABS GR. AH36 STEEL PLATE 6MX1.5MX20 ...
ABS GR. AH36 STEEL PLATE 6MX1.5MX30 ...
ABS GR AH36 STEEL BULB FLAT 60X5mm
ABS GR AH36 STEEL BULB FLAT 80X5mm
ABS GR AH36 STEEL BULB FLAT 120X7mm
PO qua...
9.114
18.650
50.521
3.390
1.840
7.350
1.060
1.413
1.060
5.270
11.060
8.210
Deliv. date
30.09.2009
Net price
951.39
871.41
883.50
866.76
854.67
833.28
823.05
867.69
862.11
1,474.05
1,036.95
868.62
Per
Matl group
99999
Plnt
SDPL
SLoc
23.05.2009
EUR
Item
10
SAP

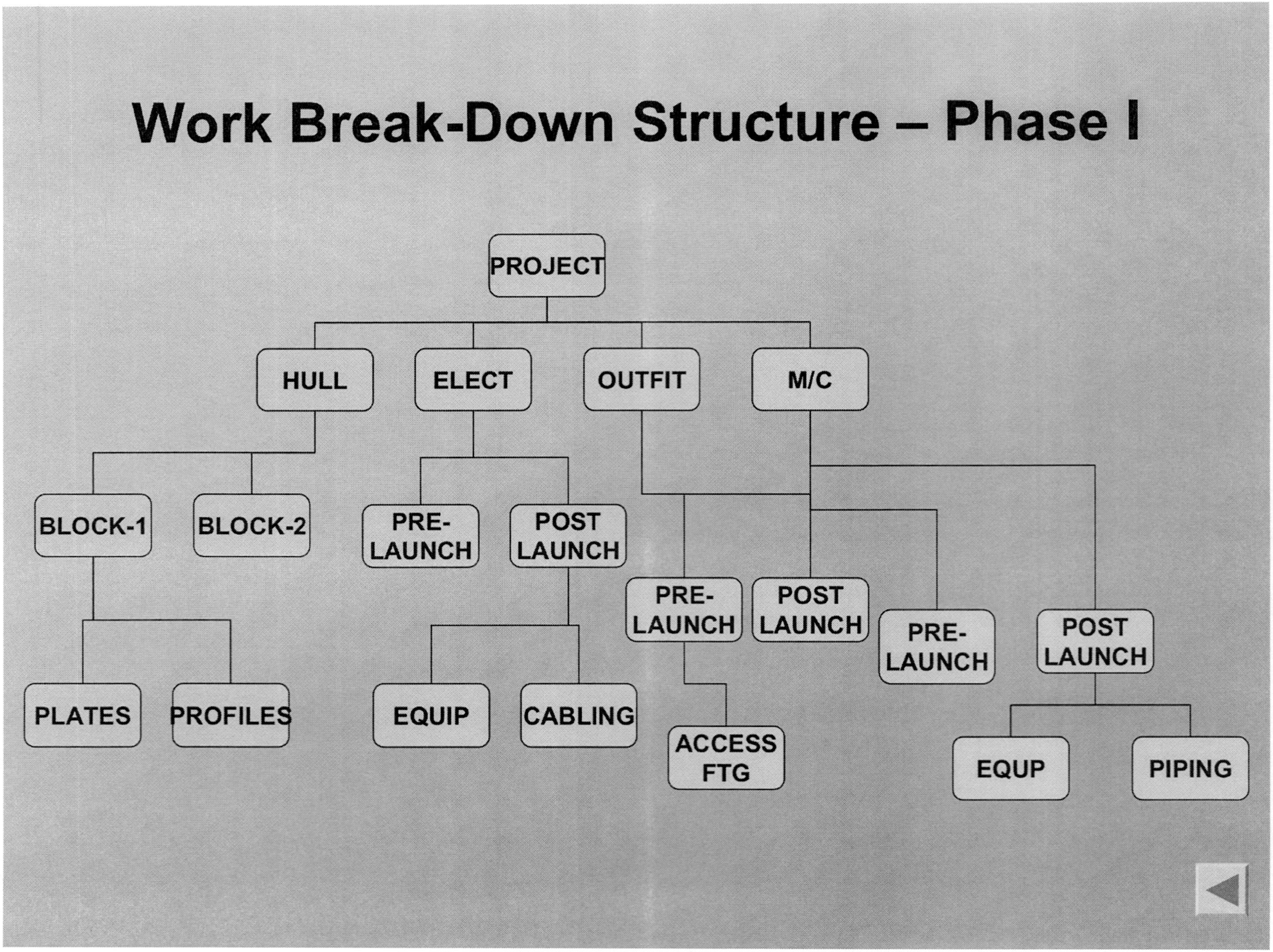

Work Break-Down Structure – Phase I
PROJECT
HULL
ELECT
OUTFIT
M/C
BLOCK-1
BLOCK-2
PRE-LAUNCH
POST LAUNCH
PRE-LAUNCH
POST LAUNCH
PRE-LAUNCH
POST LAUNCH
PLATES
PROFILES
EQUIP
CABLING
ACCESS FTG
EQUP
PIPING

PRODUCTION PLANNING – MS PROJECT

7	DRG-B8	Wed 2/7/07	1 day	Wed 2/7/07
8	FAB-B8	Tue 4/10/07	258 days	Fri 2/15/08
9	EREC-B8 (F)	Wed 3/26/08	27 days	Wed 10/1/08
10	EREC-B8 (W)	Mon 9/8/08	52 days	Sat 11/15/08
11	DRG-A9	Thu 6/14/07	1 day	Thu 6/14/07
12	FAB-A9	Thu 8/2/07	314 days	Thu 8/14/08
13	EREC-A9 (F)	Wed 10/1/08	44.6 days	Sat 11/29/08
14	EREC-A9 (W)	Fri 2/13/09	18 days	Sat 6/20/09
15	DRG-B7	Mon 9/10/07	1 day	Mon 9/10/07
16	FAB-B7	Wed 10/3/07	391.47 days	Sat 1/24/09
17	EREC-B7 (F)	Fri 10/3/08	24 days	Mon 2/2/09
18	EREC-B7 (W)	Mon 2/2/09	18 days	Tue 2/24/09

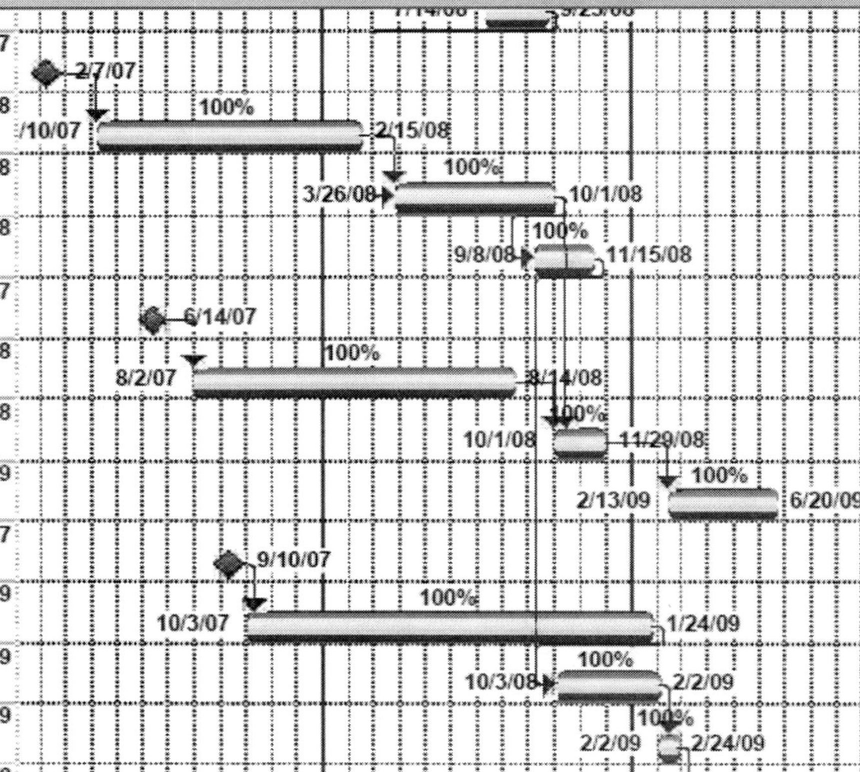

PANEL ASSY - MODEL

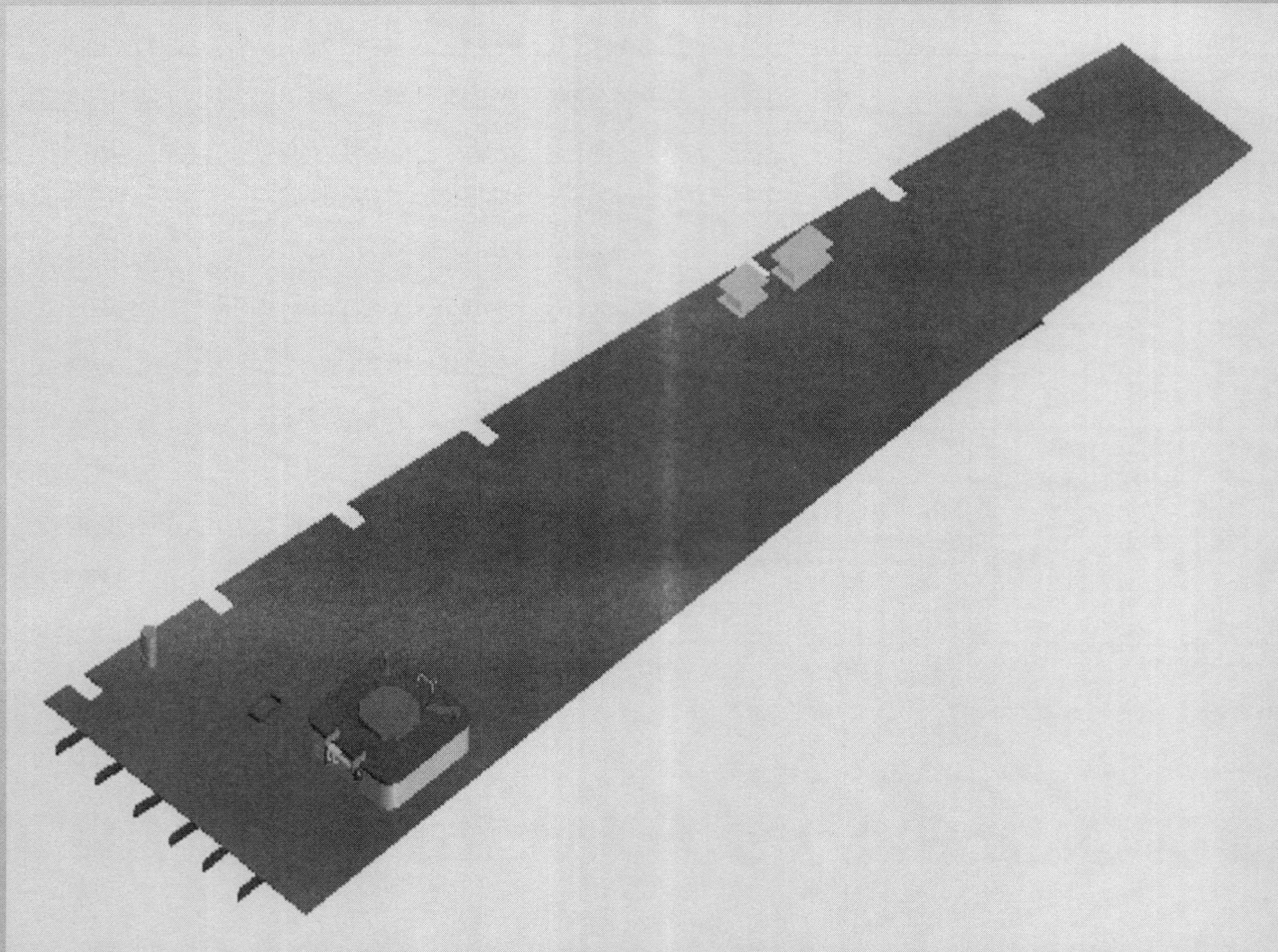

PANEL ASSY - DRAWING

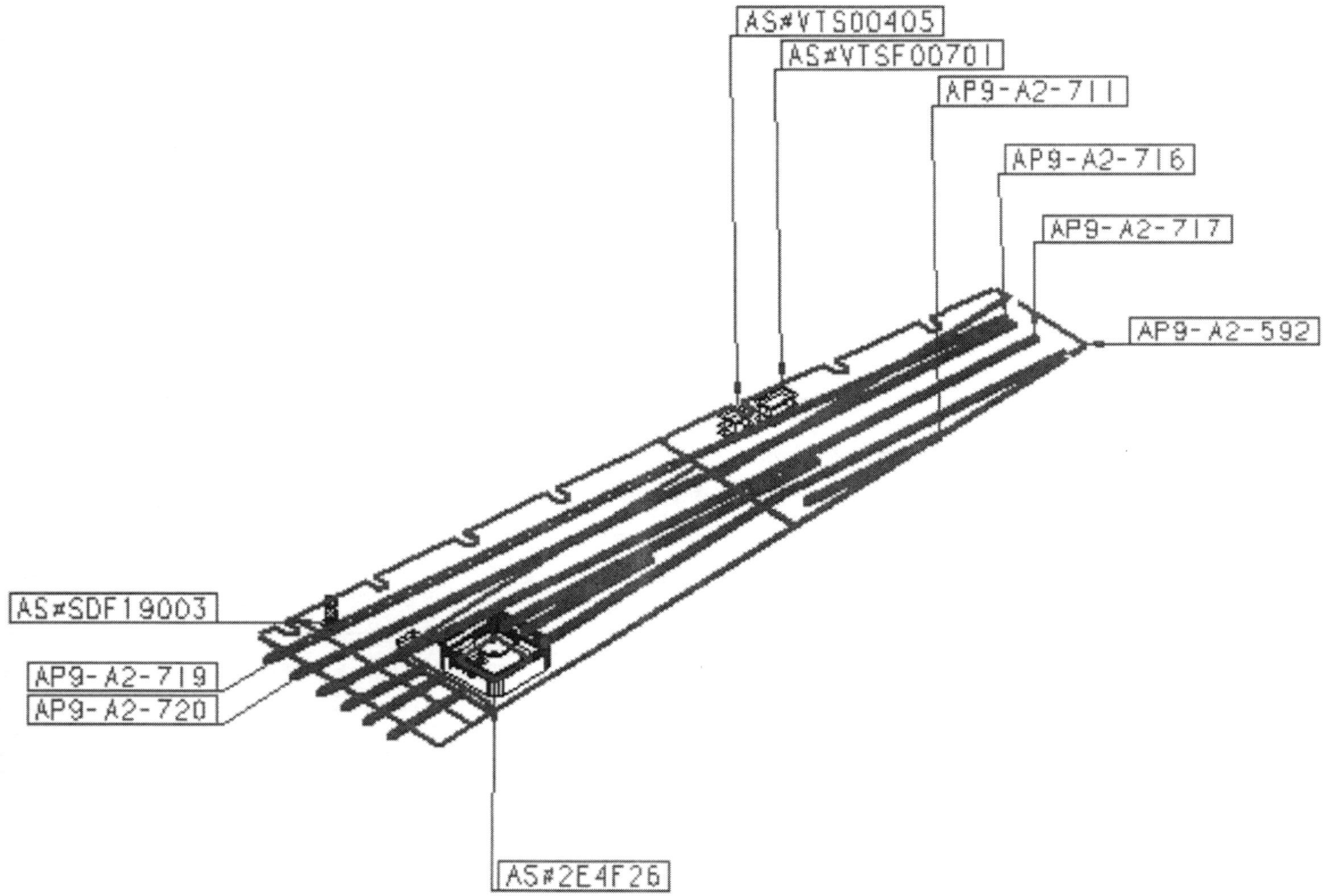

PANEL - BOM

A	B	C	D	E	F	H
SL. NO.	PART NAME	TYPE	DESCRIPTION	QTY	GR.	MATL. CODE
1	AS-VTSF00701	equipment	MS DK./BHD PIECE 300X150 FOR VENTILATION DUCT	1		590206114023
2	AS-VTS00405	equipment	MS DK./BHD PIECE 200X150 FOR VENTILATION DUCT_NEW	1		590206114025
3	AS-SDF19003	equipment	CARBON STEEL DK/BHD PC 80 NB (ONE END FLANGED & OTHER END PLAIN SLEEVED WITH CENTRE SLEEVE)	1		485115112603
4	AS-2E4_2E10D	equipment	10 MM.THK.MCT BHD.GLAND 141X238X60	1		701701010303
5	AS-2E4F26	equipment	WATER TIGHT HATCH	1		370504117071
6	ASAP9-DEC20/S6P	stiffener	HP20: 60x5.0	1	ABA	15116030002
7	ASAP9-DEC20/S5P	stiffener	HP20: 60x5.0	1	ABA	15116030002
8	ASAP9-DEC20/S4P	stiffener	HP20: 60x5.0	1	ABA	15116030002
9	ASAP9-DEC20/S3P	stiffener	HP20: 60x5.0	1	ABA	15116030002
10	ASAP9-DEC20/S2P	stiffener	HP20: 60x5.0	1	ABA	15116030002
11	ASAP9-DEC20/S1P	stiffener	HP20: 60x5.0	1	ABA	15116030002
12	ASAP9-DEC20-5P	plate	5	1	ABA	15114055337
13	ASAP9-DEC20-4P	plate	4	1	ABA	15114035233
14	ASAP9-DEC20-3P	plate	5	1	ABA	15114055337
15	ASAP9-DEC20-2P	plate	5	1	ABA	15114055337
16	ASAP9-DEC20-1P	plate	4	1	ABA	15114035233

SUB-BLOCK ASSY

BLOCK ASSY

DISTRIBUTIONS OF RESIDUAL STRAIN COMPONENTS IN WELDED STIFFENED SHIP HULL PLATES

S Y Kenno, S Das, M Hamdoon, and J B Kennedy, University of Windsor, Canada
R Rogge and M Gharghouri, Canadian Neutron Beam Centre, Canada

SUMMARY

Ship failures are often caused due to several factors, and one of the prime factors is the presence of high residual stresses resulting from the welding of stiffeners on the parent plate. A thorough experimental study using the neutron diffraction method was undertaken to determine the residual strain distributions associated with the stiffened plates typically used in ship hulls. The specimens were built of 350WT grade 9.5 mm thick steel plate stiffened by welding 127x76x9.5 angle steel stiffeners of 300W grade. This paper provides a detailed discussion on the specimens, test procedure, and the results obtained. Emphasis has been given on the resulting strain distributions when there is a stop and restart during the welding process or when a stiffener is terminated before the plate is terminated. This study shows that for these cases, the change in strain may not be a matter of concern for the ship building industry.

1. INTRODUCTION

Residual stresses are created during the welding process as the welded structure cools to the surrounding temperature, the weld material contracts and solidifies at varying rates. Residual stresses interact with external forces and stresses that a steel structure experiences during its service life. Ship hulls are comprised of multiple steel plates reinforced with stiffeners welded longitudinally and transversely and then the plates are welded together to create the hull structure. Welding is the only choice in the construction of a steel ship hull and therefore, residual stresses are inevitable within the ship hull structure. Hence, a comprehensive and highly detailed image of the residual stresses caused by welding is necessary for proper design and construction of a ship hull. Otherwise high concentrations of residual stresses can cause ultimate failure of the hull by cracks initiated transverse to the weld at these locations.

Previous studies conducted primarily focus on the longitudinal component of residual stresses and only near surface stresses. Most of these studies were conducted using conventional methods of testing such as, hole-drilling [1], sectioning method [2], and X-ray diffraction [3]. These methods of testing have limitations on the depth that the measurements can be collected and a plane strain condition is usually assumed. Other studies found in the literature considered variables such as, the welding heat input [4], the type of joints [5][6], and the weld filler material [4], but no research was completed on irregularities in a construction process.

The objective of this study was to obtain accurate representations of the residual strains, using the neutron diffraction (ND) method, from two specimens that were created to simulate typical stiffened panels found in ship hulls. The irregularities that were introduced into the specimens were a stop and start in the welding process and a stiffener terminated before the end of the plate, both of which can be typically found during the construction of a ship hull. The ND method was chosen as the most accurate and thorough method of collection with the ability to collect measurements points through the entire thickness of the plate.

2. NEUTRON DIFFRACTION

All experimental data was collected using the L3 Spectrometer at the Chalk River Laboratories at the Canadian Neutron Beam Centre in Chalk River, Ontario, Canada. The ND method is a highly specialized method of data collection based on the theories of Bragg's law, as shown in Equation 1. The beam of neutrons, with wavelength λ, passes through the test sample of material and diffracts off the lattice planes spaced at d_{hkl} in the material. The angle at which the neutron diffract is $2\theta^{B}_{hkl}$, which is called Bragg's angle, and is when the neutrons are scattered coherently and elastically by the lattice planes *hkl*. The spacing of the lattice planes of stressed material is compared with the lattice spacing of a stress-free sample of material to compute the average elastic strain in the gauge volume. [7][8]

$$\lambda = 2d_{hkl} \sin \theta^{B}_{hkl} \quad (1)$$

3. SPECIMEN SPECIFICATIONS

3.1 MATERIAL AND CONSTRUCTION

The plates for this study were 9.5 mm thick, 350WT grade [9] steel and cut using cold-cut waterjet method to eliminate the residual stresses often introduced in the cutting phase. The stiffeners were 300W grade [9] steel, angle stiffeners of dimensions 127 mm x 76.2 mm x 9.5 mm. The angles were cut using a band saw, which is also a cold-cut method.

The metals used for the plates, stiffeners, and welding wire were tested according to the ASTM standard [10] for quasi-static tension tests in the Structural Engineering Laboratory at the University of Windsor. The test results provided the yield strength, tensile strength, and modulus of elasticity for all three materials. The yield strength for the plate, the stiffener, and the weld wire material was: 405 MPa, 350 MPa, and 390 MPa, respectively. The tensile strength for the three materials was: 507 MPa, 528 MPa, and 513 MPa, respectively and for the modulus of elasticity was: 205 GPa, 204 GPa, and 209 GPa, respectively.

To weld the stiffeners onto the parent plates, the metal core arc welding process was used by a licensed welder. The specimens were fully restrained during the welding and cooling process, to simulate the plate as being part of an entire ship hull structure. The heat input that was used for both specimens was moderate and controlled by maintaining a constant current and welding speed.

3.2 SPECIMEN 1

The first specimen was a base plate 600 mm long, 400 mm wide, and 9.5 mm thick welded with one angle stiffener 600 mm long with the dimensions of 127 mm x 76.2 mm x 9.5 mm, as shown in Figure 1(a). Strain measurements were taken across the two transverse lines 1-2-3 and 4-5-6 and along the one longitudinal line 2-5. This specimen was primarily studied to find the residual strain distribution created by welding one stiffener on a parent plate. The secondary objective of this specimen was the change in the residual strain distribution with an inconsistency in the welding process. At point 5 in Figure 1(a), there was a stop and start in the welding process and a close-up detail of this inconsistency is shown in Figure 1(b). It is currently believed that a stop and start in the welding process does not dramatically affect the structural integrity of the joint but it may influence the residual strain distributions and create strain concentrations in that area. This was the focus of the strain measurements completed in this study.

All three normal strain components were measured at every location. Along line 1-2-3 which is 280 mm long, data at seven depths through the thickness of the plate were collected and along line 4-5-6 which is 133 mm long, data at three depths through the thickness were collected. The spacing of the transverse measurements was as small as 1 mm near the weld where the highest strain gradient was expected to occur and the spacing gradually increased to 40 mm as the transverse distance from the stiffener increased. On the longitudinal line 2-5, data was collected only at one depth through the thickness, closest to the welded surface of the plate, and spaced evenly at 10 mm.

3.2 SPECIMEN 2

Specimen 2 was built using the same size parent plate as Specimen 1 and the same type of angle stiffener. However, for Specimen 2, the stiffener was shortened to 400 mm long and welded in the centre of the 600 mm direction of the parent plate. This specimen was tested as a representation of locations in ship hulls where a stiffener needs to be terminated earlier than the edge of the plate. Early terminations of stiffeners can occur in ship hulls when there is an obstruction such as, mechanical or electrical equipment, and other structural apparatus.

(a)

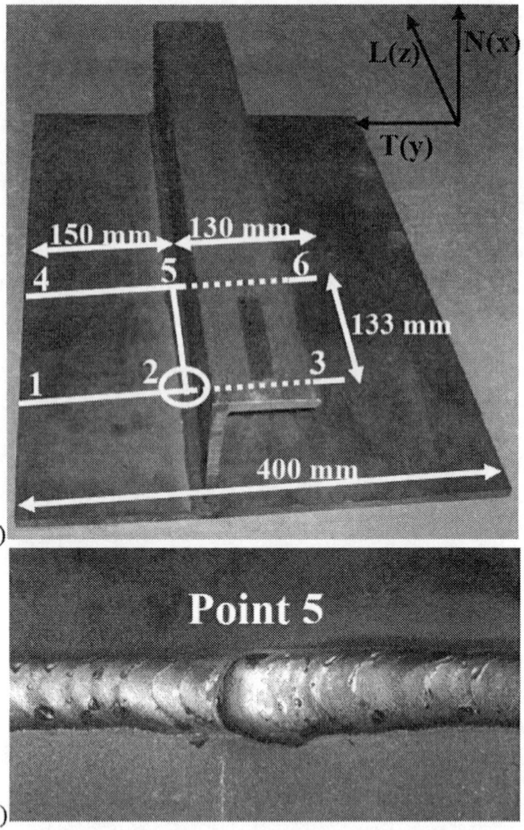

(b)

Figure 1: (a) Specimen 1 measurement details (b) close-up of inconsistency in the weld (point 5)

There were eight measurement lines for this specimen. The origin was located at the bottom edge of the stiffener at the inside edge of stiffener and it is shown by an open circle in Figure 2. The transverse measurement lines (Line 2 through Line 5) are 200 mm across the plate. Line 1 was a transverse line measured only 100 mm across the plate to avoid the high measurement times of the points underneath the stiffener. Line 1 was measured to compare to Specimen 1 and ensure that the residual strain in this location is similar to what was found transversely across the stiffener without termination and this shortened line is sufficient for comparison. Line 3 was measured as the line across the origin at the edge of the terminated stiffener. Line 2 and Line 4 were measured 20 mm on either side of Line 3, and Line 5 was measured 60 mm from Line 3 in the negative direction away from the stiffener.

The longitudinal measurement lines (Lines 6 through 8) were 70 mm in length and the measurements were taken at 10 mm intervals. All three lines extend 40 mm in the stiffener direction and 30 mm away from the stiffener from the origin Line 3. Line 6 was located through the centreline of the weld and 10 mm from the origin. Line 7 was located transversely 30 mm away from the origin and Line 8 was 50 mm away.

The location of the lines was chosen to see the change in the residual strain around the termination of the stiffener and how it compares to the residual strain where the termination of the stiffener is not affecting the residual strain distribution.

Only the transverse and longitudinal strains were measured for this specimen and these test data are presented in this paper.

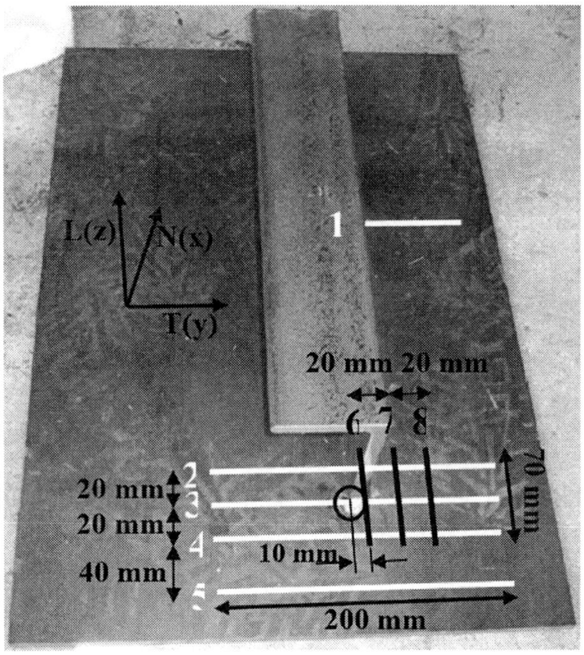

Figure 2: Specimen 2 measurement details

4. TEST RESULTS

4.1 SPECIMEN 1

Specimen 1 was measured for transverse and longitudinal residual strain measurements across two transverse lines and one longitudinal line. Figure 3 shows the longitudinal residual strain across the transverse line 1-2-3, considered the base line. The residual strain values show through the depth of the plate the strain values are consistent and increase slightly towards the welded surface of the plate at the depth of 8.9 mm. The maximum strain value is 1915 microstrain with the minimum value of -945 microstrain. The tensile peak of strain at the stiffener-weld connection is balanced by the compressive plateau as the distance from the weld increases. This pattern is typically found in welded structures and originally found by Faulkner [11] and others [2] using simply methods of stress measurement collection than neutron diffraction. The peak tensile strain is just below the yield strain of the material. Calculating the stress using all three residual strain components, the maximum tensile residual stress is above the yield strength of the material.

Figure 4 shows the longitudinal residual strain values for the transverse line 4-5-6, point 5 is the location along the weld pass where the stop and start in the welding process occurred (Figure 1(b)). These residual strain values show a similar pattern to that which was found across line 1-2-3. However, the range of strain values across this line is smaller with the maximum value at 1800 microstrain and the minimum of -855 microstrain. Given these values it appears that the strain is lower at the inconsistency than at the base weld location, contrary to what was previously assumed. To produce a more detailed look at the pattern of strain, measurements were taken along line 2-5, at the centreline of the weld from one transverse measurement line to the other. The strain values along this line are shown in Figure 5 for all three strain components. Line 2-5 is 133 mm long and the residual strain values maintain a fairly consistent level until at about 100 mm away from line 1-2-3 the strain values begin to fluctuate. The strain values peak and then drop to a minimum directly at the location of the stop and start. Due to limitations in the testing set-up, the collection of strain values past the point of the inconsistency were unattainable. Further investigation into this topic is required and planned as a continuation of this project.

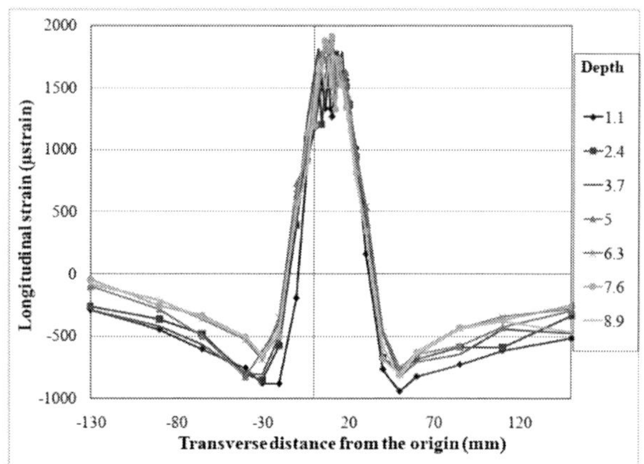

Figure 3: Longitudinal residual strain in the transverse direction across Line 1-2-3 from Specimen 1

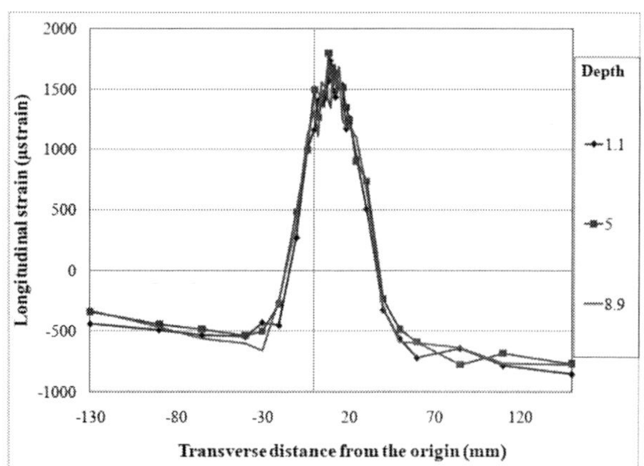

Figure 4: Longitudinal residual strain in the transverse direction across Line 4-5-6 from Specimen 1

Figure 6 shows a final comparison of the residual strain values for Specimen 1. The transverse and longitudinal strain values from both lines 1-2-3 and 4-5-6 are shown at the depth closest to the welded side of the plate (8.9 mm). The graph shows the higher maximums strain values located at line 1-2-3 as compared to line 4-5-6 for both the transverse and longitudinal strain components. The two components and two lines show similar patterns and distributions of strain across these two transverse lines.

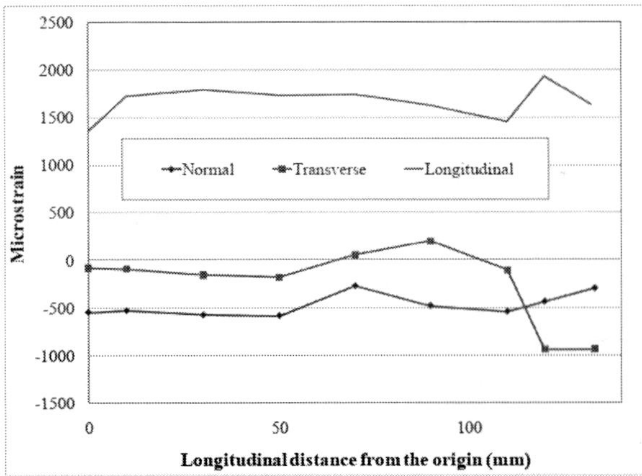

Figure 5: Residual strain distributions along the longitudinal direction for Specimen 1

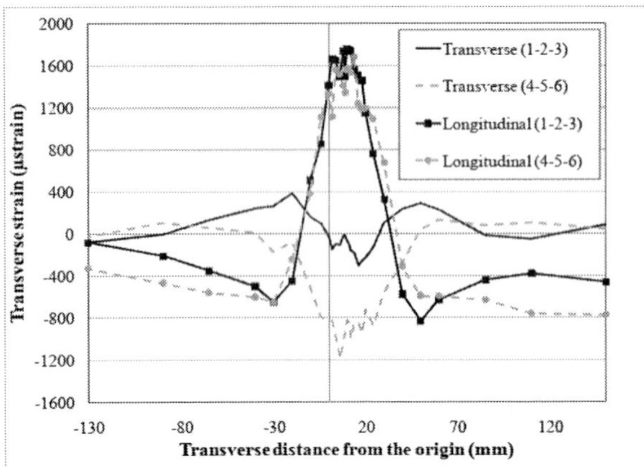

Figure 6: Comparison of transverse and longitudinal residual strains of Specimen 1

4.2 SPECIMEN 2

Specimen 2 had five transverse measurement lines at different points along the shortened stiffener on the plate. Line 1 is far away from the stiffener termination point and hence, the strains on this line are not influenced by the termination of the stiffener. Figure 7 shows the transverse strain distributions found across the transverse direction of the plate for Lines 1 through 5. It can be seen that a pattern exists among strain distributions on Lines 2 to 5 and the maximum strain occurs on Line 2 and decreases in magnitude as the measurement distance moves outside and away from the stiffener termination point and into the parent plate. Line 3 is on the stiffener termination line. For Line 2, the range of residual strain is from -240 to +618 microstrain and for Line 5 (the furthest away from the stiffener) the range is -35 to -1235 microstrain.

Figure 8 shows the transverse strain distributions along the longitudinal direction for Lines 6, 7, and 8. The graph shows that the three lines are very similar in strain ranges and patterns. Line 6 along the centreline of the weld has the least consistent strain pattern compared with the other two lines. This is most likely due to the small non-uniformity in the welding process which is unavoidable. Their plot shows how small variations in the welding process can result in a large jump in the residual strain values. All three lines show a distinct pattern at the longitudinal location of 0 where the stiffener is terminated. Within the stiffener, the residual strain values are fairly constant but when the end point of the stiffener is reached the residual strain values significantly reduce to very negative microstrain values as the distance from the end of the stiffener increases.

Where the transverse residual strain values show an excellent representation of the distributions within the plate, the longitudinal residual strain is the component that has higher and more significant values that are of most interest. Figure 9 shows the longitudinal residual strain values across the longitudinal direction of the plate for the first 5 lines. Line 1 shows a maximum residual strain value of +1880 microstrain with a similar magnitude and pattern to those found in Specimen 1 which had a maximum strain value of +1915 microstrain. The residual strain from Line 1 shows the largest range of values from -705 to +1880 microstrain. Line 2 shows a very similar residual strain pattern as Line 1 with the high tensile strain peak near the centre of the weld reducing to a compressive strain away from the stiffener. Observing the pattern of residual strain as the distance from the termination of the stiffener increases, the residual strain values are also decreasing. This is especially apparent at Line 5, which is 60 mm away from the stiffener termination point, where the strain values range from -41 to +410 microstrain, significantly lower than the other four transverse lines.

Figure 10 shows the longitudinal residual strain in the longitudinal direction for Lines 6 through 8. Line 6 appears incomplete due to restrictions in the testing set-up limiting the collection the strain values due to the position of the stiffener in relation to these points. For this line strain measurements for a 15 mm length into the stiffener were not collected. However, the general trend of all three lines is apparent in this graph. Line 6, as expected, shows the highest gradient from the termination point of the stiffener to the area past the end of the stiffener and the weld bead. Line 7 runs parallel to the stiffener at 20 mm away from the centreline of the weld. The residual strain values found along this line are consistent and vary only from +625 to +915 microstrain. This is expected of this location since extending a radius of 20 mm around the end of the weld bead termination point, still contains Line 7. Therefore, the measurement points along Line 7 are roughly 20 mm away from the weld bead at all point along the line, resulting in a plateau of stress values. There is a slight jump in the strain levels at the longitudinal origin due to the drastic stress change that occurs at the termination point of the stiffener. Line 8 is 40 mm away from the centreline of the weld bead and therefore should shows lower residual strain levels since the influence of the weld should be lesser compared to Line 7. The range of residual strain for this line is -655 to +300 microstrain. The strain levels begin to drop dramatically near and past the termination point of the stiffener and weld bead.

Figure 7: Transverse strain distribution in the transverse direction for Specimen 2

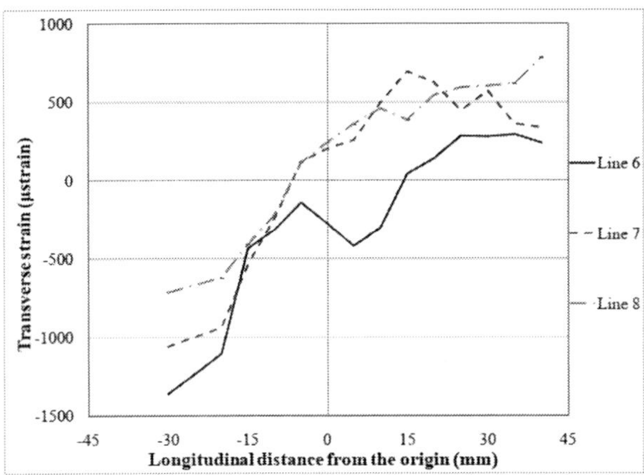

Figure 8: Transverse strain distribution in the longitudinal direction for Specimen 2

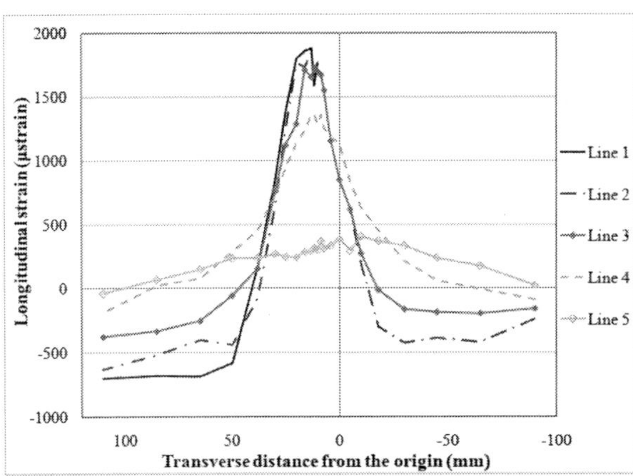

Figure 9: Longitudinal strain distribution in the transverse direction for Specimen 2

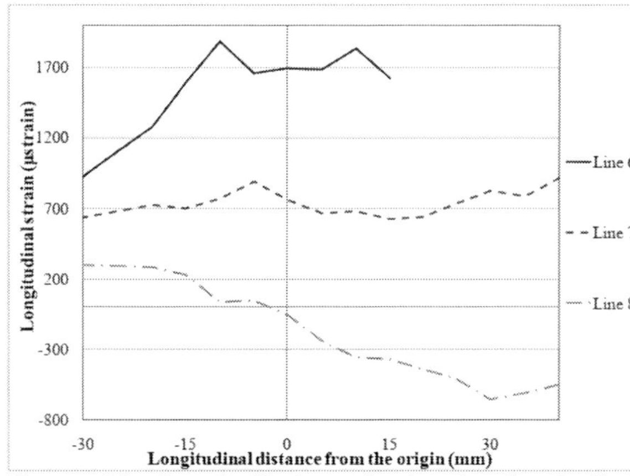

Figure 10: Longitudinal strain in the longitudinal direction for Specimen 2

5. CONCLUSIONS

The influences of stiffener termination and welding inconsistencies on residual strain in stiffened steel plate structures were examined. The neutron diffraction method was used to determine the residual strain distributions for the transverse and longitudinal components. The conclusions made here are limited to specimens similar to those used in this study:

- An inconsistency in the form of early termination of a stiffener and a stop and start in the welding process was found to reduce the residual strains in the adjacent area. Hence, such inconsistencies may not be a concern for structural integrity of ship hulls.
- The transverse and longitudinal residual strain components exhibit similar patterns with a maximum tensile peak at the stiffener balanced by a compressive plateau away from the weld.
- The longitudinal strain values are significantly higher than the transverse strains. Therefore, the longitudinal strain values are the main concern in structural design of welded structures although these values are below the yield strain of the material.
- A distinct pattern was found for the longitudinal residual strain component having a tensile strain block near the stiffener and a compressive strain plateau away from the stiffeners.

6. ACKNOWLEDGEMENTS

The authors would like to thank to Dr. David Stredulinsky and Dr. James Huang at the Defence Research and Development Canada (Atlantic) for financial and technical assistance. Sincere thanks to Mr. Ronald Donaberger and Mr. John Fox at the Canadian Neutron Beam Centre in the Chalk River Laboratories located in Chalk River, ON, Canada for their outstanding skill and assistance with the experimental work. Partial funding for the project was provided by the Natural Sciences and Engineering Research Council of Canada

7. REFERENCES

1. CHO, JR, LEE, BY, MOON, YH, and VAN TYNE, CJ., 'Investigation of residual stress and post weld heat treatment of multi-pass welds by finite element method and experiments', *Journal of Materials Processing Technology*, 2004.

2. DEXTER, RJ and PILARSKI, PJ., 'Crack propagation in welded stiffened panels', *Journal of Constructional Steel Research*, 2002.

3. GAO, H, GUO, H, BLACKBURN, JM, and HENDRICKS, RW., 'Determination of residual stress by X-ray diffraction in HSLA-100 steel weldments', *In Fifth International Conference on Residual Stress in Linkoping, Sweden*, 1997.

4. JAMES, MN, WEBSTER, PJ, HUGHES, DJ, CHEN, Z, RATEL, N, TING, S –P, and BRUNO, G., STEUWER, A., 'Correlating weld process conditions, residual strain and stress, microstructure and mechanical properties for high strength steel - the role of neutron diffraction strain scanning', *Material Science and Engineering A*, 2006.

5. PARADOWSKA, A, PRICE, JWH, IBRAHIM, R, FINLAYSON, TR, RIPLEY, M, and BLEVINS, R., 'Residual stress evaluation in multi-beads steel weldments using neutron diffraction', *In Proceedings of the IIW International Conference on Benefits of New Methods and Trends in Welding to Economy, Productivity and Quality in Prague, Czech Republic*, 2005.

6. PRICE, JWH, PARADOWSKA, A, JOSHI, S, and FINLAYSON, TR., 'Residual stresses measurement by neutron diffraction and theoretical estimation in a single weld bead', *International Journal of Pressure Vessels and Piping*, 2006.

7. HUTCHINGS, MT, WITHERS, PJ, HOLDEN, TM, and LORENTZEN, T., *Introduction to the characterization of residual stress by neutron diffraction*. Taylor & Francis Group, Boca Raton, 2005.

8. International Organization for Standardization, Geneva, Switzerland. *Technology Trends Assessment*. Versailles Project on Advanced Materials and Standards, National Institute of Standards and Technology, Gaithersburg, Maryland, USA, 2001.

9. CSA. *Canadian Standards Association: General Requirements for Rolled or Welded Structural Quality Steel/Structural Quality Steel*. Canadian Standards Association, Mississauga, 2004.

10. ASTM. *ASTM E8M: Standard Test Methods for Tension Testing of Metallic Materials*. ASTM International, West Conshohocken, 2008.

11. FAULKNER, D., 'A Review of Effective Plating for Use in the Analysis of Stiffened Plating in Bending and Compression', *Journal of Ship Research*, 1975.

9. AUTHORS BIOGRAPHY

Sara Kenno is currently a doctoral candidate in Civil Engineering at the University of Windsor, Canada. Her Master's thesis was on the residual stress in stiffened plates with one and two stiffeners.

Sreekanta Das is an associate professor in Civil Engineering at the University of Windsor, Canada. His area of expertise is structural engineering with focus on pipeline integrity and masonry structures as well as, ship structures.

Muhsin Hamdoon is a doctoral candidate in Mechanical Engineering at the University of Windsor, Canada. His area of research is in fatigue and strength of ship hulls.

John Kennedy is a professor emeritus in Civil Engineering at the University of Windsor. His research area of focus is steel structures.

Ronald Rogge is a scientist at the Canadian Neutron Beam Centre in the Chalk River Laboratories, Canada. He works with the L3 Spectrometer and the investigation of residual stresses.

Michael Gharghouri is a materials scientist at the Canadian Neutron Beam Centre in the Chalk River Laboratories, Canada.

DISTRIBUTIONS OF RESIDUAL STRAIN COMPONENTS IN WELDED STIFFENED SHIP HULL PLATES

University of Windsor

Kenno et al.

Presented by S. Das

International Conference on Ship & Offshore Technology – India 2009, Developments in Ship Design & Construction
10 – 11 December 2009, Kharagpur

Background

- Ship hull structure
  - Stiffeners welded onto steel plates

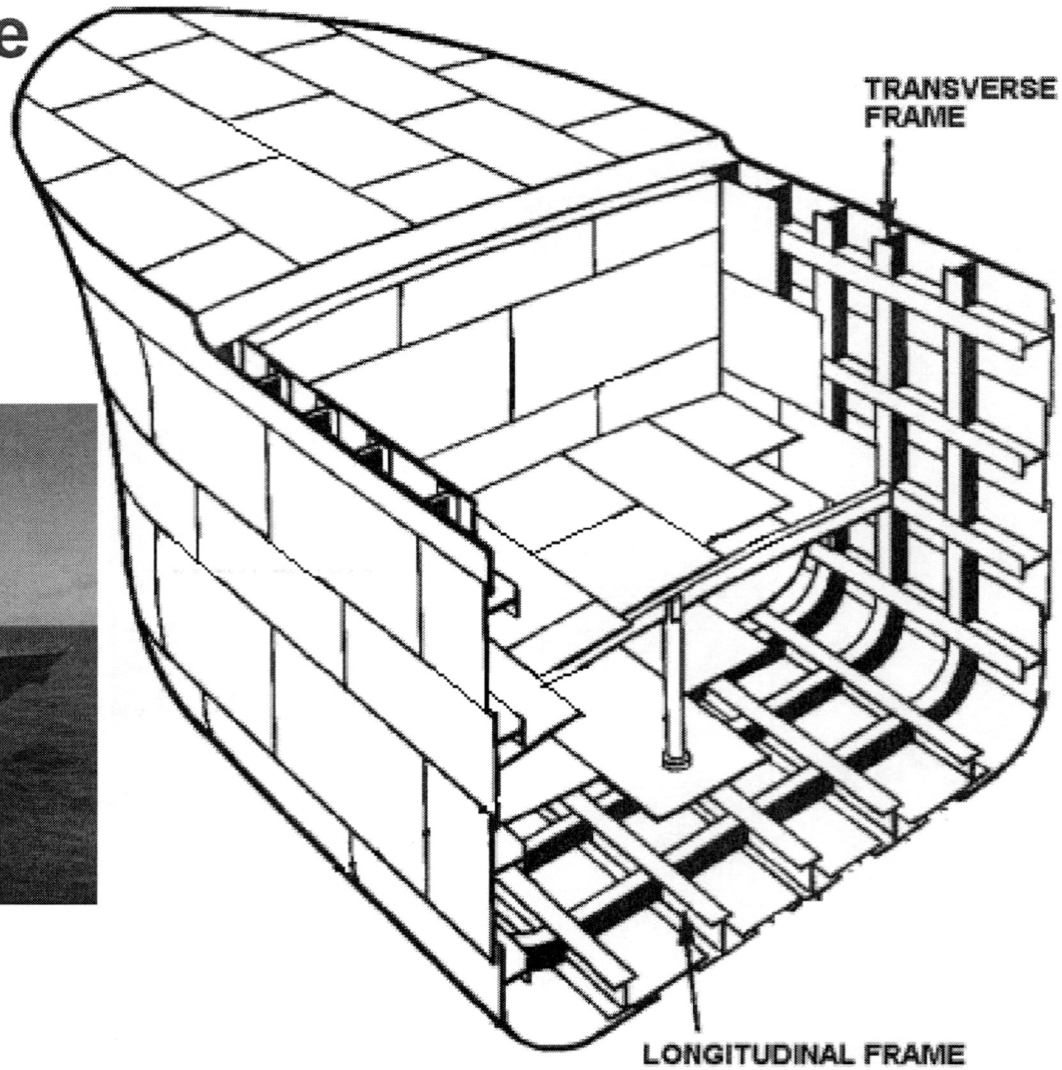

University of Windsor

Background

- Ship hull structure
  - Stiffeners welded onto steel plates
- Welding → residual stresses
  - Interact with other stresses

University of Windsor

Background

- Ship hull structure
  - Stiffeners welded onto steel plates
- Welding → residual stresses
  - Interact with other stresses
- Failure
  - Cracks at the weld

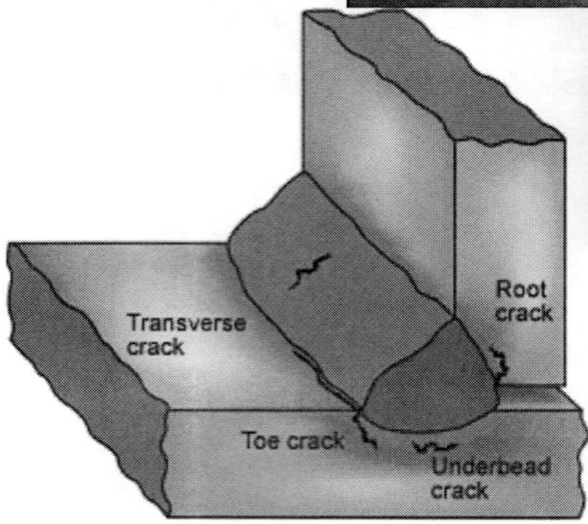

Residual Stress

University of Windsor

- Thermomechanical processing

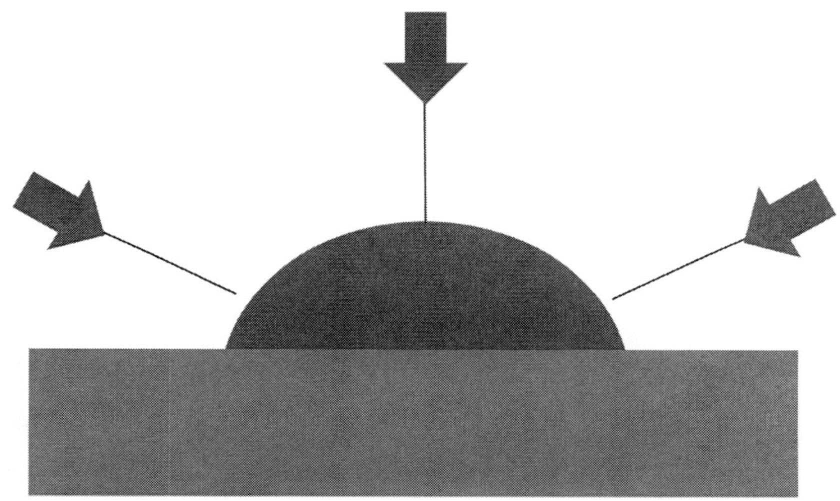

University of Windsor

Residual Stress Testing

- Destructive

University of Windsor

Residual Stress Testing

- Destructive,
- Semi-destructive, and

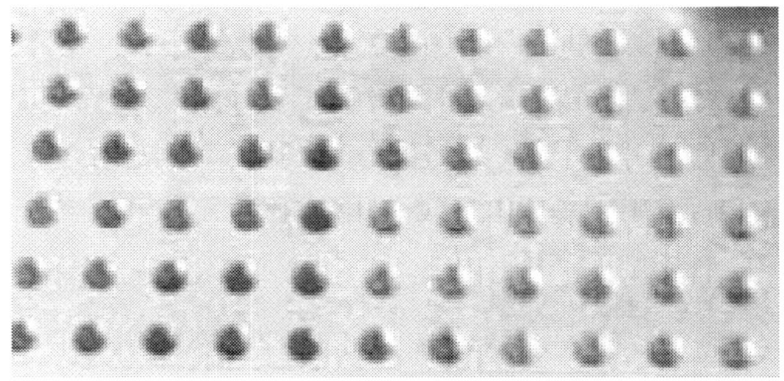

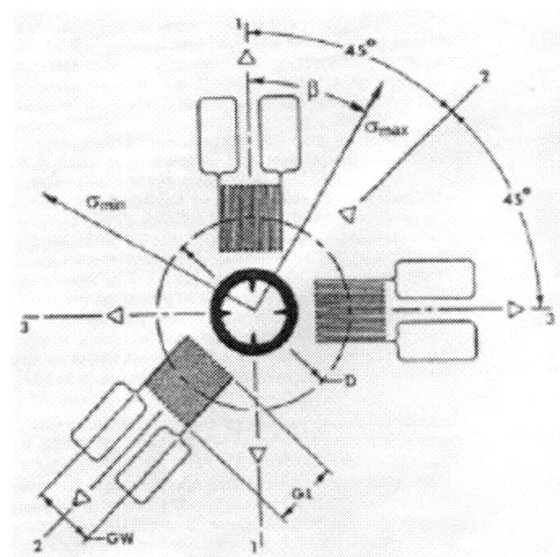

University of Windsor

Residual Stress Testing

- Destructive,
- Semi-destructive, and
- **Non-destructive**

Residual Stress Testing

University of Windsor

- Destructive,
- Semi-destructive, and
- Non-destructive

- Discrepancies between methods due to steep gradient of stress and test volume required

University of Windsor

Residual Stress Testing

- Destructive,
- Semi-destructive, and
- Non-destructive

- Discrepancies between methods due to steep gradient of stress and test volume required

- **Neutron diffraction chosen as the best method of analysis for this study for its high degree of accuracy and ability to collect data from within the volume of the specimen**

University of Windsor

Neutron Diffraction

- **The crystal lattice structure of the metal is used as an internal atomic strain gauge**
- **The lattice spacing of the stressed specimen material is compared with the lattice spacing of a sample of stress-free material**
- **Bragg's Law – neutrons with a wavelength of λ will enter the metal and diffract off the lattice planes *hkl* at Bragg's angle θ^B**

University of Windsor

Neutron Diffraction

- **Only neutrons that are scattered coherently and elastically by the metal lattice planes *hkl* are collected**
- **The shift in the diffraction angle, θ^B, determines the change in lattice spacing, d_{hkl}**

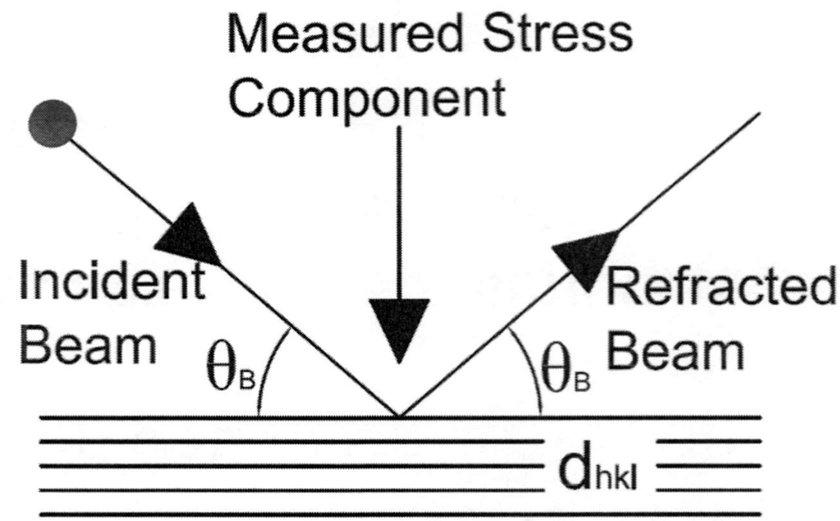

University of Windsor

Objective

- To determine the true state of residual stress distributions at a plate-stiffener connection of a ship hull with various inconsistencies

Residual Stress

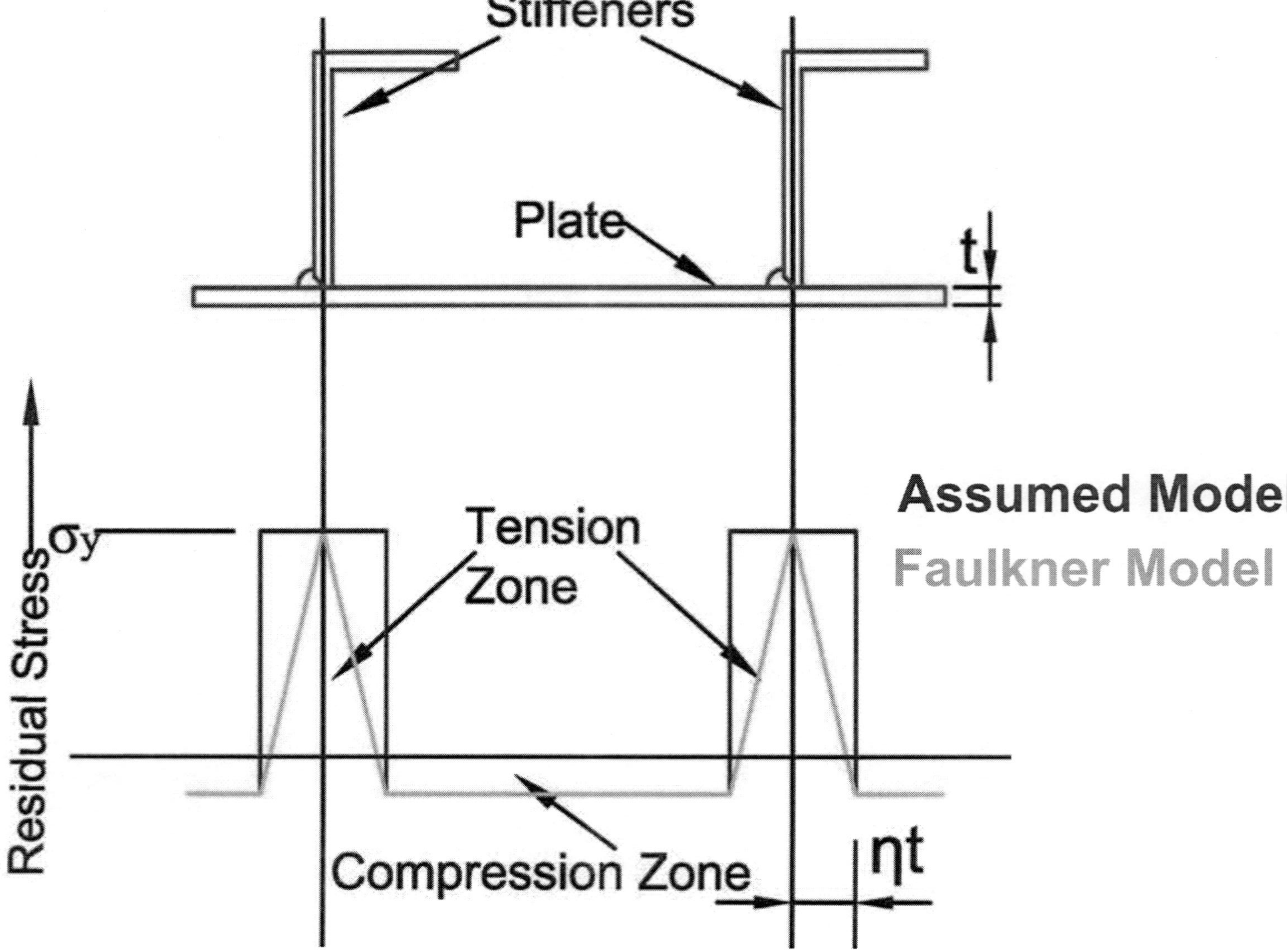

University of Windsor

Test Material

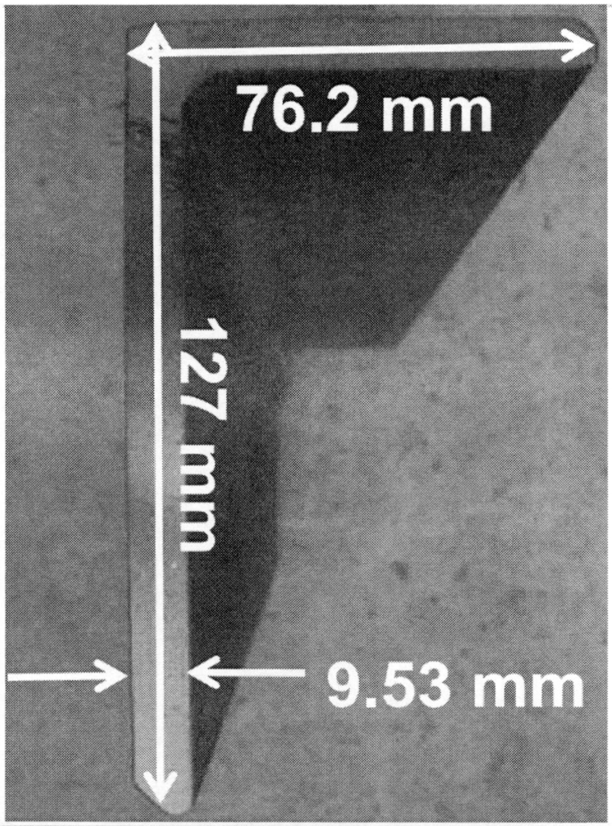

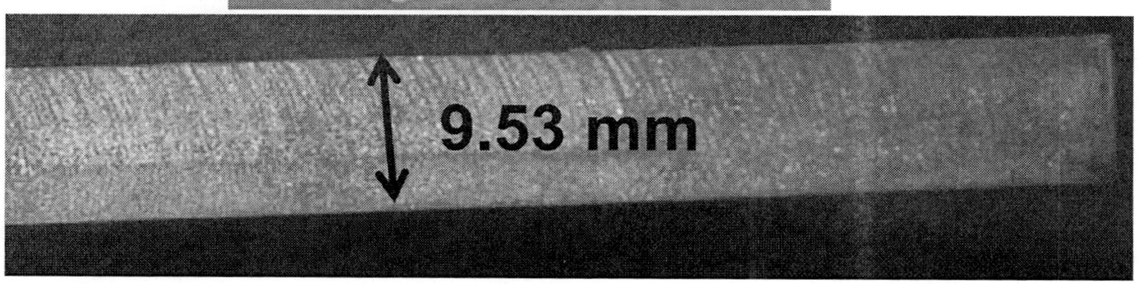

- Angle Section
  - 300W Steel
  - Band Saw cut
  - Beveled
- Parent Plate
  - 350WT Steel
  - Water-jet cut

University of Windsor

Material Properties

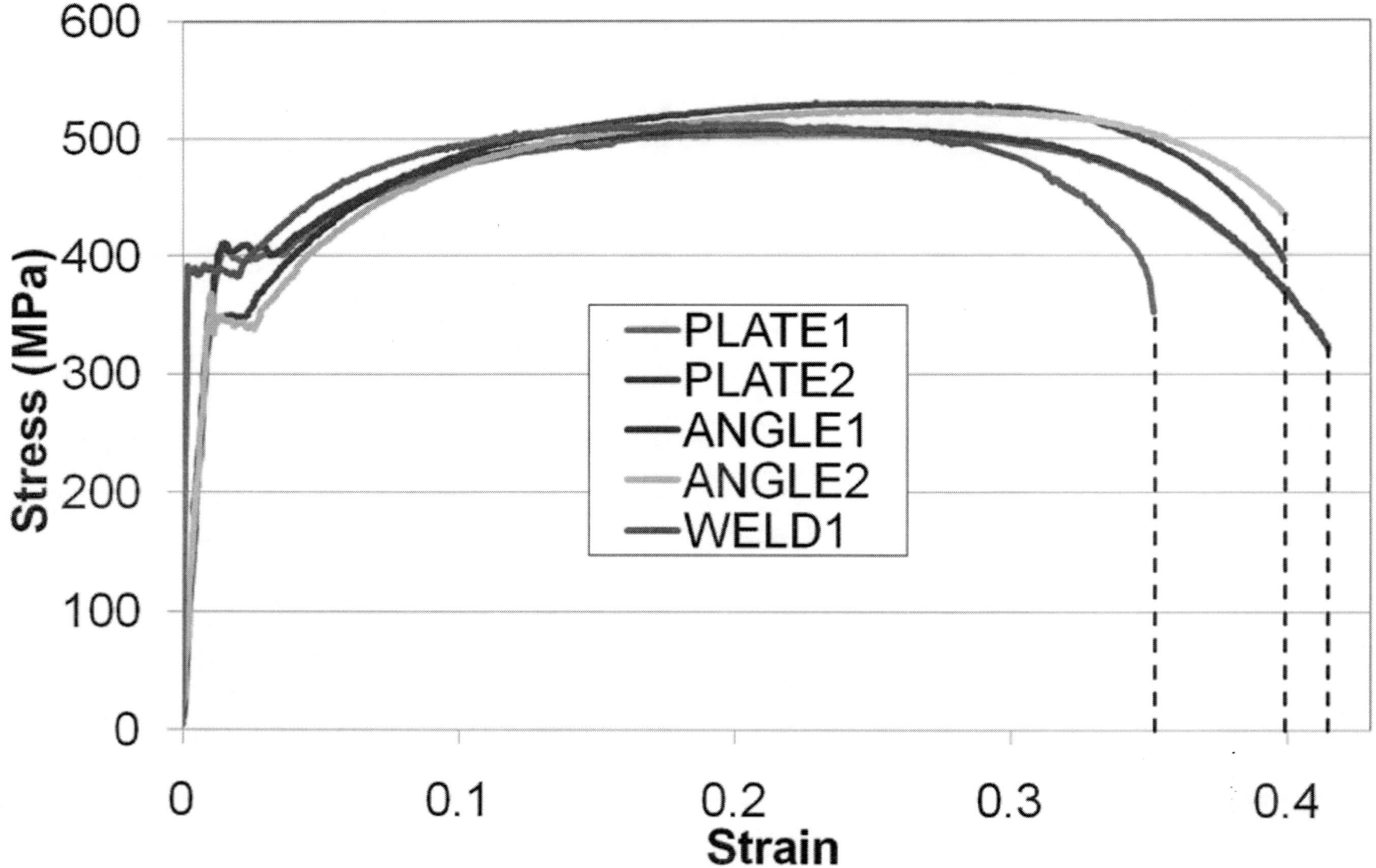

Material Properties

	Plate	Angle	Weld Wire
Grade	350WT	300W	E70C-6M H4
Average ultimate strength (MPa)	527.5	507.3	512.7
Average yield strength (MPa)	405	350	390
Elongation at rupture (%)	41	39	35
Modulus of Elasticity	208	204	209

- Quasi-static tension (pull) test according to ASTM E8M-08

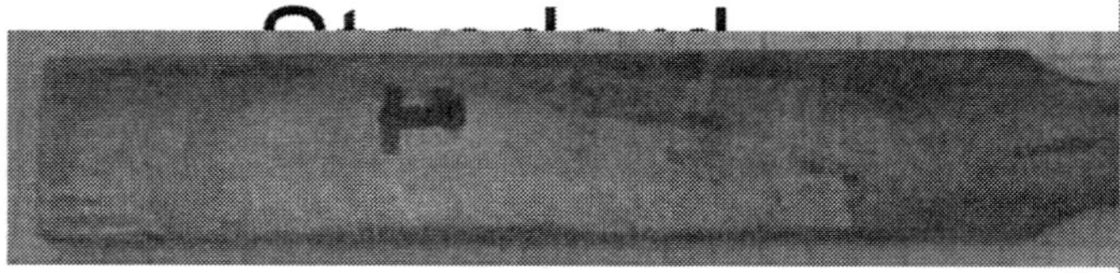

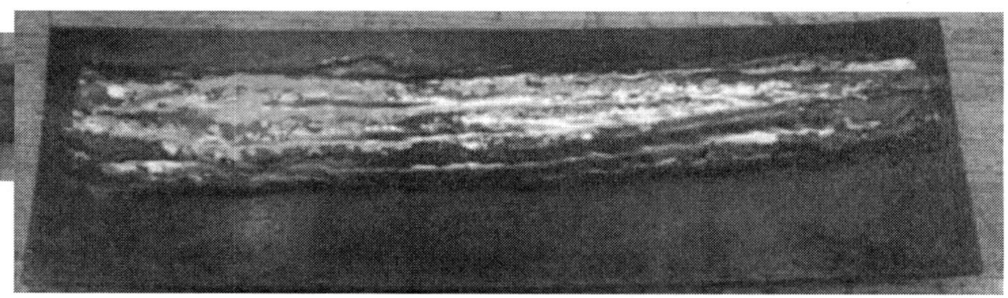

University of Windsor

Welding

- Metal Core Arc welding
- Fully restrained
- Single pass fillet welds

Specimen 1

University of Windsor

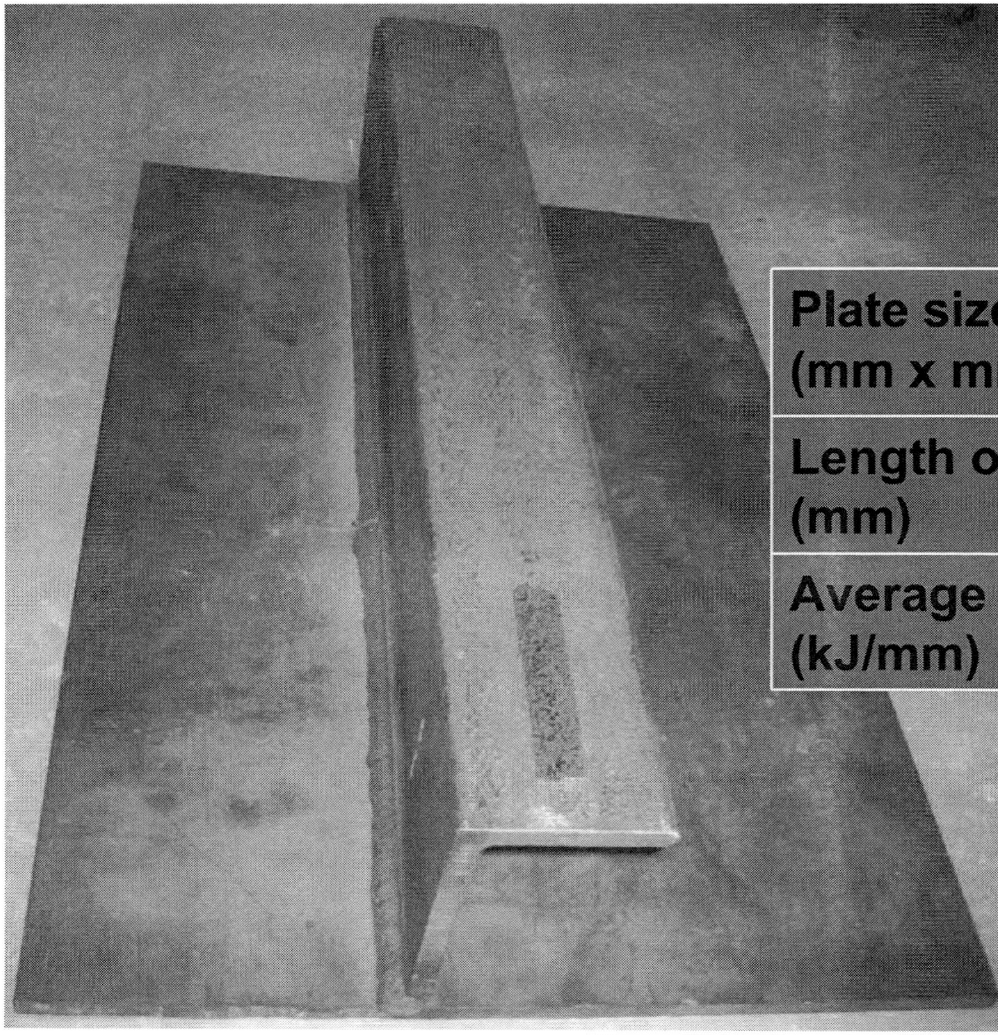

Plate size (mm x mm x mm)	600 x 400 x 9.53
Length of stiffener (mm)	600
Average heat input (kJ/mm)	2.6

Specimen 2

University of Windsor

Plate size (mm x mm x mm)	600 x 400 x 9.53
Length of stiffener (mm)	400
Average heat input (kJ/mm)	2.6

University of Windsor

Reference Samples

- Stress-free pieces of material
- Machined
- Two orientations: L and T
- Determines the stress-free lattice spacing

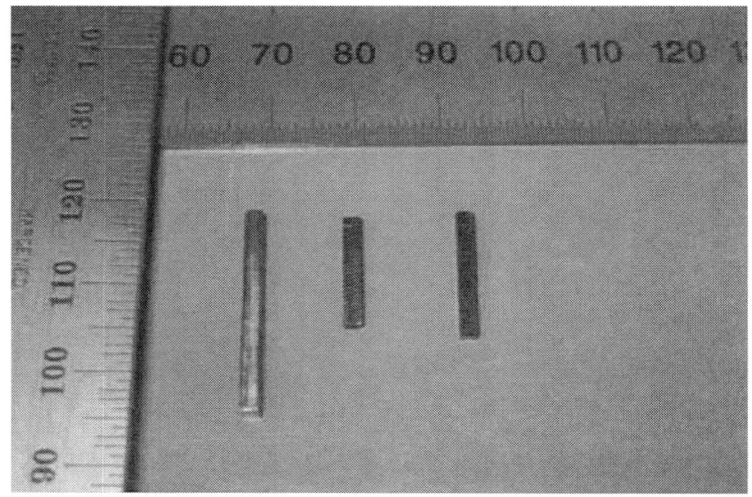

Specimen 1

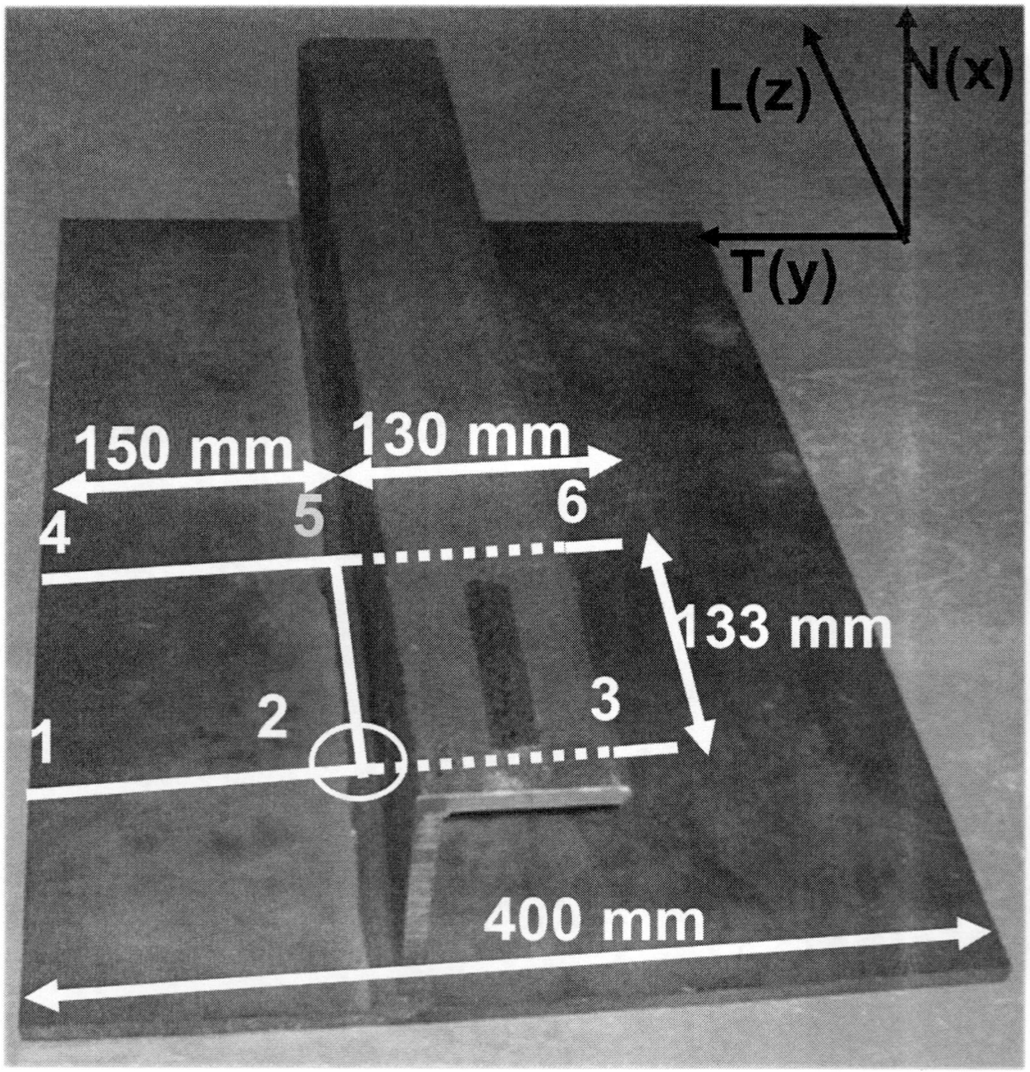

L(z)
N(x)
T(y)
150 mm
130 mm
4
5
6
133 mm
1
2
3
400 mm

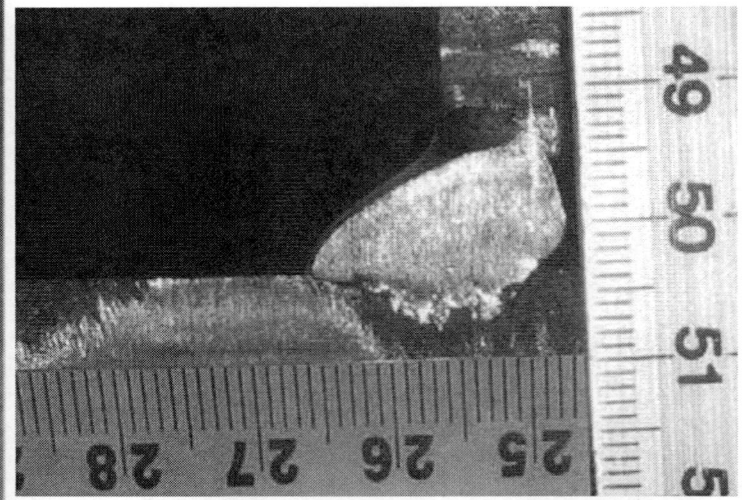

49
50
51
25
26
27
28

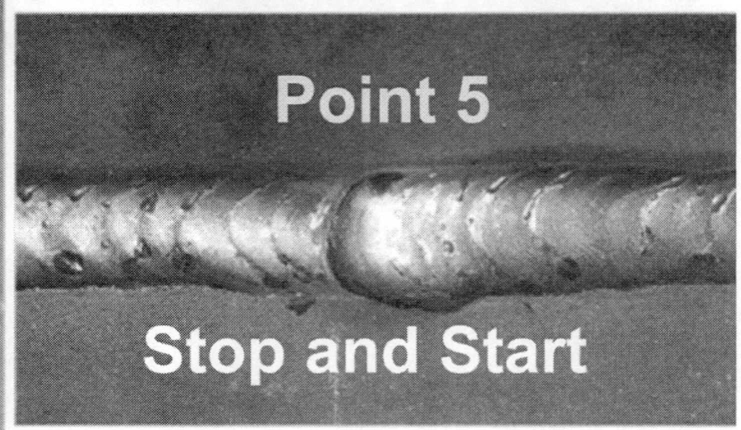

Point 5
Stop and Start

Sp 1 – Normal Comparison

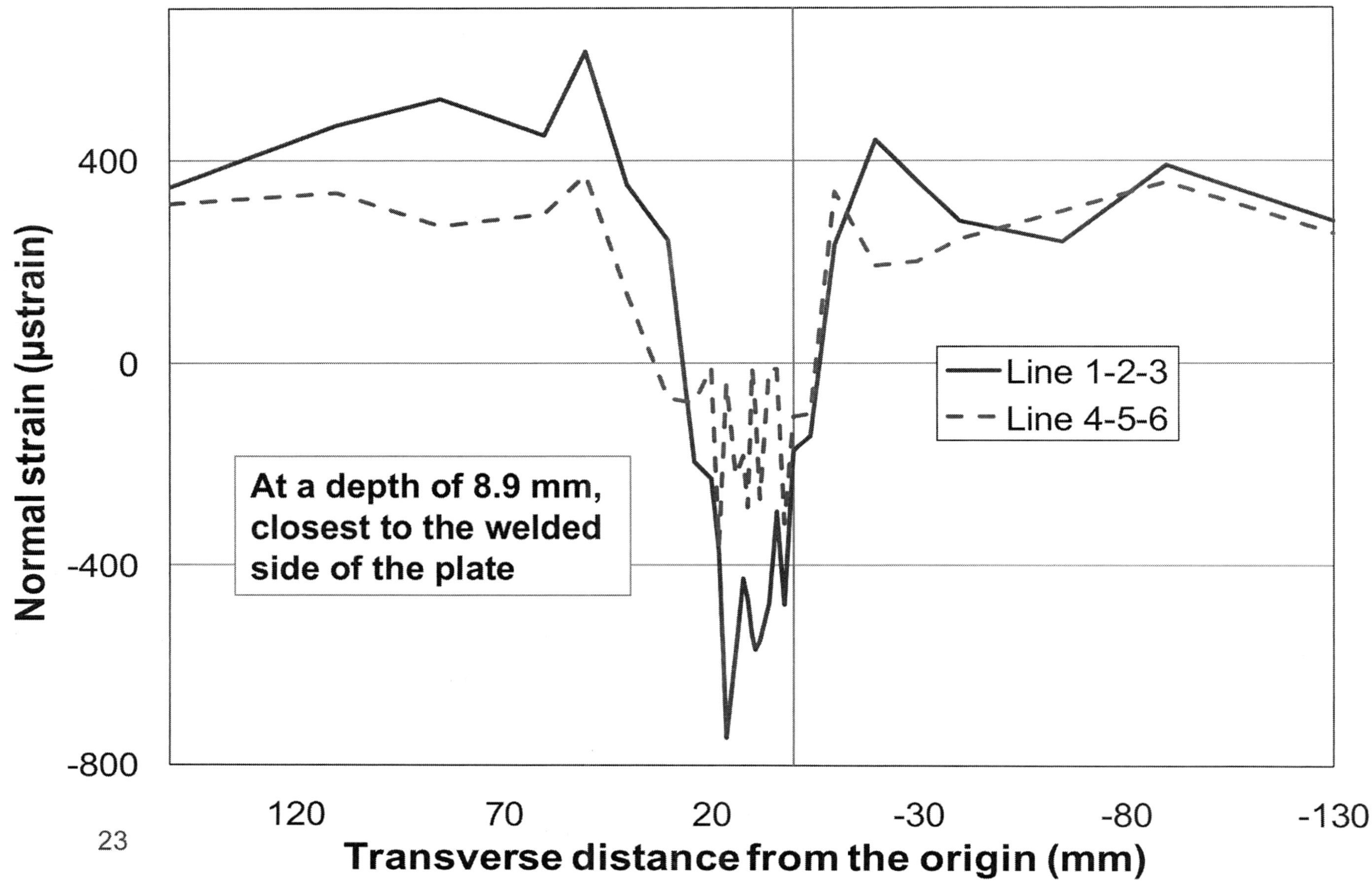

23

Sp 1 – Transverse Comparison

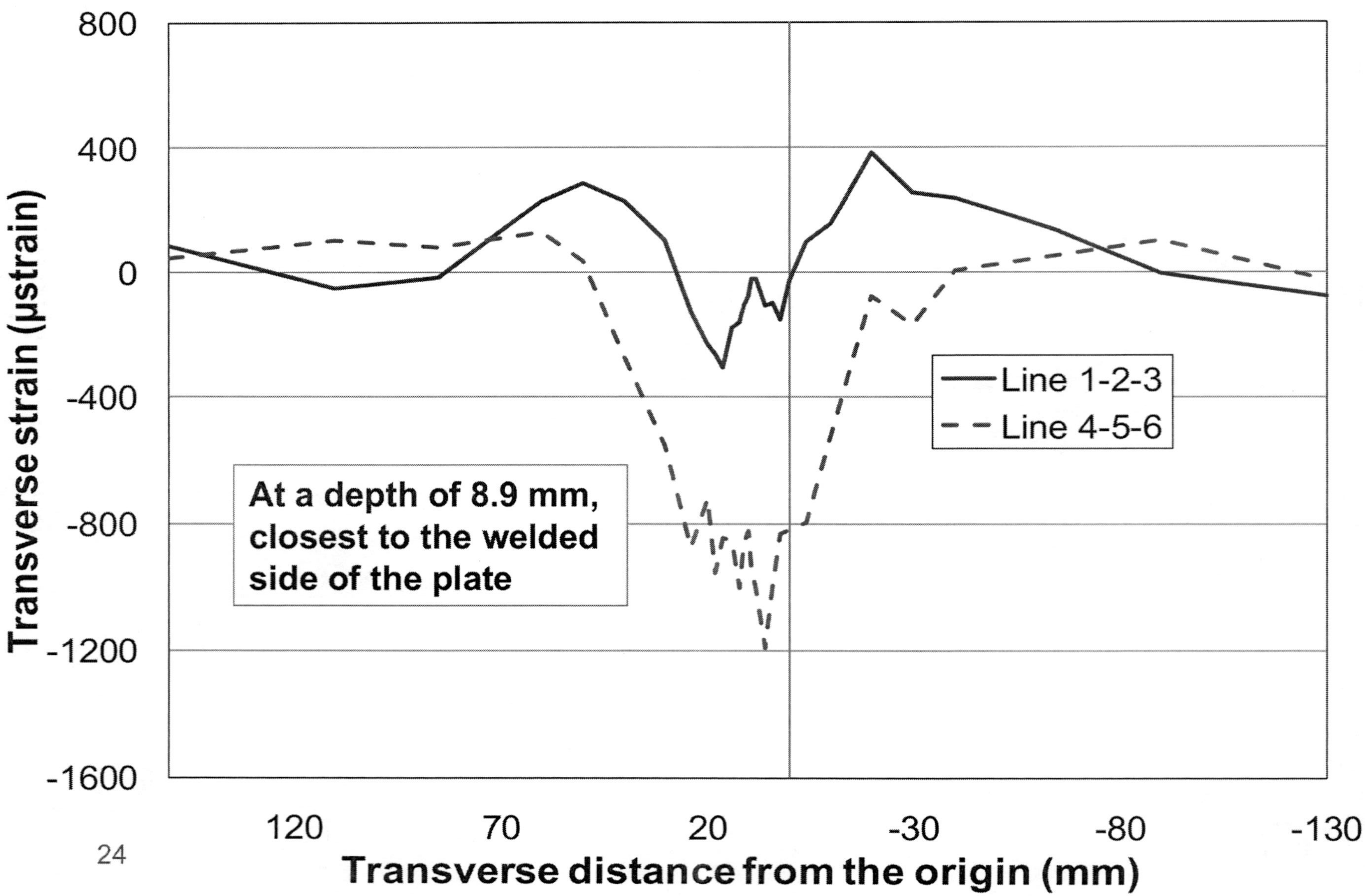

24

Specimen 1 – Line 1-2-3

University of Windsor

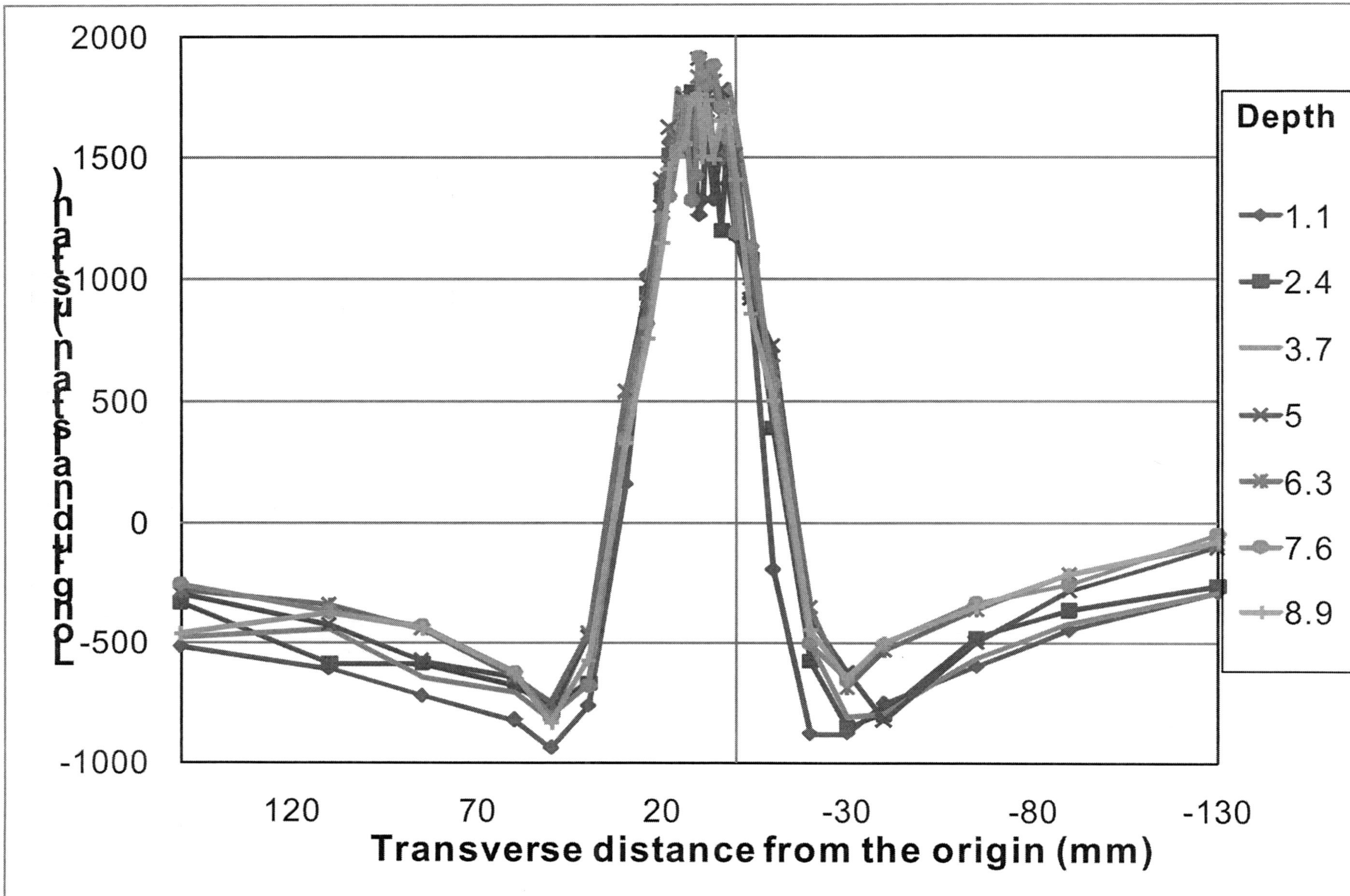

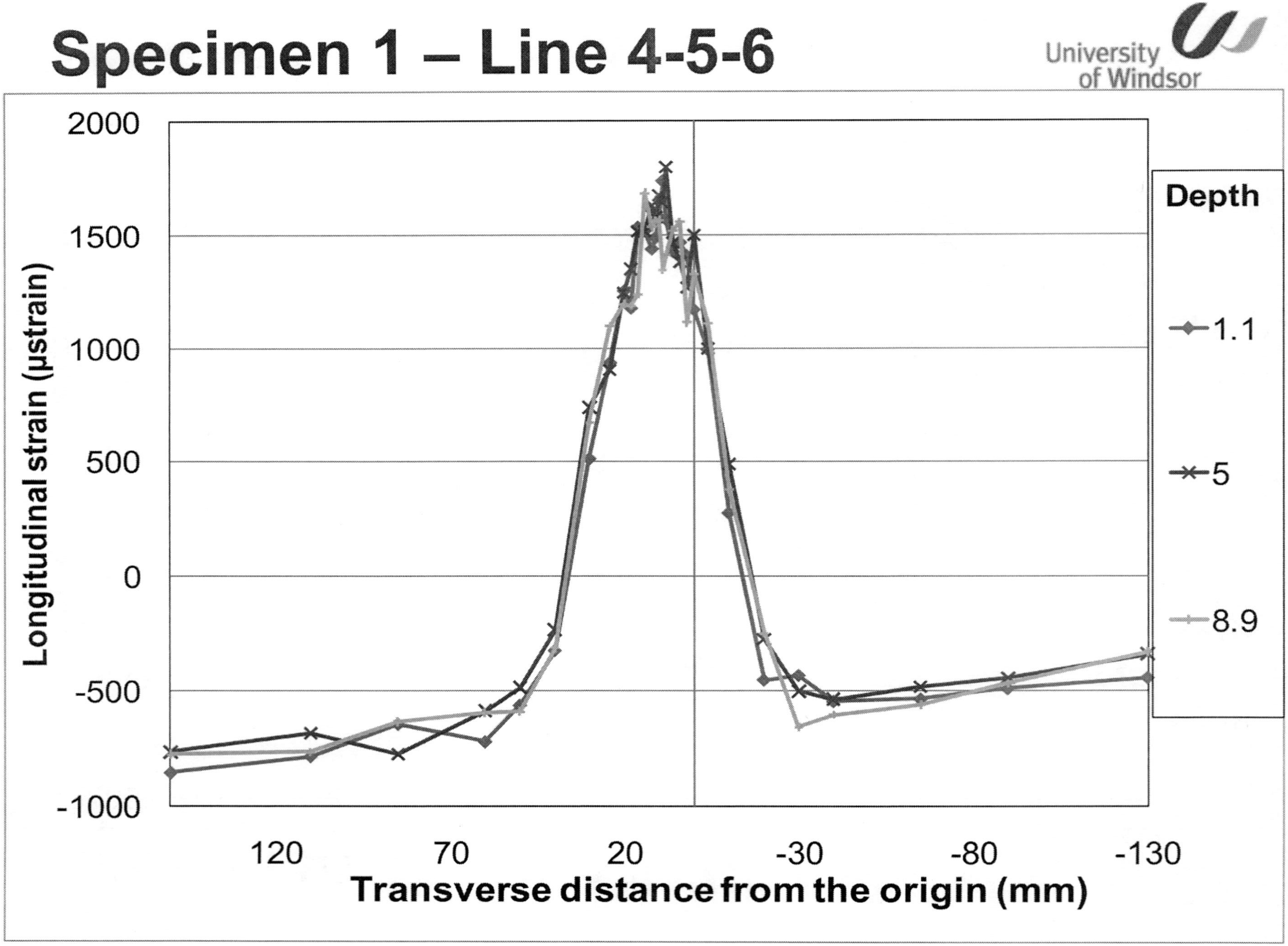

Specimen 1 – Line 4-5-6
University of Windsor
Longitudinal strain (µstrain)
2000
1500
1000
500
0
-500
-1000
120
70
20
-30
-80
-130
Transverse distance from the origin (mm)
Depth
1.1
5
8.9

Sp 1 – Longitudinal Comparison

University of Windsor

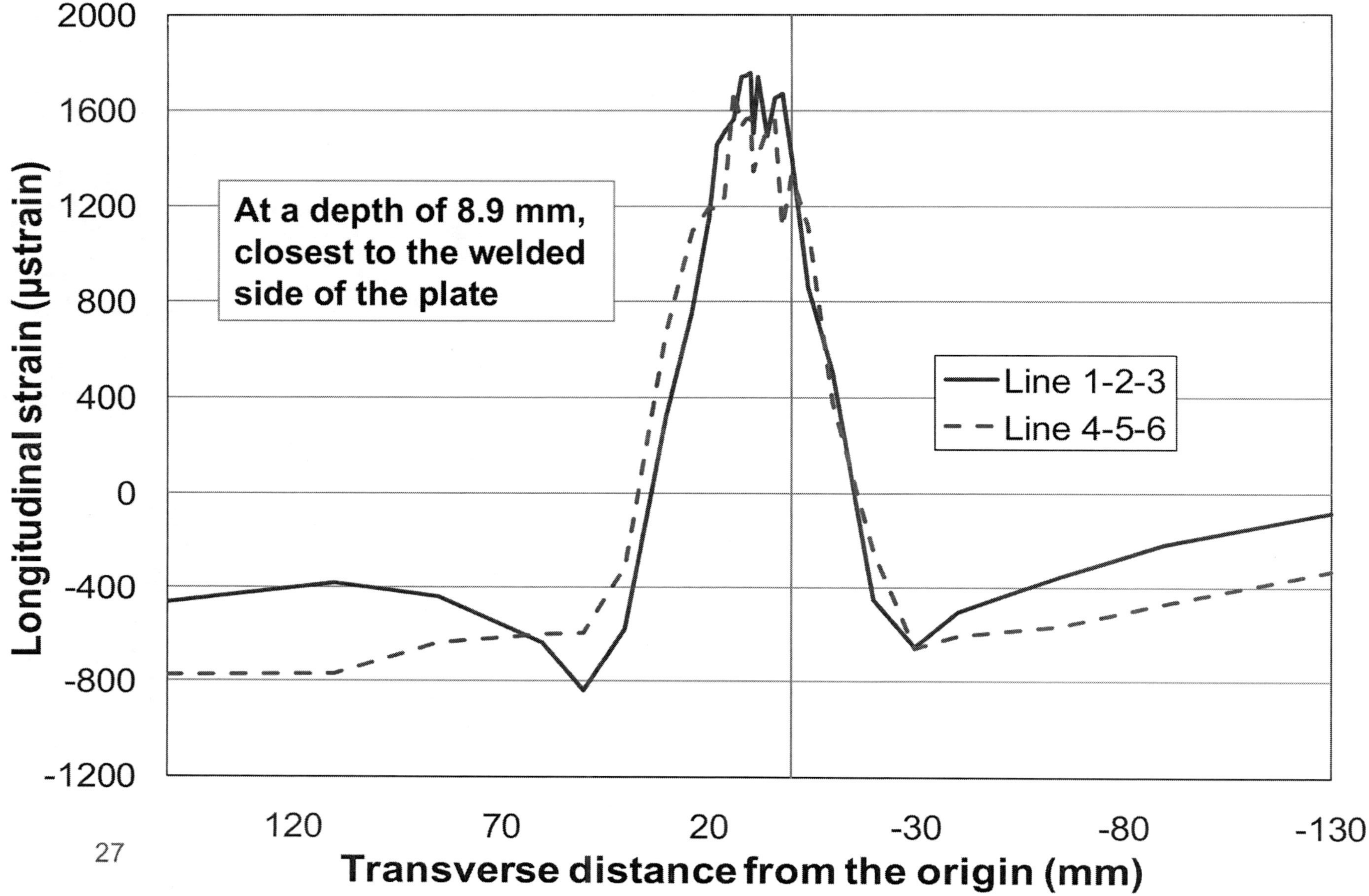

27

Specimen 1

University of Windsor

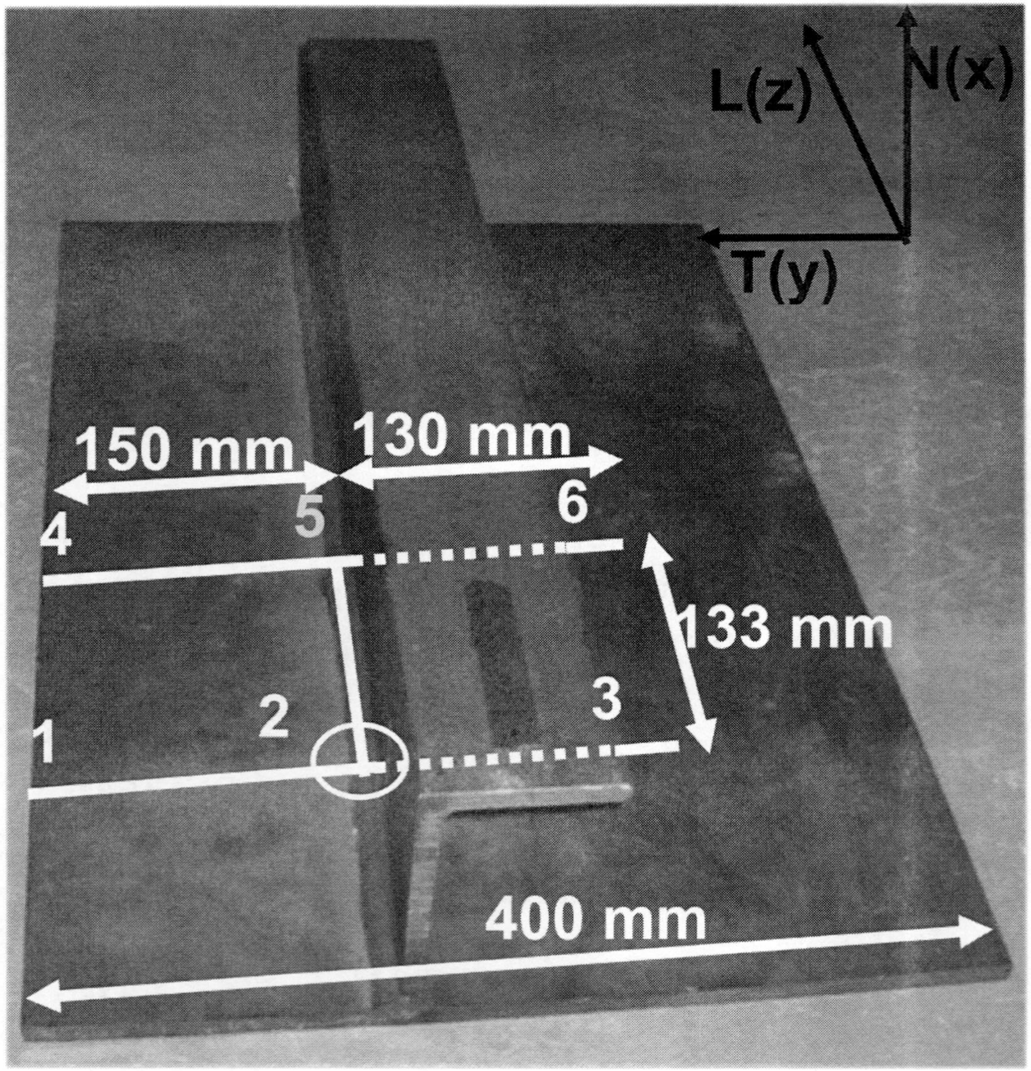

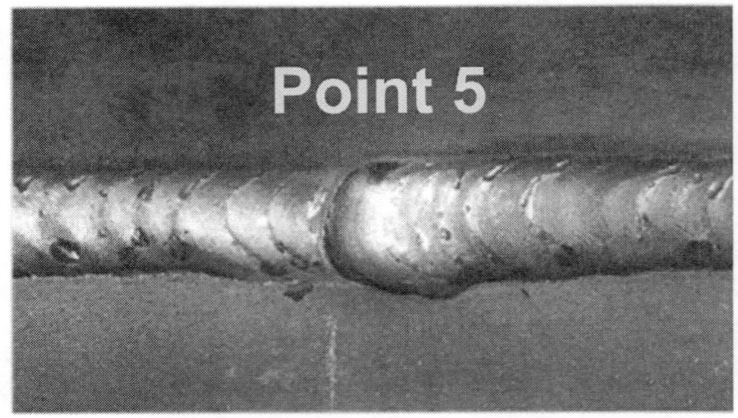

Specimen 1 – Line 2-5

University of Windsor

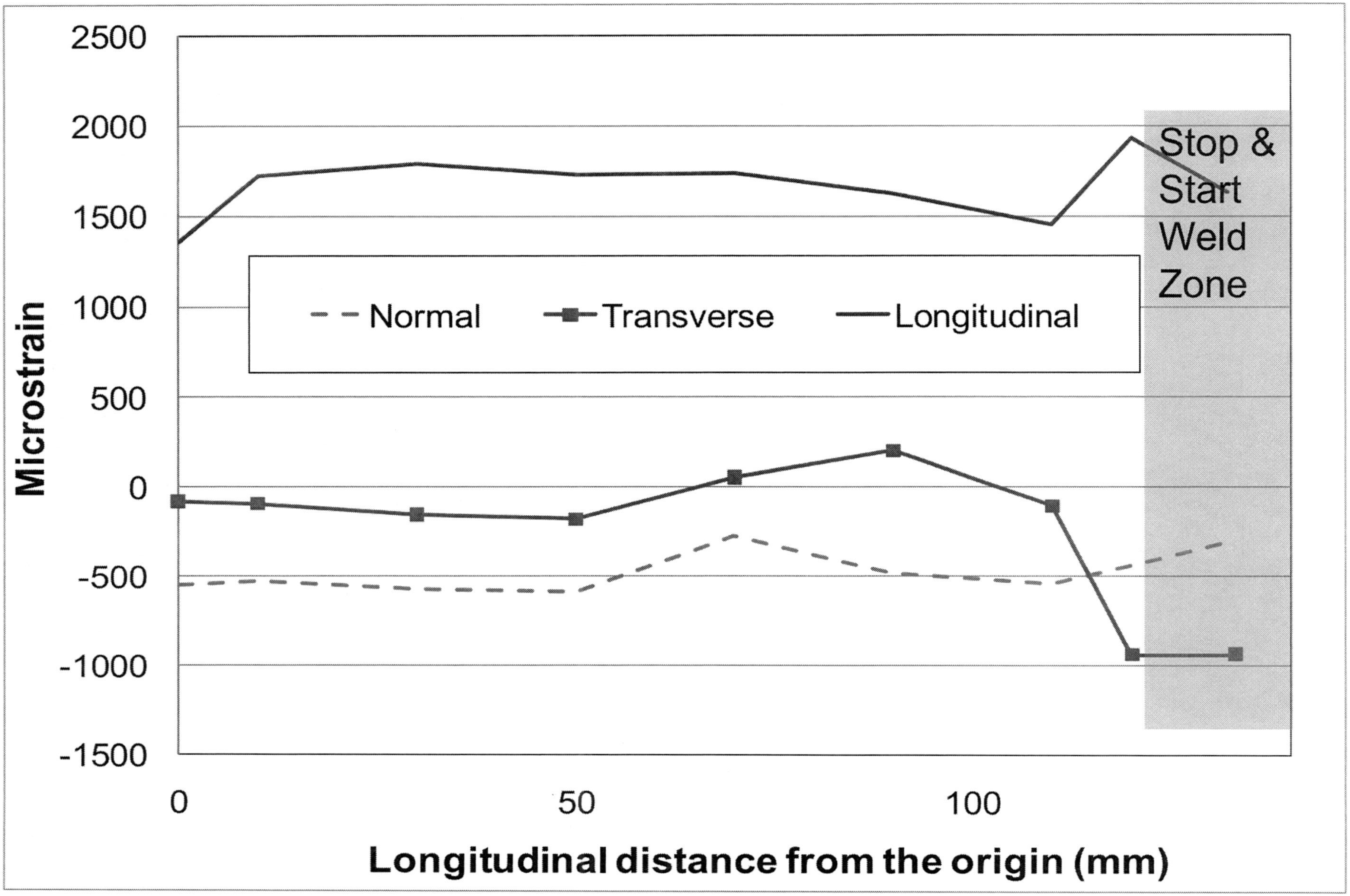

Specimen 2

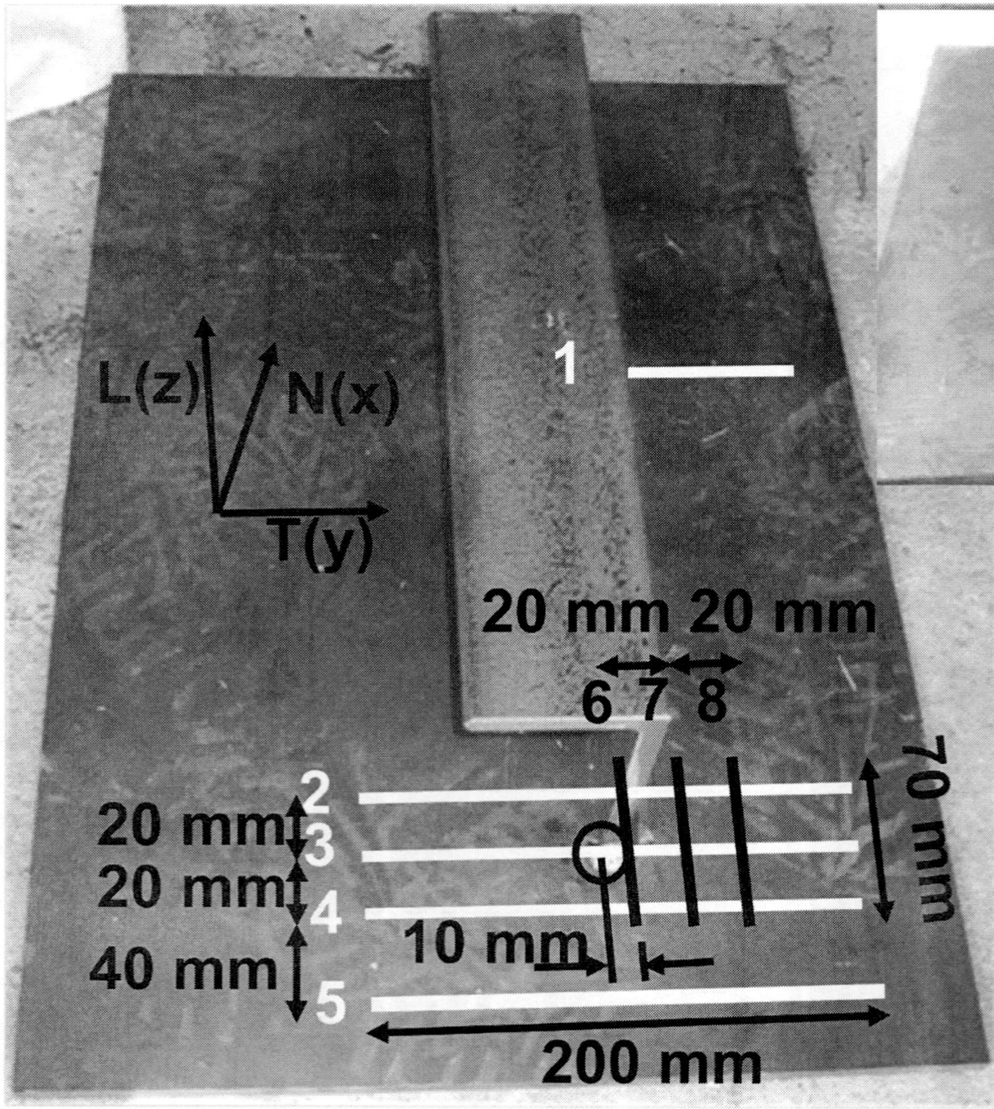

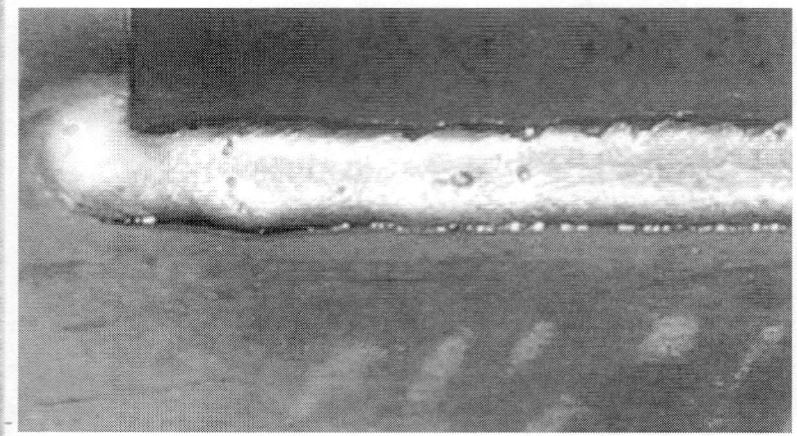

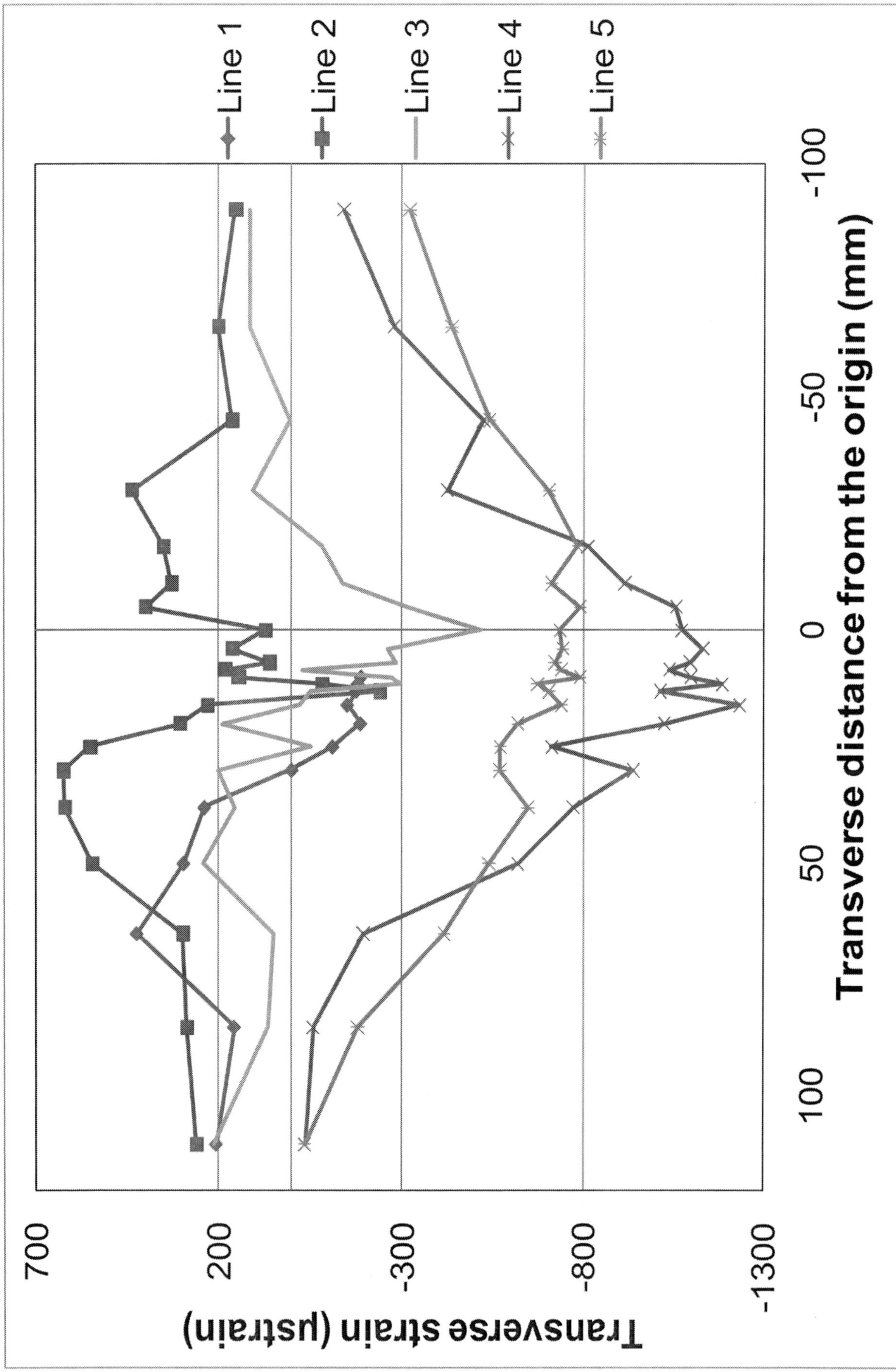
Specimen 2 – Transverse
University of Windsor
Transverse strain (μstrain)
700
200
-300
-800
-1300
100
50
0
-50
-100
Transverse distance from the origin (mm)
Line 1
Line 2
Line 3
Line 4
Line 5

Specimen 2 – Transverse

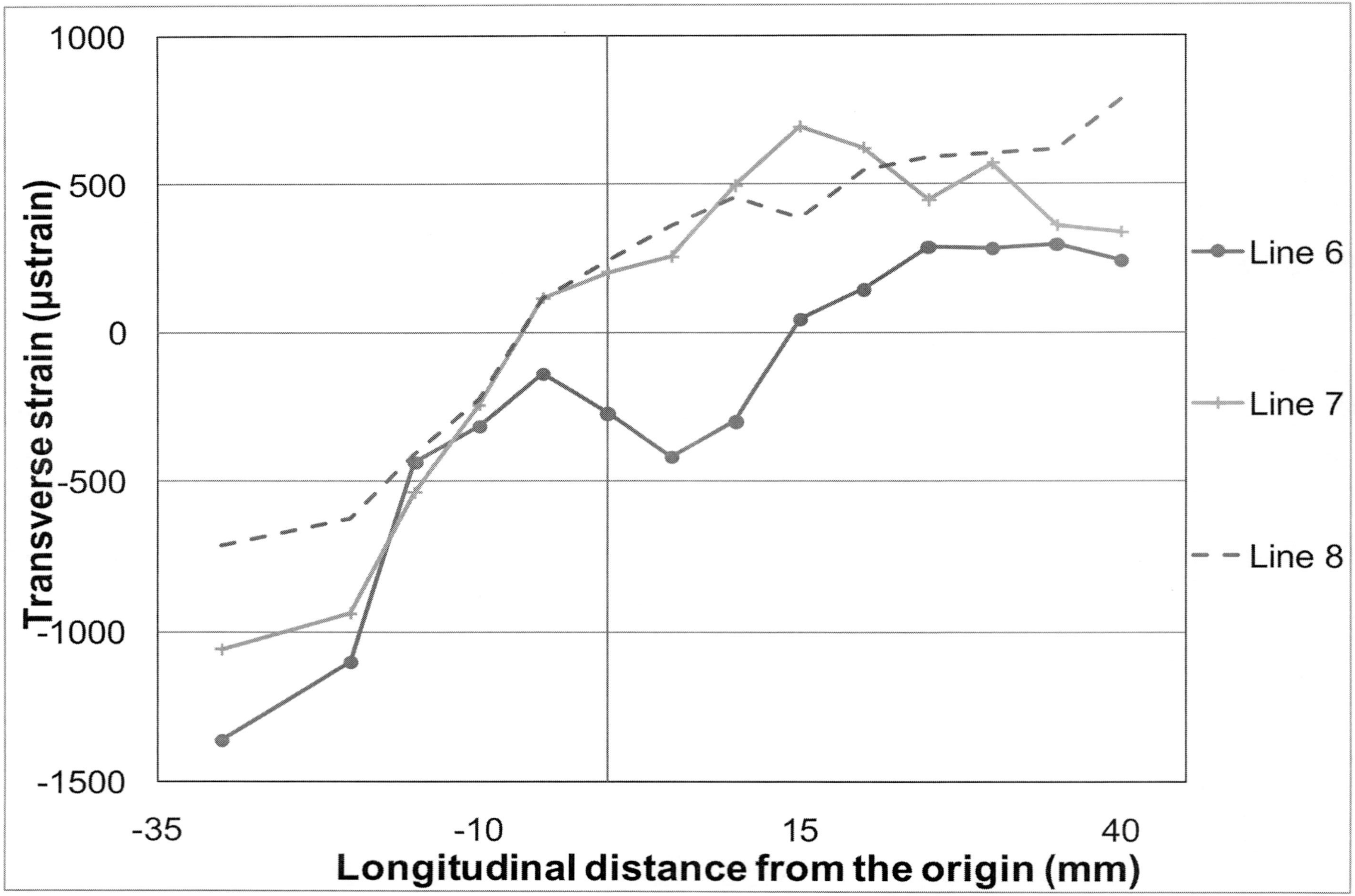

Specimen 2 – Longitudinal

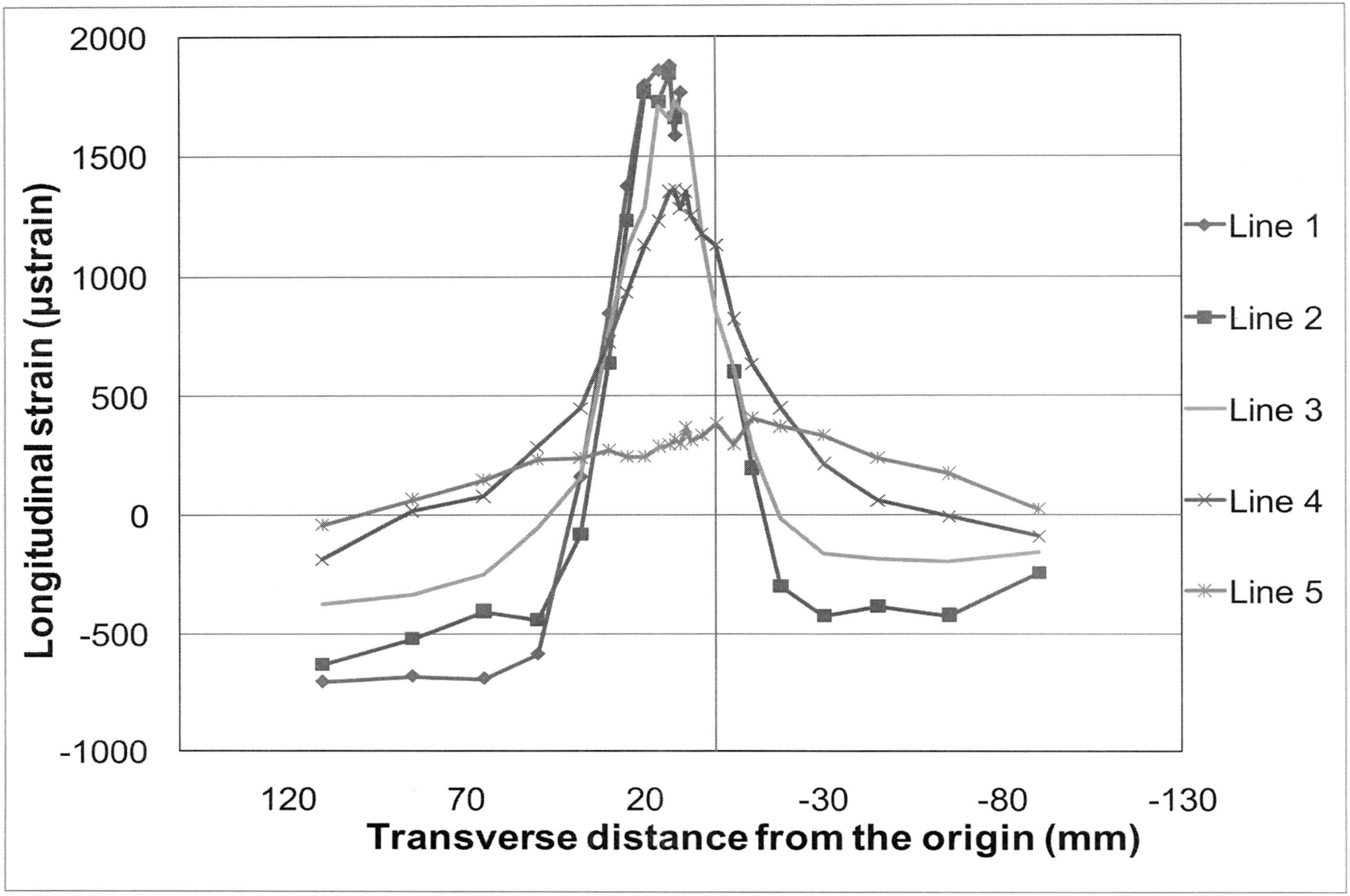

Specimen 2 – Longitudinal

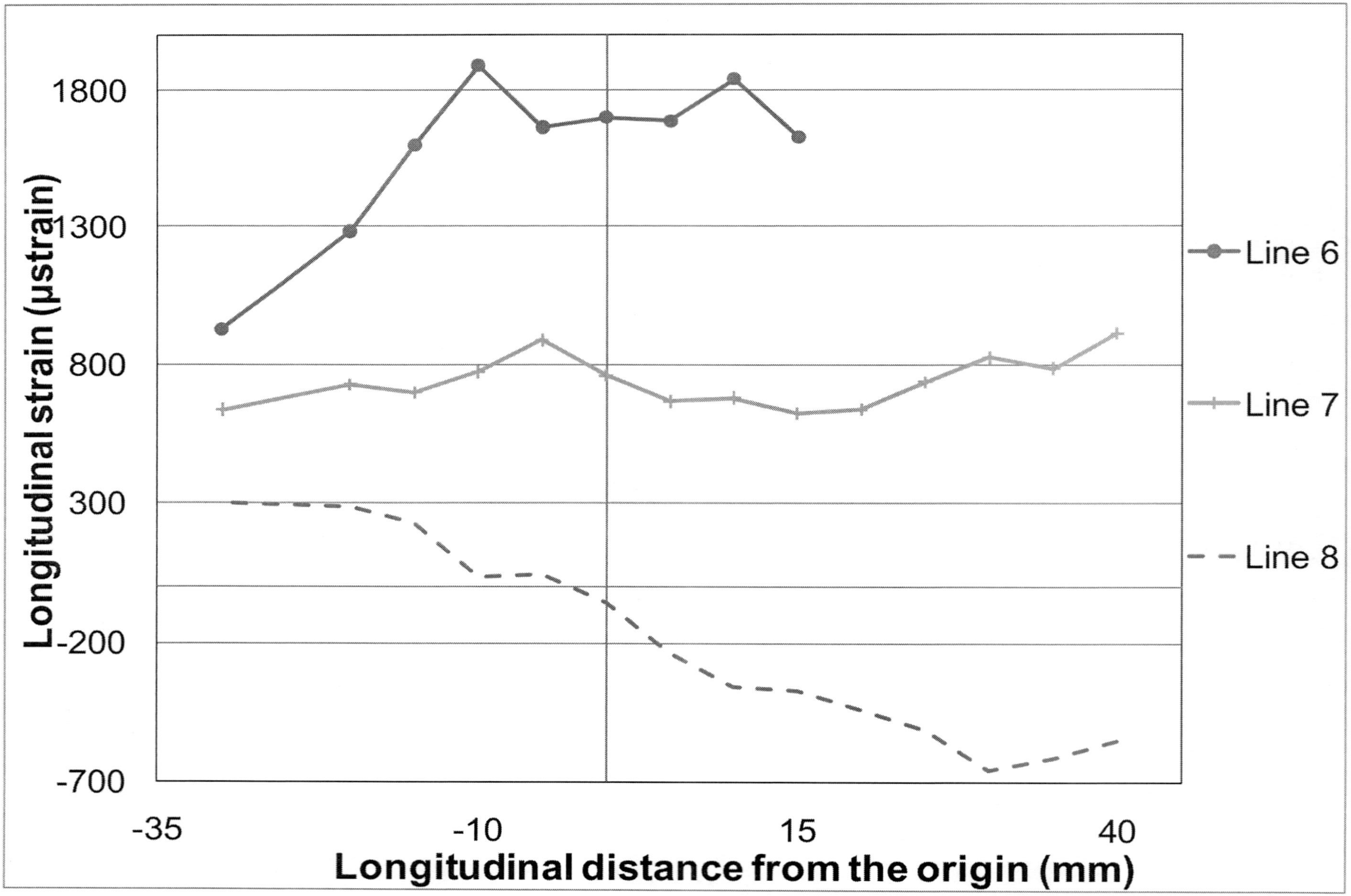

University of Windsor

Conclusions

- **Inconsistencies may reduce the residual stress hence, not a design concern**

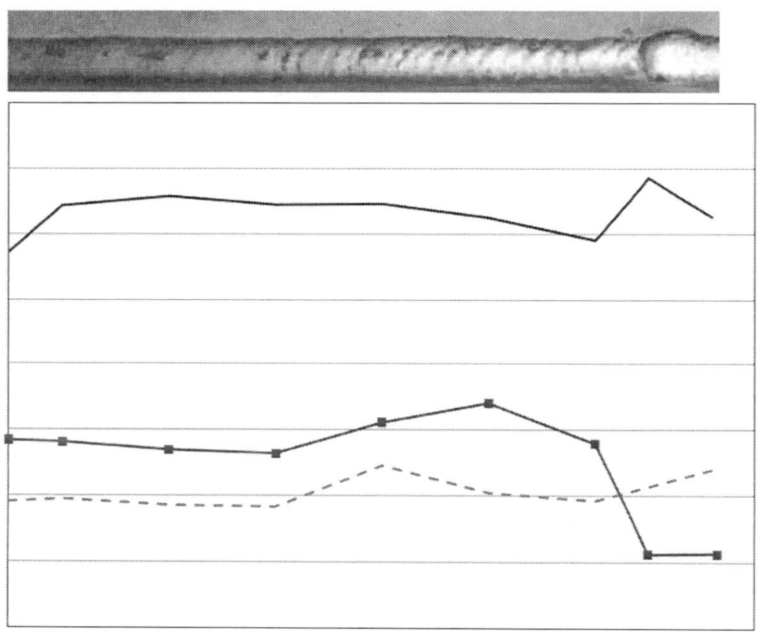

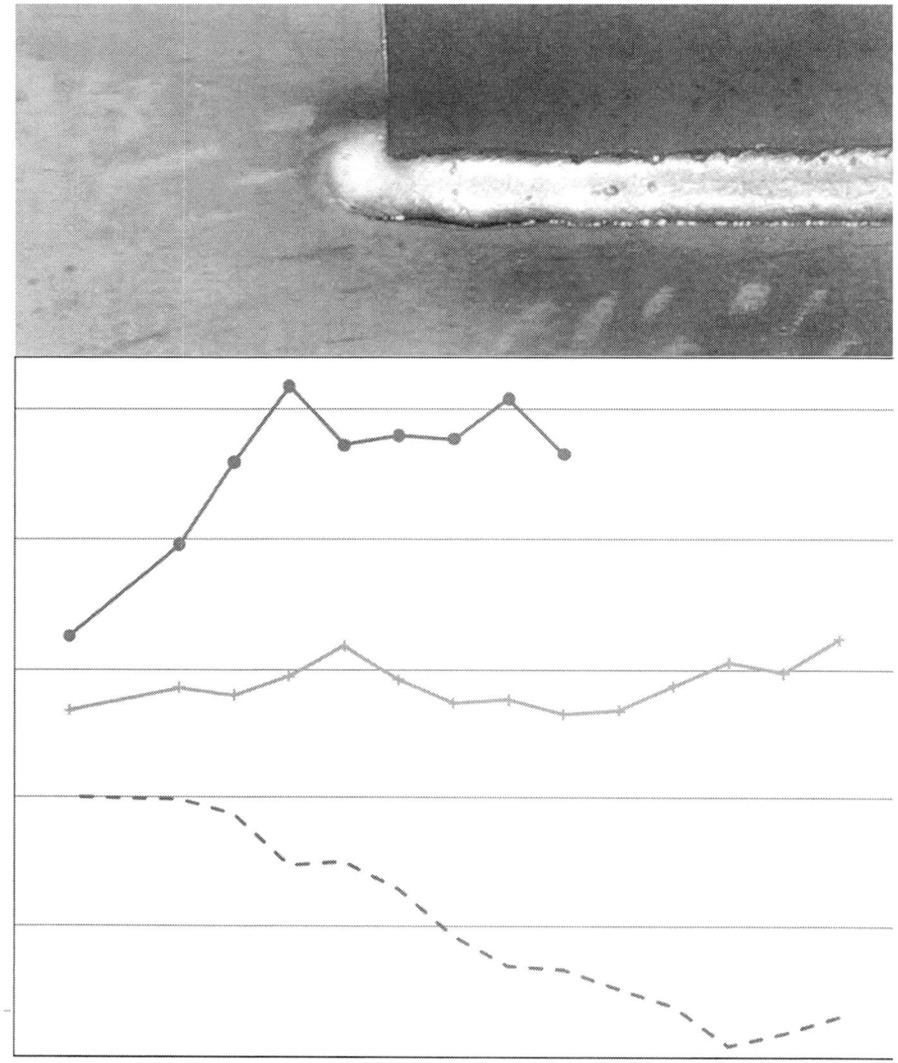

Conclusions

- Typical tensile peak & compression plateau

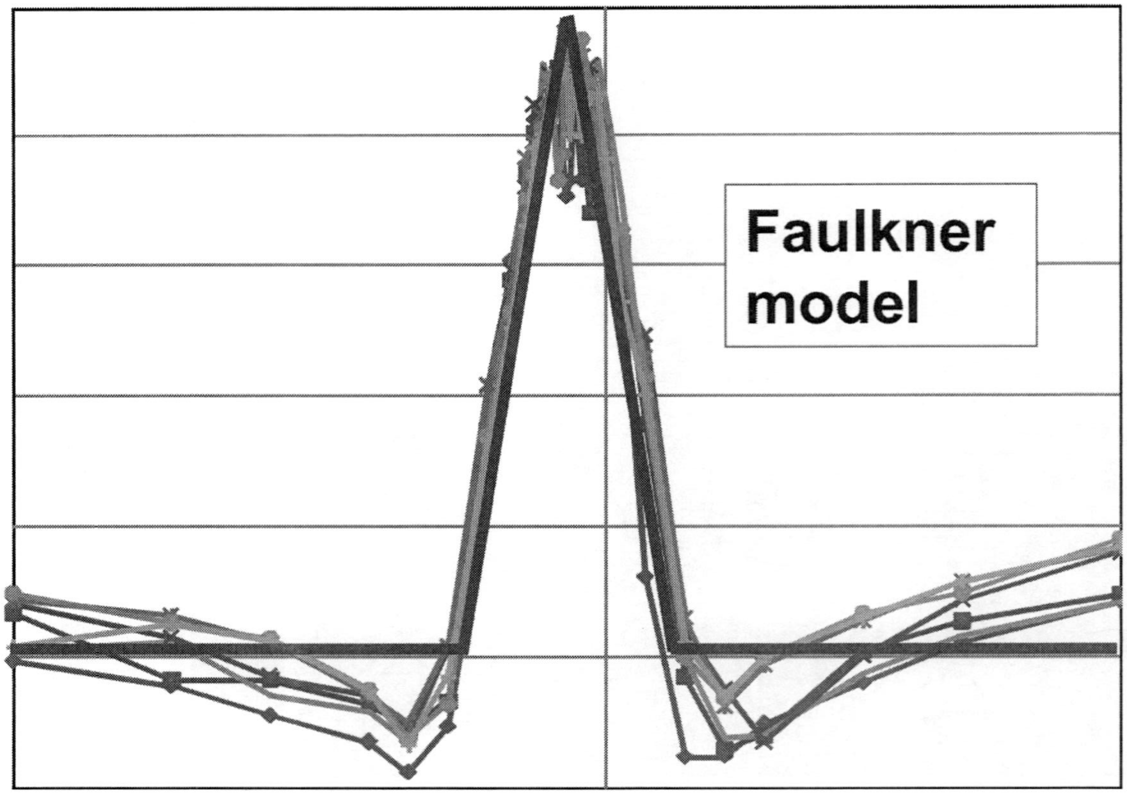

University of Windsor

Other Works

- Second phase of testing:
  - Parent plate
  - Second stiffener
  - Stiffener spacing
  - Welding heat input

Future Work

University of Windsor

- Third phase of testing:
  - Plate thickness
  - Weld metal mismatch
  - More discontinuities in welds
- Secondary Testing Methods
  - Hole-drilling: semi-destructive
  - Finite Element Modelling: parametric study

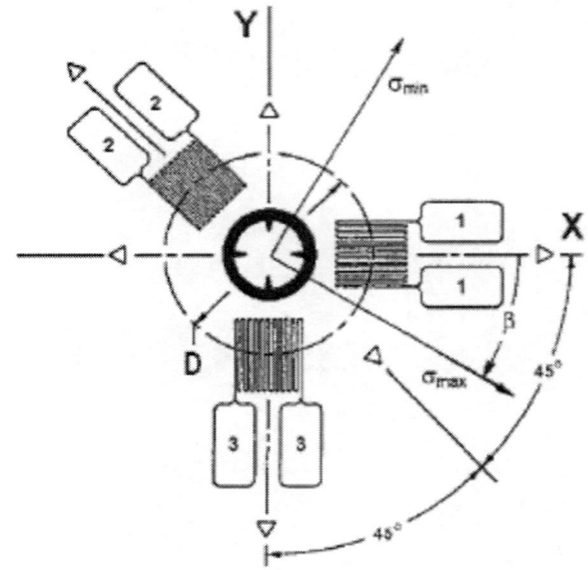

University of Windsor

Acknowledgements

- DRDC, Atlantic
  - Dr. David Stredulinsky
  - Dr. James Huang
- CNBC
  - Dr. Ron Rogge
  - Dr. Michael Gharghouri
  - Mr. Ron Donaberger
  - Mr. John Fox

LASER CLADDING- A COATING PROCESS FOR MARINE APPLICATIONS

Sanjeet Kanungo, Tolani Maritime Institute,India

Summary

This paper highlights on relative advantages seen on laser as a tool for surface engineering applications vis-à-vis other conventional methods available such as APS and HVOF techniques. The research work was carried out on mild steel plates where three different cermets were used as surface coating materials coated both by Laser cladding and Atmospheric Plasma Spray Process. The results have been summarized below to extent possible.

Introduction

During the last few decades, improved understanding of tribological behavior of different material combinations led to an intensified development of wear resistant thermal spray coatings. In the field of hard chromium replacement by thermal spraying, significant amount of work has been done and published world wide[1-3], however, the authors mainly focussed on only one tribological aspect like friction, abrasion, erosion, cavitation or corrosion.In real applications, very often more than one of these factors influence the successful use of these coatings.

The economical and ecological situation requires a continuous reflection on the existing production techniques especially in the field of surface technology. Higher demands on performance and lifetime of parts and environmental aspects influence the successful positioning in the coating market.

Alternatives to hard chromium plate applications have been widely discussed in a variety of workshops and conferences

Besides the environmental aspects, hard chromium coatings with a thickness above approximately 0.5 mm already show micro cracking. The crack density of technical coatings is between 15 to 35 and 40 to 80 cracks/mm and the crack width 0.1 to 10 μm due to phase changes (shrinkage) during production. The hardness of hard chromium coatings can reach values between 600 and 1200

$HV_{0.1}$ and the growth rate of coating is about 50 μm/h [1]. Their high hardness combined with the stable passivation layer of chromium oxide on top and the good tribological, anti-adhesive and fretting behavior make hard chromium coatings the most widely used functional surfaces, as shown in Figure1.

However, on-site repair of a damaged coating is usually not possible. The bad wettability of hard chromium by water based media and oil leads also some times to problems related to lubricant adhesion during sliding friction conditions.

So there is a need to replace chromium coatings to protect the environment. Heat treatments are suitable options environmentally but cannot be used for selective portions in the components, thus limiting their choice.

Hard Chrome platings meet the functional requirements of a wear resistant coating, but it is a highly slow process with inherent problems of coating cracks, wettability, and more importantly environmental problems. Thermal Spray Processes vis-a vis Plasma Spray Process, High Velocity Oxy Fuel, Physical Vapour Deposition Process, have already been suitably implemented and have proved their mettle. However the aforementioned processes have still certain problems with regards to porosity, bonding, oxide inclusions, coating thickness, coating areas and post machining processes.

Laser Cladding process, which is still in the laboratory phase and has found very limited use in the industry as of today due to the cost, is a potential alternative to the Galvanic and Thermal Spray Process mentioned above. The technique consists in injecting a powdery substance with a flow of inert gas over a surface, which is being radiated with a laser beam. The energy from the laser beam is responsible for melting both the powder and the substrate. This must be done carefully so that melting of the substrate is minimized.

This is in contrary to thermal spray processes mentioned above where metal powder is injected into a hot jet, wherein the powder melts inflight and splats against the substrate thereby forming a coating.

2. Experimental Procedures

2.1 Materials.

Based on the availability and literature review three cermet powders as mentioned below

Chemical Constituents of Powder chosen for study:-

1. PM-20 :50%Cr_3C_2/32%Mo/18%NiCr(custom blend MBD material)
2. Tungsten (Base), Carbon (5%), Cobalt (13%)

3.Nickel(Base),Carbon(0.80%),Iron(4.60%),Chromium(16%),Boron(3.40%),Oxygen(traces)

The powders were characterized to study their morphology and size. Each powder was coated on a low carbon steel substrate by Plasma Spray Process and Laser Cladding .In case of Laser Cladding three different scan speeds were used to evaluate the best scan speed.

The coatings obtained after deposition were characterized by X-Ray diffraction, optical microscopy and SEM/EDAX analysis to provide invaluable information concerning the morphology, phase composition, and microchemistry of the coated system.

2.2 Atmospheric Plasma Spray Process

A carrier gas is fed into a torch in which a high current arc is struck. Gases passing through the arc are dissociated and ionised resulting in a plasma jet from the nozzle. Powders are injected into the plasma jet where they are heated and ejected at the target –the injection position can be adjusted depending on their melting point. Very high temperatures are realised in the area (10000K) and heat transfer is very efficient, especially from diatomic gases. Ceramic, metal and polymer coatings are possible .The process can be carried out in vacuum to reduce atmospheric braking and oxidation ofmetallic coatings. Plasma Spray process is one of the sought after processes of thermal barrier coating. The plasma process is described here in brief. Argon is the most favored primary gas as it is the easiest of the gases to form plasma and tends to be less aggressive towards the electrode and nozzle hardware.

Hydrogen is mainly used as a secondary gas, it dramatically effects heat transfer properties and acts as antioxidant. Small amounts of hydrogen added to other plasma sources dramatically alters the plasma characteristics and energy levels and is thus used as one control for setting plasma voltage and energy.

2.3 Laser Cladding

The aim *of* laser cladding is to deposit a coating, and ensure that slight intermixture occurs only in the transitional zone in order to produce a proper metallurgical bond between base and clad materials. The coating is usually achieved by either feeding the clad material into the interaction zone, melting a pre-placed Pin on disk wear tests was performed using the Universal Wear Test Machine based on ASTM G99-90.Wear resistance of the samples was determined by Pin-on-disk method. The wear down rate was ascertained by the weight loss per unit centimeter.

Surface roughness wa s performed by a Tylor-Hobson profilometer using a cut off of 0.8mm.

powder layer or by injecting powder into the molten pool created by the laser on the substrate surface. To obtain low dilution (typically less than 5%), however, requires accurate control of the laser parameters within a narrow processing range.

In laser cladding with an injected powder, the beam melts a shallow pool on the substrate into which the cladding powder is injected forming a bead whose width is narrower than the laser spot diameter on the surface. To achieve area coverage the beads are overlapped usually by 50% *of* the bead width.

2.4 Microstructural Chararcterization

Detailed characterization of coating properties including the chemical composition, microstructure and interfaces of the different coating powders was done. Moreover, the predominant features and dissimilarities between the different coatings, which can be crucial for their performance, will be pointed out through the text. The coatings obtained after deposition were characterized by X-Ray diffraction, optical microscopy and SEM/EDAX analysis to provide invaluable information concerning the morphology, phase composition, and microchemistry of the coated system. Individual powders and their coating systems are dealt separately. For each system, the as-received powders were characterized using XRD and SEM/EDAX techniques to draw a correlation between the powder and its coatings with regard to their morphology, size and composition.

2.5. Mechanical and tribological testing

The microhardness results obtained for various coatings on the surface as well as at the cross-section for all the three powder using APS & Laser Cladding. Hardchrome plating sample results given were obtained by electroplating mild steel samples with hard chrome. The hardness results shown are average of four readings taken across the surface and cross-section. A load of 300g was used for 15 seconds to evaluate the microhardness. However, wherever the coating thickness was less than 200μm, a load of 100g was used to avoid the effect of the substrate on hardness.

3. Results and discussion

3.1 Morphology and microstructure of Powders

Characterization of As Received Powder

X-ray diffraction results of PM-20, as received powder, reveals, that Cr_3C_2 and Ni are the main components of the cermet powder. No peak for oxide phase was detected;

however, presence of trace Mo is also confirmed. Similarly x-ray diffraction results of the WC/Co powder used for the coatings, reveals that WC as the major hard phase on cobalt matrix and NiCr showed the presence of of Cr_3C_2 and Cr_2B with NiCr as the matrix. No peak of oxide phase was detected.

SEM micrograph of all the three powders showed irregular particle size between 40-50 μm of which some are completely spherical. This is shown in Figure 2.

3.2Characterization of coatings on steel:

Surface Morphology:

The micrograph in Fig 3 shows the morphology of the alloy coatings formed using APS. Layer by layer deposition of the matrix can be seen alongwith with the partially fused or fused hard particles. Due to non-homogeneous distribution of the hard phases, the system is heterogeneous. Cavities or pores are clearly seen in the photograph. Noncircular cavities seen can be due to the rebound or pull out of the unmelted hard phases as is the case normally seen in APS systems.

The micrographs in show no large scale melting and the individual particles are clearly visible. However, local melting appears to have taken place resulting in aggregation. Distribution of hard phase particles is more as compared to APS.

The principal constituents of the powder are also present in the APS coating. Peaks of $Cr_{23}C_6$ are also traced, indicating decarburisation of Cr_3C_2. The loss of carbon will contribute to formation of CO_2 which will escape as gas or can form carbide with Molybdenum. This was confirmed by XRD results of all the coated samples.

Cr_3C_2 is the major phase detected in coated samples alongwith trace amounts of Mo carbides with different scan speeds. However $Cr_{23}C_6$ phase was also detected in all the coatings formed. It appears that there is no significant difference in X-ray diffraction results of three coatings made at three different scan speeds. Even the relative intensities of Cr_3C_2 with respect to the original powder appears to be same, indicating that the Cr_3C_2 particles are intact. This indicates less formation of $Cr_{23}C_6$.

Optical micrograph shows a clear-cut interfacial line between the coating and the substrate. However, the interface line is not smooth and it can be seen that the substrate grain size does not show any change indicating a purely mechanical bonding. The near circular pores are clearly seen in the coating. These are due to the entrapment and escape of CO and CO_2 gases. Several phases are noticed which are oriented parallel to the interface line, indicating layer-by-layer splating of the molten and partly molten droplets.

One main feature of these coatings, compared to that formed by APS is that the coatings are dense and pore free. Another observation that can be noted is the varying coating thickness. The thickness of the coatings at higher scan speeds is lower than that of lowest scan speeds.

The laser coatings at a speed of 1000mm/min show large particles. It is observed that around the bigger particle, a new phase is formed which is mostly formed due to decarburisation. White matrix is that of NiCr. The interface thickness decreased with increase in translational speed. It holds good for all laser cladded coatings.

The micrograph of laser cladded PM20 with a translation speed of 1500 mm/min shows similar features as that of laser cladded PM20 with translation speed of 1000 mm/min, except for lower thickness.

It is interesting to note that as compared to APS coating all the laser cladded coatings do not show any porosity. The XRD results for all the coatings show the formation of $Cr_{23}C_6$. This indicates that decarburization has taken place, carbon and hence CO_2 has been formed. However CO_2 in the APS system remains entrapped. In laser cladding, CO_2 is released due to the turbulence formed in the molten pool. Thus the porosity is negligible in laser cladding.

The cross-sectional analysis of the coating was done by SEM/EDAX to study the thickness and dilution of the coating composition into the substrate. The laser cladded samples registered decreasing thickness with increasing translation speeds. For translation speeds of 600mm/min, 1000mm/min and 1500mm/min, the coating thickness was 600μm, 480μm and 180μm respectively for a constant powder feed rate (PFR).

The EDAX investigation was also used to check any dilution, taking place during coating formation. The results have shown that the oxygen content in the coatings in the combined form near outer layers of coatings is between 2-6%.

An analysis carried out across cross-section of the coating next to coating substrate interface showed negligible Fe transfer in APS coatings, while considerable Fe transfer was found in coating for Laser Cladding sample. And this transfer of Fe to coating from steel matrix was less at lower scan speeds than at higher scan speed. However, the dilution of Fe from matrix was restricted to within 10-50μm of the coating.

The SEM micrograph of the Plasma Sprayed WC/Co coating, showed splats, containing mainly molten metal and binder, and sparsely populated carbides. It is evident that only small carbide of less than 5 μ are trapped within the binder. The large carbide particles are either embossed or pushed out of the splat and thus can rebound easily off the substrate. As a result, only small carbide particles may be retained in the coating after deposition

The SEM micrographs of all the laser cladded samples, using of 600, 1000, 1500mm/min respectively show that as the scan speed increased the size of the particles was more. The Laser Cladding with 1000mm/min scan speed revealed clear WC particles impregnated in the matrix. As in the case APS coating, the size of the WC particles is just over 3μm or less.

X-ray diffraction of APS WC/Co shows WC as the major phase in the coating. W_2C phase was also detected in all the coatings. This indicates decarburization as original powder had no W_2C. It is evident from the diffractograms of the laser cladded samples that the intensity of W_2C with respect to WC is same indicating that there is no major change in volume fraction with increase in translation speed. Even the intensities of WC with respect to the original powder remain slightly less indicating that the WC particles are more or less intact. Reduction is proportional to conversion of WC to W_2C. No oxidation is noticed in the Laser Clad samples.

Micrograph of plasma sprayed shows a clear-cut interfacial line between the coating and the substrate. The near circular pores are clearly seen in the coating. These are due to entrapment and escape of CO and CO_2 gases. The population density of the carbide particles which is the only hard phase present in the powder is large. The non-circular black portions are the carbide particles which are either WC or W_2C particles. The interface shows turbulence indicating melting of the substrate. Laminar cracks also are seen running parallel to the interface.

The coating shows a good distribution of the WC or W_2C particles, which can identify as acicular particles. The interface line is sharp and well defined. However the interface is wavy and microstructure change is evident. This indicates melting of the substrate and thus good metallurgical bonding.

The small purple particles of size less than 1μm distributed in the grayish cobalt matrix are the WC/W_2C particles. The interface in both the cases is sharp and well defined, but the waviness indicates interface melting of the substrate.

The cross-sectional analysis of the coating was done by SEM/EDAX to determine the thickness of coating and dilution of the coating composition into the substrate.

The optical micrographs with respect to the interface region reveal a change in microstructure. The results are in line with those obtained in case of PM20.

Since, the laser cladding is a processing method wherein laser beam is used as the source to melt the alloy powder to be cladded unto the substrate, a metallurgical bonding is achieved. Minimum dilution of the substrate is required. The dilution studies showed the presence of Fe into the coating upto around 50μm.

The Plasma Sprayed NiCr SEM micrograph of coating shows a dense coating with splats being clearly visible. Small spherical particles seen are of partially melted powder. The general appearance of the coating is smooth.

Prominent boride particles can be seen in all the coatings. But careful observation shows that for the same magnification the density of the boride particles increases from as the scan speed increases.X-ray diffraction results of NiCr coatings show $Cr_3C_2\backslash Cr_7C_3$ as the major phase in all laser cladded coatings beside formation of borides. $Cr_{23}C_6$ phase was also detected in all the coatings. Black acicular particles of the hard phase are also seen. The uniform grayish background is that of oxidized matrix. The highly rough and broad interface indicates a mechanical bonding. The substrate shows the typical microstructure of low carbon steel.

The microstructure of the laser coating made at 600mm/min speed shows the black portions in the coating which are hard phases. The whitish -grayish portion is the matrix which is not uniform. The hard phase is uniformly distributed. Both large and small particles are seen. The interface is not rough, it is sharp, and indicates metallurgical bond formation between the coating and the substrate. The substrate also appears to have undergone grain refinement.The metallographic of laser coatings deposited at traverse speeds of 1000 mm/min and 1500 mm/min show that the hard phase is more uniformly dispersed and the hard particles are smaller in case of coating deposited at 1000mm/min. The photomicrograph of the coating made at 1000 mm/min is similar to the coating made at 1500 mm/min. The smooth interface in laser cladded coatings indicate melting and resolidification in the interfacial region.

As seen in Laser Cladded PM20 and WC/Co samples the coating thickness reduces with increase in translation speed, similar observations were seen in the case of Laser Cladded NiCr samples.

Since laser cladding is a processing method wherein laser beam is used as the source to melt the alloy powder to be clad unto the substrate, a metallurgical bonding is achieved. Minimum dilution of the substrate is required. The dilution study for the APS coating and Laser Clad NiCr was carried out. In the APS coated samples presence of iron was traced. It is recalled that the powder as received also contains 4.6% weight of Iron. Thus the presence of iron is seen in the coating. The Laser Cladded samples also showed the same behaviour. However the Iron concentration reduced further away from the substrate.

4.4 PERFORMANCE

Microhardness

Hardchrome plating sample results given were obtained by electroplating mild steel samples with hard chrome. The hardness results shown are average of four readings taken across the surface and cross-section. A load of 300g was used for 15 seconds to evaluate the microhardness. However, wherever the coating thickness was less than 200μm, a load of 100g was used to avoid the effect of the substrate on hardness.

It is observed that the APS coated samples register the lowest hardness compared to that of the Laser Cladded samples.

A comparison was made for each cermet powder to see the percentage increase in hardness. It was found that Laser Cladded PM20 and WC/Co powder coatings showed an improvement of 40% and 33% respectively over the corresponding APS coating. However Laser Cladded NiCr samples showed an improvement of 15% only. APS WC/Co coatings also show the formation of $CoWO_4$ which will contribute to loss in hardness.

Improvement in micro hardness in laser cladding samples is due to reduced porosity level as compared to APS coating. It is evident from readings that as the translational speed increases the hardness falls as seen Giure 3 &4). The effect is more pronounced across the crossection than on the surface. This is due to the increase in dilution with increase in translation speed for the same powder feed rate. The laser power being same for all the three translation speeds substrate melting is more pronounced at higher speeds thereby allowing iron from the substrate to mix with the cermet coating. This leads to reduction in the hardness.

In the case of PM20, presence of $Cr_{23}C_6$ will lead to lesser hardness. The Cr-C binary phase diagram is very important in understanding the PM-20 coating system. The phase diagram shows that three stable chromium carbides are formed. These are cubic $Cr_{23}C_6$, orthorhombic Cr_7C_3 and orthorhombic Cr_3C_2. These melt at 1576°C, 1766°C, and 1811°C respectively. Formation of several metastable carbides have also been reported under rapid solidification and coating conditions. In actual practice availability of carbon also decides as to which carbide is formed. It was seen from the free energy table that $Cr_{23}C_6$ has the lowest ΔF.293 value followed by Cr_7C_3 and Cr_3C_2. Thus, thermodynamically $Cr_{23}C_6$ is the most stable carbide and there is a tendency to transform from less stable to most stable carbide. However, the extent of transformation will depend on kinetic factors. X-ray patterns of the coatings investigated in the present work clearly show the formation of $Cr_{23}C_6$, including the APS coatings.

The formation of $Cr_{23}C_6$ from Cr_3C_2 proceeds by dissociation of Cr_3C_2 as follows

$$8\,Cr_3C_2 \longrightarrow 24\,Cr + 16C \qquad (1)$$

$$24\,Cr + 16C \longrightarrow Cr_{23}C_6 + Cr + 10C \qquad (2)$$

Thus, some carbon remains in liberated elemental form, as it is not needed. A small amount of Cr also remains in elemental form.

The elemental carbon may react with oxygen as the system is exposed to atmosphere and form CO and CO_2 gases. Presence of oxygen has been detected by EDAX. These gases try to escape and in the process make the coatings porous as in the case of Plasma sprayed coatings. If enough entrapped oxygen is not present then the carbon may react with molybdenum at the outer surface of carbide particles to form molybdenum carbides MoC_2, MoC_6, et cetra. The presence of molybdenum carbide has been shown by microphotographs as well as X-ray diffraction.

The surface hardness however remains more or less constant. Decarburisation of Cr_3C_2 results in formation of $Cr_{23}C_6$, which has lower hardness level, but free carbon released can form Mo_2C, which probably compensates for the loss of hardness due to the formation of $Cr_{23}C_6$.

Another important process for loss of carbide particles is the physical flight of large size particles or particle rebound especially during spraying. This reduces the concentration of carbide in the coating and consequently reduces its hardness and wear resistance. Thus, small particles of Cr_3C_2 which have not escaped due to physical flight and very small particles of $Cr_{23}C_6$ formed during the little time available will be seen in coatings as has been observed in the present investigation.

As is well known and stated previously Mo and its carbides have high hardness even at high temperatures. Thus it can be argued that the surface hardness will remain more or less constant irrespective of translation speed.

In the present work, the WC weight fraction was in the range of 55 wt. %–100 wt. % WC (87 wt %), thus, the microstructure of the clad layer consists of (WC+ (γ +WC)) (WC: primary tungsten carbide, γ +WC: eutectic structure). This can be seen in WC/Co phase diagrams. WC particles can be also seen from the SEM micrographs and has been well confirmed by the XRD results. However the particle sizes largely vary from the original size. This can be attributed to two reasons:

1. Particles rebounding as in the case of Plasma Spray, wherein larger particles rebound and fall off after impact.

2. The reduced size of the coated particle in our case can also be attributed to decomposition of carbide particle either by decarburization to W_2C or oxidation to form $CoWO_4$. This has been evident in the XRD which reveals the formation of W_2C and $CoWO_4$. Plasma flame enthalpy may be the primary driver for the formation of WC= W_2C.

The XRD results for the plasma spray process show the oxidation of WC to $CoWO_4$. Transformation of WC to tungsten rich tungsten can be established by the following equation.

$$2WC \text{ --------- > } W_2C + C \qquad \text{(equation 3)}$$

$$C + O_2 \text{ -------- > } CO_2 \qquad \text{(equation 4)}$$

W_2C formation has been reported in all the spray process vis-à-vis Plasma Spray and Laser Cladding. This is due to the intense heat in both the processes.

It is equally important to note that due to decarburization free carbon is released which can stay as free carbon or escape as gaseous carbon dioxide/monoxide. This gas in entrapped condition will lead to circular voids other wise known as pores in the coating matrix. This is evident in the plasma spray coatings. However the Laser clad coatings donot reveal any porosity as due to the molten turbulent pool the gases easily escape before the solidification starts. This contributes to the loss of hardness of WC/Co coatings.

Laser Clad NiCr coatings show a 15% increase in hardness over the corresponding APS coating. Optical micrograph of APS NiCr show a compact coating with reduced pores as compared to the corresponding PM20, WC/Co. The SEM micrographs for surface morphology show splats with heterogeneous distribution of hard phases of Carbides and Borides of Chromium. Corresponding Laser Clad samples show a homogeneous distribution of hard phases. This contributes to increase in hardness.

For NiCr, the hardness-contributing factor is chromium boride and the formation of borides is evident in the XRD as stated before. But as stated by Kaul et al boride formation is mostly at grain boundaries and thus the general fall in hardness is evident. The concentration of the boride particles in the Laser Cladded samples is substantially large in comparison to Plasma Cladded samples, and since boride is the major hardness contributing factor, the reasons are obvious. As stated earlier the coating process is heterogeneous and quite complex owing to a large number of variants governing the coating deposition on a substrate. The metallurgical and structural changes undergone are explained with the help of binary phase diagrams.

From the phase diagrams it can be seen that with very lower percentage of boron the elemental boron will exist as solid solution at room temperature. Formation of Chromium Boride (Cr_2B) is more likely as the weight fraction of Boron in the powder with respect to chromium is 0.17. Typical mole fraction for formation of Cr_2B is 0.19:0.09 and in the case Cr_5B_3 is 0.89:0.11.The XRD results show the formation of Cr_2B.

With reference to the phase diagrams Boron added combines with chromium to form chromium boride as a hard phase, which helps increase the wear resistance of the alloy. The chromium boride forms a eutectic with the nickel solid solution. Less than 0.1% by weight of boron has little effect on improvement of the wear resistance of the alloy while more than 1.5% by weight of boron reduces the toughness of the alloy. B is incorporated in the alloy as solid solution and also forming borides with Fe, Ni, besides Cr. The borides thus forming and B incorporated in the alloy is solid solution increase strength of the alloy. Further, B, when it exists with Si in the alloy, lowers melting temperature of the alloy and gives the alloy self-fluxing characteristic. If the amount of B is too small, generation of borides is insufficient, so that the alloy is given no effective increase of hardness and no effective self-fluxing characteristic. On the other hand, if the amount of B is too large, impact value of the alloy lowers too much, with simultaneous deterioration of grindability and generation of scuffing. In view of these and in accordance with the results of experiment explained later, B content should be in the range 0.5% - 3%.

As seen in the original composition only 0.8% of carbon is present in the as received powder. Thus the likelihood of formation of NiCr is reduced as can be seen in NiCr phase diagrams. Morever Chromium is also present, and with a lower free energy than nickel, the probability of formation of Nickel carbide is reduced. This is confirmed by the XRD reports. This forms chromium carbide which is a hard phase, and contributes to wear resistance.

The translation speed of the torch has an influence on the oxidation. The lesser is the translation speed the more is the oxidation. The XRD results of Laser Cladded NiCr show more oxidation in case of traverse speed of 600 mm/min and a decreasing order till 1500 mm/min in case of NiCr laser cladding. The XRD results also show prominent oxide peaks in case of clad speed of 600 mm/min. EDAX results too confirmed high oxide peaks in case of Laser Clad 600 mm/min.

It can be seen in all the optical microscopy of the Laser Cladded samples examined under the same magnification that the size of the boride particles decreases as the track speed increases. The matrix of the coating being nickel, smaller particles of borides lesser than the indentor diameter are not well supported and thus the fall in hardness.

Surface Roughness

The surface roughness of the coated samples is shown in Table 1. The laser cladded samples have a better surface finish as compared to the corresponding APS samples. The surface roughness of the Laser Cladded samples was less than the corresponding Plasma coated samples. The surface finish of Plasma coated NiCr samples was, however, smooth as compared to Cr_2C_3-NiCr and WC-12%Co.

Wear

To evaluate the wear, the initial and final weights and width of the samples was noted. The difference in height of the samples as found from the machine results was also noted. The results from the above inputs were compared. Though they were matching, the difference in weights was found to be more consistent.

It is noted that the all the coatings have a co-efficient of friction ranging from 0.3 to 0.4. The plasma coated samples for the same powders show higher co-efficient of friction. This is due to the surface roughness as shown in Table 1.

The comparative wear results are shown in Figure 6. It is seen that the laser cladded samples of PM20 and WC/Co show an improvement over the corresponding APS samples. The hard chrome sample shows a moderate wear rate. The NiCr powders show a comparative high wear irrespective of the coating process. Increase in load doesn't substantially affect the wear rate as shown in Table ---.

The co-efficient of friction is given as the sum of three terms, f_a, f_p and f_d termed as the adhesion component, the ploughing component, and the deformation component, respectively. According to Suh [5], the adhesion component for metal varies from about 0 to 0.4, depending on the presence of a contaminant layer covering the asperities.

The ploughing component is smaller than the adhesion component, usually not exceeding 0.1, except for identical metals sliding against one another, having wear particles trapped between the surfaces. The friction component due to asperity deformation, f_d, can be as large as 0.4 to 0.75 in special cases, and is believed to be responsible for the static co-efficient of friction [5].

Suh[5] defines six stages of surface interaction that characterize time dependant friction behavior of metals (Fig 8). The relative importance of f_a, f_p, f_d changes from stage to stage. Stage I is characterized by ploughing of the surface by asperities; the co-efficient of friction is largely dependant on material composition. In Stage II, the value of friction is beginning to rise. The slope is steeper if wear particles generated by asperity deformation and fracture become trapped. The swift increase in the number of wear particles gives the friction curve a step slope during Stage III; another contributor here is adhesion due to rapid increase of clean surfaces. Friction remains constant during Stage IV as adhesion and asperity deformation are constant now, as is the number of trapped wear particles. Stage V occurs when very hard stationary slider is slid against a soft specimen.

The asperities of the slider are gradually removed, creating a mirror finish of the hard surface. A decrease in friction will result in this case due to the decrease in asperity deformation and plowing. In stage VI the hard surface is mirror smooth and the value of the friction levels off. Stages V and VI do not occur if it is the hard slider that moves ad the soft specimen that is stationary [5].

Plots of Frictional force with respect to the sliding distance .The Plots were compared with the Friction Curve as shown in Fig 8. The trendline indicates that for the distance traversed the frictional forces are limited to Phase I and II and hence the co-efficient of friction is limited to 0.3-0.4. The higher value of the co-efficient of friction is attributed to wear particles trapped between surfaces.

The trend line of NiCr coated samples however showed a deviation. Plasma Sprayed NiCr samples exhibit the five phases. The stage III, IV and V are clearly visible in the figure. On the same lines the Laser Cladded NiCr samples have the same effect with however Stage III more prominent. It is to be noted at this point this phenomena of increase in Frictional Force and consequent increase in Co-efficient of friction at this stage will lead to more wear and this is evident in the results. The average co-efficient of friction of NiCr samples may be less than the other coatings.

The steady state temperature for all the three powder coatings was noted at around 1 mm from the interface. The steady state temperatures, which is largely influenced by the ambient temperatures and the thermal conductivity of the coatings was around 70°C-90°C for Cr_3C_2-NiCr coatings, 110°C for WC-Co coatings and 135°C for NiCr coatings. The temperatures are indicative of the thermal conductivity of the coatings. The higher conductive material would influence more heat flow and lower contact temperature. Pirso et al [9] hint about the influence of the binder content and type on the thermal conductivity. The results could not be conclusive and it is assumed that the contact temperature would be much higher than measured and since the temperatures mentioned will not initiate any structural changes of the cermets, the discussion was not elaborated. It can also be argued that since the co-efficient of friction reduces, it will result in a decrease in temperature.

However it can be emphasized that the applications of these coatings will be promising where the surface temperatures do not rise above 500°C to initiate thermal degradation. Thermal conductivity plays a very important role in transferring the heat to the cooled substrate.

The wear results showed improvement in the weight loss in the Laser Cladded coating as compared to Plasma Sprayed samples, with an improvement in the hardness. The SEM of WC-13%Co and Cr_3C_2-NiCr in the case of Plasma Sprayed samples revealed numerous pores, where as the corresponding Laser Cladded samples had considerably less pores, thereby improving the hardness. The wear mechanisms can thus be different.

There is no direct co-relation between hardness and the wear resistance of the coatings, although Li et al [10] have reported that the wear resistance of cermets is related to the hardness of cermet and coatings in addition to the presence of carbide particles. Rangaswamy and Herman [11] have cited that increase in the carbide content and decrease in the number of carbide particles, increase the wear resistance. H.So et al [7] have stated that the relationship between wear rate and normal load is irregular, and wear rate between in the load range of 9.8N to 78.4 N does not vary remarkably.

For sliding wear, Archard's wear rate theory has shown some similarity. The same has been tried here to verify our results.

Archard's Law of sliding wear states that the wear down is:

1. Proportional to the sliding distance

2. Proportional to the total load.

3. Inversely proportional to the hardness of the material.

It can be seen in Fig 7 that Archard's law is satisfied for cermets. Regression analysis of volume loss verses reciprocal hardness was carried out in accordance with Archard's law as shown in Fig 7. The curve shows a linear variation with a slope of 27.82. The value of the y intercept being very small is neglected. The plot clearly indicates that wear rate generally increases with decrease in the bulk hardness. The scatter below the regression line are those of Cr_3C_2-NiCr and above are of WC-13%Co coatings.

Archad's wear formula for sliding wear is:

$$W = ksFn / H \qquad (1)$$

Where 'W' is the volumetric wear in, 'K' is the dimensionless wear co-efficient, 's' is the length of distance covered, 'F_n' is the pressure applied, H is the Vickers hardness number of the material.

The wear co-efficient of both Cr_3C_2-NiCr and WC-13% Co was evaluated separately for all conditions and was found to be around 1.33E-04. The wear rate of the cermets can be thus predicted for all conditions of load and distance traveled.

However as shown in Fig 8, NiCr coatings shows the reverse trend. The mechanism of high wear rate is attributed to the boron oxide layer, which forms on the surface of the coating. In the initial stage this boron oxide layer, thus formed, is removed thereby increasing the wear rate. In addition, as seen in XRD of NiCr coatings, the matrix is predominately Ni based. When the coated sample is making a contact during initial phases of sliding, the soft ductile NiCr matrix between the sparsely populated carbides and borides of chromium, undergoes severe deformation. As the layer is removed, the underlying boride layers are exposed which retards the wear rate. It was seen in the SEM micrograph that the borides/carbides are finally pulled off due to lack of reinforcements provide by NiCr matrix. Hence the wear behavior is different. This change in wear pattern has also been reported by Pirso et al and Picas et al[9].

Dilution of Iron into the coating, influences the wear rate as it is a softer material. A controlled dilution will retard the wear rate largely. This is observed in the laser cladded sample. APS samples on the contrary though not influenced by dilution, had pores which increased the wear rate. Though Laser Cladded samples show an improvement, it is to be noted that the results are marginally better than the APS samples due to the aforementioned reasons.

Fig 8 Plot of Volume wears loss verses 1/H of all the three cermets powder coatings under investigation. The extreme points in each case represent volume loss for Plasma Sprayed Coatings and the rest are Laser Cladded coatings.

Conclusions

The objective of the research work was to search for a suitable material and process to replace existing Hard Chrome coatings.

It is observed from the preceding discussion that co-efficient of friction of hard chromium layers is less than the other deposition process. It has been observed, that Laser Cladded samples have slight higher co-efficient of friction, however the same is less than the APS samples. However, as shown in Figure , the wear resistance of Hard chromium is more than the convention thermal spray process. In our case the wear resistance of the Laser cladded samples are higher than the Plasma Spray process and the Hard Chromium Layer. It can thus be concluded that that Laser Cladded cermet coatings tide over the environmental problems of hard chromium coatings in additions to the aforementioned properties and also stand to be a quicker process.

Figures

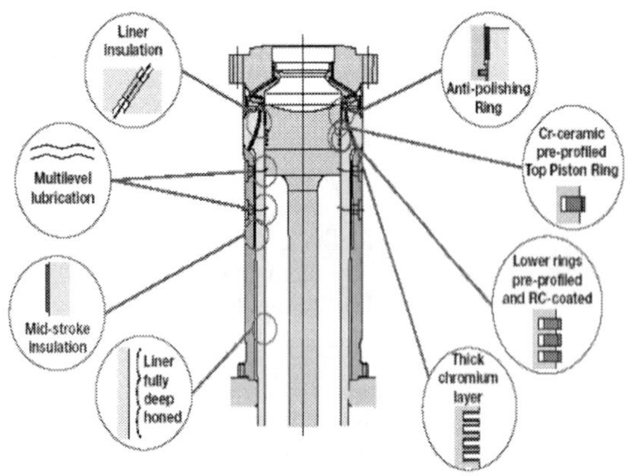

Figure 2: Morphology of as receive powders, PM20,WC/Co and NiCr respectively

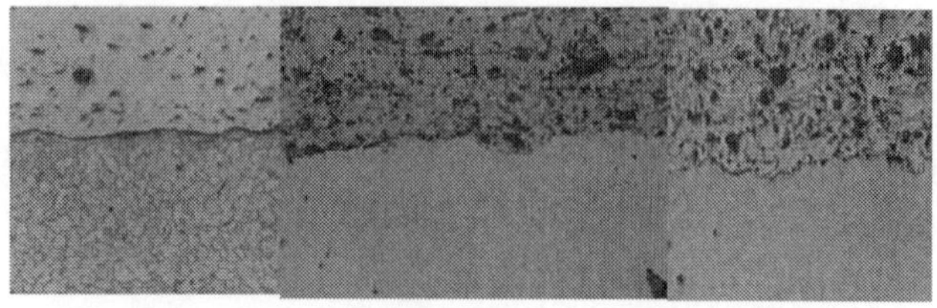

APS- NiCr APS- WC/Co APS- PM20

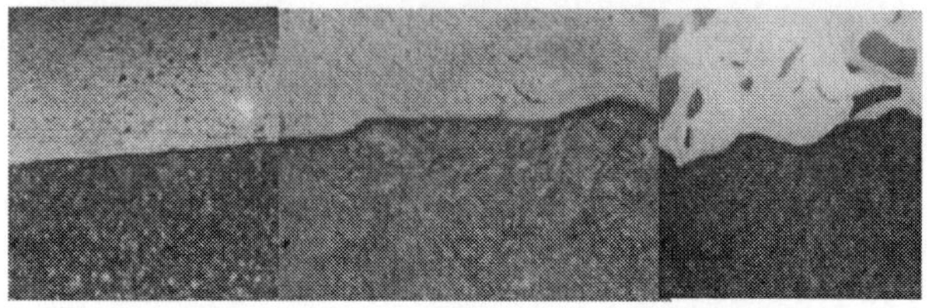

LC- NiCr LC- WC/Co LC- PM20

Figure 3: Optical Micrograph of the APS and Laser Cladded samples.

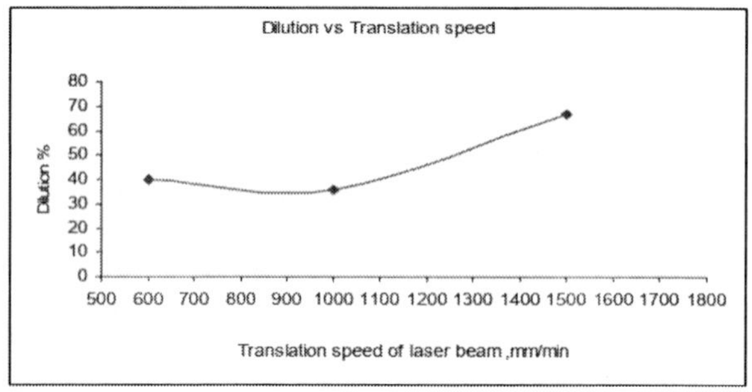

Figure 4: Coating thickness variation with laser clad speed.

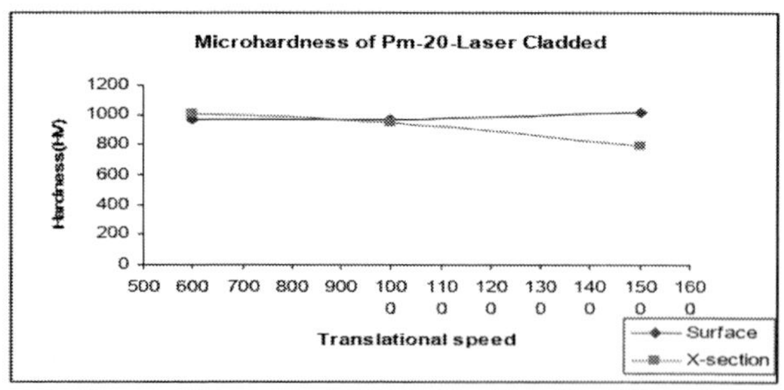

Figure 5: Micro hardness variation with clad speed.

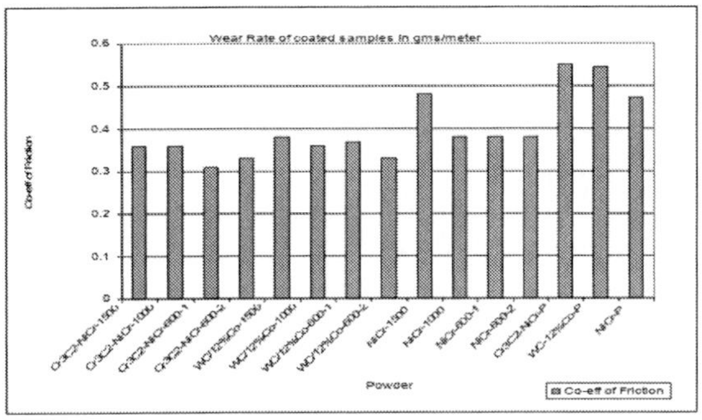

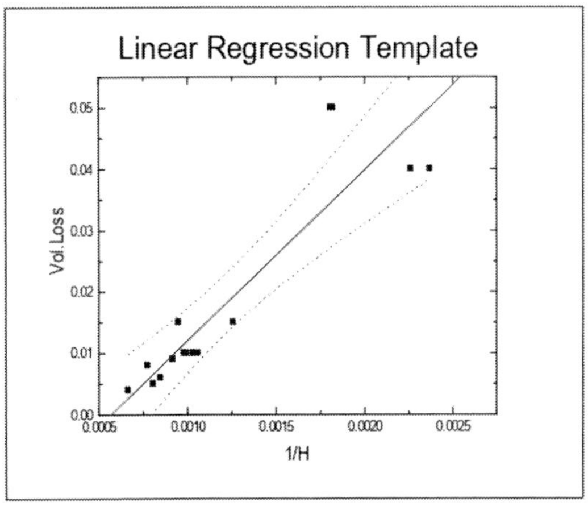

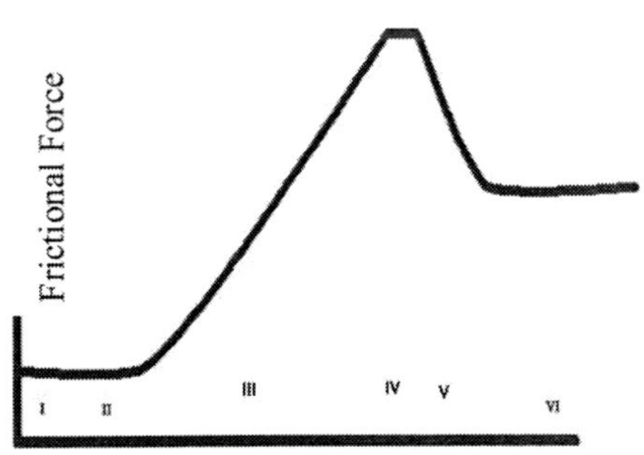

Figure8. Six stages in the frictional force versus distance skid relationship

Table 1. : Surface Roughness of as coated samples.

Sl.No.	Sample	R_a (μm)
1	APS - NiCr	7.309
2	Laser Clad-NiCr-600 mm/min	3.241
3	Laser Clad-NiCr-1000 mm/min	3.185
4	Laser Clad-NiCr-1500 mm/min	2.854
5	APS –Cr_3C_2- NiCr	10.514
6	Laser Clad-Cr_3C_2-NiCr-600 mm/min	6.584
7	Laser Clad-Cr3C2-NiCr-1000 mm/min	6.385
8	Laser Clad- Cr3C2-NiCr -1500 mm/min	6.259
9	APS – WC/Co	9.23
10	Laser Clad- WC/Co -600 mm/min	5.846
11	Laser Clad- WC/Co -1000 mm/min	5.756
12	Laser Clad- WC/Co -1500 mm/min	5.425

References:

1. D. Dowson, Engine Tribology. Elsevier, 1993, Tribology series, 26, pp. 213–240

2. M. D.Röhrle, MAHLE GmbH. Verlag Moderne Industrie. Landsberg/Lech, Germany. 1995. p. 70.

3. St.Siegmann, O.Brandt, N.Margandt, 1st International Thermal Spray Conference, 2000.

4. C. Navas, M. Cadenas, J.M. Cuetos, J. de Damborenea, Wear 260 (2006) 838–846.

5. Sue, Jiinjen Albert; Tucker, Jr., Robert Clark; Nemeth, Joseph Patrick, United States Patent 6007922, December ,1999.

6. Aram, Mehdi; Samuelsson, Per; Gong, Karin; Li, Changhai; United States Patent 7144017, December 2006.

7. E. Fern´andez , M. Cadenas , R. Gonz´alez , C. Navas, R. Fern´andez , J. de Damborenea , Wear 259 (2005) 870–875.

8.H.Chen, C.Xu, J.Qu, IM.Hutchings, P.H.Shipway, J.Liu,Wear 2599(2005)801-806.

9. Juri Pirso, Mart Viljus, Sergei Letunovits,",Wear 259(2005)44-51.

10 .J.F.Li, C.X.Ding, J.Q.Huang, P.Y.Zhang, Wear 211 (1997) 177-184.

11 .H. So, C. T. Chen and Y. A. Chen,Wear ,Volume 192, Issues 1-2, March 1996, Pages 78-84.

Dr. Sanjeet Kanungo, holds the current position of Program Chair- Practical Skills, at Tolani Maritime Institute,Pune. He is responsible the for the development of the entire skill programs in the institute. He is a Marine Engineer by profession and completed his Ph.D from IITB,Mumbai. His research interest lies in corrosion mitigation and reduction of emission in marine diesel engines.

Laser Cladding – A Coating Process for Marine Applications

Presenter: Dr. Sanjeet Kanungo

Objective

To minimize the wear of piston groove land of Marine Diesel Engines

AIM

- Chrome Lands, used presently, though functionally suiting the purpose but pose an environmental problem due to the release of Chromium hexavalent mist.
- Chrome coatings are carried out either by Chrome plating or GTAW process
- Either of the process has slow deposition rate .
- Masking operations are employed .
- Limited corrosion resistance.
- Thicker coatings lead to network of cracks due to thermal mismatch
- Post machining cost too high.

To find suitable wear resistant coatings on piston groove lands of internal combustion engines.

This was problem faced by MANB&W ,diesel engine manufacturer and a solution was sought by them to do away with the conventional chrome lands.

Where is the wear ?

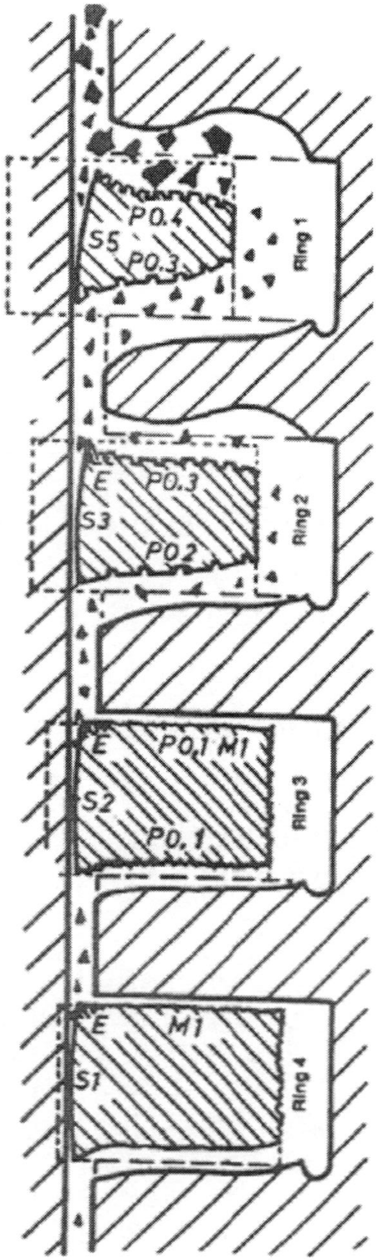

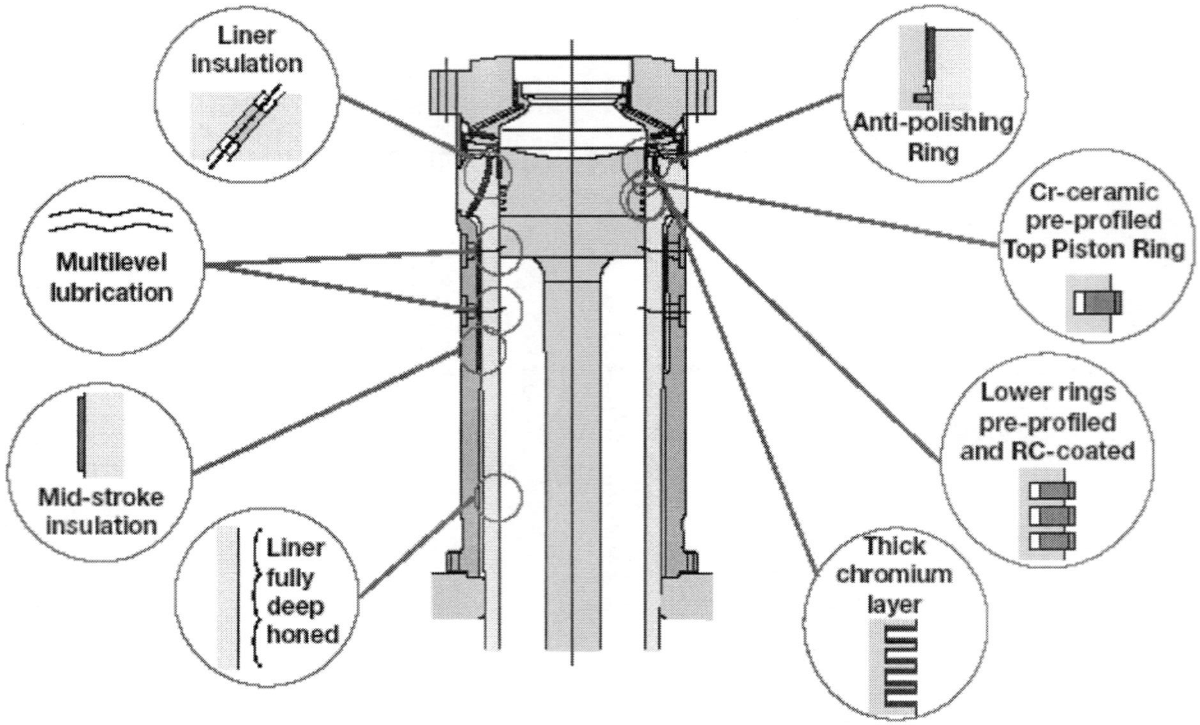

What contributes to wear ?

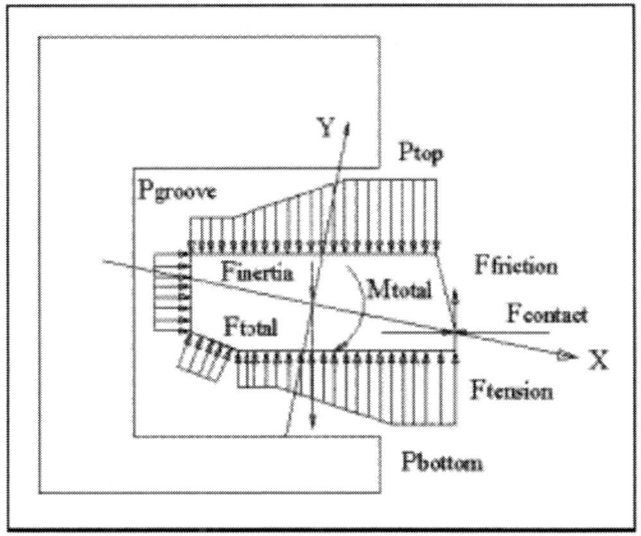

Fig. 5.7. Forces acting on the piston ring (After Ejakov et al., 1999).

- Primarily due to the combined effect of gas forces and radial motion of the ring and wear process accelerated by lack of lubrication and high temperature

How wear problems are solved?

- **Heat Treatment** is applied to the whole component at large. Selective areas cannot done
- **Hard Alloys** increase the cost of the components and thus not preffered.

Thus surface coatings are the prudent options

Methodology

- Literature Survey revealed various coating process and materials .
- Much work has been done by spray techniques like plasma spray, HVOF etc.
- Report on laser cladding technique on such components have been however limited.

Process

Plasma Spray process

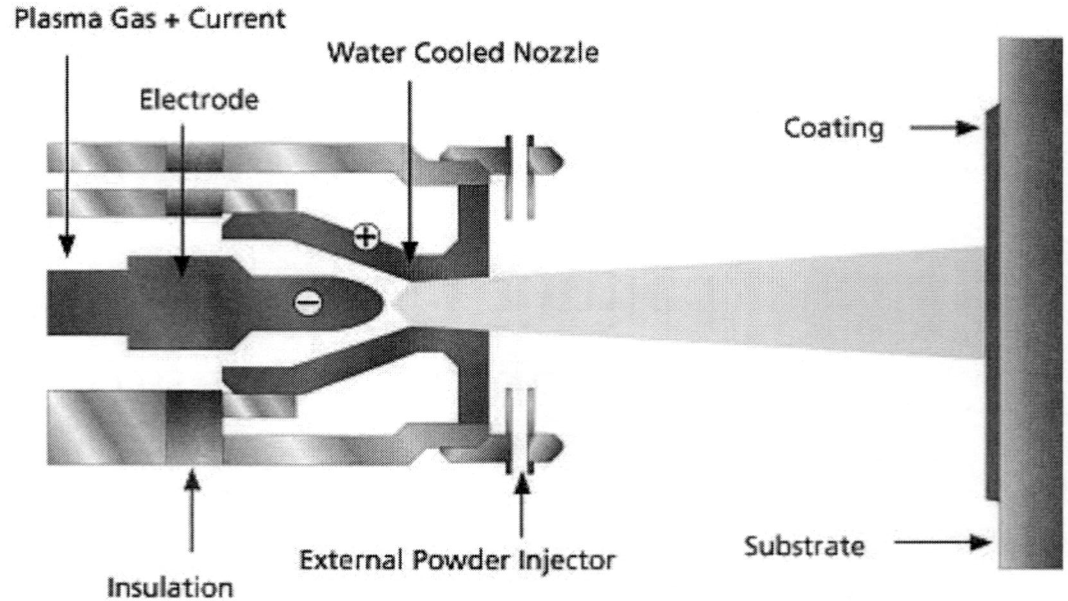

The plasma spraying process involves the latent heat of ionized inert gas (Plasma) being used to create the heat source.

The most common gas used to create the plasma is argon, this is referred to as the primary gas

Laser Cladding Process

- Insitu process,with advent of Nd:YAG laser systems using optical fibre
- High energy and localized beam
- Minimum Porosity
- Novel microstructure
- Low thermal input and controlled penetration and therefore low component distortion,
- Controlled thermal profile and therefore shape and location of the heat affected region,
- Non-contact processing
- Little or no post-machining
- Improved mechanical, physical and chemical properties
- Selective surfacing of small areas
- Chemical cleanliness and therefore minimal pollution
- Relative ease of automation

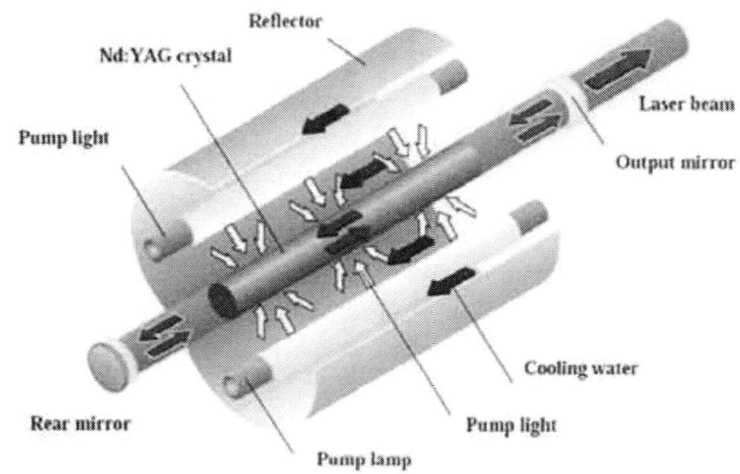

Methodology – Choice of Powders

Three cermet powders were chosen for study based on the literatures and their application in diesel engine sector.

1. ***Nickel(Base),Carbon(0.80%),Iron(4.60%), Chromium(16%),Boron(3.40%),Oxygen(traces) --- 20 μ (average)***

2. ***Tungsten (Base), Carbon (5%), Cobalt(12%) --- 20 μ (average)***

3. ***PM-20: 50%Cr_2C_3/32%Mo/18%NiCr(custom blend MBD material)-40μ-100 μ***

Characterization of NiCr powder

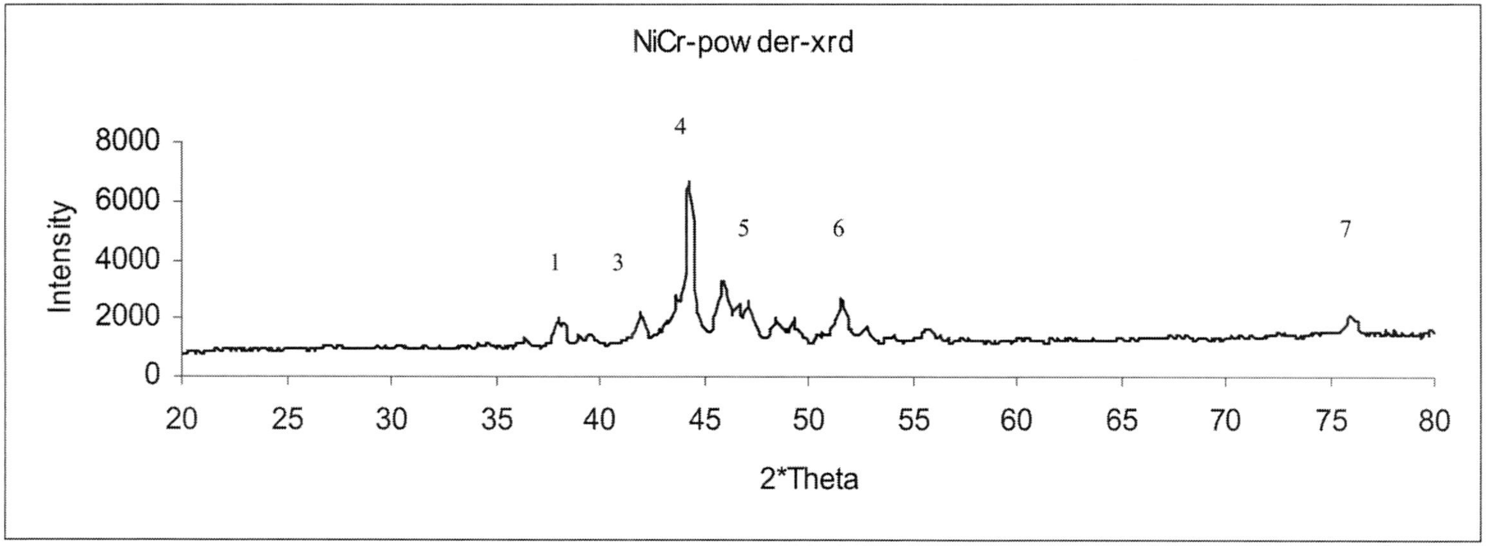

Peak 1. Cr_5B_3 *Peak 2.* $Cr_3C_2/Cr_7C_3/Cr_2B$ *Peak 3.* Cr_3C_2/Cr_7C_3

Peak 4. Cr_2B *Peak 5.* Cr_5B_3 *Peak 6 & 7. Ni*

Peak 8. CrO_2

Binary Phase Diagram of Cr-B

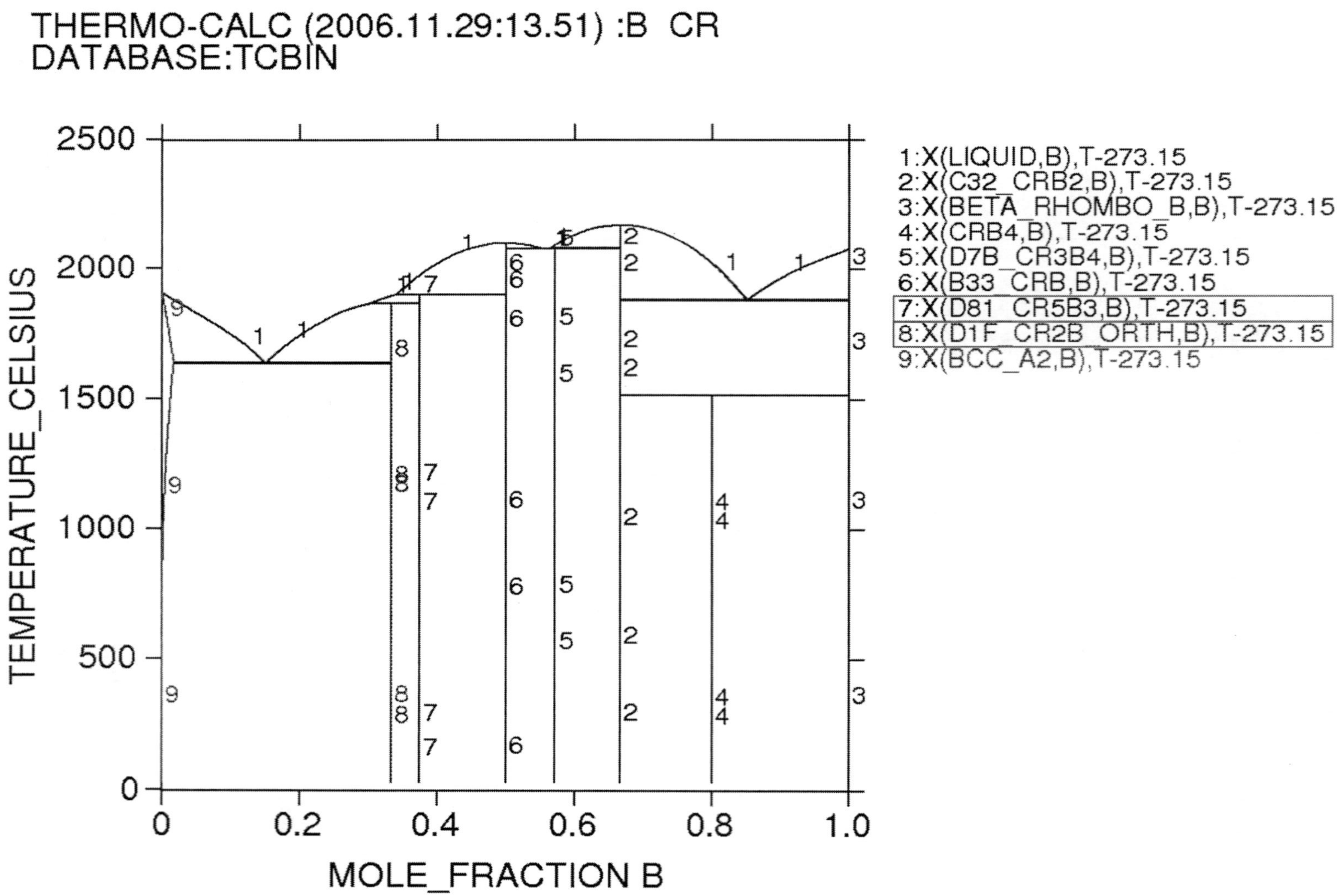

Microstructure of Plasma Sprayed Ni-Cr

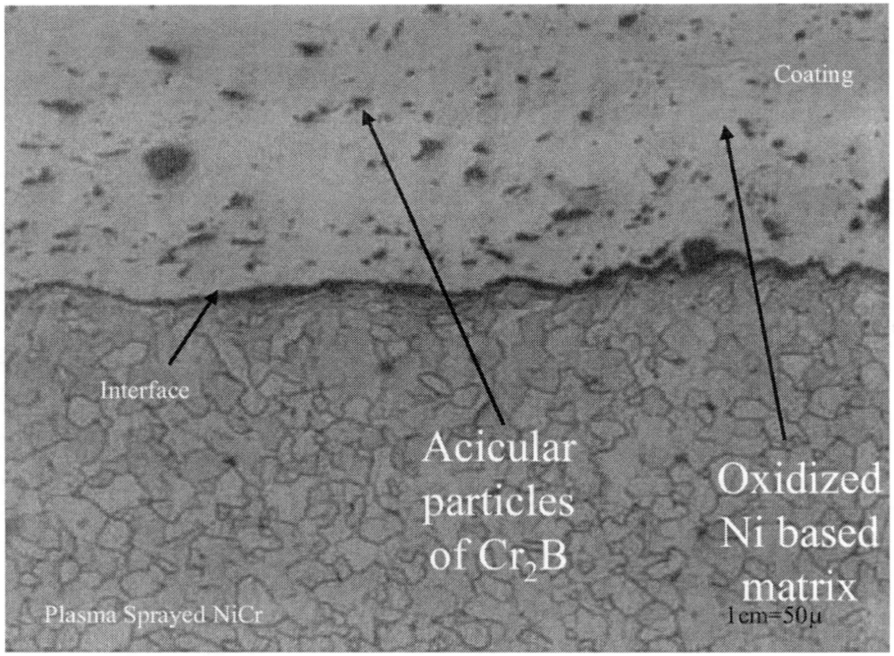

Optical (X-section)

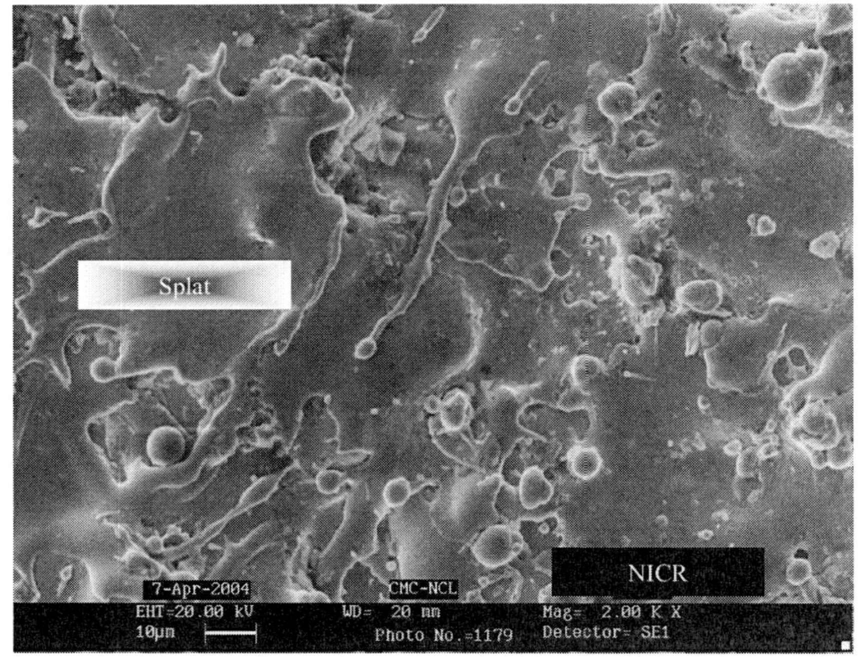

SEM(surface)

- Low Porosity
- Densely Packed
- Good Bonding
- Good Splat –hence lamellar growth of the coating
- Thickness of coating-400 microns

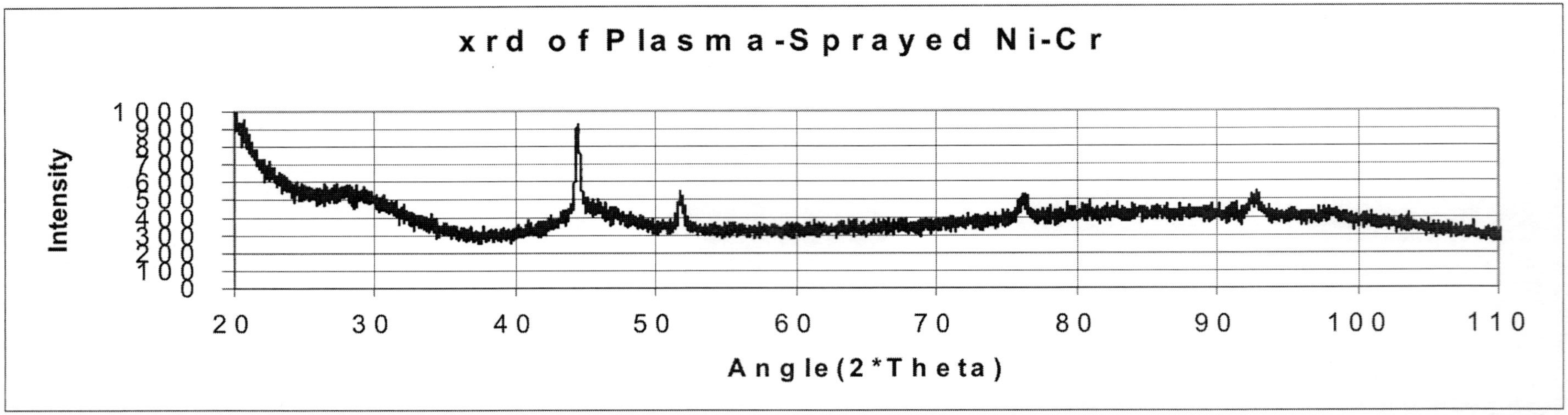

1. XRD results reveal the presence of **Cr_3C_2,Cr_2B,NiCrFe** but no presence of the **Boron.**

2. EDAX confirms the elemental presence of **O,Ni,Fe,Cr**, as per the powder composition ,but Boron is absent.Boron is presumed to have evaporated more probably as boron oxide(B_2O_3)

Microstructure of Laser Cladded Ni-Cr

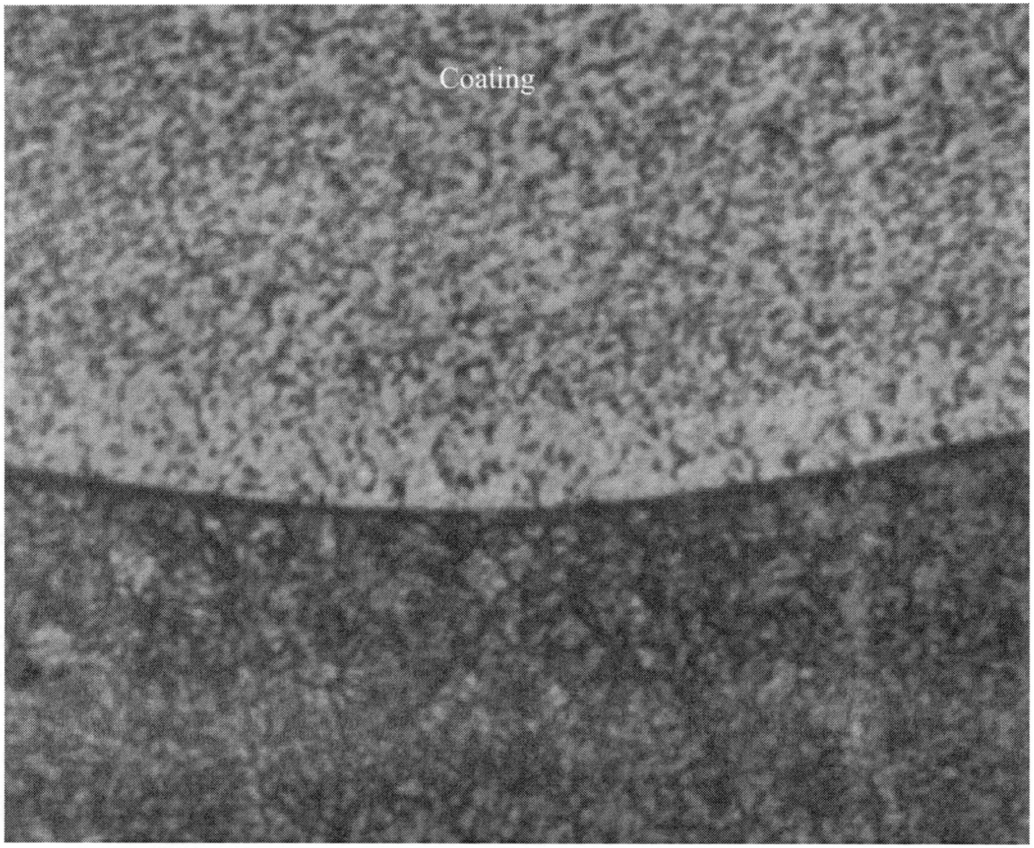

- Black portions indicate hard phase Cr_2B
- Non-uniform Whitish – grayish –Boron Oxide
- Hard Phase is uniformly distributed
- Both large and small particles seen
- Sharp interface indicating metallurgical bonding
- Substrate also appears to have undergone grain refinement.

Torch Speed –600mm/min

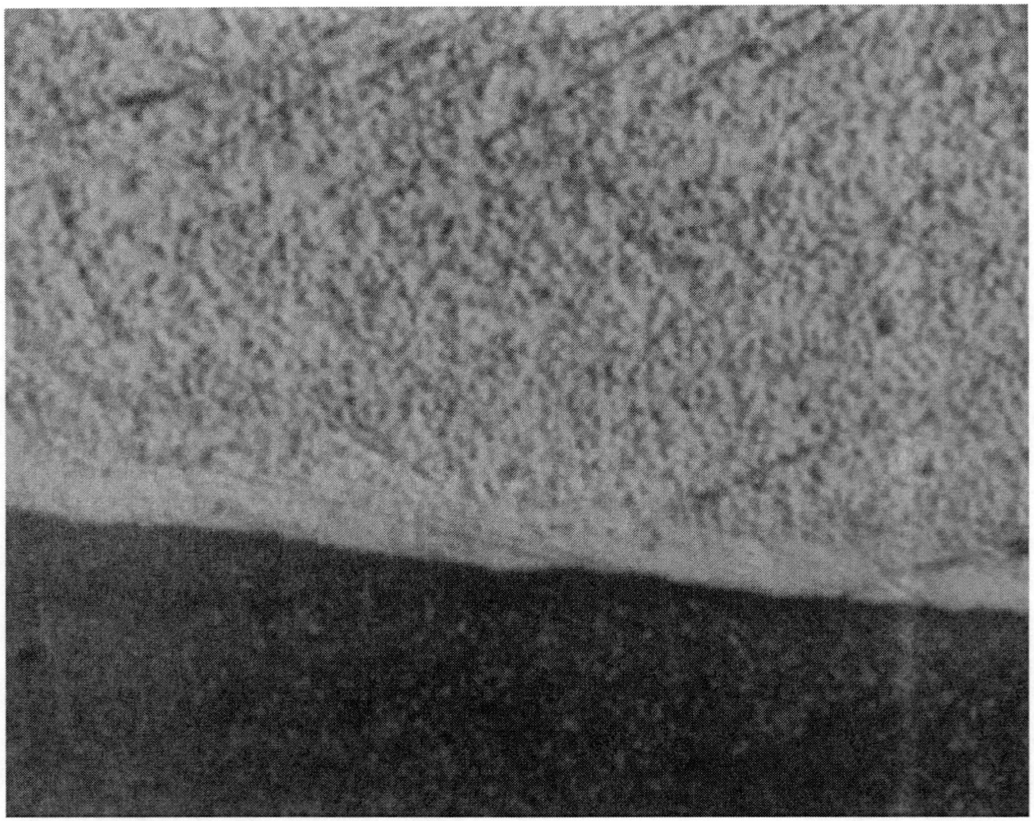

- Hard phase uniformly distributed
- Hard particles are smaller in size
- Smooth interface

Torch Speed –1000mm/min

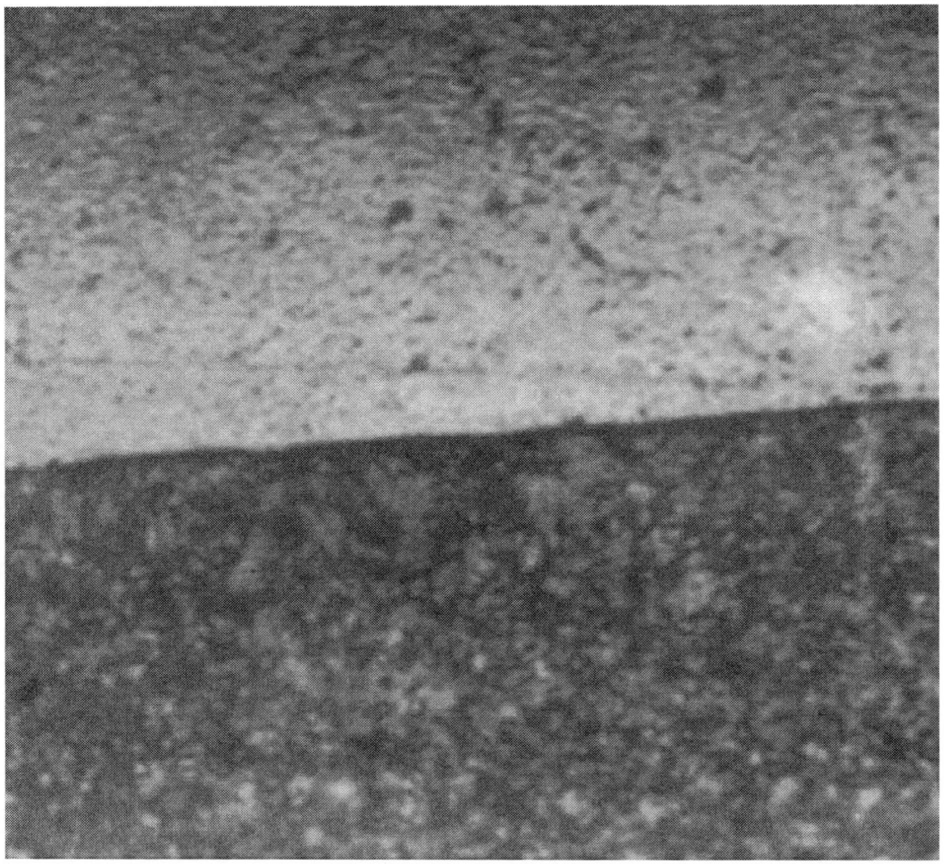

Torch Speed –1500mm/min

- Hard phase uniformly distributed
- Hard particles are further finer in size
- Smooth interface

XRD of Laser Cladded NiCr samples

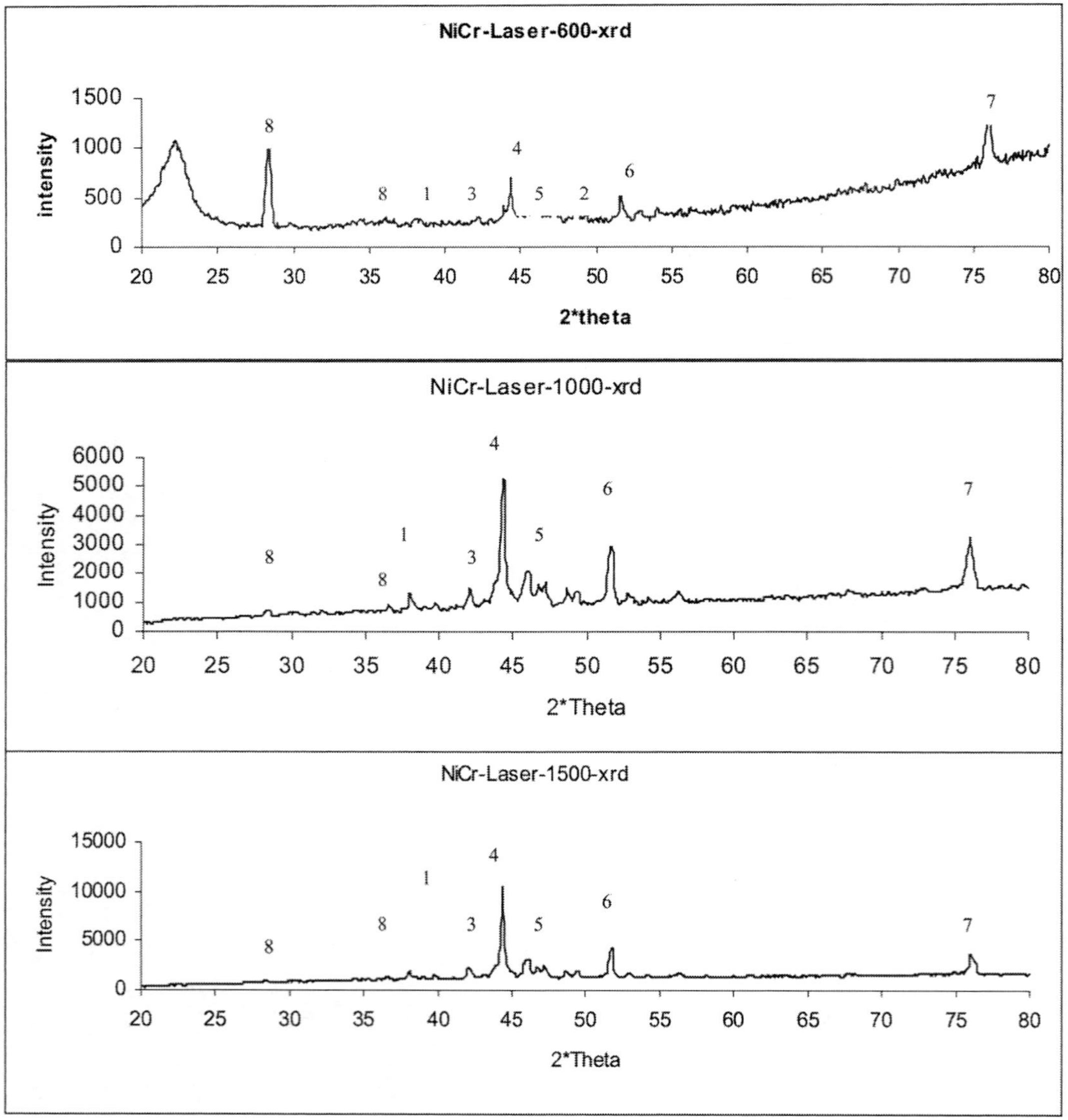

Ni,C(0.8%),Fe(4.6%),Cr(16%),B(3.4%),Oxy(traces)

Peak 1. Cr_5B_3 *Peak 2.* $Cr_3C_2/Cr_7C_3/Cr_2B$
Peak 3. Cr_3C_2/Cr_7C_3

Peak 4. Cr_2B *Peak 5.* Cr_5B_3
Peak 6 & 7. Ni

Peak 8. CrO_2

XRD of Laser Cladded NiCr samples-From 27° to 35°

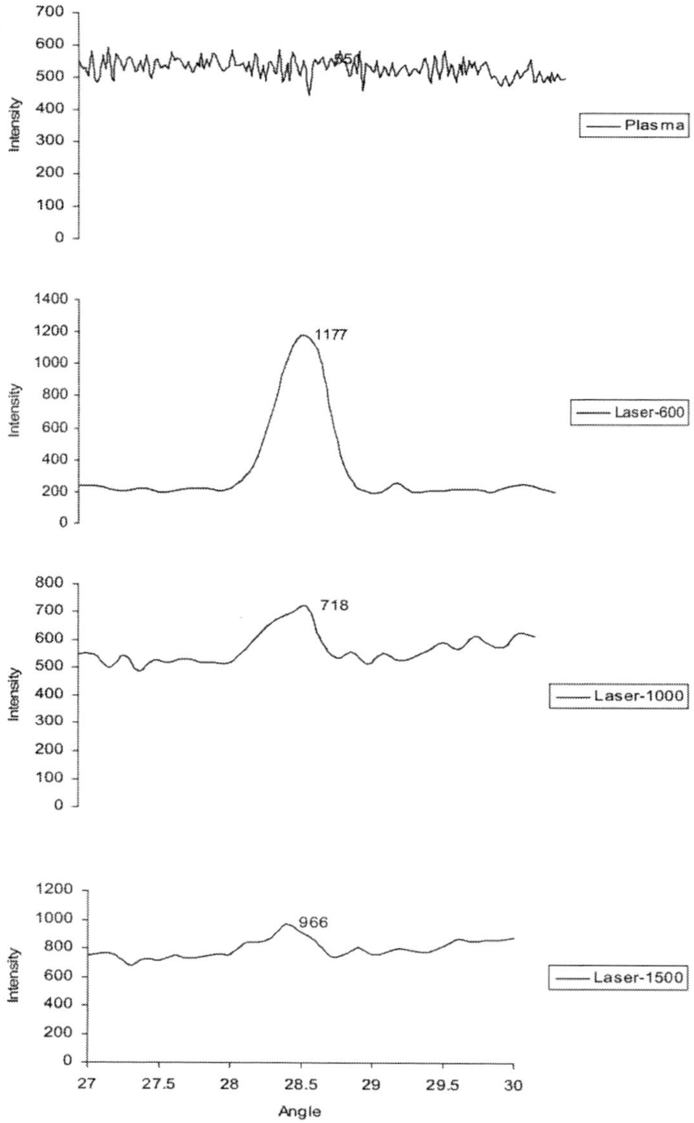

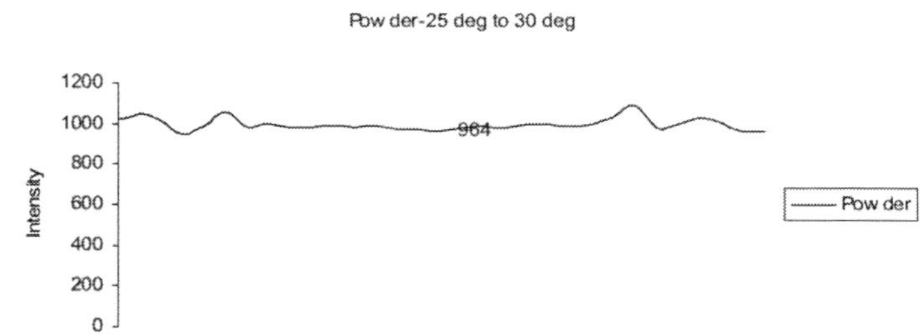

CrO_2 peaks detected in laser cladded samples though with varying intensities. Plasma Sprayed samples show no such oxidation

XRD of Laser Cladded NiCr samples-From 35° to 55°

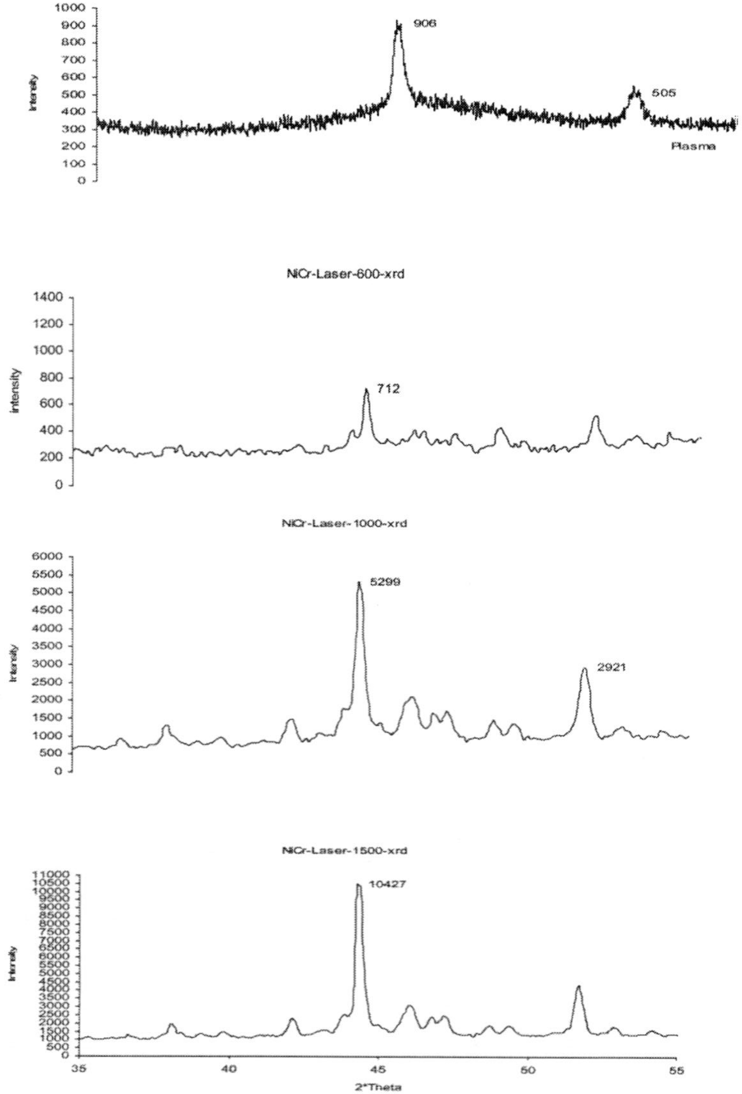

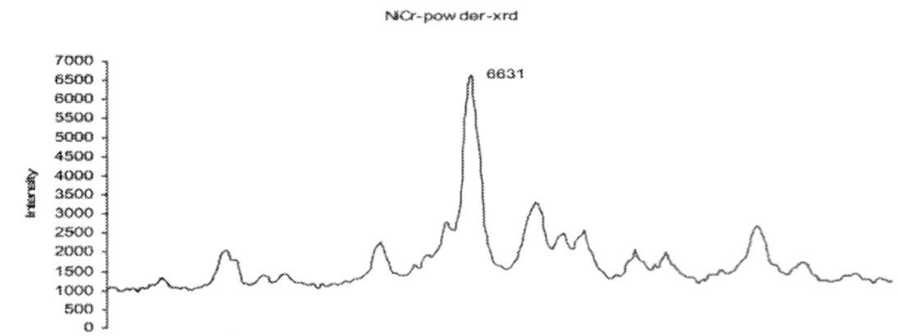

XRD peaks between 35 deg to 55 degrees. Peaks of the laser cladded samples though of varying intensities but resemble the original powder.Plasma Sprayed samples have flat peaks .

DILUTION

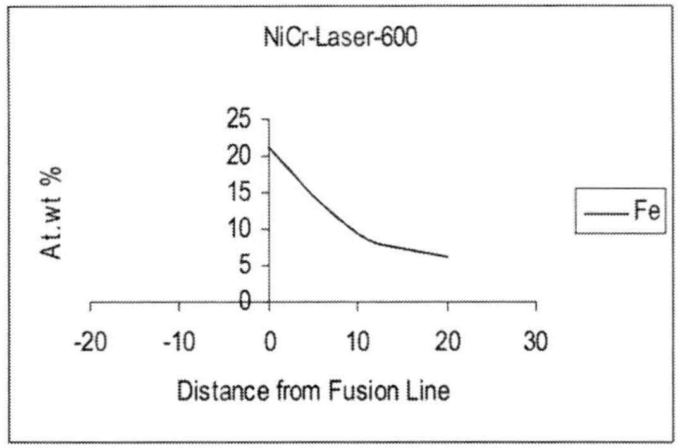

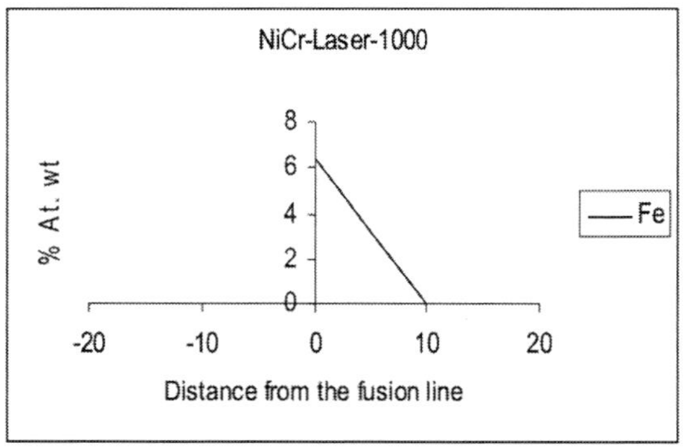

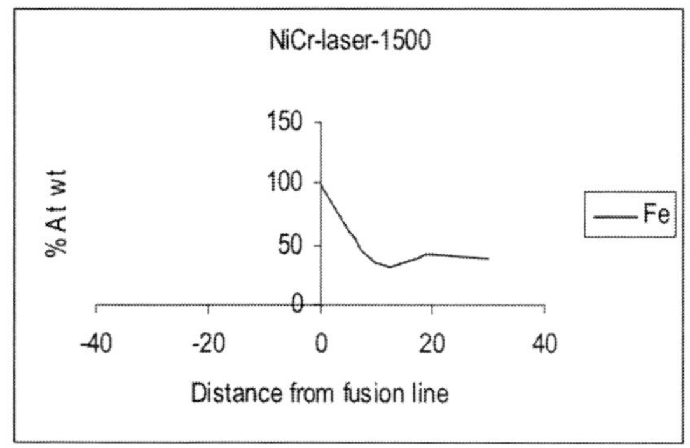

Dilution of Iron for different spraying methods and conditions

Clad Thickness

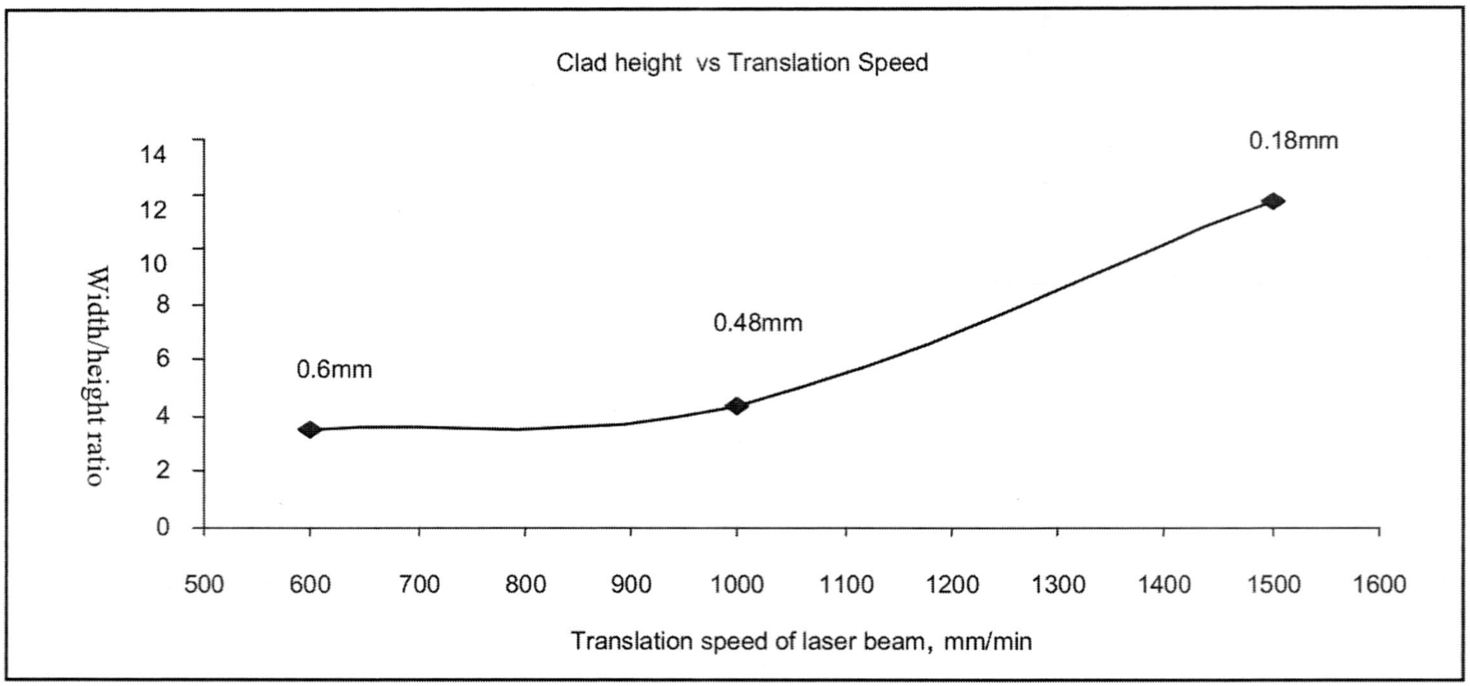

Microhardness of Laser Cladded samples

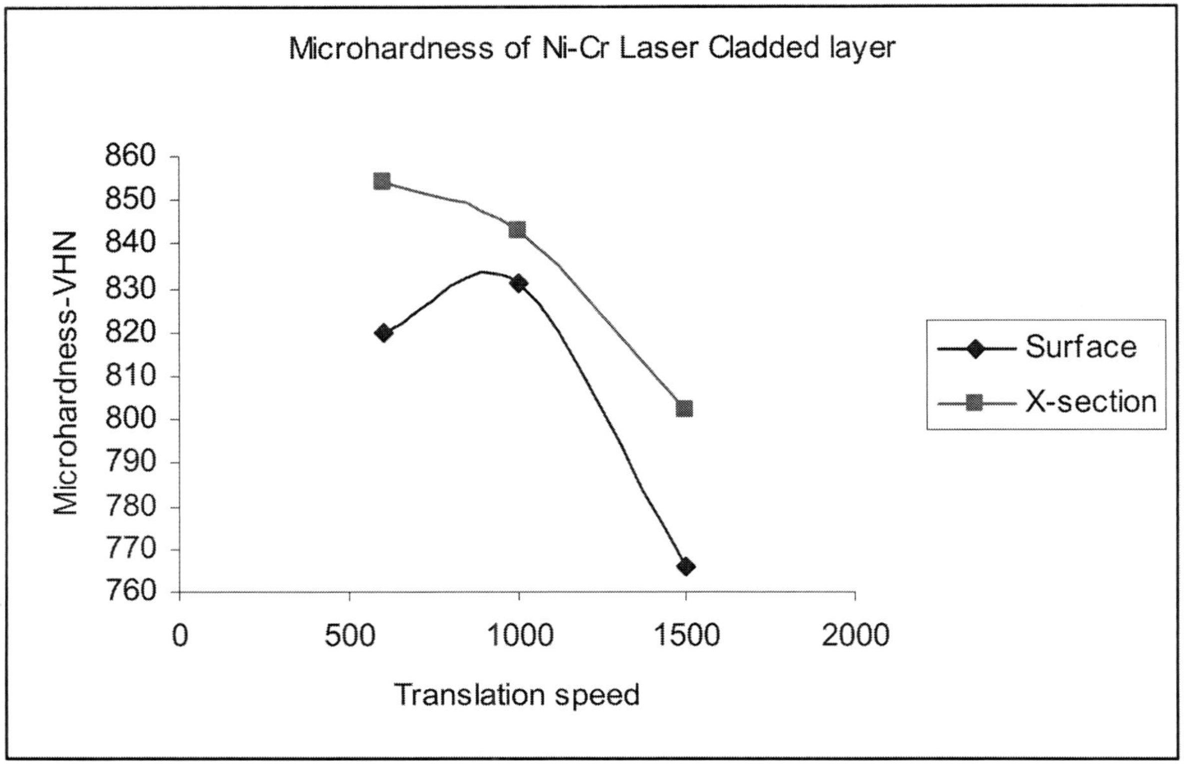

Microhardness of Plasma Sprayed samples ranged between 520-570VHN

Characterization of Cr_3C_2-NiCr powder

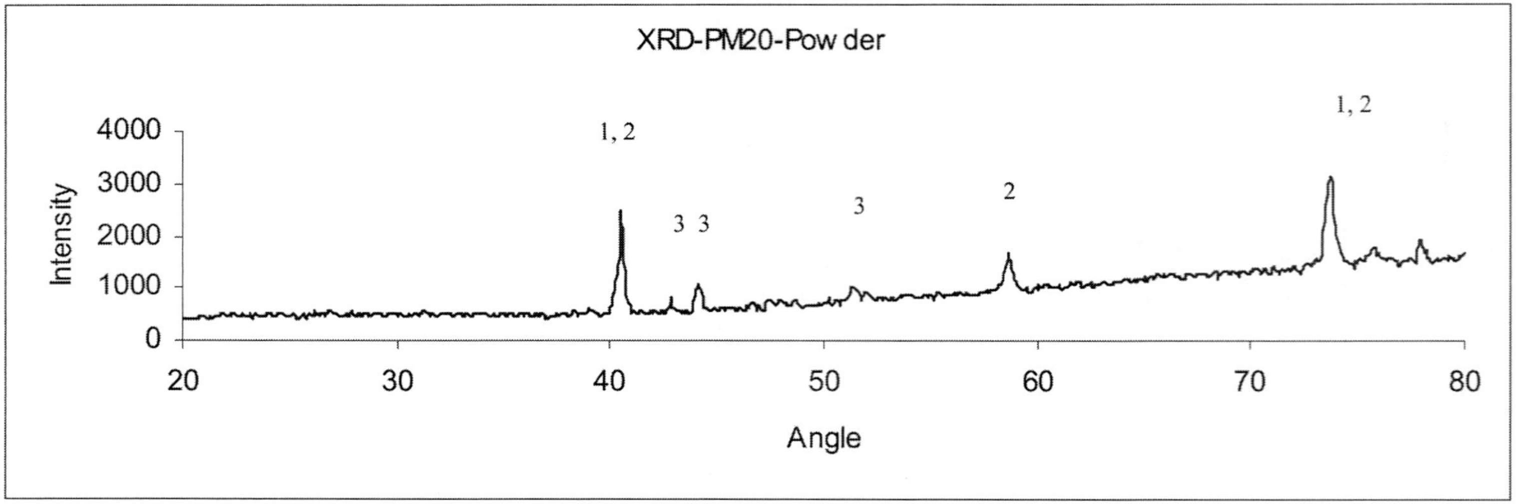

1-Cr_3C_2 2-Mo 3-Ni

Microstructure of Plasma Sprayed Cr_3C_2-NiCr

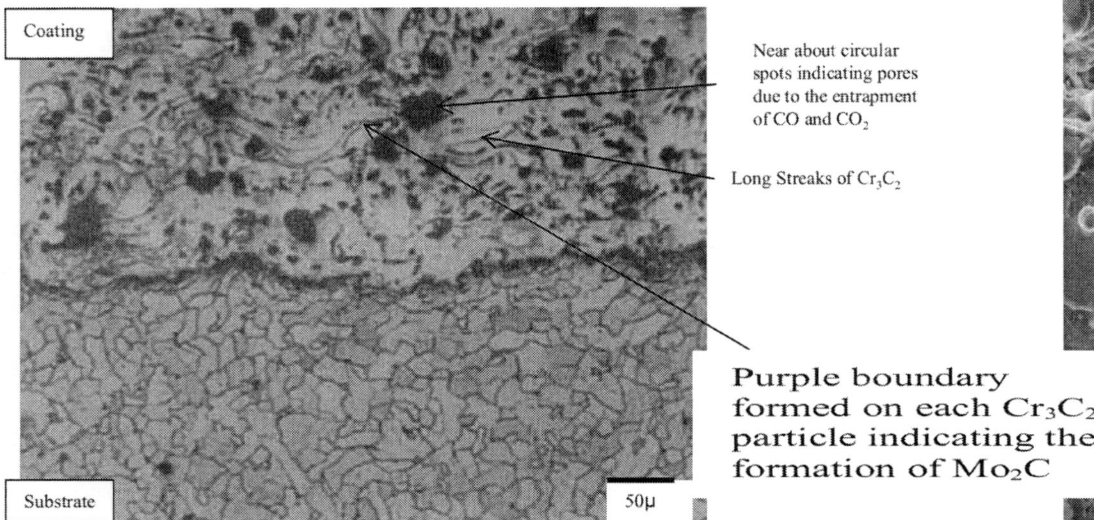

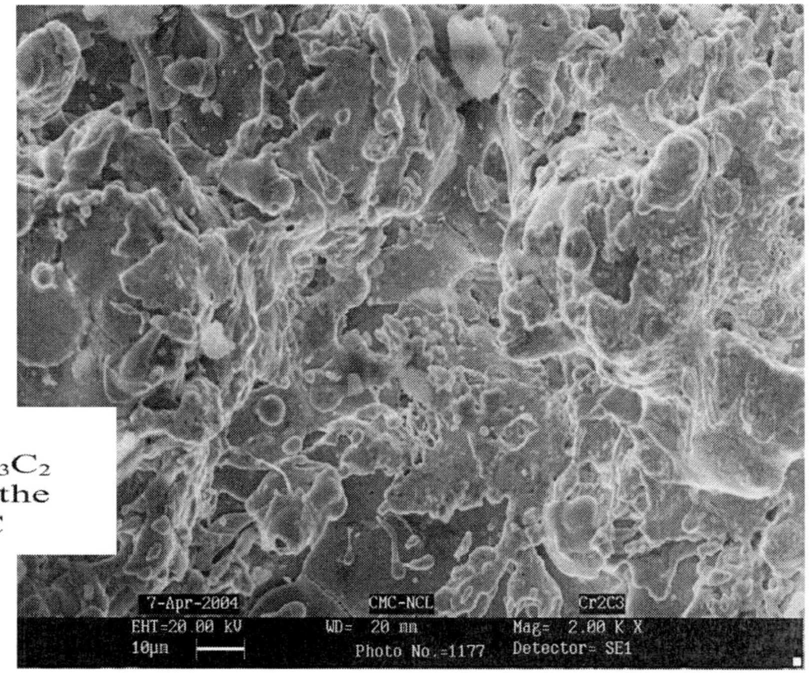

Optical (x-section)

SEM(surface)

- High porosity
- Incomplete fusion
- Good bonding and anchoring
- Thickness of Coating –290 microns

XRD of Plasma Sprayed Cr_3C_2-NiCr coating

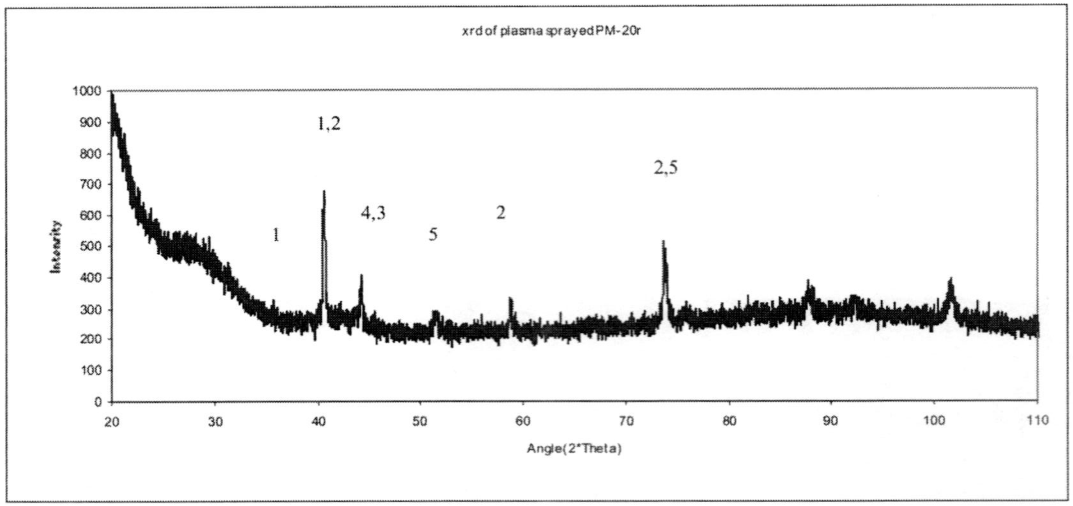

1-Cr_3C_2

2-Mo

3-Ni

4-$Cr_{23}C_6$

5- $\gamma Mo_2C/Mo_2C$

6-NiCr

1.XRD peaks consists of primarily **Cr_3C_2 and γ phase**(Nickel and chromium in solid solution)

2.EDAX shows good presence of **Cr,Fe,Ni,C and Oxygen.**

Optical Microstructure of Laser Clad Cr_3C_2-NiCr samples with varying torch speed

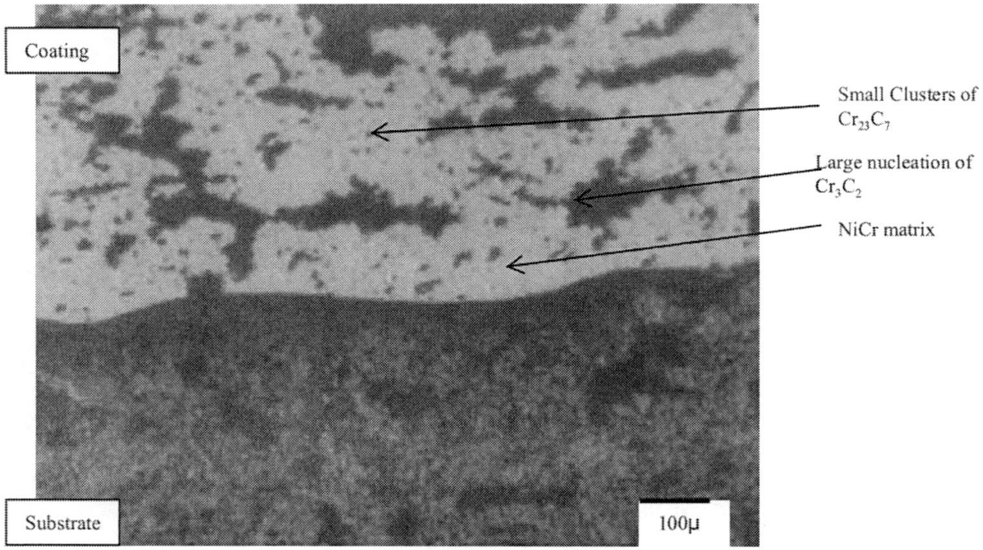

Torch Speed –600mm/min

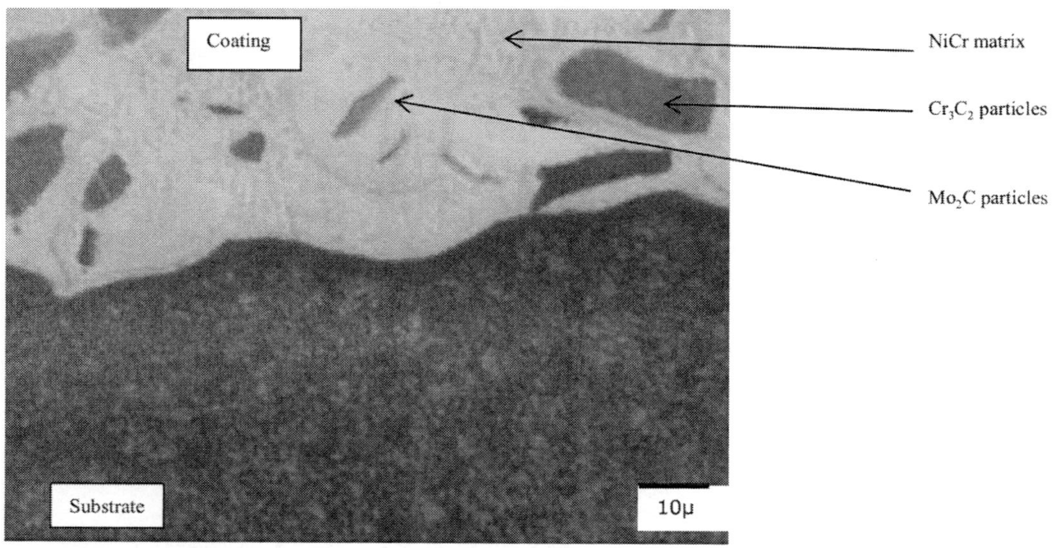

Torch Speed –1000mm/min

Optical Microstructure of Laser Clad Cr_3C_2-NiCr samples with varying torch speed .

Contd---

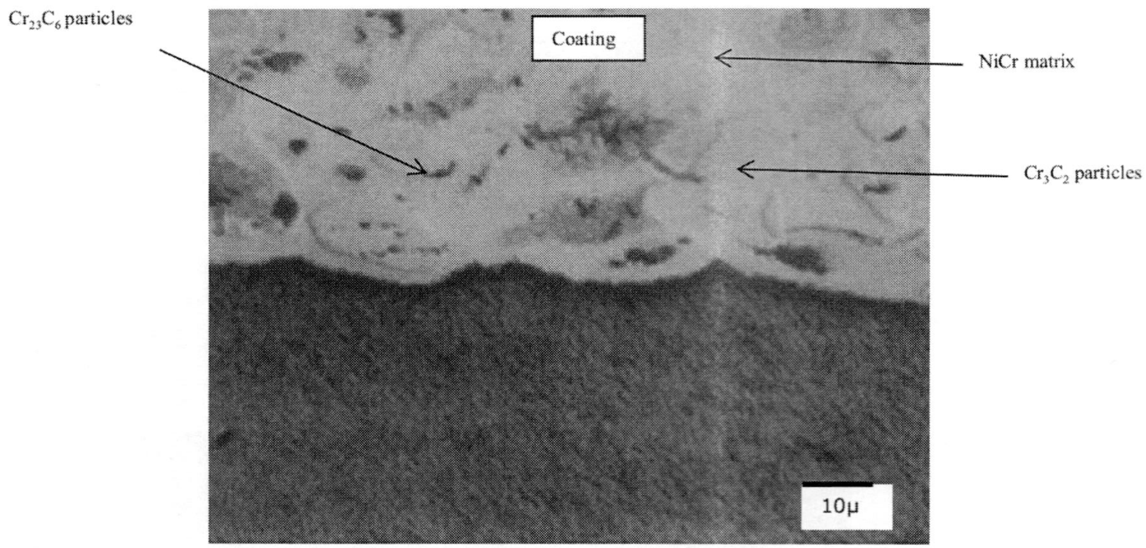

Optical Microscopy of Laser Cladded PM-20 with translation speed of 1500mm/min

XRD of Laser Cladded Cr_3C_2-NiCr coating

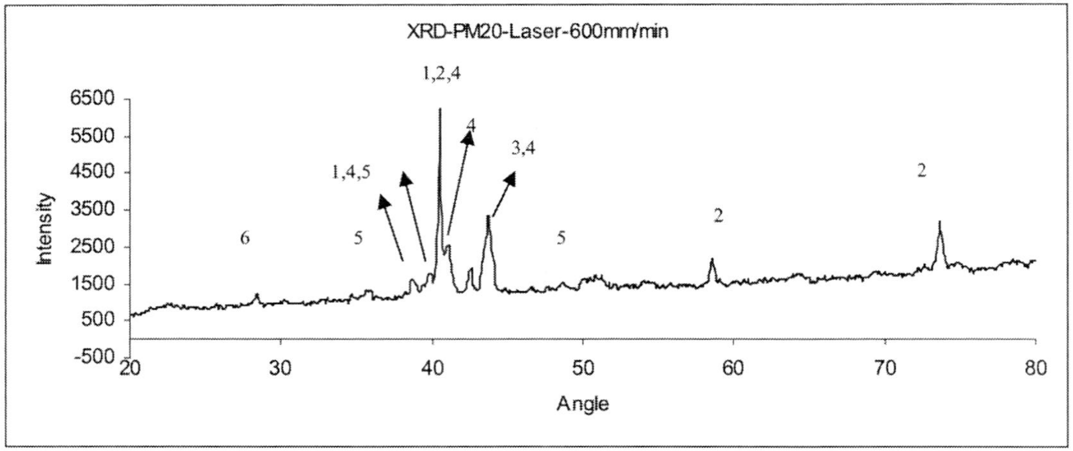

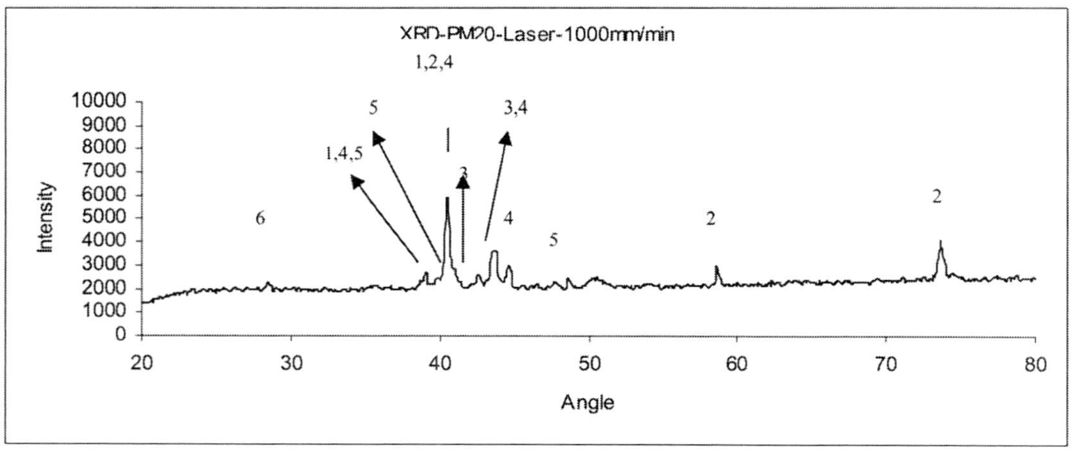

1-Cr_3C_2
2-Mo
3-Ni
4-$Cr_{23}C_6$
5- $\gamma Mo_2C/Mo_2C$
6-NiCr

XRD of Laser Cladded Cr_3C_2-NiCr coating

Contd-----

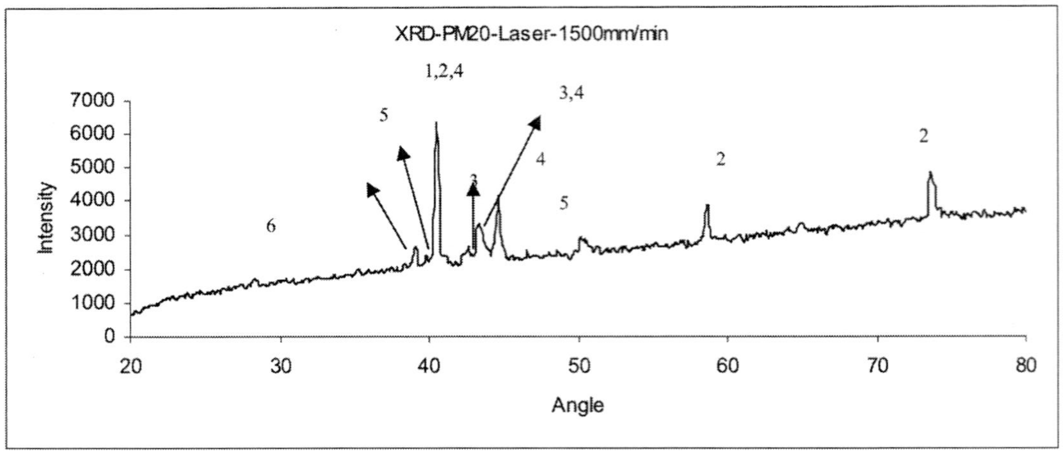

1-Cr_3C_2
2-Mo
3-Ni
4-$Cr_{23}C_6$
5- $\gamma Mo_2C/Mo_2C$
6-NiCr

SEM of Laser Cladded Cr_3C_2-NiCr samples

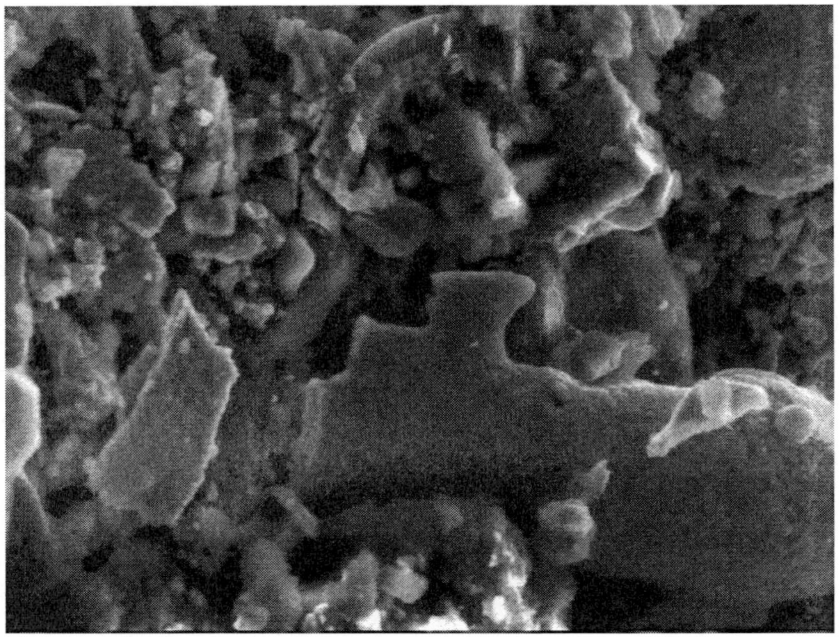

SEM of Laser Cladded PM-20 with translation speed of 600mm/min

SEM of Laser Cladded PM-20 with translation speed of 1000mm/min

SEM of Laser Cladded Cr_3C_2-NiCr samples contd----

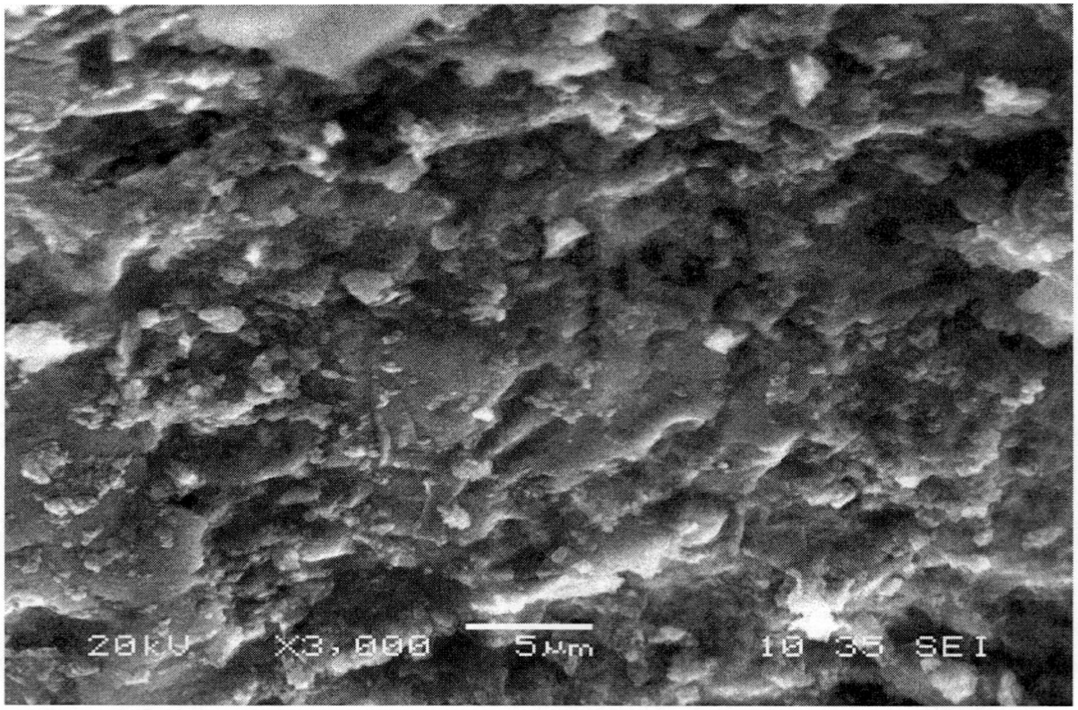

SEM of Laser Cladded PM-20 with translation speed of 1500mm/min

Dilution of Iron into the coating matrix

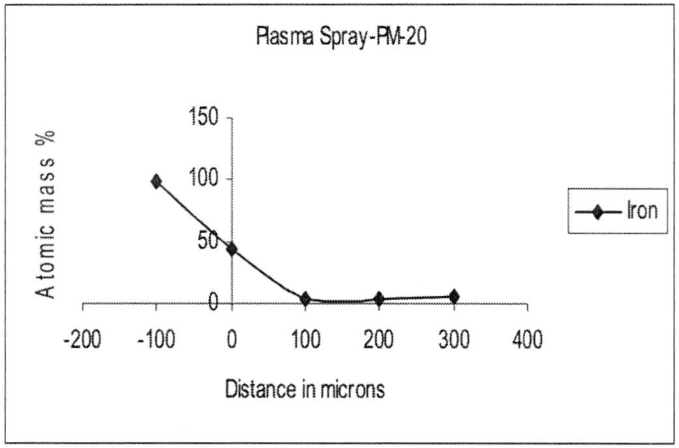

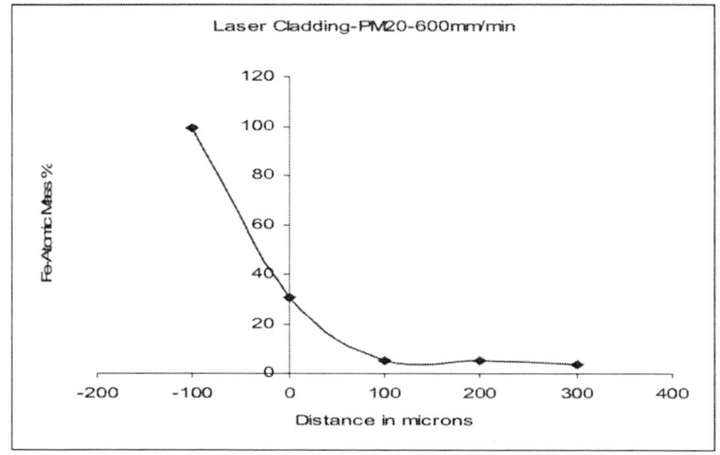

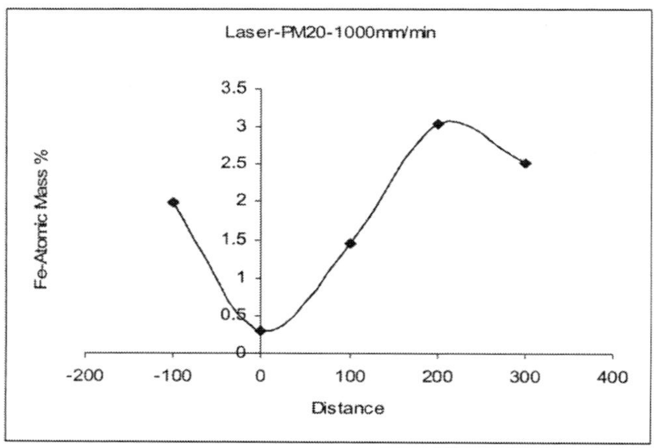

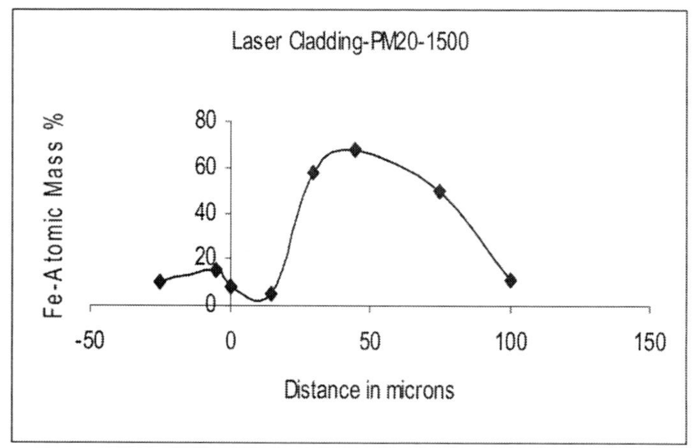

Cladding Height

The clad width to weight ratio observed to be following the same pattern as in the case of NiCr coatings i.e. the ratio increased with translation speed.

Microhardness of Laser Cladded PM-20 samples.

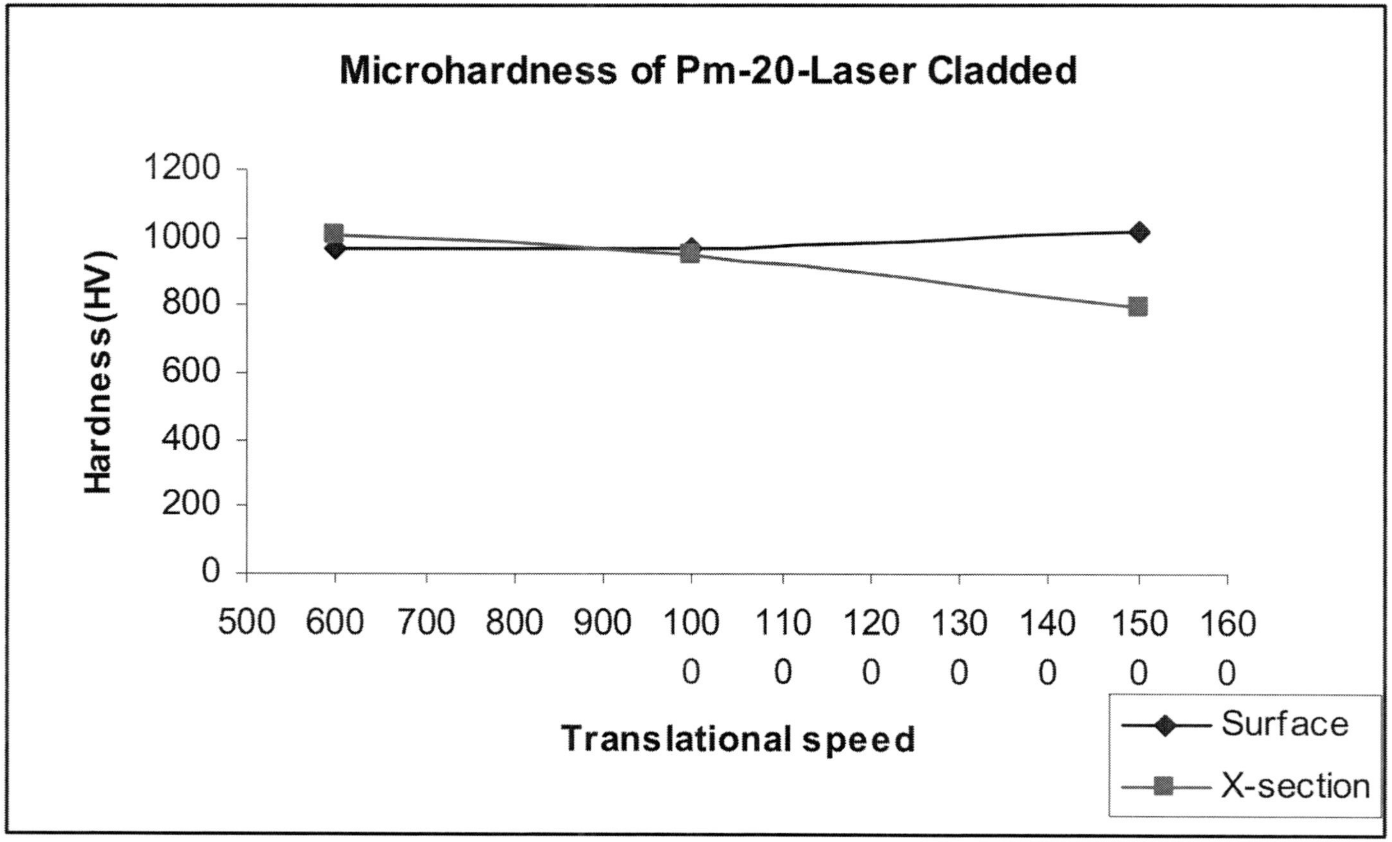

Characterization of WC-Co powder

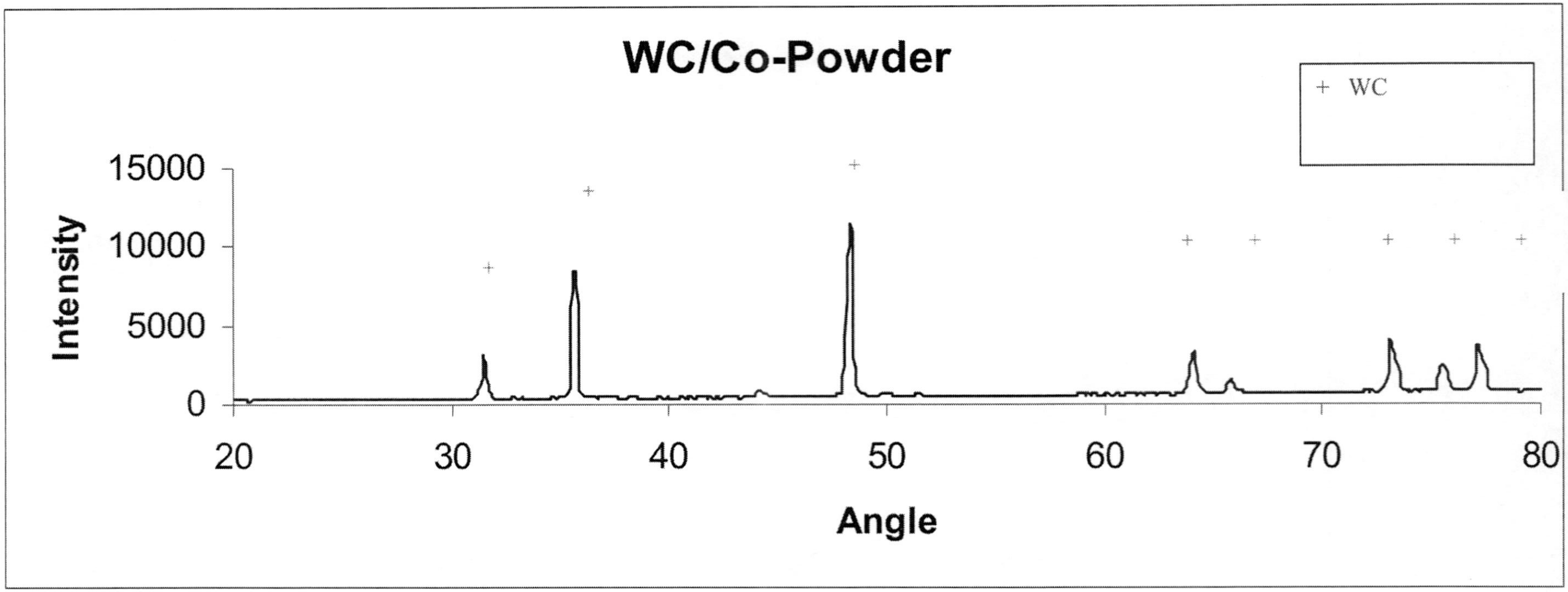

Microstructure of Plasma Sprayed WC/Co

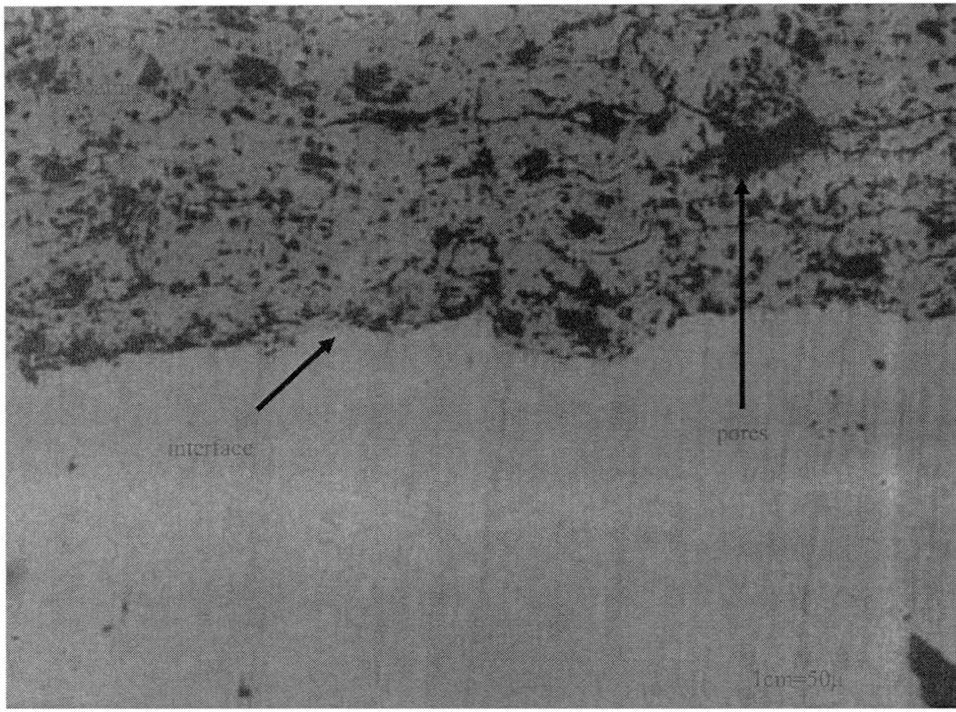

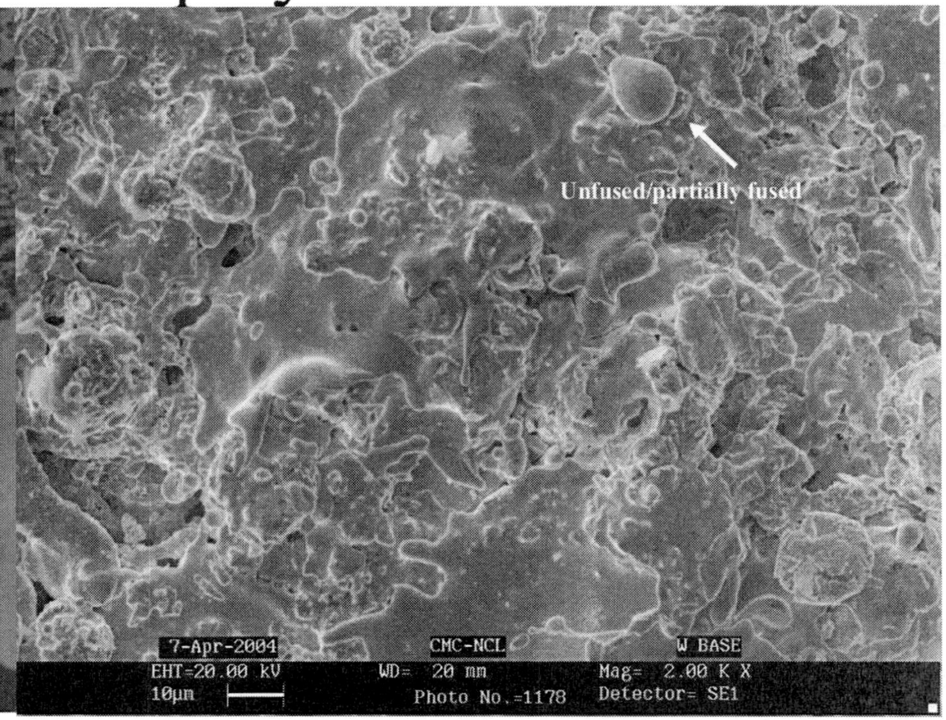

Optical (X-section)

SEM(surface)

- Medium to high Porosity
- Good substrate bonding and anchoring
- Lamination Faults :Horizontal Cracks
- Trace of incomplete fusion of powders
- Thickness of coating –490 microns

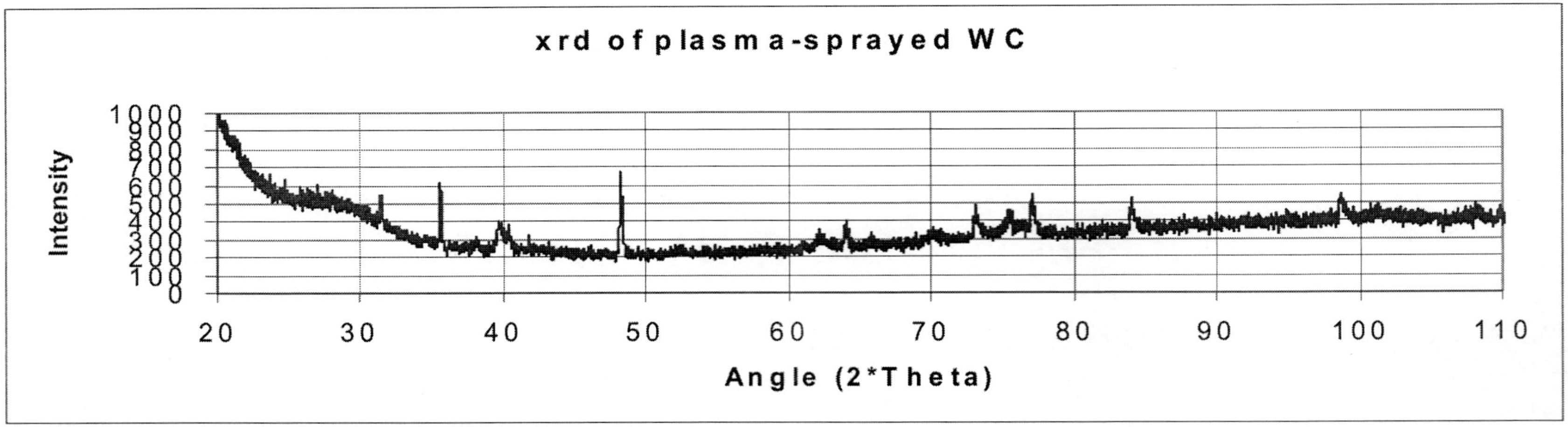

1. The XRD Peaks consists of mainly **WC, W_2C, $CoWC_4$**

2. EDAX results confirm the elemental Presence of **W, C, Co**

Microstructure of Laser Cladded WC/Co

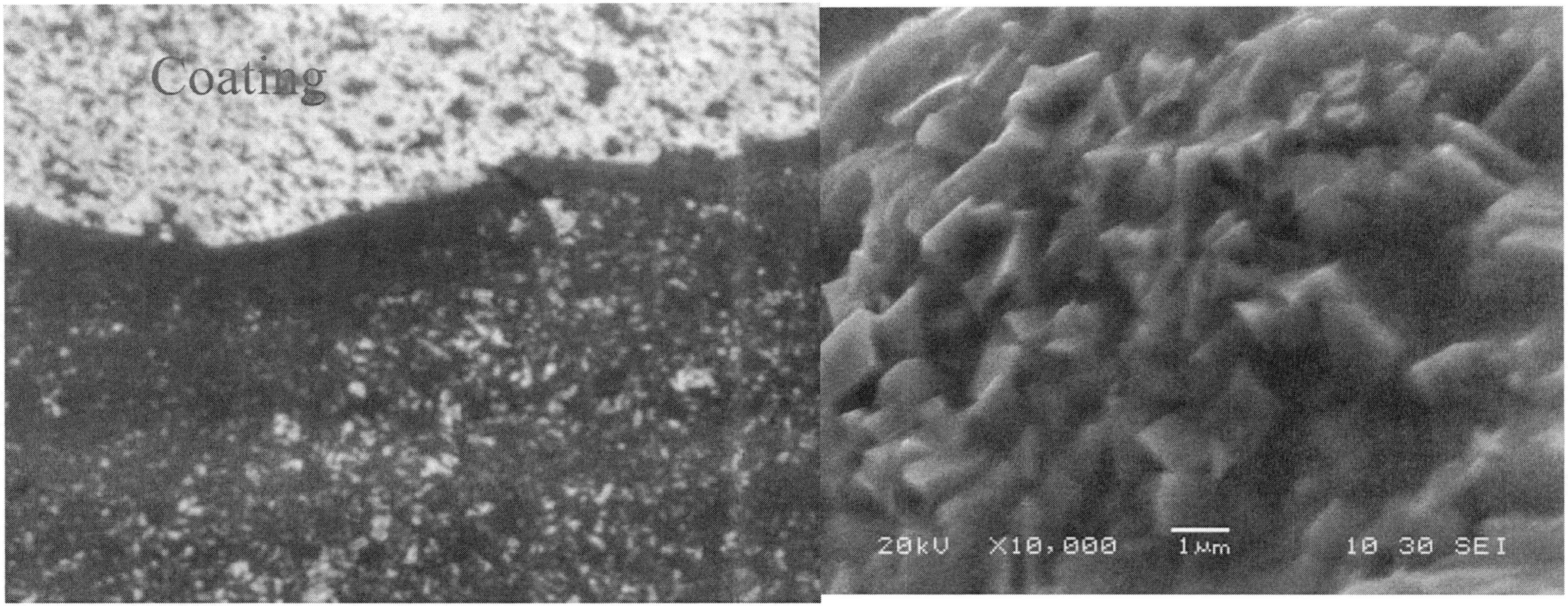

Torch Speed –600mm/min

- Dense structure
- Good Fusion at the interface
- Thickness of Coating –600 microns

Microstructure of Laser Cladded WC/CO

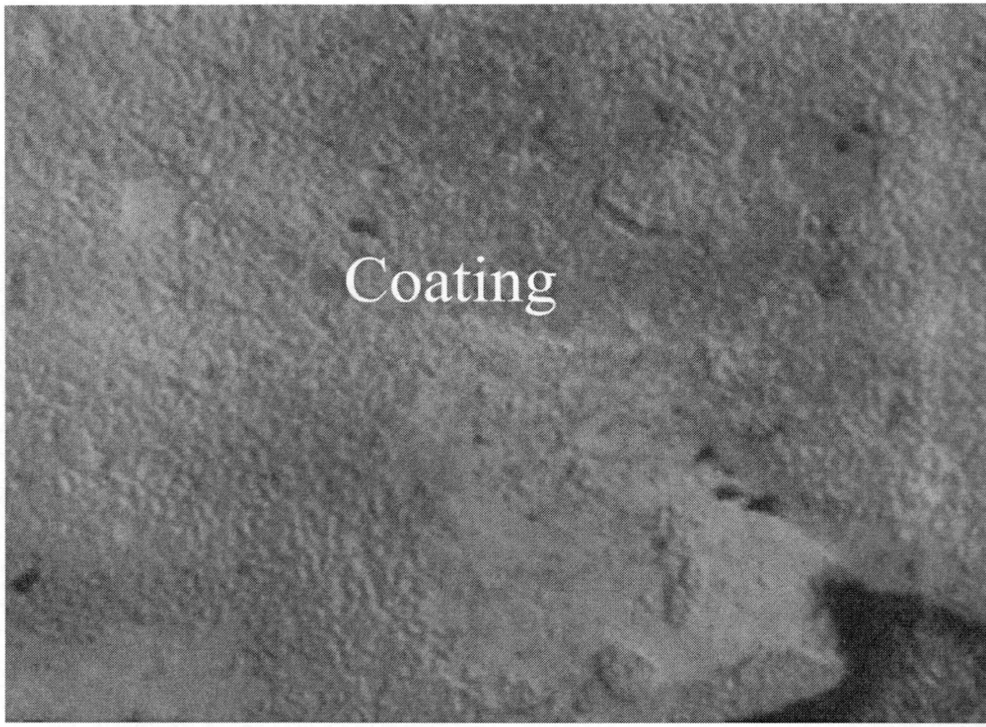

Torch Speed –1000mm/min

- Dense structure
- Good Fusion at the interface
- Thickness of Coating –480 microns

Microstructure of Laser Cladded WC/Co

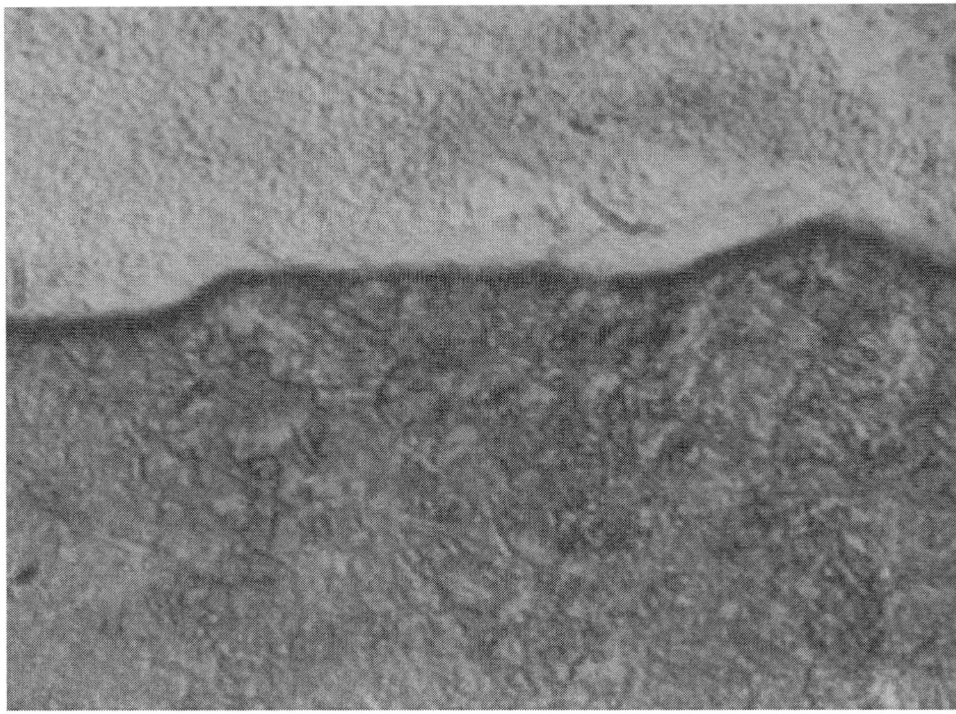

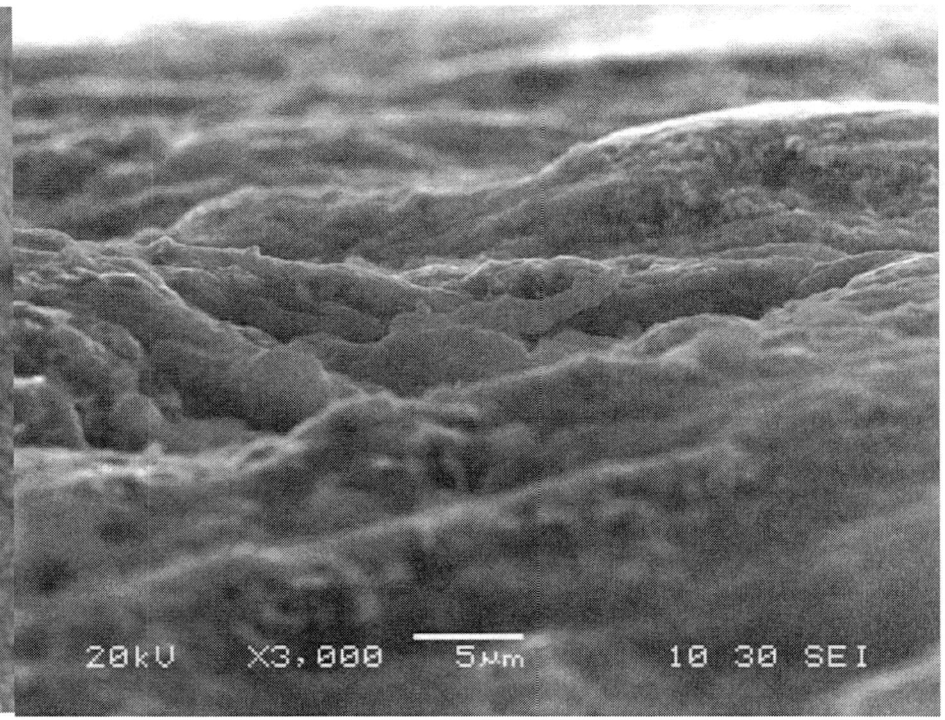

Torch Speed –1500mm/min

- Dense structure
- Good Fusion at the interface
- Thickness of Coating –180 microns

XRD Pattern of Coated -WC/CO samples

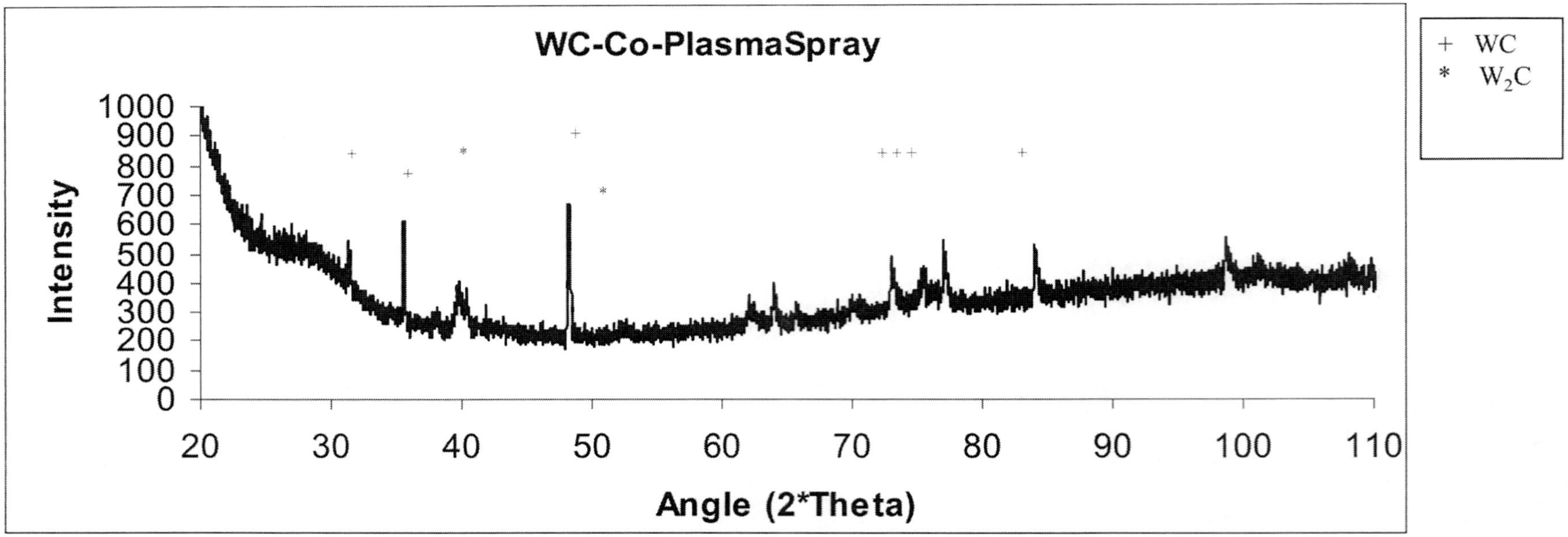

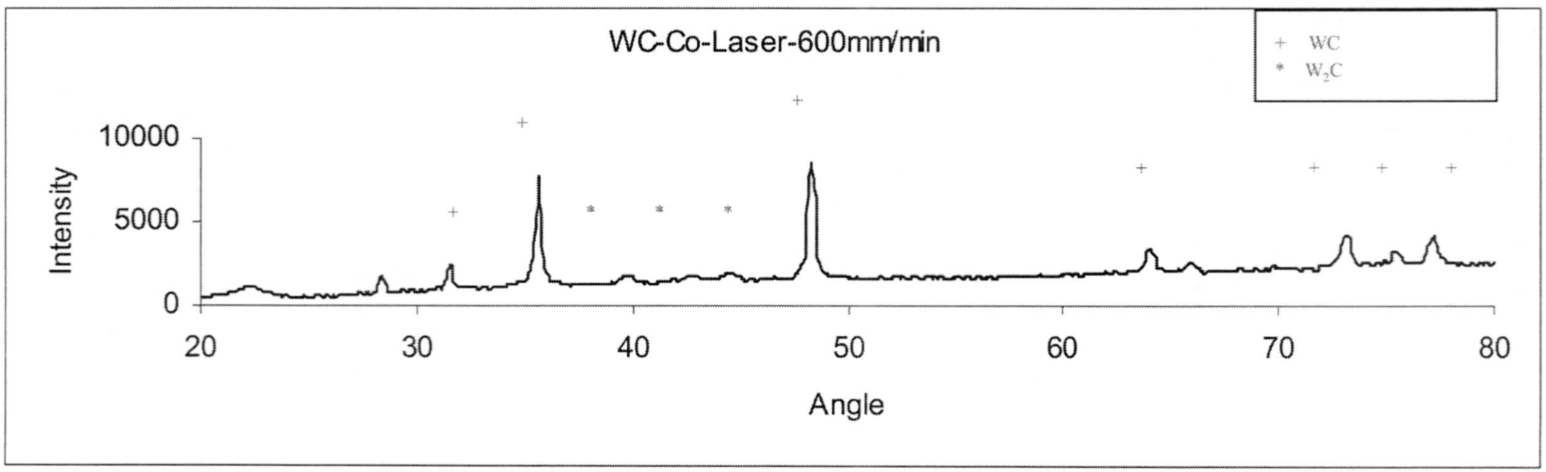

Contd…..

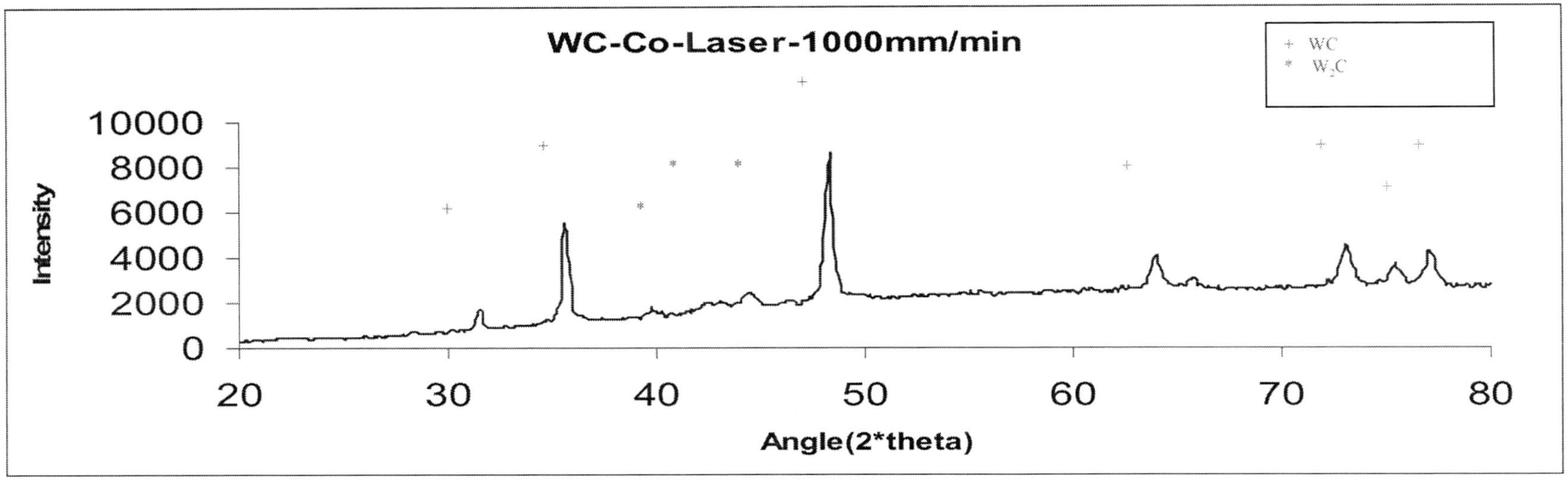
WC-Co-Laser-1000mm/min
WC
W_2C
Intensity
10000
8000
6000
4000
2000
0
20
30
40
50
60
70
80
Angle(2*theta)

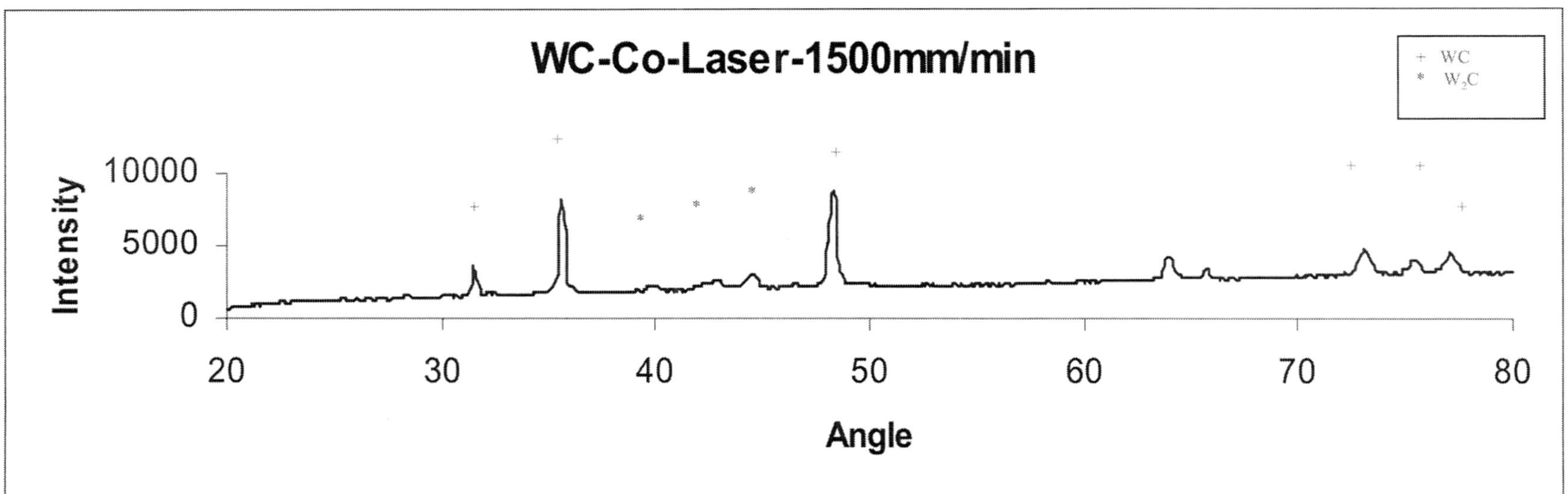
WC-Co-Laser-1500mm/min
WC
W_2C
Intensity
10000
5000
0
20
30
40
50
60
70
80
Angle

Dilution

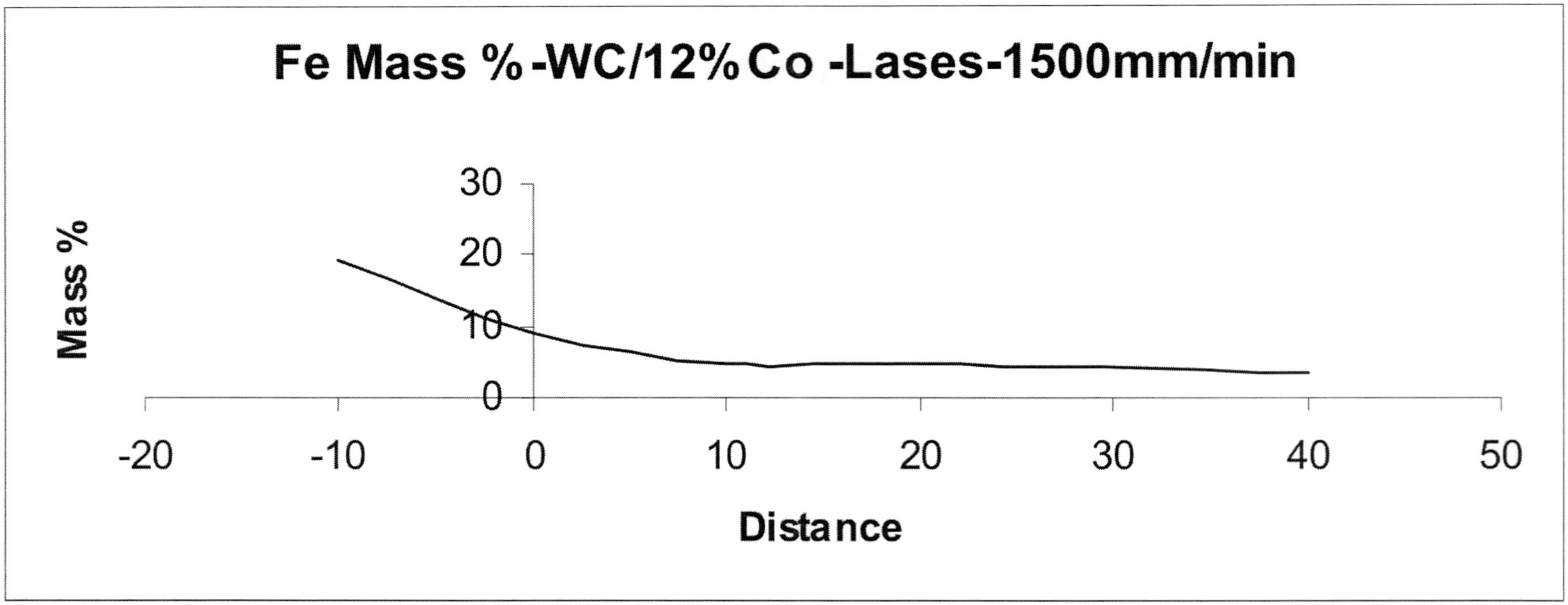

Plot for showing iron dilution. Iron entering the coating showing dilution. As we move away from the interface % of Fe reduces

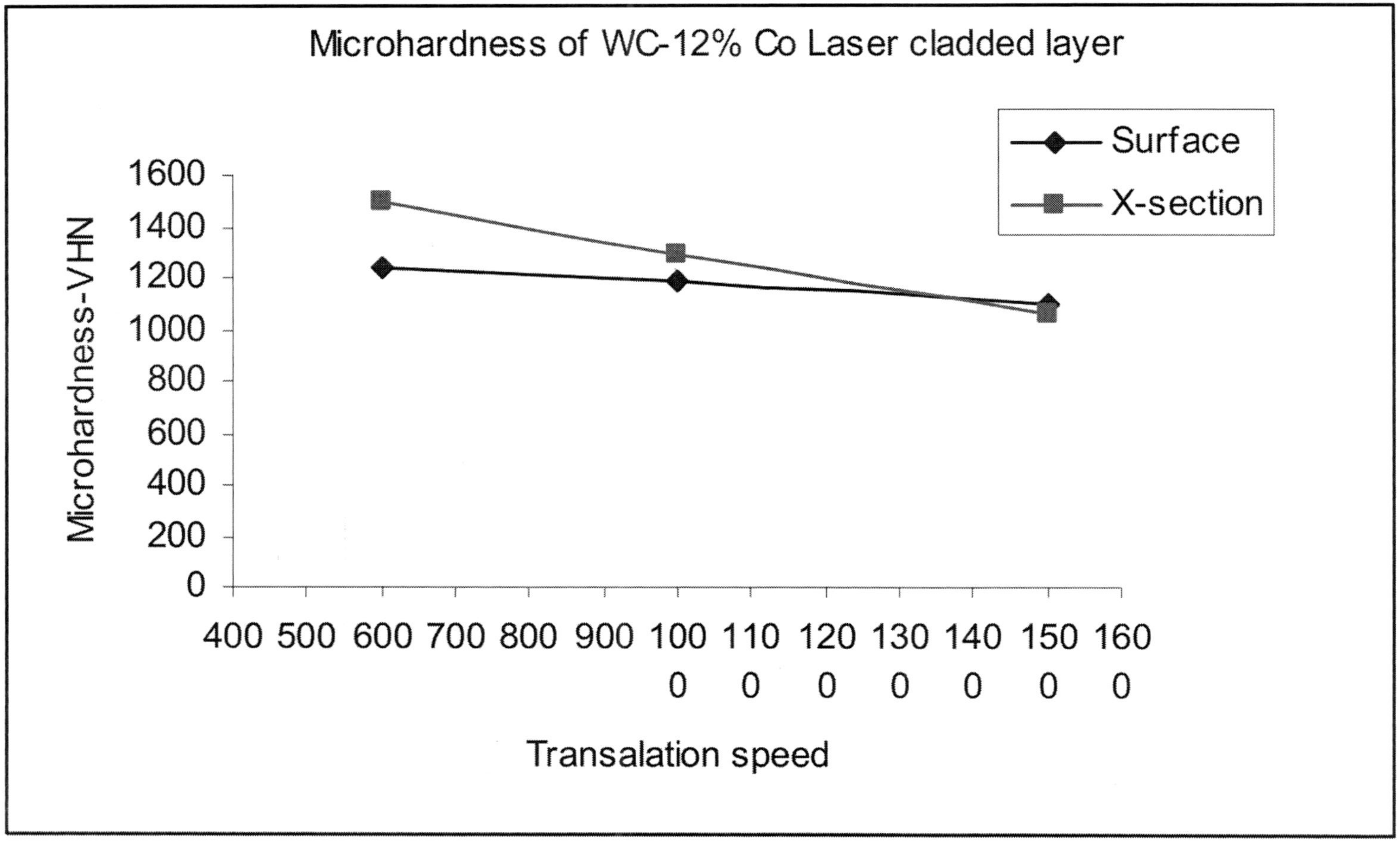

Plasma sprayed samples registered a microhardness of 540 VHN

Summary – Characterization of three coated powders

Powder	Coating Process	XRD	Hard-ness	Comments /Inferences	General Remarks
Nickel(Base), ***Carbon(0.80%)***, ***Iron(4.60%)***, ***Chromium(16%)*** ***,Boron(3.40%)***, ***Oxygen(traces)--- 20 μ (average)***	Plasma Spray Process	Cr_2B	580VHN	Hard phase peak less than original powder. Contributes to loss of hardness	Compact coating with good as sprayed surface finish and no oxidation
	Laser Cladding-translation speed of 600mm/min	Cr_2B Cr_2O_3 Cr_3C_2	850 VHN	Hard phase peaks less due to evaporation of boron oxide, as oxidation traced.	General appearance good, Interface well defined indicating metallurgical bonding.
	Laser Cladding-translation speed of 1000mm/min	Cr_2B Cr_2O_3 Cr_3C_2	830 VHN	Hard phase peaks prominent, Oxidation slight.	General appearance good, Interface well defined indicating metallurgical bonding.
	Laser Cladding-translation speed of 1500mm/min	Cr_2B, Cr_2O_3 Cr_3C_2	800VHN	Hard phase peaks prominent. Hardness fall due dilution with increasing translation speed	Grain refinement taking place leading very fine boride particles. Interface well defined indicating metallurgical bonding.

Contd…

Chromium Carbide Coatings

Powder	Coating Process	XRD	Hard-ness	Comments /Inferences	General Remarks
50%Cr_3C_2/32%Mo/18%NiCr 40μ-100 μ	Plasma Spray Process	Cr_3C_2 $Cr_{23}C_6$ Mo_2C	422VHN	Hard phase peak less than original powder. Formation of $Cr_{23}C_6$ indicates decarburization. Formation and entrapment of CO/CO_2. This contributes to loss of hardness	Lack of Fusion with abundant porosity. Good bonding at substrate
	Laser Cladding-translation speed of 600mm/min	Cr_3C_2 $Cr_{23}C_6$ Mo_2C	966VHN	Hard phase remains very close to the original powder. Partial decarburization evident which leads to formation of Mo_2C, contributing to hardness.	General appearance good, Interface well defined indicating metallurgical bonding.
	Laser Cladding-translation speed of 1000mm/min	Cr_3C_2 $Cr_{23}C_6$ Mo_2C	966VHN	Hard phase remains very close to the original powder. Partial decarburization evident which leads to formation of Mo_2C, contributing to hardness.	General appearance good, Interface well defined indicating metallurgical bonding.
	Laser Cladding-translation speed of 1500mm/min	Cr_3C_2 $Cr_{23}C_6$ Mo_2C	800VHN	Hard phase remains very close to the original powder. Partial decarburization evident which leads to formation of Mo_2C, contributing to hardness. Drop in hardness due to more dilution, allowing Fe to enter the coating.	General appearance good, Interface well defined indicating metallurgical bonding.

Tungsten carbide Coatings

Powder	Coating Process	XRD	Hard-ness	Comments /Inferences	General Remarks
Tungsten (Base), Carbon (5%), Cobalt(12%) --- 20 μ (average)	Plasma Spray Process	WC W_2C	550VHN	Hard phase peak less than original powder. Decarburisation evident.	Medium to high porosity Lamination faults. Incomplete Fusion
	Laser Cladding-translation speed of 600mm/min	WC W_2C	1200 VHN	Hard Phase proportion remains constant	General appearance good, Interface well defined indicating metallurgical bonding.
	Laser Cladding-translation speed of 1000mm/min	WC W_2C	1100 VHN	Hard Phase proportion remains constant	General appearance good, Interface well defined indicating metallurgical bonding.
	Laser Cladding-translation speed of 1500mm/min	WC W_2C	1000VHN	Hard phase peaks prominent. Hardness fall due dilution with increasing translation speed	Grain refinement taking place leading very fine boride particles. Interface well defined indicating metallurgical bonding.

Performance Evaluation

Determination of Wear and Friction Of Plasma Coated Samples By Pin on Disc Method

Carried

out at

NMRL,Ambernath

Specifications

- Wear test carried out by Pin-on –Disc Method
- Disc Material –Bearing Steel 5210, with 780HV
- The sample size of all specimens kept at 10 mm dia.
- The total track length was kept approximately constant at 3000mts
- The time of contact thus varied, thereby changing the sliding velocity
- The load on all specimens was kept at 7kg.

Wear Sample Holder

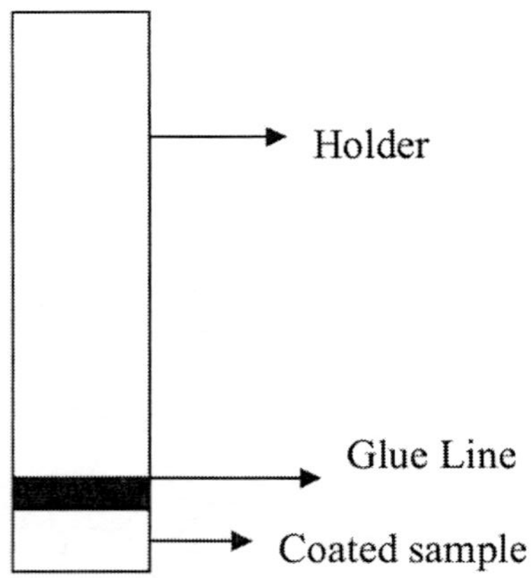

Old Design

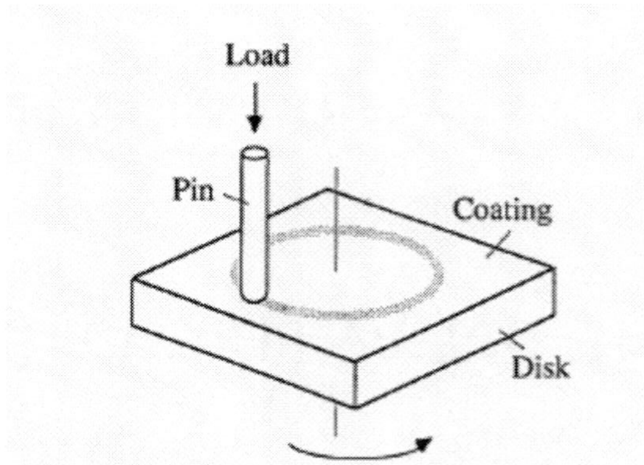

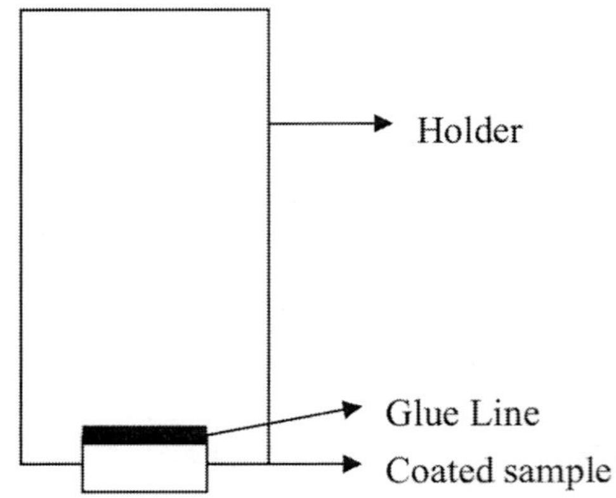

Improved Design

Due to shear at the holder and sample interface the sample came out and hence the improved design.

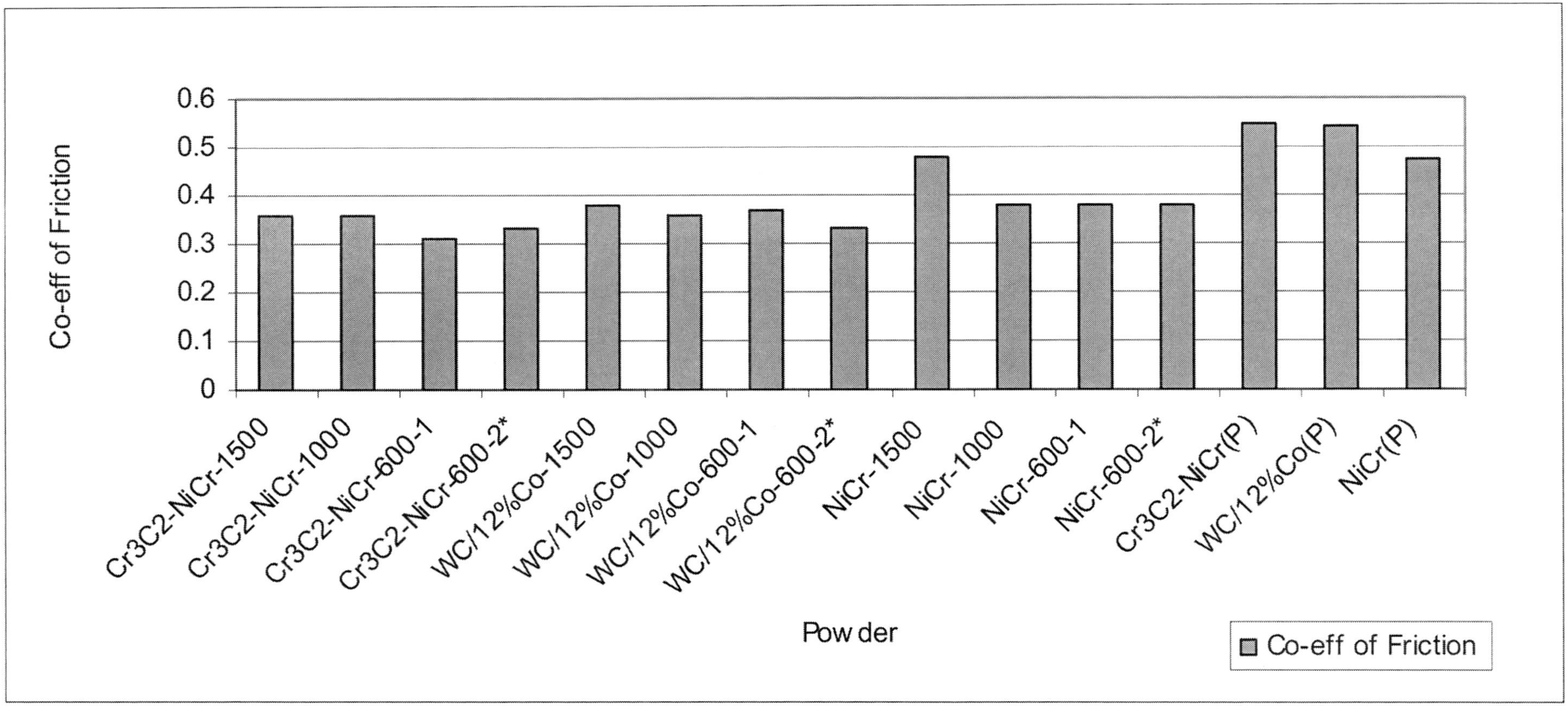

Histogram of all the coated samples showing the co-efficient of friction for a constant sliding velocity and fixed load of 69N, with the exception of the some samples where the load was increased to 147N, marked in asterix.

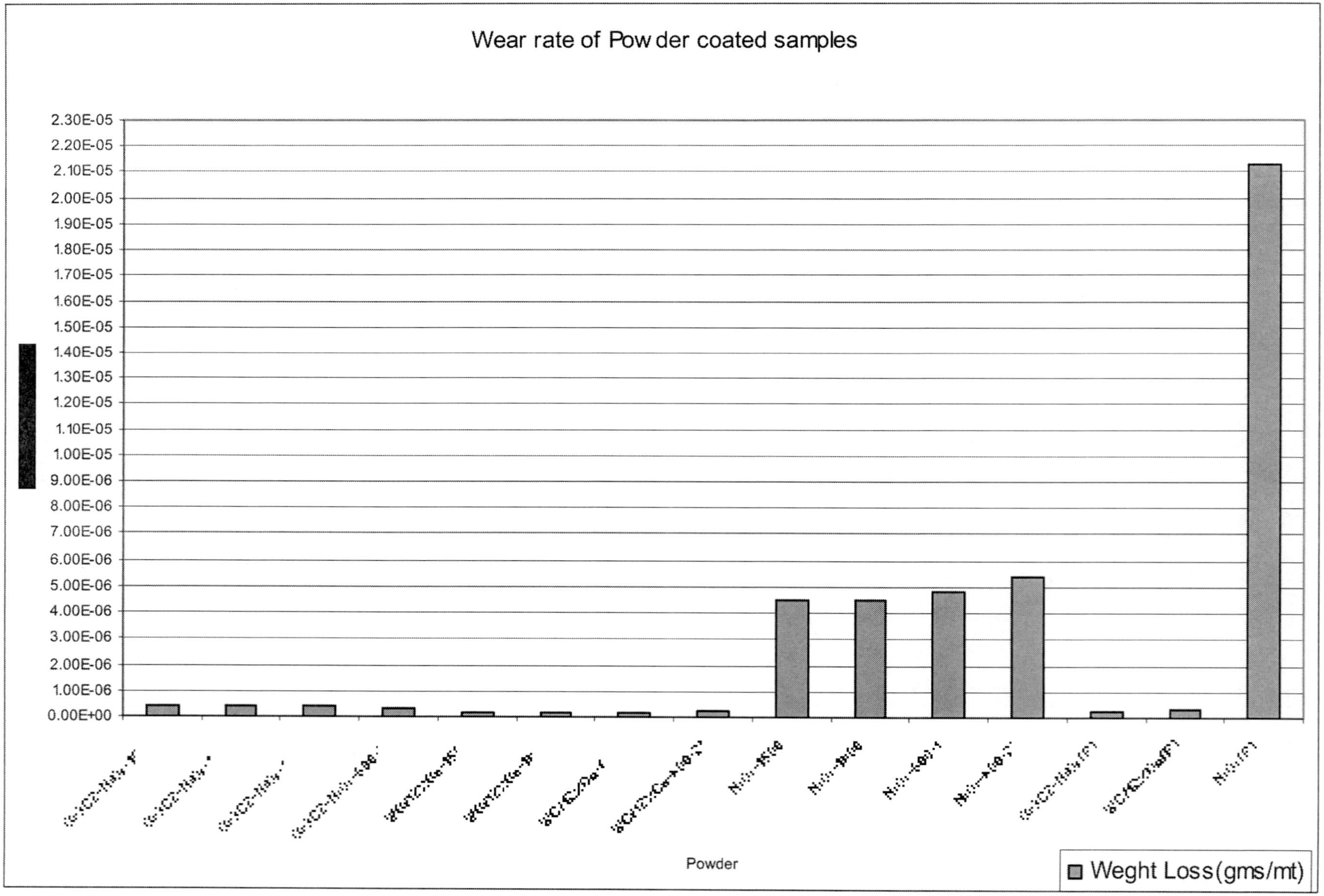

Histogram of all the coated samples showing the weight loss for a constant sliding velocity and fixed load of 69N, with the exception of the some samples where the load was increased to 147N, marked in asterix

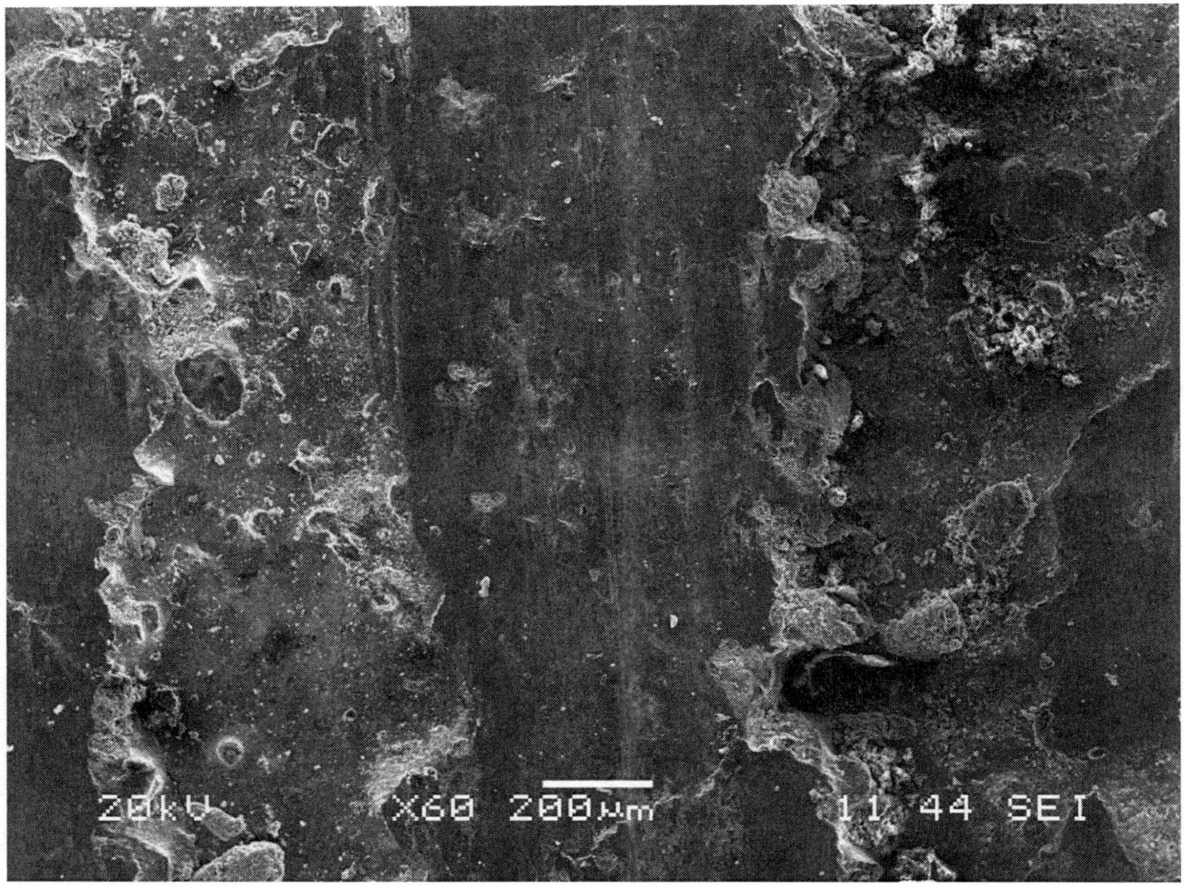

SEM of Wear Track of PM-20 sample

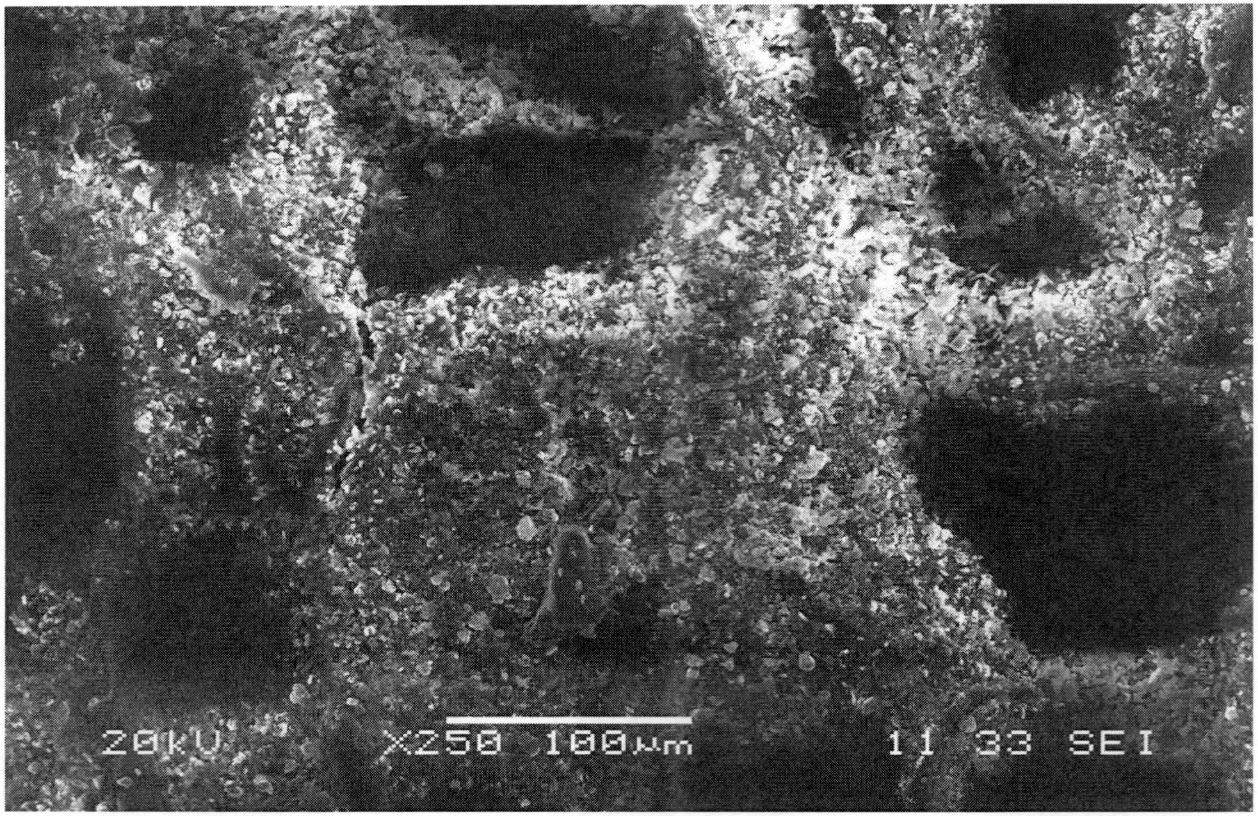

SEM of Wear Track of WC-12%Co sample

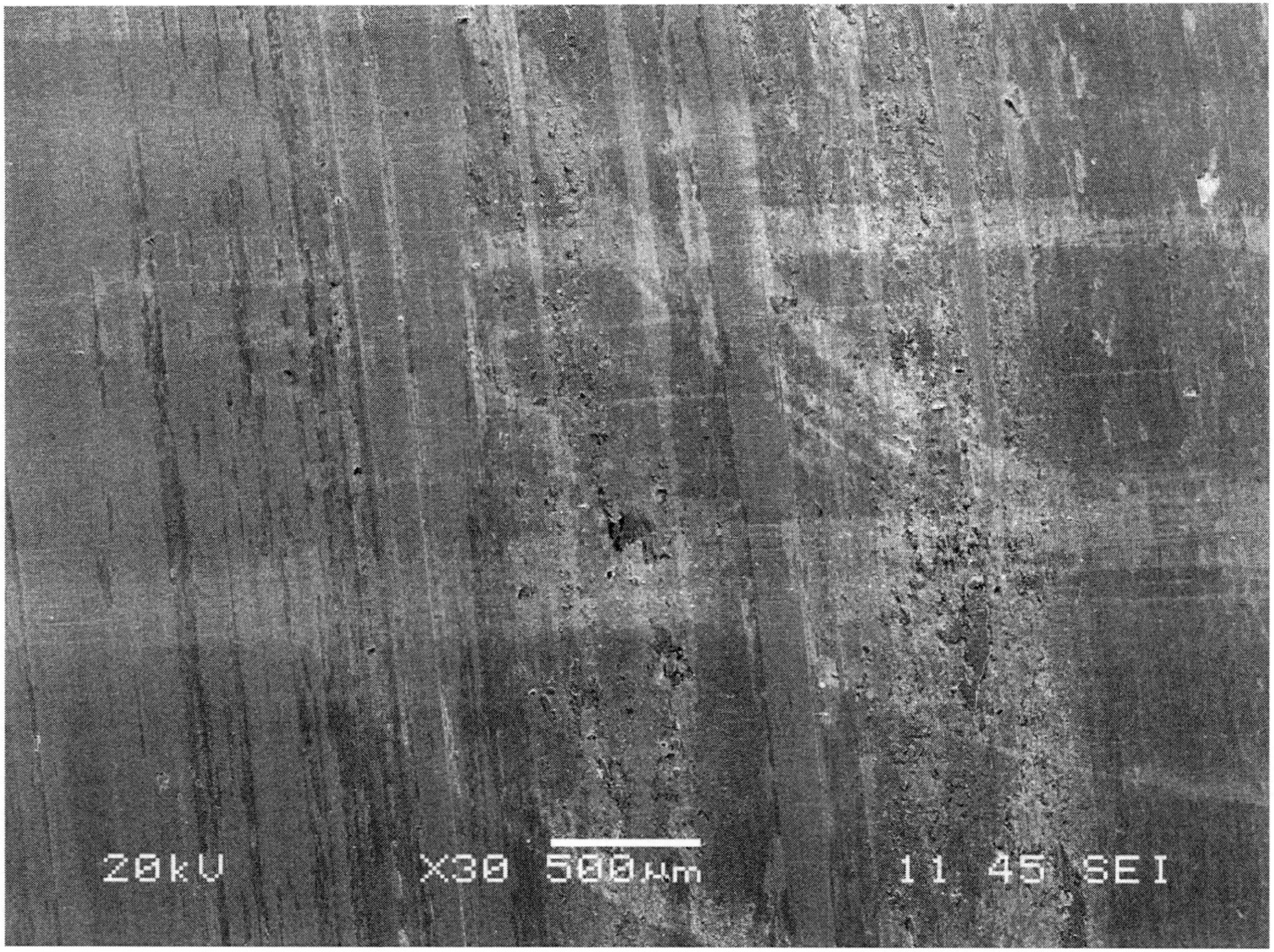

SEM of Wear Track of NiCr sample

Conclusion

- The objective of the research work was to decide upon a suitable cermet material having good wear resistance and hardness properties to replace the Chrome Coatings and a suitable process to achieve the coating.
- Laser Cladding results have been promising especially wrt hardness. Microstructure reveals coatings with reduced porosity.All the coatings exhibited a well distributed hard phase in their respective matrix.
- Laser Cladding ,translation speed of 1000mm/min has yielded the best results wrt oxidation ,thickness ,hardness and dilution and finally structural integrity .
- WC/Co and PM-20 samples have improved results as compared to NiCr samples.

Conclusion

- NiCr coatings lacked good wear resistance due to yielding of the matrix as is well known for Nickel.

THANK YOU

OPTIMUM CONCEPTUAL DESIGN OF AN ULTRA-LOW MOTION SEMI-SUBMERSIBLE FLOATING OIL AND GAS PRODUCTION SYSTEM

R Sharma, Advanced Institutes of Convergence Technology, Republic of Korea
Tae-wan Kim, Seoul National University, Republic of Korea
O P Sha, Indian Institute of Technology, Kharagpur, India

SUMMARY

Economic and efficient energy resources are keys to a nation's development. Because of their low cost and advancement in drilling and exploration technologies oil and gas based energy systems are most widely used in practice throughout the world. The inexpensive oil and gas based energy systems are used for everything from transportation of goods and people to the harvesting of crops for food to being able see at night. As the energy demand continues to rise, this is pushing further the demands for inexpensive energy solutions. Since, available reservoirs of energy rich fossil fuels at shallow and moderately deep waters have been exploited already, the fossil fuel exploration and production is being forced to move into extremely deep waters. At deep water the expenses associated with fixed production platforms are no longer feasible, and that makes a floating production platform design a far more economical choice. This paper presents a critical parameter driven optimum conceptual design of an ultra-low motion semi-submersible floating oil and gas production system. The design is carried to satisfy the given parameter of weights and dimensions of the platform topsides required for production and drilling, water depth location, and low motion requirements. The proposed design process is highly iterative process of altering the key dimensions of a 'n-column ring' pontoon, and it meets applicable regulations and requests of the order placing agency while minimizing costs and satisfying the chosen parameters. The chosen parameters are classified into groups depending upon the scientific and technological requirements, and they are: GAaOHD - general arrangement and overall hull design, WaCL - wind and current loading, LaGL - local and global loading, WBS - weight/buoyancy/stability, HMaL – hydrodynamic motion and loading, SaSD - strength and structural design, M – mooring, and EaC – economics and cost. Overall, this paper introduces a novel design process for an ultra-low motion semi-submersible that covers the complete life cycle of the structure. Finally, we discuss one design example motivated by real world applications to show the effectiveness, usability and efficiency of our proposed model.

1. INTRODUCTION

An offshore platform is a large structure (floating or fixed) which is used to house workers and machinery needed to drill wells in the ocean bed, extract oil and/or natural gas, process the produced fluids, and ship or pipe them to shore. A semi-submersible is a type of offshore platform and technically it is considered as a special type of 'watercraft' that has much of its hull form (70 – 85 %) underwater. Since, for a semi-submersible the waterplane is low, the semi-submersible is less affected by the waves than a normal ship. However, it needs proper and continuous ballasting to achieve desired trim and stability. Though, semi-submersibles show far less motions in waves than mono-hull vessels making them suitable for tasks that demand very strict motion requirements, but their actual motion control is very difficult and complex. In modern engineering environment semi-submersibles are used as crane vessels, drilling vessels, production platforms and accommodation facilities. A semi-submersible offers better motion characteristics; equal resistance to wave, wind, and current from any direction; the ability to support a mooring system; and a large deck area.

Modern engineering structures are becoming more complex with higher demands of economy and efficiency and these demands are forcing designers to design a structure that satisfies the requirements of the order placing agency within the affordable cost. Furthermore, with the ever increasing requirements for working in deeper waters the recent semi-submersibles are often fitted with a 'Dynamic Position System (DPS)' to control the craft more efficiently. A technical analysis of semi-submersible is shown in Table 1, more details about critical analysis of semi-submersibles can be found in our other works, e.g. [3-5]. For exploration purposes, a semi-submersible is the design of choice for development drilling. The design of semi-submersible is a process that employs model tests and numerical simulations in the field of motions, mooring loads, wave drift forces and Dynamic Positioning (DP). The critical issues are the accurate estimation of the impact of 'green water', 'slamming' and 'Vortex Induced Motions (VIM)' on semi-submersibles. Semi-submersibles are designed for extreme wave conditions and the motions characteristics in extreme wave conditions become important. Also, wave impact loads on semi-submersible pontoons during transport in extreme weather conditions on the deck of a barge or heavy-load vessel are important parameters in design. The trials and monitoring are done with scaled and full-scale measurements on semi-submersibles. In offshore platforms it is preferred to have a mooring system for field development at any water depth. Apart from the issue of mooring system, there are other design considerations that affect the applicability of semi-submersible for deepwaters.

Table 1: A technical analysis of semi-submersible.

Type	Deck load	Deck area	Maneuverability	Stability (under motion)	Storage Volume	Production unit installation	Building cost	Operating cost
Semi-submersible	Low	High	Low	High* (*Though, actual steering of it in bad weather is very difficult)	Low	Not possible	Low	High
Our analysis: Since for them initial investment is low, for moderately deeper water (water depth < 2250 m) semi-submersible is and will remain an attractive option. However, for ultra deep water (water depth > 3500 m) large size drillships will become the more favorable option in future.								

Though a fully automatic optimum design is possible at least theoretically but in practice a fully automatic process is rarely used mainly because a fully automatic process computes a unique and rigid result and flexible choices in design are rarely achieved. In practice it is desired that the optimization process computes flexible choices of design and then these sets of choices are discussed. Since, there are many parameters that affect the final choice of selection the flexible choices of design serve the final choice of selection efficiently. Since, fossil fuel and ore resources are depleting at shallow water depths because of the extensive utilization in the last century, economically there is a growing need for deepwater solutions. The exploration and production requirement in deep sea demands that the motion characteristics of the structure are further lower as compared to the inherently low motion characteristics of the semi-submersible. In this work we explore a novel design concept for ultra-low motion semi-submersible for operations in deepwaters. In our design, highly constrained ultra low motion can be achieved with novel use of the concepts from 'seismic engineering', column sizes and spacing, floaters spacing and arrangement, and application of dynamic position systems (DPS).

This paper presents a critical parameter driven optimum conceptual design of an ultra-low motion semi-submersible floating oil and gas production system. The critical parameters are chosen based upon their scientific and technical merits. The chosen critical parameters are: Technical design – critical parameter 1 (hydrodynamic stability); Technical design – critical parameter 2 (hydrodynamic intact and stability for societal safety regulations); Technical design – critical parameter 3 (structural analysis for societal safety regulations); Technical design – critical parameter 4 (mooring and motion analysis for constrained design regulations); Technical design – critical parameter 5 (environmental load analysis, global mooring analysis, regulatory compliance analysis, and design objective test analysis); Techincal design – critical parameter 6 (global dimensional test); Technical design – critical parameter 7; (cost analysis); and Technical design – critical parameter 8 (global optimization test). The critical parameters are utilized in sequence and iteratively the design is refined. The design is carried to satisfy the given parameter of weights and dimensions of the platform topsides required for production and drilling, water depth location, and low motion requirements. The proposed design process is highly iterative process of altering the key dimensions of a 'n-column ring' pontoon, and it meets applicable regulations and requests of the order placing agency while minimizing costs and satisfying the chosen parameters. The iterative process is employed at each critical parameter selection, and critical parameters are designed sequentially from 'Technical design – critical parameter 1' to 'Technical design – critical parameter 2', …, 'Technical design – critical parameter 8'. The chosen parameters are classified into groups depending upon the scientific and technological requirements, and they are: GAaOHD - general arrangement and overall hull design, WaCL - wind and current loading, LaGL - local and global loading, WBS - weight/buoyancy/stability, HMaL – hydrodynamic motion and loading, SaSD - strength and structural design, M – mooring, and EaC – economics and cost. Overall, this paper introduces a novel design process for an ultra-low motion semi-submersible that covers the complete life cycle of the structure. Finally, we discuss several design examples motivated by real world applications to show the effectiveness, usability and efficiency of our proposed design model.

The remaining of this paper is organized as follows: Section 2 describes the basic ideas of the design of the ultra-low motion semisubmersible; Section 3 presents the optimization model and Section 4 concludes the paper with identifications of future areas of research and application. Some technical details about the terms used in the present paper and design examples are given in Appendix 1.

2. DESCRIPTION OF ULTRA-LOW MOTION DESIGN CONCEPT

Our idea of ultra-low motion design concept for semi-submersible is new, novel and has been derived from the fundamentals of 'Seismic Engineering', [1]. In our design we combine the earthquake resistance technology with basic flotation fundamentals to maximize their combined effects on floatation and minimize the motion characteristics. We use dampers, which absorb energy and reduce its effects on the semi-submersible. Our idea is shown conceptually in Fig. 1.

Additional, control on motion characteristics can be achieved with pendulum driven control of the centralized trusses of the semi-submersibles. The pendulum acts as mass damper and preferably it is used in extreme weather conditions. In normal and rough weather control system at the columns is sufficient. This is shown in Fig. 2.

We use 'Magneto-rheological Fluid (MR fluid)' ([6]), in the column for damping control to constrain motion characteristics. MR fluid is a liquid that changes to a near-solid when exposed to a magnetic force, and then back to liquid once the magnetic force is removed. In columns a 'tube-in-tube' design is used, and in the inside tubular column dampers filled with MR fluids are used to control or stabilize the semi-submersible during rough and extreme weather conditions. The control design concept for the column of the semi-submersible is shown in Fig. 3. During rough and extreme weather, MR fluid inside the dampers will change from solid to liquid and back as weather induced vibrations or movements activate a magnetic force inside the damper. Using this damper in semi-submersible not only the motion characteristics can be reduced but also a smart semi-submersible that automatically react to weather conditions can be designed. This will limit the amount of damage caused by extreme weather environments.

Technically, our MR fluid damper is a 'semi-active device' that change its damping level by varying the amount of current supplied to an internal electromagnet that controls the flow of MR fluid. Each damper sits on the floor of the inner tubular column and is attached to the braces that are welded into a steel cross beam. As the semi-submersible is influenced with motion or vibration, the dampers move back and forth to compensate for the motion/vibration of the force. When it's magnetized, the MR fluid increases the amount of force that the dampers can exert.

3. DESCRIPTION OF OPTIMZATION MODEL

Now, we describe the optimization approach. An optimization approach needs clearly defined parameters: free variables, technical design parameters, constraints and objective function. In computational implementation free variables are changed during the computation of the optimum solution while the parameters have fixed values defined at the input stage. The constrained parameters are the requirements that need to be fulfilled, and the objective function needs to be optimized. The objective is user selected and it can be linear, non-linear and stochastic. Because of many possibilities in the very definition of objective function it is desired that a powerful optimization technique is adopted. Some examples are linear programming technique (for linear objective function and constraints), gradient minimization technique (for non-linear objective function and linear constraints), sequential quadratic programming (SQP) technique (for non-linear objective function (for linear objective function and constraints)) and evolutionary programming technique (for stochastic objective functions and constraints). The schematic diagram of generic optimization approach is shown in Fig. 4. In Fig. 4, the main program reads the parameters values from the input set and runs the optimization process. The iterative optimization process computes the optimum values of free variables ($O(S_{fv})$) via iteration. The process is initialized with some starting values (initial guess S_{fv_0}), and then these initial values are entered in the computing zone where constraints and objective functions are checked and satisfied. The check and satisfaction of the constraints and objective functions lead to modification in the set values of S_{fv}.

The modified set values (S^*_{fv}) are used again and the process is continued till all the constraints and objective functions are satisfied, and the optimum $O(S_{fv})$ is computed. The definition of free variables depend upon the chosen primary structural configuration, e.g. 3 column semi-submersible or 4 column semi-submersible, or n -column semi-submersible where n = 3, 4, 6, 8, etc. Once a configuration is chosen then other free variable are chosen. In general nine free variables are enough, e.g. for upper deck: width and length; for columns: width, height and spacing; for floaters: width, height and length; and for column additions: size of heave plates (if designed with water-entrapment plates at the keel). The technical design parameters consist of two categories: owner related technical design parameters (e.g. SA – sea state: environmental conditions for operations and survival, VDL: variable drilling load, TVLiT: total variable load in transit, HM: heave motion, PM: pitch motion, RM: roll motion, SL: safety level, IS: intact stability, DS: damaged stability, ES: equipment specifications, CP: conversion possibilities, and DC: dismantling cost) and shipbuilding yard related technical design parameters (e.g. CSA: cross sectional size of the deck box, HDB: height of deck box, NB: number of bracings, SBC: size of bracings, CSHP: cross sectional size of heave plate). The air gap requirements are determined from simulated results of experimental tests that are conducted on similar scaled models. Again, the constraints are categorized: MSL - maximum stability level, MEC - maximum operating conditions, MTVLiTs - total variable load in transit, and MRFB - minimum required freeboard.

MSL - maximum stability level: A semi-submersible is classified by a professional agency, and within classification society rules and regulations the design aim is to achieve maximum stability levels. In general the semi-submersible has a significant buoyant upper hull structure and this feature improves IS. For damaged condition there is a restricted limit of the heeling angle and because of that the amount of water that can accumulate is limited. However, a small amount of water endanger more with free surface effects, and to achieve better damaged stability levels columns are strengthened with double hull, etc. In this configuration the vessel's restoring moment counteract the inflow of water. A linear or non-linear approximation relationship between restoring moment, displacement, GM value and angle can be used, e.g. linear approximations are fairly accurate up to 15 - 17.5 degrees of angle, and non-linear relationship is needed only for higher angles (17.5 - 30). In both linear and non-linear relationships GM value is important.

MHM – minimum heave motion, MPM – minimum pitch motion and MRM – minimum roll motion: The motion characteristics affect the workability conditions and therefore it is desired that they are minimum under maximum drilling and operating conditions. The minimum HM/PM/RM are determined with specifying allowable drill and operation compensator limit. The HM/PM/RM depend upon natural periods and submergence of floaters. There exist various formulations that link the values. In the optimization process a natural period is selected first and then second/third/fourth hump responses are computed.

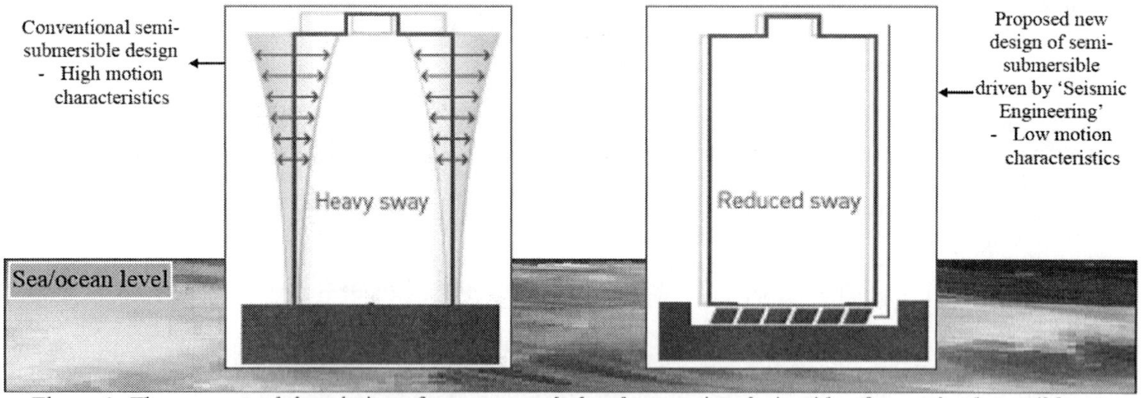

Figure 1: The conceptual description of our proposed ultra-low motion design idea for semi-submersibles.

Figure 2: The additional pendulum based control of centralized trusses of semi-submersible.

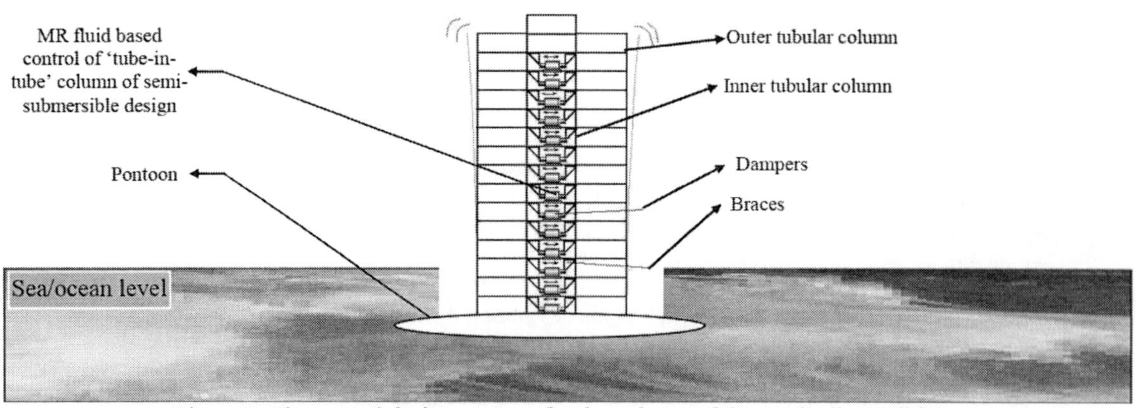

Figure 3: The control design concept for the column of the semi-submersible.

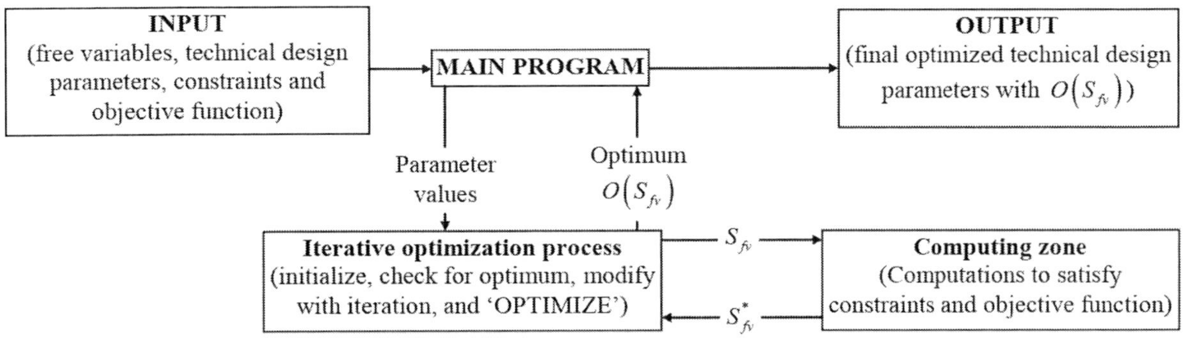

Figure 4: The schematic diagram of generic optimization approach.

An optimum value is computed with assuming second/third/fourth hump responses as constraints. A motion based analysis ensures that workable conditions are determined within a given set of environmental conditions.

MRFB - minimum required freeboard: Minimum freeboard is determined at transit draft. The floater size determines total variable load and freeboard in the transit condition. The freeboard is a flexible criterion and it is both a requirement and constraint. It is a requirement because freeboard needs to be positive and it acts as a margin for weight increase during construction. It is a constraint because the owner demands a minimum freeboard as higher freeboard means higher material consumption and cost. These conflicting requirements are settled using optimization techniques, and an optimally selected freeboard and total variable load determine efficient size of floaters and displacement of the vessel.

Production related constraints (PRS): Flat/single curved panels have a low cost of production. Hence, from cost point of view it is desired that the semi-submersible be designed with flat/single curved panels. This also adds some constraints, and they are:

- Length of floater ≥ upper hull length.
- Width of floater ≥ column width.
- Total width of vessel ≤ width of building dock/repair dock of the shipbuilding yard.

Objective function: The central idea is to optimize the overall performance of the semi-submersible. In general the performance definition is owner driven meaning that total investment cost (to cover total design cost, construction cost, installation cost, operation and maintenance cost and dismantling cost) is minimized. The design cost is around 4 – 7.5 % (i.e. low for designs already in practice and high for new and innovative designs) of the construction cost. The construction cost consists of various sub-costs like drilling cost (material cost = 27.5 – 32.5 %, labor cost = 2 – 3 %), structure cost (material cost = 4 – 6 %, labor cost = 7 – 8 %), machinery cost (material cost = 13 – 17 %, labor cost = 2 – 3 %), hull outfitting cost (material cost = 8 – 11 %, labor cost = 1.5 – 3.5 %), electrical machinery and fitting cost (material cost = 5.5 – 7 %, labor cost = 2 – 4.5 %), painting cost (material cost = 0.5 – 2 %, labor cost = 2 – 3 %), others (labor cost = 5 – 7 %), and general cost for assembling etc. (labor cost = 5 – 8 %). The installation cost is around 3 – 4 % of construction cost. The operation and maintenance cost is dynamic and it increases with time. However, for implementation purpose a linear variation over the life cycle can be assumed, and at the design stage the operation and maintenance cost is assumed between 10 – 20 % of construction cost. Of late environmental regulations are getting tighter every year. This has affected the dismantling cost of semi-submersibles. At present, assuming a life cycle of 30 years, the dismantling cost of a semi-submersible is around 8 – 12.5 % of construction cost.

Implementation of optimization approach: As mentioned previously, our optimization model is driven by critical parameters. The detailed schematic diagram of our proposed critical parameter driven iterative optimum conceptual design process model is shown in Fig. 5.

In our work the objective is to minimize the CX, also referred to as objective function herein. The objective function is defined as:

$$CX = f(\lambda) \tag{1}$$

where λ is collective set of all the design variables. For more robust implementation, we neglect the fixed costs, e.g. engineering cost. The main components of the cost function are: hull and deck fabrication cost; mooring manufacturing cost; riser fabrication cost and installation cost. The hull and deck related costs are computed as a sum over its individual parts of the estimated cost density c_i of the part times its mass (M_i). Obviously, mass is a function of λ. The installation costs are treated as a function of the size of the semi-submersible. To avoid complexity, again, we do not consider the installation costs of the mooring, riser and semi-submersible.

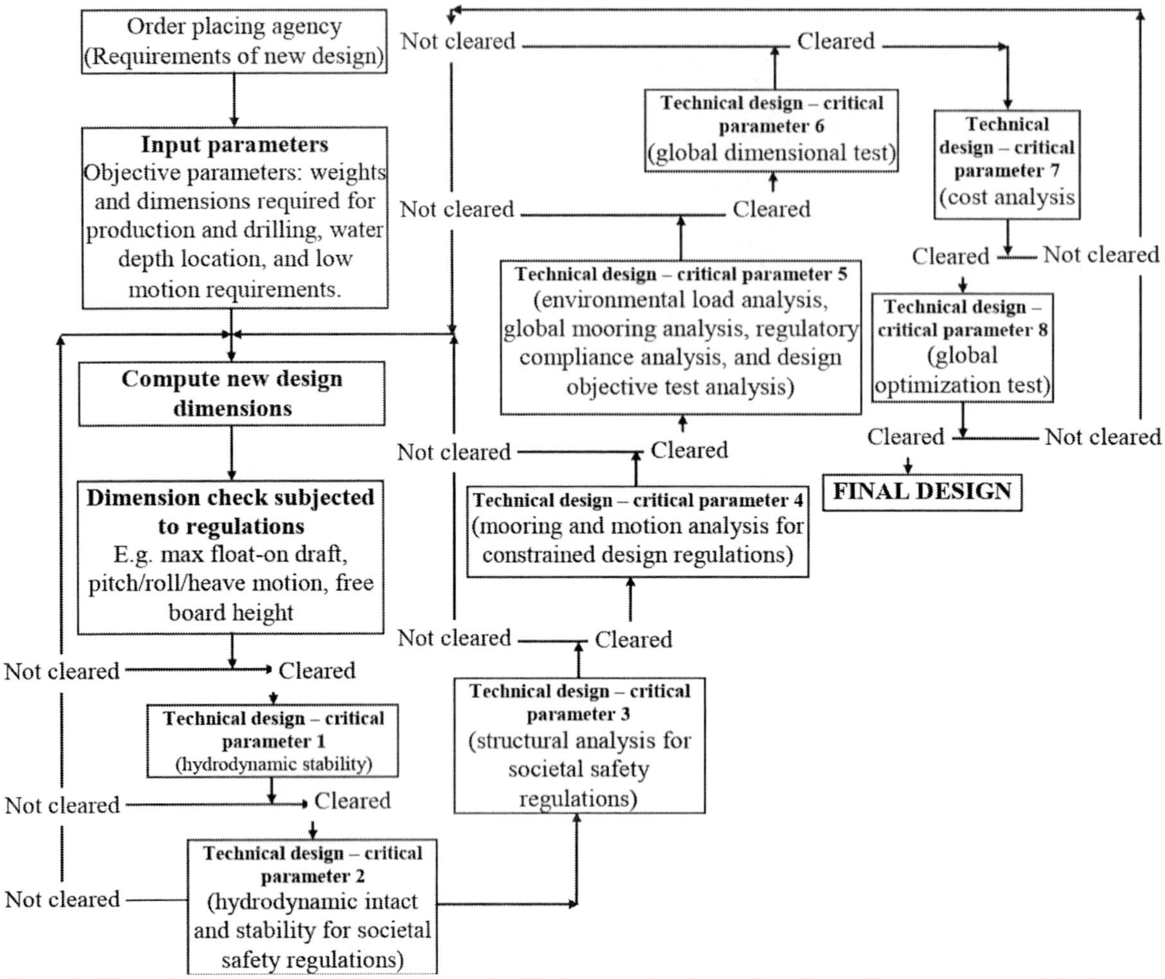

Figure 5: The schematic diagram of our proposed critical parameter driven iterative optimum conceptual design process model.

Now we formulate the optimization problem as follows:

$$\text{Min}_{\lambda} \left\{ \sum_i c_i . M_i + C_{moor} + C_{riser} + C_{instal} \right\}, \qquad (2)$$

where i-th represents the parts of the semi-submersible (e.g. tubular columns (both inner and outer), pontoons, bottom plates, deck, and equipments, etc.).

The minimization is subjected to physical constraints. These physical constraints include the requirement to balance weight and semi-submersible displacement:

- $\rho\nabla = \sum_i M_i + M_{bal}$, where $\rho\nabla$ is the operational displacement and M_{bal} is the ballast weight. Furthermore, this payload constraint is verified with no ballast at the defined transit draft. The optimization is also constrained by the requirements of the hydrodynamic static and dynamic stability. We consider both intact and damaged stability. Then the solution results in a stable and efficient structure for conditions of towing, operation and damaged.
- $GM \geq \beta_{meta}$, where β_{meta} is the minimum meta centeric height. The hydrodynamic behaviour of semi-submersible is considered with MHM – minimum heave motion, MPM – minimum pitch motion and MRM – minimum roll motion.
- $T_{HM} \geq \gamma_{HM}$, $T_{PM} \geq \gamma_{PM}$, $T_{RM} \geq \gamma_{RM}$, where γ_{HM}, γ_{PM}, γ_{RM} are minimum heave/pitch/roll periods.
- $MOT_{HM} \leq MHM$, $MOT_{PM} \leq MPM$, $MOT_{RM} \leq MRM$, where MHM, MPM, MRM are minimum heave/pitch/roll motions.
- $Wa \leq h(Te, Lo_B) \leq Wb$, where Lo_B is the maximum breaking load of the given set of the mooring, and Te is the tension in the mooring.

All equality and inequality constraints are summarized as:

$$\mathrm{H}(\lambda) = 0, \mathrm{G}(\lambda) = 0\,. \qquad (3)$$

To solve the above mentioned optimization problem, we use technique from constrained optimization ([2]), and the final constrained problem is defined as:

$$\mathrm{Min}_{\lambda}\left(f(\lambda)+\phi(P,\mathrm{g}(\lambda))+\phi(P',\mathrm{h}(\lambda))\right), \quad (4)$$

where g is the inequality constraint function and h is the equality constraint function as defined in Equation (3); P and P' are the penalty parameters; and ϕ and ϕ' are the penalty functions. The penalty functions are defined as:

$$\phi=\begin{cases}0\ if\ \mathrm{g}(\lambda)\geq 0\\ u(P,\mathrm{g}(\lambda))\ \text{otherwise}.\end{cases}, \text{ and} \quad (5)$$

$$\phi=\begin{cases}0\ if\ |\mathrm{h}(\lambda)|\leq \varepsilon\\ u(P,\mathrm{h}(\lambda))\ \text{otherwise}.\end{cases} \quad (6)$$

The parameters play an important role in the optimization and to gain a better insight into the design process it is desired that an optimization program is developed. The program allows a variation of the main design criteria, and helps in determining the effects of design criteria on overall size and shape of the semi-submersible. Initially, the design exercise starts with the market survey. The market survey coupled with the analysis by the design team helps in determining the main design criteria. Once, the main design criteria are determined then they are divided into owner's and shipyard's requirements.

We present the implementation details of an optimization program developed by us. In general the owner's requirements are: variable drilling load in operational, survival and transit condition (e.g. ≥ 5500 tonnes), total variable load in transit condition including mooring spread (e.g. ≥ 10000 tonnes), workability conditions in operations (e.g. up to 9 m significant waves and riser disconnection once every storm year), workability in survival condition (e.g. 100 years condition, maximum waves of 40 m height), classification societies (e.g. ABS/DNV/LRS etc.).

The shipbuilding yard requirements are: optimizing parameter with respect to shipyard's constraints (e.g. shipbuilding yard's fabrication system, reduction of joints in the structure, and use plates of maximum size, etc.), maximum construction width (depends upon the available width of building dock, e.g. $\leq$ 80 m), and minimize steel weight consumption.

These requirements form the basic design set and from this set workability in operation and survival conditions are converted to the motional behaviour input data (natural heave/pitch/roll period and second/third hump response). The significant wave height and storm conditions are combined with permissible strokes of the drill string compensator and the telescopic joint. The combination of these values computes the heave/pitch/roll motion curve in irregular seas. The experimental and analytical studies are used to determine the combination of natural period and response values at the second/third hump which satisfies the motion requirements. The height of air-gap is approximated to around 10-15 m.

The algorithms presented above have been implemented with Matlab[#] 7.1, and in C++ using its object-oriented features on a Silicon Graphics[TM##] Origin[TM##] 200 workstation. To clearly demonstrate the applicability of the presented optimization process we present a design example. Let the full set of input details for the conceptual design of a semi-submersible be: Free variables: 65 m < deck width < 80 m – Due to available building dock width in the shipbuilding yard, 0 m < column width, floater length, floater/column width, and floater height.

Parameters: VDL = 5500 tonnes, TVL = 10, 000 tonnes, Height of airgap = 13 m, Height of deck box = 10 m, Number of bracings = 6, and Diameter of bracings = 3.0 m.

Constraints: GM operating = 4.50 m, GM transient = 0.338 m, Transient freeboard = 0.65 m.

Design example 1: The results of the first final optimized design are: Rules and regulations – ABS/DNV/LRS; Principal dimensions; L of main deck – 80.16 m, B of main deck = 71.15 m, D of main deck – 48.38 m, D of flush bottom deck – 39.38 m, L of lower hull – 133.85 m, B of lower hull = 15.475 m, D of lower hull – 10.25 m, L of column – 15.475, B of column = 15.475 m, Longitudinal column space – 64.675 m, Transverse space of column – 64.675 m, T of operation – 25.75 m, T of survival – 21.375 m, T of transit – 9.5625 m; Capacities; Bulk mud and cement – 950 m^3, Liquid mud – 825 m^3, Drill water – 2950 m^3, Fuel oil – 3450 m^3, Potable water – 500 m^3, Ocean/sea water - 500 m^3, Sack storage – 213.45 m^3, Drill pipe storage – 458.825 m^3, Casing pipe storage – 376.25 m^3, Riser storage – 706.367 m^3; Variable load; Deck and column – 5500 tonnes for operating/survival/transit condition, Lower hull – 6750 tonnes for operating/survival condition, Lower hull – 2250 tonnes for transit condition, Mooring – 2500 tonnes; Design criteria; Water depth – 1125 m for operating/survival condition, Wind (60 seconds) – 92 knots for operating condition, Wind (60 seconds) – 124 knots for survival condition, Wind (3600 seconds) – 82 knots for operating condition, Wind (3600 seconds) – 110 knots for survival condition, Significant wave height/period – 10 m/13 seconds for operating condition, Significant wave height/period – 23 m/21 seconds for survival condition, Maximum wave height – 19 m for operating condition, Maximum wave height – 43 m for survival condition, Surface current – 3.325 knots for operating/survival condition, Temperature – (-) 25 ^{0}C for operating/survival condition; Machinery/equipment/out fittings; Anchors – (11 – 21) tonnes, Mooring lines – 12 (100 mm diameter chain * 1275 m plus in length, 132 mm diameter wire * 2550 m plus in length), Cranes – (3* 82.5) tonnes, Accommodation – for 120 persons in 2 persons per cabin, Ballast pump room – 6, Main generators – 6*3500 kW, and Azimuth thrusters – 4*3125 kW.

Design example 2: The results of the second final optimized design are: Water depth: 2000 m, Oil throughput: 75, 000 b/d, Gas throughput: 90 MMscfd, Water throughput: 45,000 barrels per day, Number of production wells 6 (sub-sea), Production risers: numbers - 2, Payload (deck/facilities/risers - including future) 15, 750 tons, Mooring 12-leg mooring system, Ropings: polyester rope, Hull dimensions: 85m X 85m X 232 ft x 232 ft x 60m, Hull columns dimensions: 15.75 m X 15.75 m, Draft: 36 m, Pontoon dimensions: 13.175m X 9.050m, Deck dimensions: 48.950 m X 62.875m (two levels), 47.475m X 13.750 (future module). The graphical design output 1 is shown in Fig. 6. And other graphical design output 2 of the design listed in the Appendix 1 is shown in Fig. 7.

Figure 6: The graphical design 1 of ultra-low motion semi-submersible floating oil and gas production system.

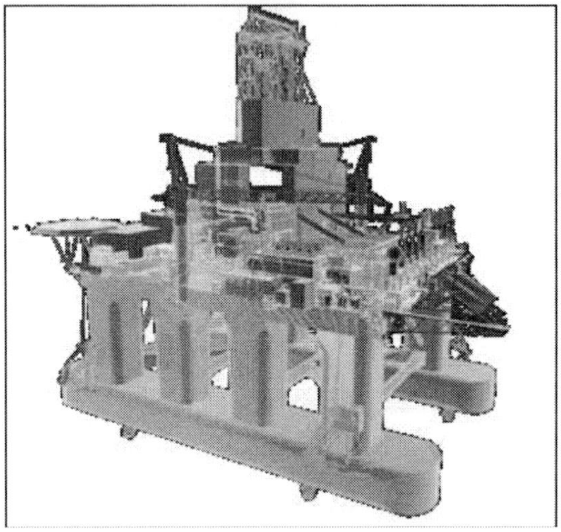

Figure 7: The graphical design 2 of ultra-low motion semi-submersible floating oil and gas production system.

Some other designs that have been optimized with our program are listed in the Appendix.

The steel weight is roughly linearly dependent upon VDL. The operating conditions affect the design and though the reduction of maximum working conditions brings the second/third hump and period of heave/pitch/roll forward thereby reducing the steel weight but it increases the chances of riser disconnection. In proper design, these conflicting choices are settled with the preference given by the owner. Similarly, the drilling condition affects the steel weight. A higher value of drilling condition is preferred because it results in better earnings and additional cost of steel is low as compared to the enhanced earnings because of higher drilling condition. The owner's requirements strongly affect the final design, and parameters like VDL, TVL, GM all affect the steel weight, e.g. higher the VDL/TVL/GM higher the steel weight with TVL/VDL/GM being the order of decreasing influence on steel weight. The fabrication parameters like preference for flat panels, single curved panels, reducing number of main construction elements and continuous use of bulkheads from floaters to columns to upper hull result in significant cost savings (i.e. up to 20 – 30 % of the cost of production of structure). The construction width affects the steel weight and final design. The optimum width balances the design with proper floaters, columns and deck dimensions within the set of input design data. The optimum ratio of L / B for semi-submersibles is around 1.125 – 1.625.

4. CONCLUSIONS

In this work, we have presented a critical parameter driven optimum conceptual design of an ultra-low motion semi-submersible floating oil and gas production system. The design has been carried to satisfy the given parameter of weights and dimensions of the platform topsides required for production and drilling, water depth location, and low motion requirements. Since, the proposed design process is highly iterative process and the key dimensions of a 'n-column ring' pontoon are refined in step wise manner with user input the proposed model can be efficiently used in practice.

The proposed model meets applicable regulations and requests of the order placing agency while minimizing costs and satisfying the chosen parameters. The model is based upon some chosen parameters which might change from user to user depending upon variable set of requirements placed by the order placing agency. We have discussed one design example motivated by real world applications and shown the effectiveness, usability and efficiency of our proposed model.

We believe that there is an urgent need to introduce optimization driven design to ensure competitive utilization of various technical parameters in the design of ships and offshore structures. Since, a ship and offshore structure use many independent systems (e.g. CAD/CAM/CAE system, CFD system, and FEM systems) for design; a dedicated interface to exchange data will facilitate the integration process. For this purpose expandable (flexible) middle data format that synchronizes CAD/CAM/CAE DB, CFD DB, and FEM DB, might hold the key in future research. Our future work shall go in this direction, and currently this is under investigation.

5. ACKNOWLEDGEMENTS

The first author is supported by the internal research grants of AICT, SNU, Korea. Partially, this research was supported by the Korea Science and Engineering Foundation (KOSEF), Ministry of Education, Science and Technology (MEST), Government of Korea via grant number: 2009-0081219.

6. REFERENCES

1. BETBEDER-MATIBET, J., '*Seismic Engineering*', Wiley-ISTE, USA, pp. 141-145, 2008.
2. DENNIS, J. E. and SCHANBEL, R. B., '*Numerical Methods for Unconstrained Optimization and Nonlinear Equations*', Society for Industrial and Applied Mathematics, USA, pp. 201-212, 1987.
3. SHARMA, R., KIM, TAE-WAN, SHA, O. P. and MISRA S. C., 'Semisubmersible Design Faces Challenges', *Offshore Marine Technology*, 3rd Quarter 2009, pp. 22 – 30, 2009.
4. SHARMA, R., MISRA, S. C., and SHA, O. P., 'Deepwater Drilling Designs - System Integration', *Marine Engineers Review*, March 2009, pp. 41 – 44, 2009.
5. SHARMA, R., MISRA, S. C., and SHA, O. P., 'Drillships or Semi-submersibles for Deep Waters", *Marine Engineers Review*, Feb. 2009, pp. 36 – 41, 2009.
6. STANWAY, R., 'Smart Fluids: Current and Future Developments', *Materials Science and Technology*, 20, pp. 931 - 939, 2004.

TRADEMARKS AND COPYRIGHTS

Appendix 1

Design example 3: This example deals with the drilling semisubmersible for extreme weather conditions, e.g. ultra deep waters, harsh environment, artic waters, and great distance from existing infrastructures, etc. The optimized features are: Large payload and storage capacity; excellent motion characteristics; high efficiency for drilling and completion; double rig (main and auxillary); vertical riser and casing racking, independent handling of BOP and trees; maximum parallel activities; large working space for safe and efficient operations; protected working and storage areas; sufficient bed capacity; environment friendly design with low emissions; diagnostic capabilities for remote drilling; efficient riser mud return system; and prepared for full extreme weather conditions of wind, winter, and waves etc. MAIN DETAILS: HYN D 2500 DP-UR; MAIN DESIGN: , L = 103.84 m (main hull excluding helideck), B = 74.15 m (main hull excluding anchor racks), D = 38 m (bottom line to upper hull main deck),), D = 30.25 m (bottom line to flush bottom), Upper hull = 7.75 m deep (1.5 m inner bottom), Columns = 4×21.46 m height, Pontoon level = 13.75 m square, Operating draft = 13.75 m (circular), Longitudinal spacing = 40.28 m (80.56 m overall), Transverse spacing = 60.91 m, Pontoons – 107.35 m (L) × 13.75 m (B) ×

8.78 m (D) (fore/aft); 46.83 m (L) × 13.75 m (B) × 8.78 m (D) (port/stbd); DRAFT AND DISPLACEMENT: Operating draft = 19.75 m, Transit draft = 8.25 m, Survival draft = 14.82 m, Operating displacement = 50031 tonnes, Transit displacement = 35973 tonnes, Survival displacement = 45416 tonnes; OPERATING PARAMETERS: Water depth = 3000 m, Maximum drilling depth = 9150 m, Transit speed = 7 knots, Survival conditions - Wave = 27.45 m, Wind = 105 knots, Current – 2.625 knots, Drilling conditions - Wave = 13.75 m (13 second), Wind = 55 knots, Current – 4.125 knots, Assisted mooring = 2×5000 HP thruster, Design conditions – Air temperature = +38 to -20 ^{0}C, Water temperature = +32 to -2 ^{0}C; MAIN MACHINERY: $6 \times$ diesel engines, 21000 kW/28000 kVA from 6 engine generators, Power distribution = 2 main 6000 VAC buses, 480 VAC, 10 steel catenary risers, Emergency power = 800 kW/1000 kVA emergency generator, Propulsion = 2×5000 HP variable speed azimuthing thrusters; CAPACITIES: Variable drilling load = 9225 tonnes (operating), Variable drilling load = 6150 tonnes (transit), Riser rack area = 915 m^2, Casing rack area = 472 m^2, Other rack area = 1210 m^2, Total rack area = 2597 m^2, Free deck area = 1388 m^2, Drill pipe in horizontal racker = 9150 m, Sacks = 6000 each, Drill water = 2650 m^3, Potable water = 595 m^3, Fuel oil = 2891 m^3, Bulk amount of bentonite = 348.30 m^3, Bulk amount of cement = 348.3 m^3, Amount of liquid mud/salt water = 3490 m^3, Ballast water = 19690 m^3; Drilling equipments: 52 m× 20 m× 13.45 m main hoist 1000 tonnes and auxiliary hoist 350 tonnes, Setblock = 440 tonnes (i.e. landing strings, casings, etc.), Drawworks = 4000 HP, auxiliary – 1100 HP, Rotary = main and auxiliary (16 m× 1.6 m), Top drive = main (775 tonnes) and auxiliary (AC), Travelling block = main (1000 tonnes) and auxiliary (350 tonnes), Horizontal pipe handling = 12200 m (430 stands), Vertical pipe handling = 950 m (325 stands), Riser tensioners = 12 mounts up to 3000 kips capability, Motion compensator – at main hoist (static – 1050 tonnes and dynamic – 525 tonnes with 7.65 m stroke), Active Hv. compensation = installed; MUD SYSTEM: Cementing = 2 × 600 HP (space availability = 2200 HP), Mud pumps = 4× 2200 HP 52 MPa, Mud processing equipment = 11425 litres/minute (circulating rate), Desander - 3× 0.35 m cones, Desilter = 24× 0.15 m cones, Centrifuge = 4× decanting type and 2 additional, Shakers = 14× linear motion (6 – scalpers, 6 – finishing and 2 mud conditions), Pumps = 6 × 8 × 100 HP, 10 × 8 × 100 HP (for mix processing) and 14 × 8 × 100 HP (for mud transferring), Hot swap mud system = yes; SUBSEA SYSTEMS: BOP = 0.34 m (105 MPs, 5 ram, 2 annular, flexi joint and adapter), BOP handling system = transporter of 300 tinnes, Control system = multiplex hydraulic, Riser Details = 2500 m, 0.575 m outside diameter and 27.50 m joint (buoyancy – 98 %), Telescopic joint = 16.825 m stroke, Diverter – 1.54 m inside diameter, Drill pipe = outside diameters of 0.09 m, 0.128 m, 0.140 m and 0.170 m, Drill collars = outside diameters of 0.120 m, 0.165 m, 0.210 m, 0.240 m, TV system = ROV with 2500 m reach, Positioning = dynamic positioning system with lower flex joint angle, Choke and kill lines = 0.08 m inside diameter with 105 MPa and trim compensation, Tree handling system = 175 tonnes transporter with skidding system and carts with maximum 6 trees in simultaneous operation, Subsea gantry = 100 tonnes × 9 m span for BOP and 90 tonnes × 11.4 span for trees; MOONPOOL: 45 m × 7.75 m; CRANAGE: 3× 100 tonnes × 55 m (boom range); MOORING SYSTEM: Winches = 12× traction winches, Wire/chain = 12× 3100 m × 0.09 m wire and 12× 1550 m × 0.085 m chain, Anchors = 12× 15 tonnes HHP, Thruster assisted mooring = 2 × 5000 tonnes HP thruster; Accommodation = 160 people; DIVING SPEED = ROV installed in upper hull; ADDITIONAL INFORMATION: Motion characteristics = Heave period (20-25 seconds), Roll/pitch period (40 – 45 seconds), Life boats = 6 × 54 persons, Life rafts = 8× 20 persons, Rescue boats = 1× 15 persons and Fire fighting systems – helideck (aqueous film-forming foams (AFFF)), test areas (impreza inter-cooling water spray heat shield), Quarters (automatic fire sprinkling system), and Fire and gas detection (automatic fire and gas alarm systems).

7. AUTHORS BIOGRAPHY

R. Sharma holds the current position of Senior Researcher at the Advanced Institutes of Convergence Technology, Seoul National University, Seoul, Republic of Korea. He is responsible for the research in the areas of computer aided geometric design, computational geometry, visualization, and their applications in design and manufacturing; dynamic data driven forecasting systems; and participatory/democratic economy.

Tae-wan Kim holds the current positions of Associate Professor in the Department of Naval Architecture and Ocean Engineering, College of Engineering; Director, Planning and Coordination Office, Advanced Institutes of Convergence Technology, Associate Head, Future IT-Convergence Program, and Vice President, Korean Society of CAD/CAM Engineers, Seoul National University, Seoul, Republic of Korea. He is responsible for the teaching and research in the areas of computer-aided design (CAD), ship production management, system modeling and simulation, information technology application for shipbuilding industry, and PDM/PLM (Product Development/Lifecycle Management).

O. P. Sha holds the current position of Professor in the Department of Ocean Engineering and Naval Architecture at the IIT, Kharagpur, India. He has been teaching various courses at the graduate and postgraduate level for the last more than two decades. Currently he is involved in many research projects, and specializes in the areas of CAD/CAM, computational geometry, and computer aided design of high-speed hull forms for marine crafts.

OPTIMUM CONCEPTUAL DESIGN OF AN ULTRA-LOW MOTION SEMI-SUBMERSIBLE FLOATING OIL AND GAS PRODUCTION SYSTEM

R Sharma, Indian Institute of Technology, Chennai, India
Tae-wan Kim, Seoul National University, Republic of Korea
O P Sha, Indian Institute of Technology, Kharagpur, India

Introduction

- Since, fossil fuel and ore resources are depleting at shallow water depths because of the extensive utilization in the last century, economically there is a growing need for deepwater solutions.

- The exploration and production requirement in deep sea demands that the motion characteristics of the structure are further lower as compared to the inherently low motion characteristics of the semi-submersible.

- FUTURE DEMAND – Low motion semi-submersibles and drill ships.

Introduction

- An offshore platform is a large structure (floating or fixed) which is used to house workers and machinery needed to drill wells in the ocean bed, extract oil and/or natural gas, process the produced fluids, and ship or pipe them to shore.

- A semi-submersible is a type of offshore platform and technically it is considered as a special type of 'watercraft' that has much of its hull form (70 – 85 %) underwater.

- Since, for a semi-submersible the waterplane is low, the semi-submersible is less affected by the waves than a normal ship.

Introduction

- However, it needs proper and continuous ballasting to achieve desired trim and stability.

- Though a fully automatic optimum design is possible at least theoretically but in practice a fully automatic process is rarely used mainly because a fully automatic process computes a unique and rigid result and flexible choices in design are rarely achieved.

- In practice it is desired that the optimization process computes flexible choices of design and then these sets of choices are discussed.

Introduction

- In this work we explore a novel design concept for ultra-low motion semi-submersible for operations in deepwaters.

- In our design, highly constrained ultra low motion can be achieved with novel use of the concepts from 'seismic engineering', column sizes and spacing, floaters spacing and arrangement, and application of dynamic position systems (DPS).

- MOTIVATIONS: 'Seismic Engineering', and we combine the earthquake resistance technology with basic flotation fundamentals.

Introduction

- AIMS – To maximize the combined effects on floatation and minimize the motion characteristics.

- TECHNICAL DETAILS: We use dampers, which absorb energy and reduce its effects on the semi-submersible.

- DAMPERS - ‘Magneto-rheological Fluid (MR fluid)’ based dampers.

Description of ultra-low motion design concept

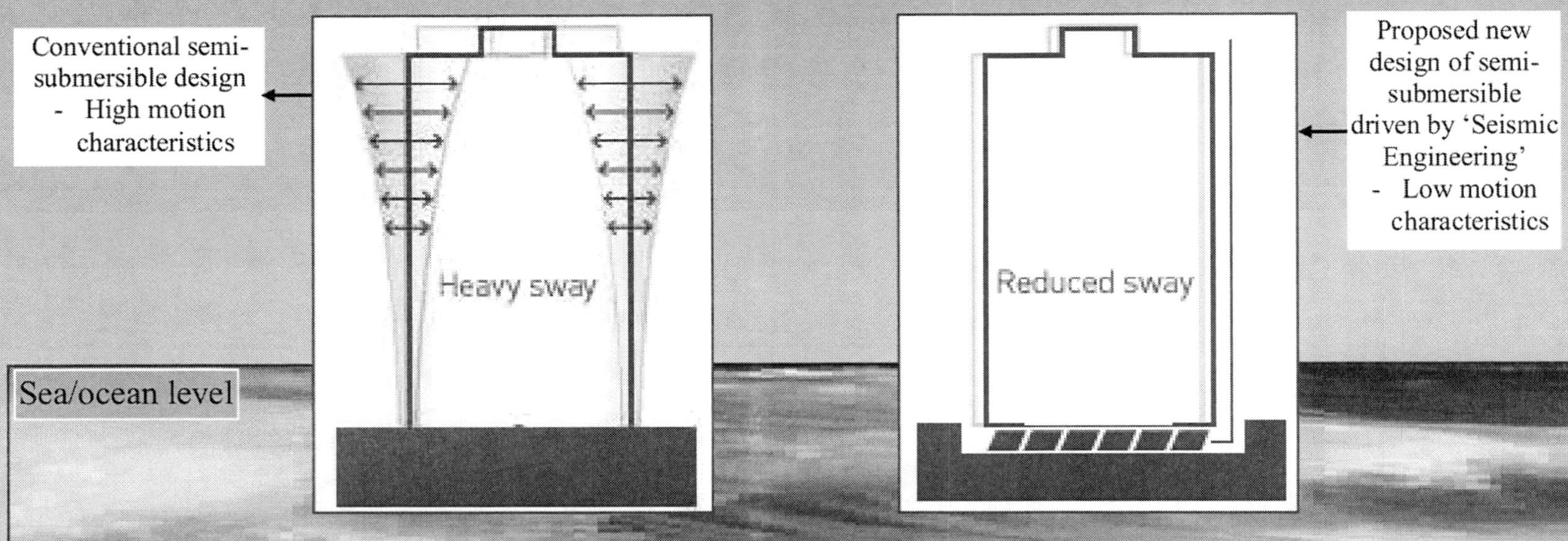

Figure 1: The conceptual description of our proposed ultra-low motion design idea for semi-submersibles.

Description of ultra-low motion design concept

- Dampers in the column for damping control to constrain motion characteristics.

- MR fluid is a liquid that changes to a near-solid when exposed to a magnetic force, and then back to liquid once the magnetic force is removed.

- In columns a 'tube-in-tube' design is used, and in the inside tubular column dampers filled with MR fluids are used to control or stabilize the semi-submersible during rough and extreme weather conditions.

- The control design concept for the column of the semi-submersible is shown in Fig. 2.

Description of ultra-low motion design concept

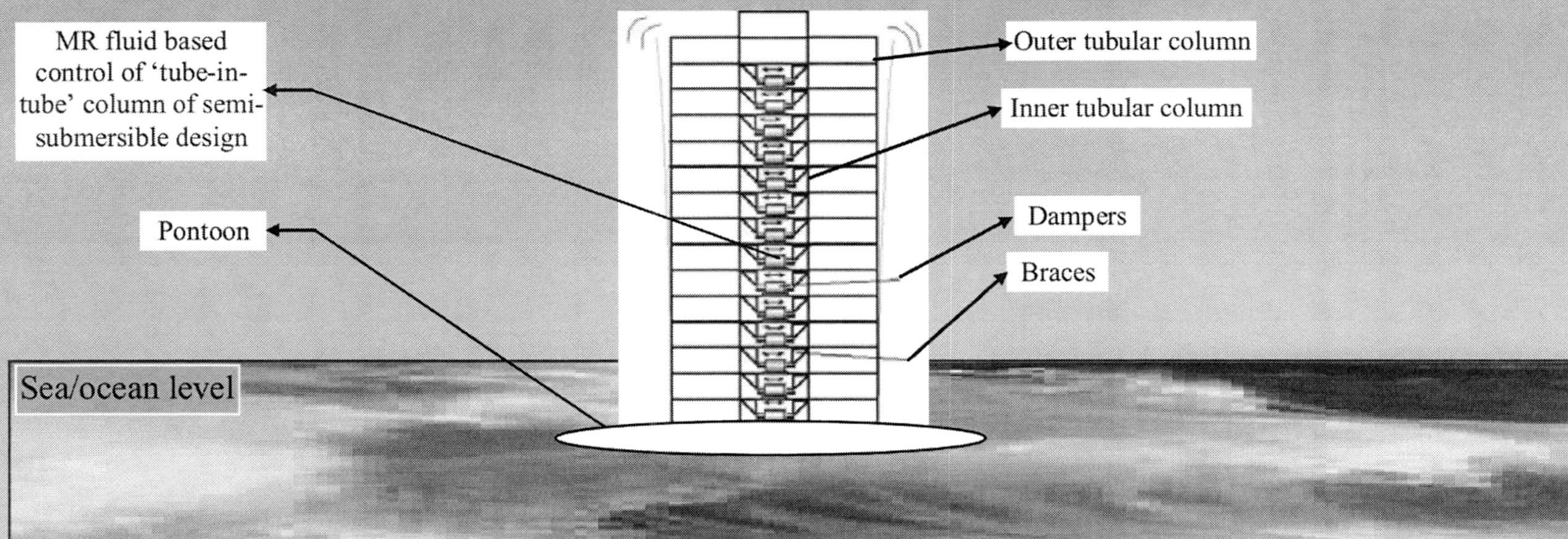

Figure 2: The control design concept for the column of the semi-submersible.

Description of ultra-low motion design concept

- ADDITIONAL CONTROL: On motion can be achieved with pendulum driven control of the centralized trusses of the semi-submersibles.

- The pendulum acts as mass damper and preferably it is used in extreme weather conditions.

- In normal and rough weather control system at the columns is sufficient.

- This is shown in Fig. 3.

Description of ultra-low motion design concept

Figure 3: The additional pendulum based control of centralized trusses of semi-submersible.

Description of ultra-low motion design concept

- OPERATIONAL DETAILS: During rough and extreme weather, MR fluid inside the dampers will change from solid to liquid and back as weather induced vibrations or movements activate a magnetic force inside the damper.

- Using this damper in semi-submersible not only the motion characteristics can be reduced.

- This idea can also be extended to a smart semi-submersible that automatically react to weather conditions can be designed. This will limit the amount of damage caused by extreme weather environments.

Description of ultra-low motion design concept

- Technically, our MR fluid damper is a ‘semi-active device’ that change its damping level by varying the amount of current supplied to an internal electromagnet that controls the flow of MR fluid.

- Each damper sits on the floor of the inner tubular column and is attached to the braces that are welded into a steel cross beam.

- As the semi-submersible is influenced with motion or vibration, the dampers move back and forth to compensate for the motion/vibration of the force.

Description of optimization model

- An optimization approach needs clearly defined parameters: free variables, technical design parameters, constraints and objective function.

- In computational implementation free variables are changed during the computation of the optimum solution while the parameters have fixed values defined at the input stage.

- The constrained parameters are the requirements that need to be fulfilled, and the objective function needs to be optimized.

Description of optimization model

- The objective is user selected and it can be linear, non-linear and stochastic.

- Because of many possibilities in the very definition of objective function it is desired that a powerful optimization technique is adopted.

- The schematic diagram of generic optimization approach is shown in Fig. 4.

- In Fig. 4, the main program reads the parameters values from the input set and runs the optimization process.

Description of optimization model

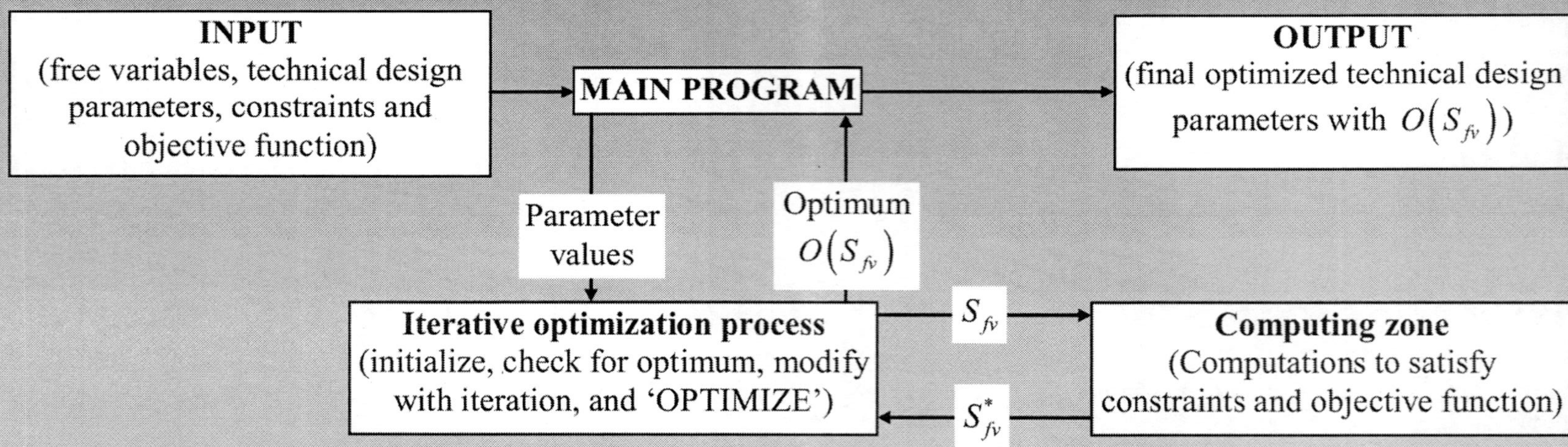

Figure 4: The schematic diagram of generic optimization approach.

Description of optimization model

- The objective is user selected and it can be linear, non-linear and stochastic.

- Because of many possibilities in the very definition of objective function it is desired that a powerful optimization technique is adopted.

- The schematic diagram of generic optimization approach is shown in Fig. 4.

- In Fig. 4, the main program reads the parameters values from the input set and runs the optimization process.

Description of optimization model

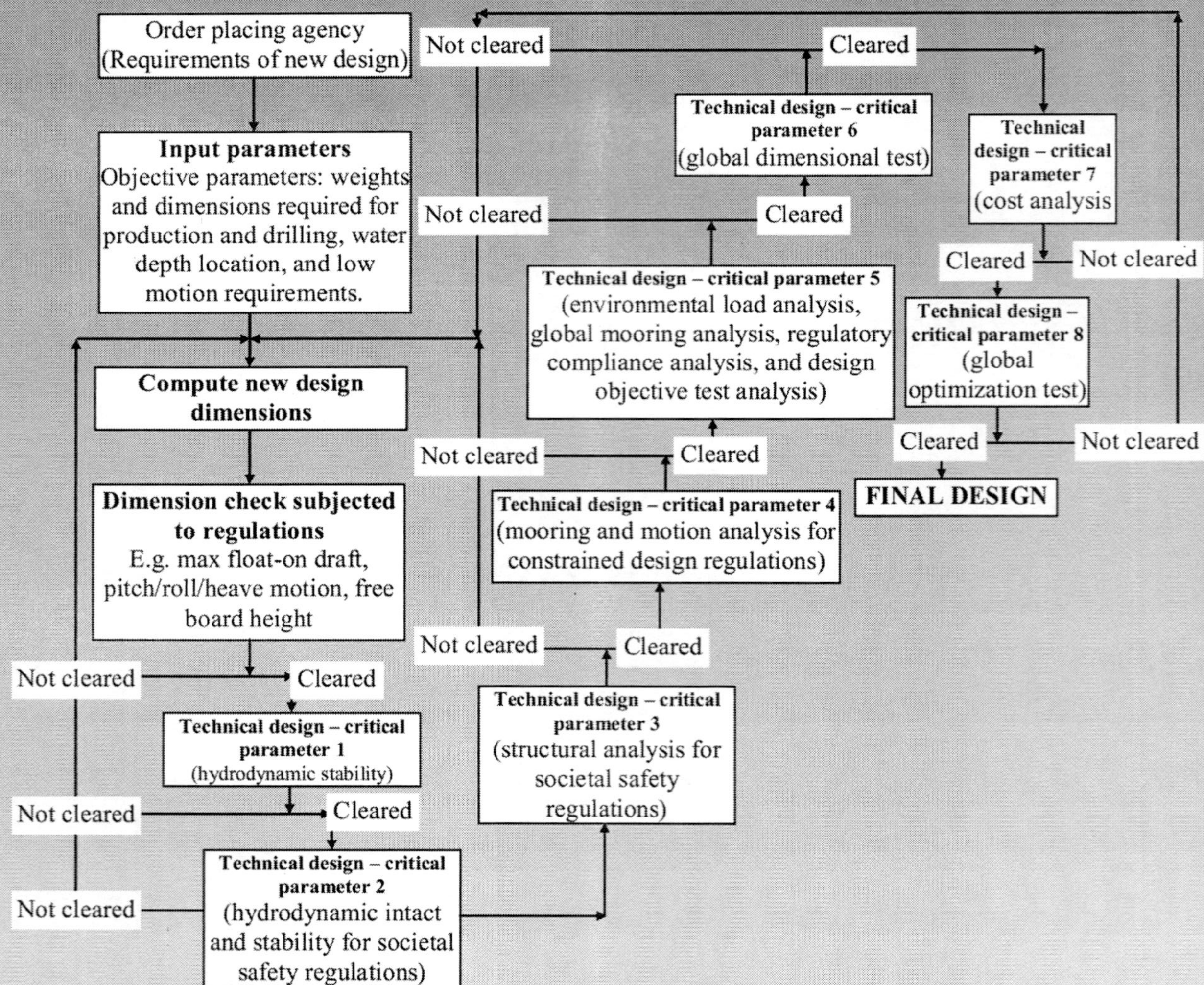

Figure 5: The schematic diagram of our proposed critical parameter driven iterative optimum conceptual design process model.

Results and discussion

- IMPLEMENTATION: The algorithms presented above have been implemented with Matlab# 7.1, and in C++ using its object-oriented features on a Silicon Graphics™## Origin™## 200 workstation.

- To clearly demonstrate the applicability of the presented optimization process we present a design example.

- Let the full set of input details for the conceptual design of a semi-submersible be: Free variables: 65 m deck width 80 m – Due to available building dock width in the shipbuilding yard, 0 m column width, floater length, floater/column width, and floater height.

Results and discussion

- Parameters: VDL = 5500 tonnes, TVL = 10, 000 tonnes, Height of airgap = 13 m, Height of deck box = 10 m, Number of bracings = 6, and Diameter of bracings = 3.0 m.

- Constraints: GM operating = 4.50 m, GM transient = 0.338 m, Transient freeboard = 0.65 m.

Results and discussion

Figure 6: The graphical design 1 of ultra-low motion semi-submersible floating oil and gas production system.

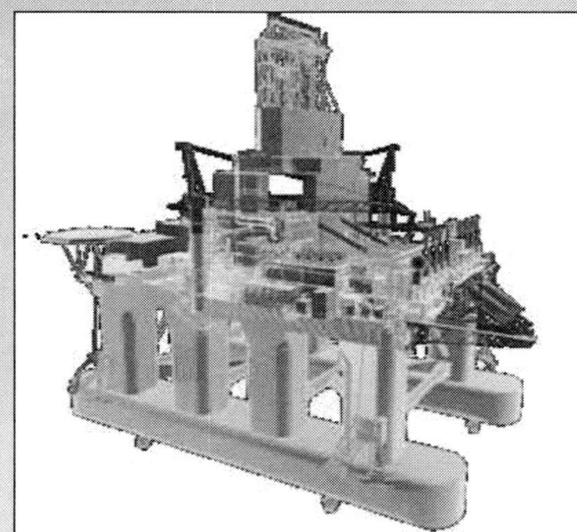

Figure 7: The graphical design 2 of ultra-low motion semi-submersible floating oil and gas production system.

Conclusions

- In this work, we have presented a critical parameter driven optimum conceptual design of an ultra-low motion semi-submersible floating oil and gas production system.

- The design has been carried to satisfy the given parameter of weights and dimensions of the platform topsides required for production and drilling, water depth location, and low motion requirements.

- Since, the proposed design process is highly iterative process and the key dimensions of a '*n*-column ring' pontoon are refined in step wise manner with user input the proposed model can be efficiently used in practice.

Conclusions

- The proposed model meets applicable regulations and requests of the order placing agency while minimizing costs and satisfying the chosen parameters.

- The model is based upon some chosen parameters which might change from user to user depending upon variable set of requirements placed by the order placing agency.

- We have discussed one design example motivated by real world applications and shown the effectiveness, usability and efficiency of our proposed model.

Conclusions

- We believe that there is an urgent need to introduce optimization driven design to ensure competitive utilization of various technical parameters in the design of ships and offshore structures.

- The presented design is currently being investigated with a experimental prototype to validate the design.

- OPEN ISSUE: Small size dampers with large numbers or large size dampers with small numbers.

ACKNOWLEDGEMENTS

The first author was supported by the internal research grants of AICT, SNU, Korea. Partially, this research was supported by the Korea Science and Engineering Foundation (KOSEF), Ministry of Education, Science and Technology (MEST), Government of Korea via grant number: 2009-0081219.

Thank you for your kind attention.

QUANTIFICATION OF UNCERTAINTY IN MANOEUVRING CHARACTERISTICS FOR DESIGN OF UNDERWATER VEHICLES

Amit Ray, Indian Institute of Technology Delhi, India
Debabrata Sen, Indian Institute of Technology Kharagpur, India
SN Singh and **V Seshadri,** Indian Institute of Technology Delhi, India

SUMMARY

The prediction of manoeuvring characteristics of underwater vehicles during design involves approximations at various stages. This paper attempts to quantify some of the uncertainties involved in the manoeuvring characteristics of underwater vehicles. The first source of uncertainty is in idealization of mathematical model selected for trajectory simulation. This is illustrated for alternative mathematical models in trajectory simulation programs. Next, the values of the hydrodynamic coefficients (HDCs) in the equations of motion have their own levels of uncertainty, depending upon the methods used to determine them. The sensitivity of trajectory simulation results to uncertainty levels in various HDCs is examined. Finally, the level of uncertainty in full-scale measurements of manoeuvres of underwater vehicles is discussed and estimated. It emerges that the cumulative errors in the prediction process during design need to be curtailed, in order to reduce their levels of uncertainty below those of the validation process.

NOMENCLATURE

$\vec{F} = \{X, Y, Z\}$	External force on the body and its components along x, y and z directions
$\vec{M} = \{K, M, N\}$	Moment of the external force and its components along x, y and z directions
$\{x_G, y_G, z_G\}$	Coordinates of the body's center of gravity (CG) with respect to the body-fixed system $oxyz$
$\{x_G, y_G, z_G\}$	Coordinates of the body's center of buoyancy
m	Mass of the vehicle
I_{ij}	Moments of inertia about the body-fixed system $oxyz$, where the indices i,j correspond to x,y,z coordinates.
u, v, w	Straight-line velocities of the body along x, y and z directions in the body-fixed frame of reference
p, q, r	Angular velocities of the body about x, y and z directions in the body-fixed frame of reference
W	Weight of the body in excess of buoyancy

1. INTRODUCTION

The study of manoeuvring characteristics of underwater vehicles includes assessment of motion stability, controllability and trajectory simulation in six degrees of freedom. These have important operational implications for all types of underwater vehicles. There are a variety of approaches (empirical/analytical, experimental and numerical/computational techniques) for manoeuvring studies of underwater vehicles [1]. The problem of mathematical modelling of the motion of marine vehicles for trajectory simulation has been studied extensively over many decades [2-6]. The subject has gained renewed interest after the proliferation of Unmanned Underwater Vehicles (UUVs), which include Remotely Operated Vehicles (ROVs) and Autonomous Underwater Vehicles (AUVs) used for many commercial and scientific applications [7, 8].

Manoeuvring characteristics of underwater vehicles are difficult to predict accurately during initial stages of design. Frequently, recourse is made to empirical or semi-empirical methods for the purpose. Even after more detailed information is available from time-taking studies using tools like Computational Fluid Dynamics (CFD), or from expensive model tests, there are inherent uncertainties in the prediction process for manoeuvring qualities. All too often, attention is concentrated on particular aspects of the problem, such as model testing, CFD, control system design, system identification, or trajectory simulation. It is necessary to estimate the magnitude of uncertainty for the overall process of prediction of manoeuvring characteristics.

In order to quantify these uncertainties, it is necessary to understand the choices being made at each stage of the prediction process and their implications on the ultimate result. These aspects or stages for manoeuvring prediction / evaluation may be outlined as follows.

- Formulation of mathematical model for trajectory simulation
- Estimation of values of Hydrodynamic Coefficients (HDCs) for the body
- Simulation of trajectories for various manoeuvres
- Full-scale trials for standard manoeuvres and comparison with predictions

Each of these aspects has inherent uncertainties. The ultimate benchmark for evaluating any set of predictions should be comparing them with full-scale trail results for a set of standard manoeuvres. These are the only 'true' values against which the 'errors' in each stage of manoeuvrability prediction must be judged. However, not only are the full-scale measurements rarely conducted or reported, but the full-scale measurement process itself is beset with uncertainties. Therefore, it is important to define realistic limits for error bounds in each stage of manoeuvring prediction, in order to ensure practically reliable results.

This paper presents glimpses of the choices available to the designer / analyst at each of the stages for the manoeuvrability prediction process for underwater vehicles. Cases are presented illustrating the effects of these decisions and the effects of variation in parameters on the trajectory simulation for typical underwater vehicles. Based on these simulations, the effect of variation in values of HDCs can be appreciated. Finally, the paper presents error bounds likely during full-scale measurements, which suggest the level of accuracy desirable for the preceding stages of manoeuvring prediction and analysis.

2. FOCUS OF THIS STUDY

Any study on the sensitivity of manoeuvring characteristics to the many variables in the modelling process needs to be limited to certain cases in order to keep the scope of the study within reasonable length. In this study, therefore, two underwater vehicle hull forms have been considered and various trajectory simulations have been undertaken for one particular type of definitive manoeuvre.

2.1 UNDERWATER VEHICLES CONSIDERED

Trajectory simulations were carried out for two underwater vehicle geometries in this study:

- Axisymmetric body of revolution with cruciform control surfaces at stern [9, 10]. The length of the body is 10 metres and weight is 196.5 kg.
- SUBOFF body with appendages (fin and cruciform control surfaces) as described in [11]. Length of model is 4.26 m and weight is 18.1 kg.

Mass properties and HDC values determined by model testing (Planar Motion Mechanism) were used in the trajectory simulation programs developed.

2.2 MANEOUVRE CONSIDERED

Definitive manoeuvres are carried out to characterise and compare the handling qualities of underwater vehicles. These include zigzag/overshoot, meander, spiral, pullout and turning circle manoeuvres [2, 12]. The depth-changing and depth-keeping abilities are particularly important for underwater vehicles due to the relatively narrow band of water (in depth) within which they require to operate. Therefore, the zigzag/overshoot manoeuvre in the vertical plane is considered for this study.

In the zigzag manoeuvre, starting from a level submerged course, the control surfaces are set at constant angle δ_1 as quickly and as smoothly as possible until the pitch angle (θ) becomes equal to the execute pitch angle (θ_e) decided for the manoeuvre. The planes are then deflected to an angle $-\delta_1$ until the original depth at the start of the manoeuvre is reached, or the pitch angle in the opposite sense ($-\theta$) is reached. The parameters of interest are the time to reach execute (t_e), time to check pitch (t_c), time to check depth (t_d), overshoot pitch angle (θ_O) and overshoot change of depth (z_O), as defined in Figure 1. These values will depend upon the specified plane deflection, execute pitch angle and the speed at which the manoeuvre is conducted.

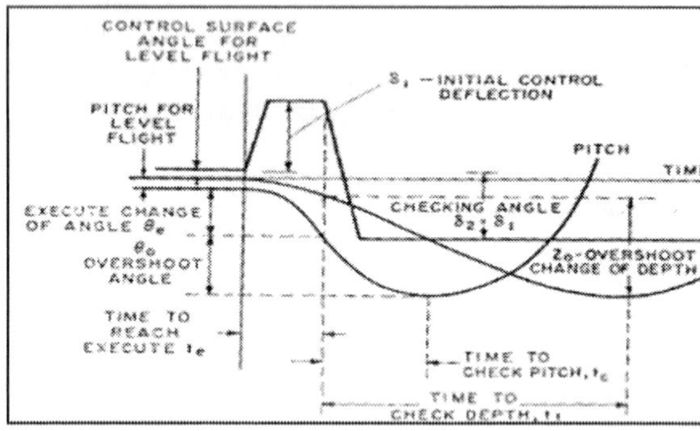

Figure 1 Definition of parameters for Zigzag manoeuvre in vertical plane [2]

The variation in values of the above parameters for standard 10/10 zigzag manoeuvre (i.e. δ_1=10 degrees, θ_e =10 degrees) and 15/5 zigzag at speed 5 knots was examined in this study for various mathematical models and HDC values.

3. MATHEMATICAL MODELS FOR TRAJECTORY SIMULATION

The motion of an underwater vehicle can be described in terms of Newton–Euler laws of motion, where the rate of change of momentum of a rigid body is equated to the external forces/ moments causing the change. For an underwater vehicle such as a submarine or an AUV, controlled motion is possible with six degrees of freedom (6-DOF).

3.1 RIGID BODY EQUATIONS OF MOTION

The trajectory of the body at any instant of time is described by its linear velocities u, v, w, and by its angular velocities p, q, r, in the body-fixed frame of reference, and by its position and orientation in an inertial frame of reference (Figure 2).

Let $\vec{F}$ denote the external force on the body, and $\vec{M}$ denote the moment of the external force about the origin o of the body-fixed system $oxyz$. Their components are denoted as, $\vec{F} = \{X,Y,Z\}$ and

$\vec{M} = \{K, M, N\}$. The mass of the vehicle is m and the moments of inertia about the body-fixed system $oxyz$ are denoted as I_{ij} where the indices i,j correspond to x,y,z coordinates. The coordinates of the body's center of gravity (CG) with respect to the body system $oxyz$ are $\{x_G, y_G, z_G\}$.

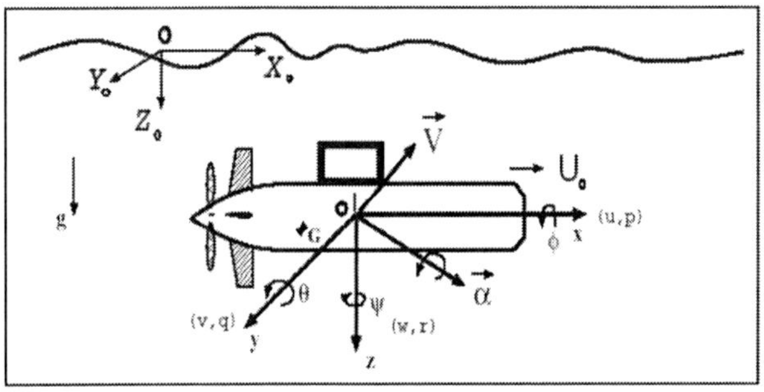

Figure 2 Definition diagram of symbols for orientation; linear and angular velocities

The governing 6-DOF equations of motion can be written as follows for the body-fixed frame of reference [5, 6, 13]:

Surge:

$$m[\dot{u} - vr + wq - x_G(q^2 + r^2) + y_G(pq - \dot{r}) + z_G(pr + \dot{q})] = X \quad (1)$$

Sway:

$$m[\dot{v} - wp + ur - y_G(r^2 + p^2) + z_G(qr - \dot{p}) + x_G(qp + \dot{r})] = Y \quad (2)$$

Heave:

$$m[\dot{w} - uq + vp - z_G(p^2 + q^2) + x_G(rp - \dot{q}) + y_G(rq + \dot{p})] = Z \quad (3)$$

Roll:

$$I_{xx}\dot{p} + (I_{zz} - I_{yy})qr - (\dot{r} + pq)I_{xz} + (r^2 - q^2)I_{yz} + (pr - \dot{q})I_{xy} + m[y_G(\dot{w} - uq + vp) - z_G(\dot{v} - wp + ur)] = K \quad (4)$$

Pitch:

$$I_{yy}\dot{q} + (I_{xx} - I_{zz})rp - (\dot{p} + qr)I_{yx} + (p^2 - r^2)I_{zx} + (qp - \dot{r})I_{yz} + m[z_G(\dot{u} - vr + wq) - x_G(\dot{w} - uq + vp)] = M \quad (5)$$

Yaw:

$$I_{zz}\dot{r} + (I_{yy} - I_{xx})pq - (\dot{q} + rp)I_{zy} + (q^2 - p^2)I_{xy} + (rq - \dot{p})I_{zx} + m[x_G(\dot{v} - wp + ur) - y_G(\dot{u} - vr + wq)] = N \quad (6)$$

The above six dynamic motion equations can be written in matrix form [13] as:

$$[A]\{\dot{V}\} - \{F_I\} = \{F_E\} \quad (7)$$

where the external generalised forces are denoted by

$$\{F_E\} = \{X, Y, Z, K, M, N\}^T \quad (8)$$

and the terms on the left hand side of equation (7) are matrices of size 6x6, defined as follows.

The extended mass matrix (including added mass terms) is given by:

$$[A] = [\mathfrak{M}] + [\mathfrak{M}'] \quad (9)$$

where mass matrix is given by:

$$[\mathfrak{M}] = \begin{bmatrix} m & 0 & 0 & 0 & mz_G & -my_G \\ 0 & m & 0 & -mz_G & 0 & mx_G \\ 0 & 0 & m & my_G & -mx_G & 0 \\ 0 & -mz_G & my_G & I_{xx} & -I_{xy} & -I_{xz} \\ mz_G & 0 & -mx_G & -I_{yx} & I_{yy} & -I_{yz} \\ -my_G & mx_G & 0 & -I_{zx} & -I_{zy} & I_{zz} \end{bmatrix} \quad (10)$$

and added mass matrix is taken as:

$$[\mathfrak{M}'] = \begin{bmatrix} -X_{\dot{u}} & 0 & 0 & 0 & 0 & 0 \\ 0 & -Y_{\dot{v}} & 0 & -Y_{\dot{p}} & 0 & -Y_{\dot{r}} \\ 0 & 0 & -Z_{\dot{w}} & 0 & -Z_{\dot{q}} & 0 \\ 0 & -K_{\dot{v}} & 0 & -K_{\dot{p}} & 0 & -K_{\dot{r}} \\ 0 & 0 & -M_{\dot{w}} & 0 & -M_{\dot{q}} & 0 \\ 0 & -N_{\dot{v}} & 0 & -N_{\dot{p}} & 0 & -N_{\dot{r}} \end{bmatrix} \quad (11)$$

The subscript notation represents partial differentiation, so $X_{\dot{u}} = \partial X / \partial \dot{u}$, etc.

Acceleration vector is given by:

$$\{\dot{V}\} = \{\dot{u} \quad \dot{v} \quad \dot{w} \quad \dot{p} \quad \dot{q} \quad \dot{r}\}^T \quad (12)$$

The inertial force terms, independent of modelling of external hydrodynamic forces, are given by

$$\{F_I\} = \begin{Bmatrix} m[vr - wq + x_G(q^2 + r^2) - y_G pq - z_G pr \\ m[wp - ur + y_G(r^2 + p^2) - z_G qr - x_G qp \\ m[uq - vp + z_G(p^2 + q^2) - x_G rp - y_G rq \\ (I_{yy} - I_{zz})qr + I_{xz}pq + I_{yz}(q^2 - r^2) - I_{xy}pr - my_G(vp - uq) + mz_G(ur - wp) \\ (I_{zz} - I_{xx})rp + I_{yx}qr + I_{zx}(r^2 - p^2) - I_{yz}qp - mz_G(wq - vr) + mx_G(vp - uq) \\ (I_{xx} - I_{yy})pq + I_{zy}rp + I_{xy}(p^2 - q^2) - I_{zx}rq - mx_G(ur - wp) + my_G(wq - vr) \end{Bmatrix} \quad (13)$$

Thus far, there are no approximations in the mathematical model, but the external forces and moments are yet to be modelled. Simplification of equations (9) and (12) will follow if centre of gravity is coincident with the origin, and if the axes system is along the principal axes of the body.

3.2 VARIANTS OF MATHEMATICAL MODEL

The external forces and moments due to hydrodynamic (and hydrostatic) loads are complicated functions of many factors, including water density, viscosity, surface tension, pressure, vapour pressure, and motions of the body. Rather than attempting to obtain hydrodynamic forces exactly from the Navier-Stokes equations, the components of the hydrodynamic forces and moments acting on the hull and on each appendage are usually treated as a function of the motion state variables, in terms of 'hydrodynamic coefficients' (HDCs) [2, 4, 5].

In general, the most important part of developing the dynamic motion equations is in expressing the external force and moment vector $\{F_E\}$ properly. These forces /moments may be split into physically meaningful components corresponding to hydrostatic (gravity-dependent) forces, fluid inertia (i.e. added mass, or acceleration-dependent terms), fluid damping forces (velocity-dependent terms), propulsion forces, and control surface forces, which are the most crucial for initiating any manoeuvre of the body.

The HDCs may be denoted using another form of subscripts, where subscripts denote the motion variable they are multiplied with. For example, for a small sway velocity v, the hydrodynamic force component arising from that motion is expressed as $Y_v v$, where Y_v is the HDC for the motion variable v in the equation of motion for the Y-force. All coefficients can be made non-dimensional using appropriate combinations of vessel length and initial axial velocity.

In order to generalise the various mathematical models available for expressing $\{F_E\}$ in terms of HDCs, we can consider $\{F_E\}$ as the product of various HDCs (as matrix [B] of size 6 x n) with respective motion variables (as matrix [C] of size n x 6), where n depends on the particular model and may typically vary from 6 to 15.

Thus,

$$\{F_E\} = diag\,([B].[C]) \tag{14}$$

i.e. the vector $\{F_E\}$ is formed by the diagonal terms of the 6 x 6 matrix $[B].[C]$, which also includes terms corresponding to hydrostatic forces and control forces.

Various models developed for specific applications can be conveniently expressed in the above form. For example, only linear terms may be retained in some models, or cross-coupling between various motions may be neglected, or special terms may be added to account for cross-flow drag, or empirical corrective terms may be included.

Three mathematical models for external hydrodynamic force have been considered in this study, as described below.

3.2 (a) First Mathematical Model

For the first model [9, 13], we consider the following structure of matrices representing the external hydrodynamic forces.

$[B_1] =$

$$\begin{bmatrix} X_{uu} & X_{vv} & X_{ww} & Z_{\dot{w}} & Z_{\dot{q}} & -Y_{\dot{r}} & -Y_{\dot{v}} & X_{\delta s\delta s} & X_{\delta r\delta r} \\ X_{\dot{u}} & Y_v & Y_{v|v|} & -Z_{\dot{w}} & -Z_{\dot{q}} & Y_r & Y_{\delta r} & 0 & 0 \\ Z_w & Z_{w|w|} & Y_{\dot{v}} & -X_{\dot{u}} & Z_q & Y_{\dot{r}} & Z_{\delta s} & 0 & 0 \\ K_{vw} & K_p & K_{v\dot{q}} & K_{\dot{q}r} & K_{w\dot{r}} & K_{\delta s} & K_{\delta r} & z_G W & 0 \\ M_{u\dot{w}} & M_w & M_{w|w|} & -N_{\dot{r}} & -Y_{\dot{r}} & -Z_{\dot{q}} & M_q & M_{\delta s} & -z_G W \\ N_{uv} & N_v & N_{v|v|} & Z_{\dot{q}} & M_{\dot{q}} & Y_{\dot{r}} & N_r & N_{\delta r} & 0 \end{bmatrix} \tag{15}$$

where,

$$K_{vw} = Z_{\dot{w}} - Y_{\dot{v}} \qquad K_{vq} = Y_{\dot{r}} + Z_{\dot{q}}$$

$$K_{qr} = N_{\dot{r}} - M_{\dot{q}} \qquad K_{wr} = -Y_{\dot{r}} - Z_{\dot{q}} \tag{16}$$

$$M_{uw} = X_{\dot{u}} - Z_{\dot{w}} \qquad N_{uv} = Y_{\dot{v}} - X_{\dot{u}}$$

The relations of (14) above hold good for any model, since the combination of some acceleration-dependent HDCs are merely being represented in terms of the state variables with which they are multiplied in the respective force / moment equation.

In the first model considered, we have

$[C_1] =$

$$\begin{bmatrix} u^2 & ur & w & vw & uw & uv \\ v^2 & v & w|w| & p & w & v \\ w^2 & v|v| & vp & vq & w|w| & v|v| \\ wq & wp & uq & qr & rp & wp \\ q^2 & pq & q & wr & vp & pq \\ r^2 & r & pr & \delta s & uq & ur \\ vr & \delta r & \delta s & \delta r & q & r \\ (\delta s)^2 & & & c\theta.s\phi & \delta s & \delta r \\ (\delta r)^2 & & & & s\theta & \end{bmatrix} \tag{17}$$

where $c\theta$ denotes $cos\theta$, $s\theta$ denotes $sin\theta$, and δr and δs denote rudder and stern plane deflection, respectively.

3.2 (b) Second Mathematical Model

The second model considered here is a partly linearised version of the first model, dropping squared velocity terms and retaining only those second order terms which are cross-products of velocity components. Thus, we have:

$[B_2] =$

$$\begin{bmatrix} 0 & 0 & 0 & Z_{\dot{w}} & 0 & 0 & -Y_{\dot{v}} & 0 & 0 \\ X_{\dot{u}} & Y_v & 0 & -Z_{\dot{w}} & -Z_{\dot{q}} & Y_r & Y_{\delta r} & 0 & 0 \\ Z_w & 0 & Y_{\dot{v}} & -X_{\dot{u}} & Z_q & Y_{\dot{r}} & Z_{\delta s} & 0 & 0 \\ K_{vw} & K_p & K_{vq} & K_{qr} & K_{wr} & K_{\delta s} & K_{\delta r} & z_G W & 0 \\ M_{uw} & M_w & 0 & -N_{\dot{r}} & -Y_{\dot{r}} & -Z_{\dot{q}} & M_q & M_{\delta s} & -z_G W \\ N_{uv} & N_v & 0 & Z_{\dot{q}} & M_{\dot{q}} & Y_{\dot{r}} & N_r & N_{\delta r} & 0 \end{bmatrix} \quad (18)$$

and

$[C_2] =$

$$\begin{bmatrix} 0 & ur & w & vw & uw & uv \\ 0 & v & 0 & p & w & v \\ 0 & 0 & vp & vq & 0 & 0 \\ wq & wp & uq & qr & rp & wp \\ 0 & pq & q & wr & vp & pq \\ 0 & r & pr & \delta s & uq & ur \\ vr & \delta r & \delta s & \delta r & q & r \\ 0 & 0 & 0 & c\theta.s\phi & \delta s & \delta r \\ 0 & 0 & 0 & 0 & s\theta & 0 \end{bmatrix} \quad (19)$$

3.2 (c) Third Mathematical Model

The third model considered retains only strictly linear terms. Thus,

$[B_3] =$

$$\begin{bmatrix} 0 & 0 & 0 & 0 & 0 & 0 & 0 & 0 & 0 \\ 0 & Y_v & 0 & 0 & 0 & Y_r & Y_{\delta r} & 0 & 0 \\ Z_w & 0 & 0 & 0 & Z_q & 0 & Z_{\delta s} & 0 & 0 \\ 0 & K_p & 0 & 0 & 0 & K_{\delta s} & K_{\delta r} & z_G W & 0 \\ 0 & M_w & 0 & 0 & 0 & 0 & M_q & M_{\delta s} & -z_G W \\ 0 & N_v & 0 & 0 & 0 & 0 & N_r & N_{\delta r} & 0 \end{bmatrix} \quad (20)$$

and

$[C_3] =$

$$\begin{bmatrix} 0 & 0 & w & 0 & 0 & 0 \\ 0 & v & 0 & p & w & v \\ 0 & 0 & 0 & 0 & 0 & 0 \\ 0 & 0 & 0 & 0 & 0 & 0 \\ 0 & 0 & q & 0 & 0 & 0 \\ 0 & r & 0 & \delta s & 0 & 0 \\ 0 & \delta r & \delta s & \delta r & q & r \\ 0 & 0 & 0 & c\theta.s\phi & \delta s & \delta r \\ 0 & 0 & 0 & 0 & s\theta & 0 \end{bmatrix} \quad (21)$$

Other mathematical models such as those given in [14, 15] can also be cast in this form, requiring additional HDCs to be evaluated / estimated. These models have retained many additional (cross-coupled and non-linear) HDC's, in addition to those obtained strictly by Taylor series expansions, based on the best-fit for measured forces in model testing.

4. TRAJECTORY SIMULATION

4.1 ALGORITHM FOR SIMULATION

The dynamic equations summarised in equation (7) can now be recast as:

$$\{\dot{u}\ \dot{v}\ \dot{w}\ \dot{p}\ \dot{q}\ \dot{r}\}^T = [A]^{-1}\{\{F_E\} + \{F_I\}\}$$
$$= [A]^{-1}\{\{diag[B].[C] + \{F_I\}\} \quad (22)$$

The following transformation relations exist between the velocities in the body system *oxyz* and the inertial system $Ox_O y_O z_O$:

$$\dot{x}_0 = c_2 c_3 u + (-c_1 s_3 + s_1 s_2 c_3)v + (s_1 s_3 + c_1 s_2 c_3)w$$
$$\dot{y}_0 = c_2 s_3 u + (c_1 c_3 + s_1 s_2 s_3)v + (-s_1 c_3 + c_1 s_2 s_3)w$$
$$\dot{z}_0 = -s_2 u + s_1 c_2 v + c_1 c_2 w \quad (23)$$

$$\dot{\phi} = p + (s_1 s_2 / c_2)q + (c_1 s_2 / c_2)r$$
$$\dot{\theta} = c_1 q - s_1 r \quad (24)$$
$$\dot{\psi} = (s_1 / c_2)q + (c_1 / c_2)r$$

where

$$s_1 = \sin\phi; \quad s_2 = \sin\theta; \quad s_3 = \sin\psi$$
$$c_1 = \sin\phi; \quad c_2 = \sin\theta; \quad c_3 = \sin\psi \quad (25)$$

Equations (22, 23, 24) are thus 12 ordinary linear differential equation whose integration in time will determine the values of vehicle position X_O, Y_O, Z_O and orientation ϕ, θ, ψ with time. There are numerous integration schemes for integration of such ordinary linear differential equations, starting from simplest 1st order explicit schemes to high order implicit schemes such as Runge-Kutta schemes, Predictor-corrector schemes, etc. These (the dynamic motion) equations are usually very robust and do not lead to numerical instability due to integration scheme if the time step size is sufficiently small.

By substituting the HDC values in the above equations of motion, the equations of motion can thus be integrated for known control inputs and speed, to calculate the accelerations, velocities and displacements (position) of the body as function of time. Thus, the trajectory of an underwater vessel can be simulated [5, 13].

4.2 TYPICAL RESULTS

A flexible trajectory simulation program was created, with scope for modifying the external force matrices $[B]$ and $[C]$, thus enabling the mathematical model to be readily changed.

Results of a 15/5 zigzag manoeuvre at 5 knots for the axisymmetric body and for SUBOFF geometry, both using the first mathematical model (3.2(a) above) are shown in Figures 3 and 4. For the trajectory simulations shown in Figures 3 and 4, the main parameters of interest are shown in Table 1.

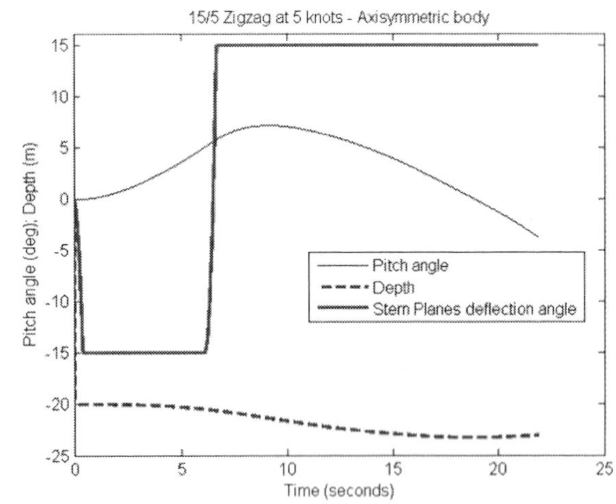

Figure 3 Simulation of 15/5 Zigzag at 5 knots – Axisymmetric body

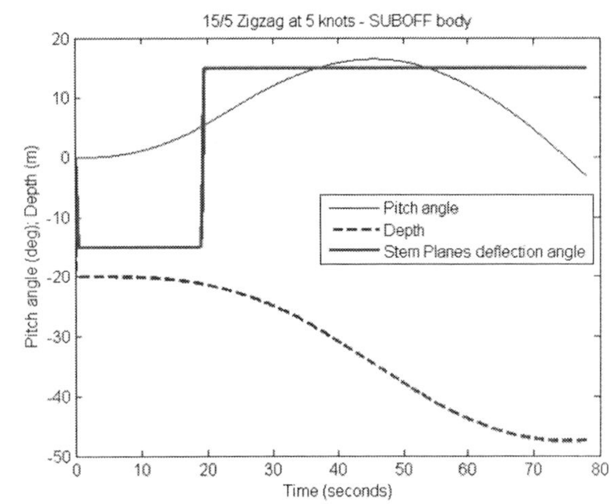

Figure 4 Simulation of 15/5 Zigzag at 5 knots – SUBOFF body

To examine the effect of varying the mathematical model, the same manoeuvre was simulated using different mathematical models (described in sub-section 3.2 above). Results are plotted for two of the models in Figure 5 (for axisymmetric body) and the parameter values obtained for all three mathematical models are given in Table 2.

Parameter	Axisym-metric body	SUBOFF body
Overshoot pitch angle (θ_O); degrees	2.16	11.43
Overshoot change of depth (z_O); metres	2.81	26.34
Time to reach execute (t_e); seconds	6.12	18.96
Time to check pitch (t_c); seconds	9.16	45.48
Time to check depth (t_d); seconds	18.94	74.94

Table 1 Parameters estimated by trajectory simulation for 15/5 zigzag at 5 knots using first mathematical model (linear terms only)

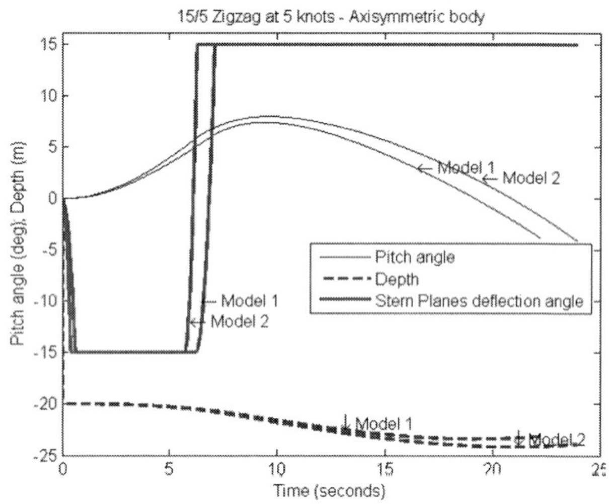

Figure 5 Effect of variation in mathematical model on trajectory simulation for axisymmetric body

Para-meters of Zigzag →	θ_O (deg)	z_O (m)	t_e (sec)	t_c (sec)	t_d (sec)
	(Difference from Model 3 values given in %)				
Model 1	2.40	2.91	6.28	9.44	19.14
	11.1%	*3.6%*	*2.6%*	*3.1%*	*1.1%*
Model 2	2.95	3.74	5.76	9.66	20.86
	36.6%	*33.1%*	*5.9%*	*5.5%*	*10.1%*
Model 3 (fully linear)	2.16	2.81	6.12	9.16	18.94

Table 2 Effect of variation in mathematical model on trajectory simulation results (15/5 zigzag at 5 kts) for axisymmetric body.

Examining the results shown in Table 2, it is seen that changing the mathematical model from strictly linear to non-linear terms causes variation in parameter values in the range of 1 to 37% for the manoeuvre and the body considered. In the sequel, the significance of this level of uncertainty will be examined in light of the other uncertainties in the manoeuvring prediction process.

5. VALUES OF HYDRODYNAMIC COEFFICIENTS

The main unknowns in the procedure of prediction of manoeuvring characteristics are the numerous HDC's. It is desirable that once the geometry of the vessel is defined in the early stages of the design, these derivatives are estimated by empirical methods based on similar vessels, or other methods, so that necessary checks can be made as to whether the design satisfies the various manoeuvring requirements.

As described in Section 3, the choice of mathematical model dictates which HDCs that need to be evaluated. Of the 36 first-order HDC's possible, many can be neglected depending on symmetry of the body [4, 16]. The model test data available in [11] includes only 21 linear HDCs. The number of second-order HDCs and various correction term may vary from model to model. The mathematical models described in sub-sections 3.2(a) (Model 1) and 3.2(b) (Model 2) contain 55 and 24 HDCs, respectively.

The values of the hydrodynamic coefficients (HDCs) in the equations of motion have their own levels of uncertainty depending upon the model testing methods and facilities used to determine them. For the SUBOFF body, [11] the reported levels of uncertainty are listed in Table 3.

HDC	Estimated Uncertainty
Static HDCs: Z_w', M_w', Y_v', N_v'	4 – 5 %
Rotary HDCs: Z_q', M_q', Y_r', N_r'	10%
Control HDCs	6 – 10%
Added mass HDCs: $Z_{\dot{w}}'$, $M_{\dot{q}}'$, $Y_{\dot{v}}'$, $N_{\dot{r}}'$	7%

Table 3 Uncertainty estimated for various HDCs for SUBOFF body obtained by model testing [11]

The error level in one coefficient may be masked by errors in others, such that the margins of stability computed give misleading results, or such that the trajectories computed are similar for different combinations of HDC values.

Sensitivity studies have been carried out for HDCs of underwater vehicles in order to determine their relative importance and hence, the error bounds permissible for their values. In general, sensitivity has been defined as the ratio of the relative change in some output variable, O, and the relative change in an input parameter, I, each compared to nominal values (O_{nom}, I_{nom}). For each case the sensitivity, S, of the response to the variation in parameter can be calculated as [10]:

$$S = \frac{(O - O_{nom}) / O_{nom}}{(I - I_{nom}) / I_{nom}} \tag{26}$$

The influence of various hydrodynamic coefficients on the predicted manoeuvrability of submerged bodies has been examined [9, 10] and it is reported that for a submarine-like body, trajectories are most sensitive to linear damping coefficients. For a bare hull axisymmetric body, the linear inertial coefficients were found to be the most significant. In another study [17], the sensitivity of geometrical characteristics of an AUV vis-à-vis its added mass HDCs have been explored. Sensitivity analysis using Genetic Algorithms [18] for various manoeuvres has revealed that difference in the model geometry caused HDCs to have different tendencies in sensitivity change. Also, the nature of sensitivity changes depending on the trials executed.

Considering the focus of this study on vertical plane manoeuvres, the most significant coefficients for axisymmetric body have been identified [9] as: $M_{\delta s}, Z_{\dot{w}}, Z_{\delta s}, M_{\dot{w}}, M_{\dot{q}}, M_q, Z_{\dot{q}}$. For submarine-like body, these are: $M_q, M_{\delta s}, Z_{\delta s}, Z_q, Z_{\dot{w}}, Z_w, Z_{\dot{q}}, M_{\dot{w}}, M_{\dot{q}}$. The difference in relative sensitivity has been attributed to the difference in the mathematical model used for the two bodies, but may also be the result of the difference in their geometries. In order to ensure that the vertical plane trajectories do not vary by more than 10%, it is reported [9] that the HDCs that need to be determined within about 10% accuracy are $Z_{\dot{w}}, M_{\delta s}, Z_{\dot{q}}, Z_{\delta s}, M_{\dot{q}}, M_{\dot{w}}$ for axisymmetric body. Based on the study using Genetic Algorithm [18], M_w and M_q were found to be the two most significant HDCs.

With this background, the HDCs considered most significant for the axisymmetric body were varied, one at a time, by ±25 to 50% of their initial values and the trajectory simulation was repeated using the linear mathematical model (Model 3). To consider the extreme case, all the HDC values were simultaneously changed by ±100% from initial values and trajectory simulation repeated. Results of these sensitivity studies are summarised in Table 4. These results are graphically shown in Figure 6, wherein the cumulative effect of changes in each HDC is shown for the five parameters of the vertical plane zigzag manoeuvre.

It is seen that the three most important HDCs for the case considered are $M_{\delta S}$, $Z_{\delta S}$ and M_q. Change in value of these HDCs by ±50% causes change in at least one of the vertical plane zigzag manoeuvre parameters by more than 50%. The marked influence of HDCs $M_{\delta S}$, $Z_{\delta S}$ and M_q is clearly evident from Figure 6.

Change in HDC by ↓		% Change in values of Zigzag Parameters				
		θ_O	z_O	t_e	t_c	t_d
$M_{\delta s}$	+50%	56.3	3.2	-23.9	-13.5	-18.0
	-25%	-48.5	-26.1	29.1	10.5	2.2
$Z_{\dot{w}}$	+50%	8.0	-4.6	-2.9	-1.5	-6.5
	-50%	-18.2	21.8	7.2	3.9	28.5
$Z_{\delta s}$	+50%	-50.6	-45.2	11.1	-6.6	-20.1
	-50%	68.7	55.5	-7.2	14.6	12.0
$Z_{\dot{q}}$	+50%	-2.9	-1.7	0.7	-0.2	-0.2
	-50%	3.0	1.8	-0.7	0.2	0.3
$M_{\dot{w}}$	+50%	-2.1	-2.2	-0.7	-1.1	-1.3
	-50%	0.5	1.4	0.3	0.7	1.0
$M_{\dot{q}}$	+50%	-6.9	15.3	21.2	19.7	18.9
	-50%	0.5	-26.5	-25.8	-26.9	-25.9
M_q	+50%	-38.3	-28.7	9.8	-3.1	-8.7
	-50%	78.2	110.8	-8.2	18.3	44.2
M_w	+50%	-26.5	-6.5	7.5	-0.7	10.6
	-50%	17.6	-5.5	-5.6	-2.4	-10.7
All	+50%	-31.9	-34.9	3.9	-7.0	-17.8
	-50%	163.6	197.5	-6.5	69.0	50.4

Table 4 Results of sensitivity studies of HDC values for vertical plane zigzag - axisymmetric body

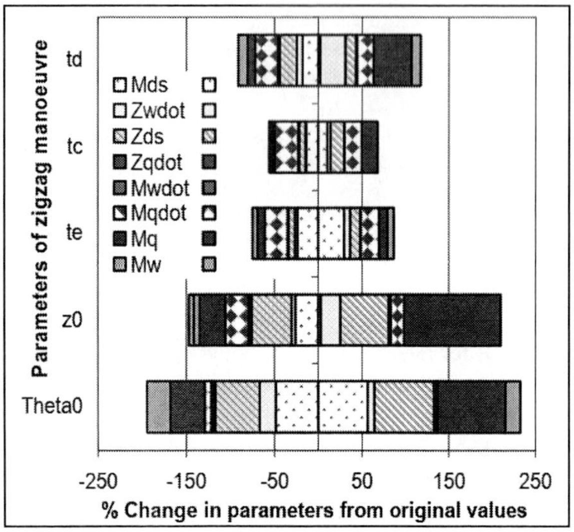

Figure 6 Graphical representation of results of sensitivity analysis showing percentage change in zigzag parameters due to ±50% change in HDCs

Figure 6 also shows that the effect of variation in each HDC may affect the value of overshoot angle (θ_O) by +275% to -250%. However, as indicated in Table 3, the effect of simultaneous change in HDC values by ±50% is less than the sum of the changes in any parameter value due to each of the HDCs by ±50%. Around 200% increase in z_O and 160% increase in θ_O is expected in case of simultaneous reduction in all listed HDC values by 50%.

It is also seen that the parameters time to reach execute (t_e) and time to check pitch (t_c) and depth (t_d) are relatively less influenced (about ±100%) by change in HDC values, as compared to the parameters θ_O and z_O. The effect of change in trajectory of the body due to simultaneous change of ±50% in all listed HDCs is shown in Figure 7. The original response curves are in the middle, while quicker responses are for the case when HDCs were increased by 50%. Responses were delayed when all HDC values were reduced by 50%.

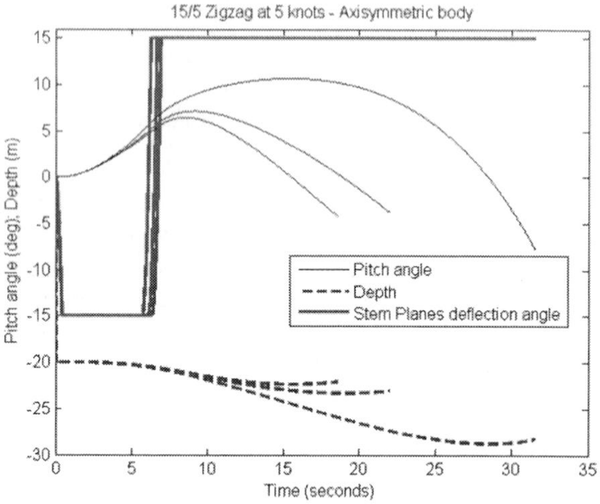

Figure 7 Change in trajectory simulation results for axisymmetric body due to change in HDCs by ±50%

These studies were repeated for the submarine-like SUBOFF body, varying all HDCs varied for axisymmetric body, as well as Z_q, again using the linear mathematical model (Model 3). Results of these sensitivity studies are summarised in Table 5. Only those HDCs which had an effect of more than 2% on the zigzag parameters have been included in the Table. The most significant HDCs for the SUBOFF body simulation emerge as $Z_{\dot{w}}, Z_{\delta s}, M_w$. It is thus seen that even for the same mathematical model, the body geometry has a significant influence on the relative importance of the HDCs.

Change in HDC by ↓		% Change in Zigzag Parameters				
		θ_O	z_O	t_e	t_c	t_d
$M_{\delta s}$	+50%	-3.6	-7.7	-5.6	-5.6	-5.3
	-50%	4.4	9.4	6.6	6.5	6.1
$Z_{\dot{w}}$	+50%	15.7	19.1	5.3	11.5	11.4
	-50%	-17.4	-19.4	-6.4	-12.0	-11.4
$Z_{\delta s}$	+50%	17.3	0.3	-11.4	-6.6	-6.6
	-50%	-20.8	1.3	20.5	13.1	13.8
M_w	+50%	13.6	-2.4	-11.9	-9.5	-10.3
	-50%	-18.4	3.9	21.6	16.6	18.5
Z_q	+50%	-2.33	-2.58	0.11	-1.01	-0.93
	-50%	2.43	2.72	-0.11	1.01	0.96
All	+50%	54.9	16.0	-20.0	-5.9	-6.0
	-50%	-37.9	11.0	57.1	39.0	41.7

Table 5 Results of sensitivity studies of HDC values for vertical plane zigzag - SUBOFF body

The change in zigzag manoeuvre parameters due to HDC value variations does not exceed 21% when HDCs are varied one at a time. This is much lesser than the effect of variations in case of axisymmetric body. Even when all relevant HDCs are varied simultaneously by ±50% from original values, the variation in zigzag parameters is up to 55% only. In actual terms, the magnitude of variation is much greater, but percentage change is relatively less. The change in trajectory for simultaneous variation in HDCs by ±50% for SUBOFF body is shown in Figure 8. As in Figure 7, the quickest responses correspond to 50% increased HDC values.

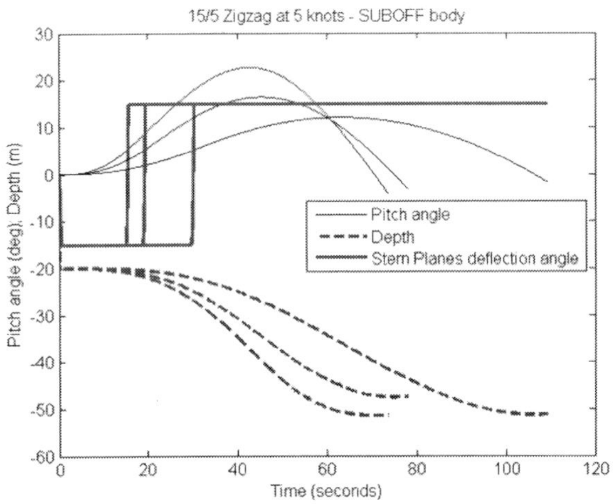

Figure 8 Change in trajectory simulation results for SUBOFF body due to change in HDCs by ±50%

6. UNCERTAINTY IN FULL-SCALE TRIALS

For each stage of the trajectory simulation process described above (formulation of mathematical model, program for trajectory simulation and estimating values of HDCs), the ultimate check for the manoeuvrability prediction procedure is to compare the predicted trajectory with the measured values during full-scale trials.

However, such comparisons are not only rare [2, 7, 19], but are also fraught with the uncertainties involved with the uncertainties due to noise in the data obtained during full-scale trials, since laboratory conditions (say, of model testing for obtaining HDCs) cannot be replicated.

It is therefore necessary to quantify typical levels of uncertainty in the full-scale measurement process in order to arrive at realistic estimates of the accuracy required of the entire process of manoeuvrability prediction.

6.1 SOURCES OF ERRORS

As per uncertainty analysis described in [11], the errors in measurement can be of two types: bias errors (which are constant throughout an experiment, affecting all measurements in the same sense) and precision errors (which are random scatter in results during an experiment).

Focusing our attention on zigzag manoeuvres in the vertical plane for underwater vehicles, we may identify the possible sources of bias errors in full-scale trials as:

- Errors in measuring instruments for depth, pitch angle and speed of the vehicle
- Set in inertial navigation or other motion / position-recording instruments
- Hydrostatic imbalance of the vehicle
- Changes in temperature of water with depth (affecting density and viscosity)
- Ocean currents at different depths

Sources of precision errors may be listed as follows:

- Errors in recording and processing data, depending on frequency of measurements
- Unknown and random sea water disturbances
- Errors due to variation in time for applying control forces

6.2 UNCERTAINTY ESTIMATES

Most of the above errors are impossible to quantify exactly, but the expected variations can be estimated, albeit somewhat crudely.

6.2(a) Instrumentation Bias

During full-scale sea trials on manned underwater vehicles reported in [19], the least count for on-board instrumentation of various parameters was as follows.

- Depth: 1 metre
- Pitch angle: 1 degree
- Time: 0.5 seconds
- Plane angle: 1 degree
- Speed: 0.5 knots

Based on the above values, the uncertainty in each of the parameters of interest in the zigzag manoeuvre may be quantified for assumed sample values, as shown in Table 6. These are close to values for axisymmetric body parameters (Table 1) for 15/5 zigzag.

Parameter	Typical Value	Instrument Uncertainty
Overshoot pitch angle (θ_O)	2 degrees	0.50
Overshoot change of depth (z_O)	5 metres	0.20
Time to reach execute (t_e)	12 seconds	0.017
Time to check pitch (t_c)	30 seconds	0.008
Time to check depth (t_d)	60 seconds	0.006

Table 6 Typical bias error estimated based on least count of measuring instruments

6.2(b) Sea Water Density Variation

The variation in density of sea water of tropical waters may be from 1.015 to 1.030 tons/m^3. For a vehicle of displacement 2000 tons, this implies a variation in buoyancy by 30 tons. The most straightforward way of countering this is to adjust variable ballast to achieve neutral buoyancy at a certain depth, at zero or near-zero forward speed. The bias error in this case is in lack of achieving neutral buoyancy, which may be masked due to non-zero speed of the vehicle. Additionally, change in buoyancy of the vehicle may result due to change in consumables (variable weights) or hull compressibility. Even if neutral buoyancy is achieved at a certain depth, the change of depth in a region with sharp temperature gradient may cause imbalance between weight and buoyancy, which may be as much as 0.25% of the vehicle displacement. The effect of this magnitude of change during the course of a typical zigzag manoeuvre is shown for two body geometries in Table 7.

Parameters of Zigzag →	t_e	θ_O	z_O
Axisymmetric body			
(a) Original weight	5.76 sec	2.95 deg	3.74 m
(b) For Weight increased by 0.25%	5.76 sec	2.96 deg	3.76 m
SUBOFF body			
(a) Original weight	18.96 sec	11.43 deg	18.80 m
(b) For Weight increased by 0.25%	18.96 sec	11.43 deg	18.81 m

Table 7 Effect of change in weight on simulated 15/5 zigzag manoeuvre for two body geometries

It may be seen that variation in depth overshoot and pitch angle overshoot are marginal, while change in time to reach execute pitch angle is insignificant. The uncertainty due to this source of error may thus be roughly estimated (on the higher side) as 0.003 for θ_O and 0.005 for z_O.

6.2(c) Errors in Data Recording

Precision errors in recording and processing data may be estimated by repeating a particular manoeuvre at different points of time, in the same location in the ocean and without changing the vehicle trim. It is near-impossible to meet these requirements in order to estimate this source of precision error in isolation. However, estimated values are of the order of the least count of the instruments for depth, pitch angle, time and speed, leading to additional uncertainty values similar to those listed in Table 6.

6.2(d) Error in Control Force Application

Precision errors due to control force application result from inexact or asynchronous application of plane angles. Particularly in case of manned submarines, the time required to apply plane angle when ordered may vary from operator to operator and may not exactly coincide with the moment the execute pitch angle is reached. Typical variations are of the order of 2 seconds, leading to uncertainties in all time parameters of the zigzag manoeuvre by the same amount, and a variation in the overshoot angle by as much as 1-2 degrees, and variation in overshoot change of depth by 2-5 metres. These values are typical for low speeds (3 to 5 knots) and may be somewhat higher for greater speeds. The consequent uncertainties in the parameters of interest, for sample values listed in Table 6, may be (lower values): 0.5 for θ_O , 0.4 for z_O, 0.167 for t_e, 0.067 for t_c, and 0.033 for t_d.

6.3 TOTAL UNCERTAINTY

In the worst case scenario, all the errors quantified above can occur simultaneously and hence the uncertainties estimated due to various sources of error would add up. Since all possible sources of error have not been quantified, this estimate is not necessarily conservative (i.e., the actual error may be even greater). Since all parameters of interest in the zigzag manoeuvre are measured directly in the full-scale trials, simple addition of the uncertainties due to the sources of error would suffice. Thus, summing up the estimated uncertainty values for various sources of error described above, the results are given in Table 8.

Parameter	Assumed Value	Total Uncertainty
Overshoot pitch angle (θ_O)	2 degrees	1.503
Overshoot change of depth (z_O)	5 metres	0.805
Time to reach execute (t_e)	30 seconds	0.250
Time to check pitch (t_c)	60 seconds	0.133
Time to check depth (t_d)	90 seconds	0.050

Table 8 Typical uncertainty levels estimated based on various sources of bias and precision errors

It may be noted that these estimates are only meaningful for values of parameters close to those assumed as 'typical'. For different speeds, and certainly for different vehicles, these values would vary and accordingly the uncertainty levels would also change. In general, it is likely that for larger bodies and / or greater speeds (or plane deflections), the values of parameters measured would be greater. In such a scenario, the associated uncertainties are likely to be lower.

To illustrate this aspect, the estimated total uncertainty for another set of parameters (for a case of higher speed or a larger vehicle or a greater control surface deflection) is shown in Table 9. It may be noted that these values are closer to those for the SUBOFF body in 15/5 zigzag (Table 1), which is a larger and less stable body than the axisymmetric body considered.

Parameter	Assumed Value	Total Uncertainty
Overshoot pitch angle (θ_O)	8 degrees	0.378
Overshoot change of depth (z_O)	15 metres	0.272
Time to reach execute (t_e)	15 seconds	0.200
Time to check pitch (t_c)	40 seconds	0.100
Time to check depth (t_d)	70 seconds	0.043

Table 9 Typical uncertainty levels estimated for another set of parameter values (for higher speeds / larger body/ greater plane deflection)

7. CONCLUSIONS

The process of manoeuvrability prediction of underwater vehicles is fraught with uncertainties at every stage. In view of the criticality of the application, it is important to have an overall perspective of the subject in order to ensure that estimates at every stage are reasonably accurate, while also optimising the time and effort required in the process.

This study has considered two body geometries. A generalised trajectory simulation program was developed, using which simulations have been carried out using three mathematical models, for a variety of cases. Most results have been presented for a typical 15/5 vertical plane zigzag manoeuvre at 5 knots. Conclusions have been drawn regarding the effect of various choices on the values of main parameters obtained from simulating this manoeuvre.

It was seen that choice of mathematical model (linear / non-linear) can influence the zigzag parameters by 1% to 37%, with the greater effect on the pitch overshoot and depth overshoot, which are measures of the depth-keeping ability of the underwater vehicle.

Estimation of HDC values (whose number is determined by choice of mathematical model) is usually the most time-consuming aspect of manoeuvrability prediction. Experimental uncertainty in estimation of HDC values has been reported in the range of 5 to 10%.

Based on a sensitivity analysis of effect of HDC values on vertical plane zigzag parameters, it was seen that for the axisymmetric body considered, $M_{\delta s}$, $Z_{\delta s}$ and M_q are most significant, causing more than by 50% change in at least one of the zigzag parameters when any one HDC is varied by ±50%. However, for the SUBOFF body (which is more akin to a submarine form), the most significant HDCs emerge as $Z_{\dot{w}}, Z_{\delta s}, M_w$, for which individual ±50% change does not cause change of zigzag parameters by more than 21%. If all relevant HDCs are varied by ±50% simultaneously, the zigzag parameters change by up to 55% for SUBOFF and up to almost 200% for the axisymmetric body. Thus, the sensitivity of HDCs depends not only on the mathematical model adopted, but also on the manoeuvre considered and the body geometry being studied.

Summing up the minimum and maximum possible uncertainties in the stages of manoeuvrability prediction thus discussed, a figure of uncertainty of 60% to 250% may be suggested for the parameters of the manoeuvre considered in this study.

Uncertainty estimation for various sources of error has been discussed for full-scale sea trials, which are the ultimate benchmark for any prediction process. It is seen that the uncertainty levels in the measurement process depend on many unknown factors, and also on the values of manoeuvre parameters. Based on two typical cases, it is estimated that error magnitude of various parameters during full-scale measurements may vary from 5% to 150%. This value, though large, appears smaller than the cumulative uncertainty (60-250%) in the overall prediction process.

It needs to be remembered that predictions need to be only as accurate as can be measured and verified. The quantification of uncertainties at various stages can offer insight into margins for improvement at various stages of the prediction process. Further studies may expand the approach adopted for this study to other definitive manoeuvres, other bodies and other mathematical models.

8. ACKNOWLEDGEMENT

This study was undertaken as part of a project supported by the Naval Research Board (NRB) of the Defence Research & Development Organisation (DRDO), India, which is gratefully acknowledged by the authors.

9. REFERENCES

1. RAY, A., SESHADRI, V., SINGH, S.N., SEN, D., 'Manoeuvring Studies of Underwater Vehicles – A Review', *Transactions of RINA*, 2008.

2. ARENTZEN, A. and MANDEL, P., 'Naval Architectural Aspects of Submarine Design', *SNAME Transactions*, 1960.

3. SPENCER, J.B., 'Stability and Control of Submarines', *Journal of the Royal Navy Scientific Service (JRNSS)*, Vol. 23, No. 3, 1967.

4. ABKOWITZ, M.A., *Stability and Motion Control of Ocean Vehicles*, MIT Press, 1969.

5. FOSSEN, T.I., *Guidance and Control of Ocean Vehicles*, John Wiley, 1994.

6. LEWANDOWSKI, E.M., *The Dynamics of Marine Craft*, World Scientific, 2004.

7. PRESTERO, T., 'Verification of a Six-Degree of Freedom Simulation Model for the REMUS Autonomous Underwater Vehicle', *M.S. Thesis, Massachusetts Institute Of Technology & Woods Hole Oceanographic Institution*, 2001.

8. NAHON, M., 'A Simplified Dynamics Model for Autonomous Underwater Vehicles', *The Journal of Ocean Technology*, Vol. 1, No. 1, 2006.

9. REDDY, D.N., 'Theoretical and Experimental Studies on Motion and Control of Submerged Bodies', *M.S. Thesis, Indian Institute of Technology Kharagpur*, 1996.

10. SEN, D., 'A Study on Sensitivity of Maneuverability Performance on the Hydrodynamic Coefficients for Submerged Bodies', *Journal of Ship Research*, Vol. 44 (3), pp. 186–196, Sep 2000.

11. RODDY, R. F. 'Investigation of the stability and control characteristics of several configurations of the DARPA SUBOFF model (DTRC Model 5470) from captive-model experiments', *David Taylor Research Center Report No.DTRC/SHD-1298-08*, September 1990.

12. LEWIS, E.V. (ed.), 'Principles of Naval Architecture, Vol. III', *SNAME*, 1989.

13. SEN, D., 'A Generalized Hydrodynamic Model for Underwater Bodies', *2nd Indian National Conference on Harbour & Ocean Engineering (INCHOE-97)*, Dec 1997.

14. GERTLER, M. and HAGEN, G.R., 'Standard Equations of Motion for Submarine Simulation', *Report 2510, Naval Ship Research and Development Center*, 1967

15. FELDMAN, J.P., 'DTNSRDC Revised Standard Submarine Equations of Motion'; *DTNSRDC Report SPD-0393-09*, June 1979.

16. PETERSON, R.S., 'Evaluation of Semi-Empirical Methods for Predicting Linear Static and Rotary Hydrodynamic Coefficients', *Naval Coastal Systems Center, NCSC TM-291-80*, June 1980.

17. PERRAULT, D., BOSE, N., O'YOUNG, S., WILLIAMS, C.D., 'Sensitivity of AUV added mass coefficients to variations in hull and control plane geometry', *Ocean Engineering*, Vol. 30, 2003.

18. YEOA, D.J., RHEE, K.P., 'Sensitivity analysis of submersibles' manoeuvrability and its application to the design of actuator inputs', *Ocean Engineering*, Vol. 33, 2006.

19. RAY, A., 'Manoeuvring Trials of Underwater Vehicles', *Journal of Ship Technology*, Vol.3, No. 2, July 2007.

10. AUTHORS BIOGRAPHY

Amit Ray is a Naval Architect posted at Integrated Headquarters MoD(Navy), New Delhi, India. His field of interest and specialization has been submarines, throughout his B.Tech (Hons.), PG Diploma at IIT Delhi and M.Tech. at IIT Kharagpur. He has been involved in various assignments pertaining to submarine repairs and trials, as well as teaching submarine design.

Debabrata Sen is a Professor in the Department of Ocean Engineering and Naval Architecture at the Indian Institute of Technology Kharagpur, India. He is a Naval Architect with over 20 years of teaching and R&D experience at IIT. His area of expertise is marine hydrodynamics, including seakeeping and manoeuvring of marine vehicles.

SN Singh is a Professor in the Department of Applied Mechanics at the Indian Institute of Technology Delhi, India. He is an Aeronautical engineer and has over 20 years of teaching and R&D experience at IIT. His area of expertise is fluid mechanics, CFD, flow instrumentation and allied fields.

V Seshadri is a Chair Professor in the Department of Applied Mechanics at the Indian Institute of Technology Delhi, India, with over 30 years of teaching and R&D experience. His areas of specialisation include pipeline engineering, CFD, development and calibration of fluid devices, and Biofluid mechanics.

QUANTIFICATION OF UNCERTAINTY IN MANOEUVRING CHARACTERISTICS FOR DESIGN OF UNDERWATER VEHICLES

Amit Ray, *IIT Delhi*

Debabrata Sen, *IIT Kharagpur*

SN Singh *&* ***V Seshadri***, *IIT Delhi*

AIM

- To estimate the magnitude of uncertainty for prediction of manoeuvring characteristics during design of underwater vehicles.

- To compare this with uncertainty in full-scale trials undertaken for validation, and thus identify desired levels of accuracy.

SCOPE

- Introduction
- Mathematical models
- Trajectory simulation
- Sensitivity studies for Hydrodynamic coefficients
- Uncertainty in full-scale trials
- Conclusion

INTRODUCTION

- Manoeuvring characteristics of underwater vehicles:
  - Motion stability, controllability, trajectory simulation.
  - Difficult to predict accurately during initial stages of design
- Techniques: empirical/analytical, experimental and numerical/computational.
- Attention is usually concentrated on one particular aspect of the problem (model testing, CFD, control system design, system identification, or trajectory simulation).
- Ultimate benchmark for evaluating any set of predictions should be comparison with full-scale trail results for standard manoeuvres.

ASPECTS OF MANOEUVRING PREDICTION

- Formulation of mathematical model (for trajectory simulation)
- Estimation of values of Hydrodynamic Coefficients (HDCs) for the specific body
- Simulation of trajectories for various manoeuvres
- Full-scale trials for standard manoeuvres and comparison with predictions

UNDERWATER VEHICLES

- Axisymmetric body of revolution with cruciform control surfaces at stern.
  - Length 10 metres
  - Weight 196.5 kg

- SUBOFF body with appendages (fin and cruciform control surfaces).
  - Length 4.26 m
  - Weight 18.1 kg

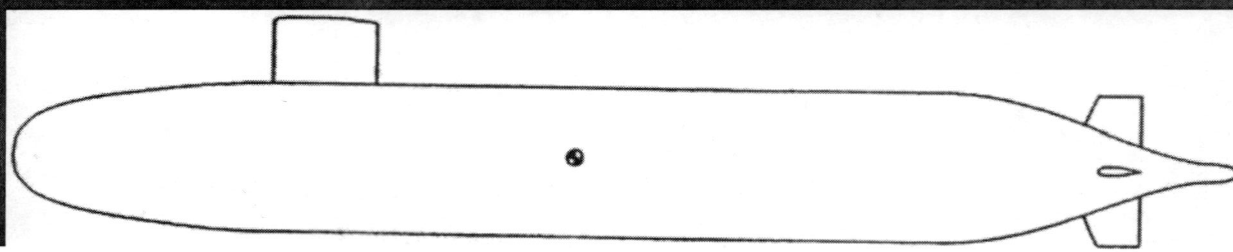

MANEOUVRE CONSIDERED

- Zigzag manoeuvre in vertical plane
  - 15/5 (i.e. δ_1=15 degrees, θ_e =5 degrees)
  - Speed 5 knots

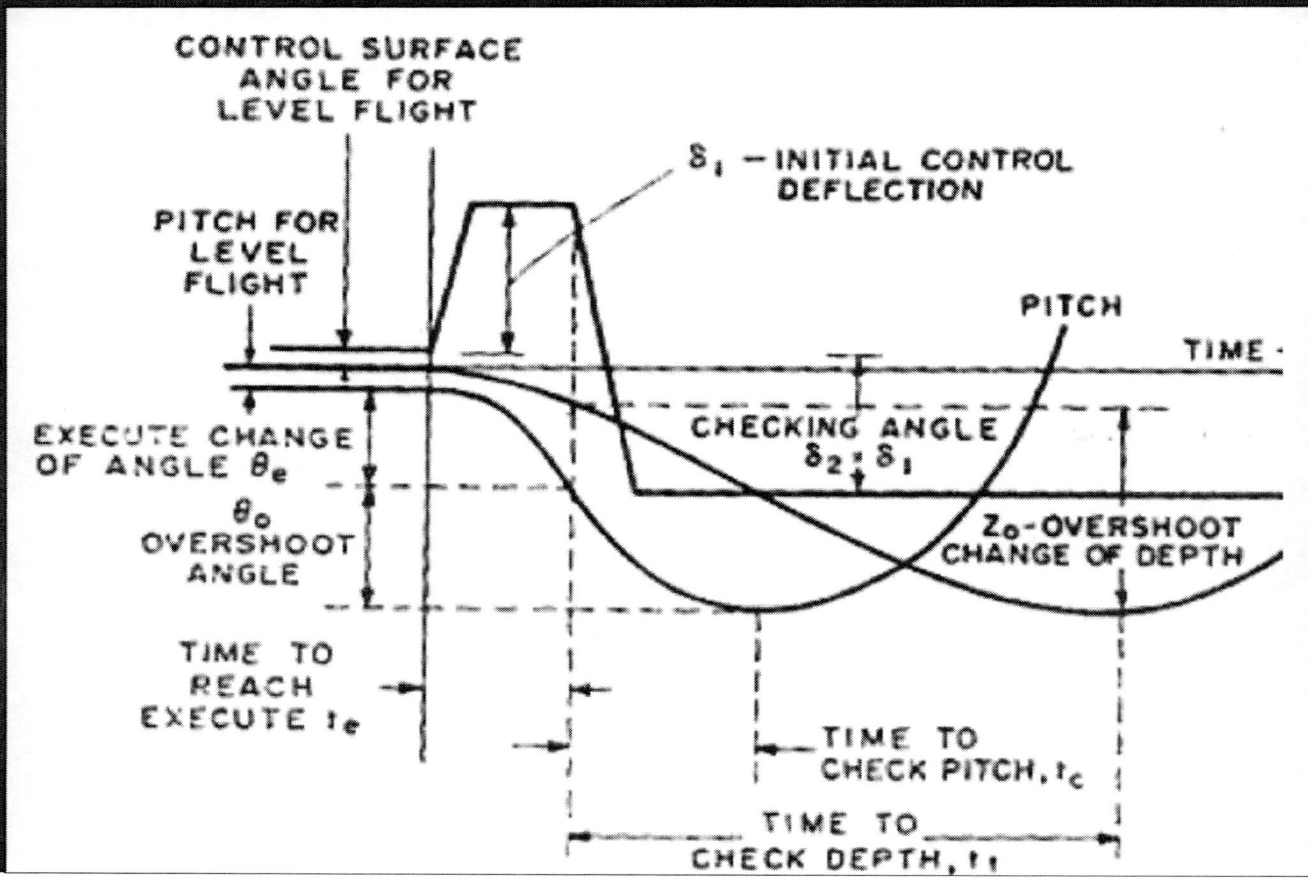

MATHEMATICAL MODELS FOR TRAJECTORY SIMULATION

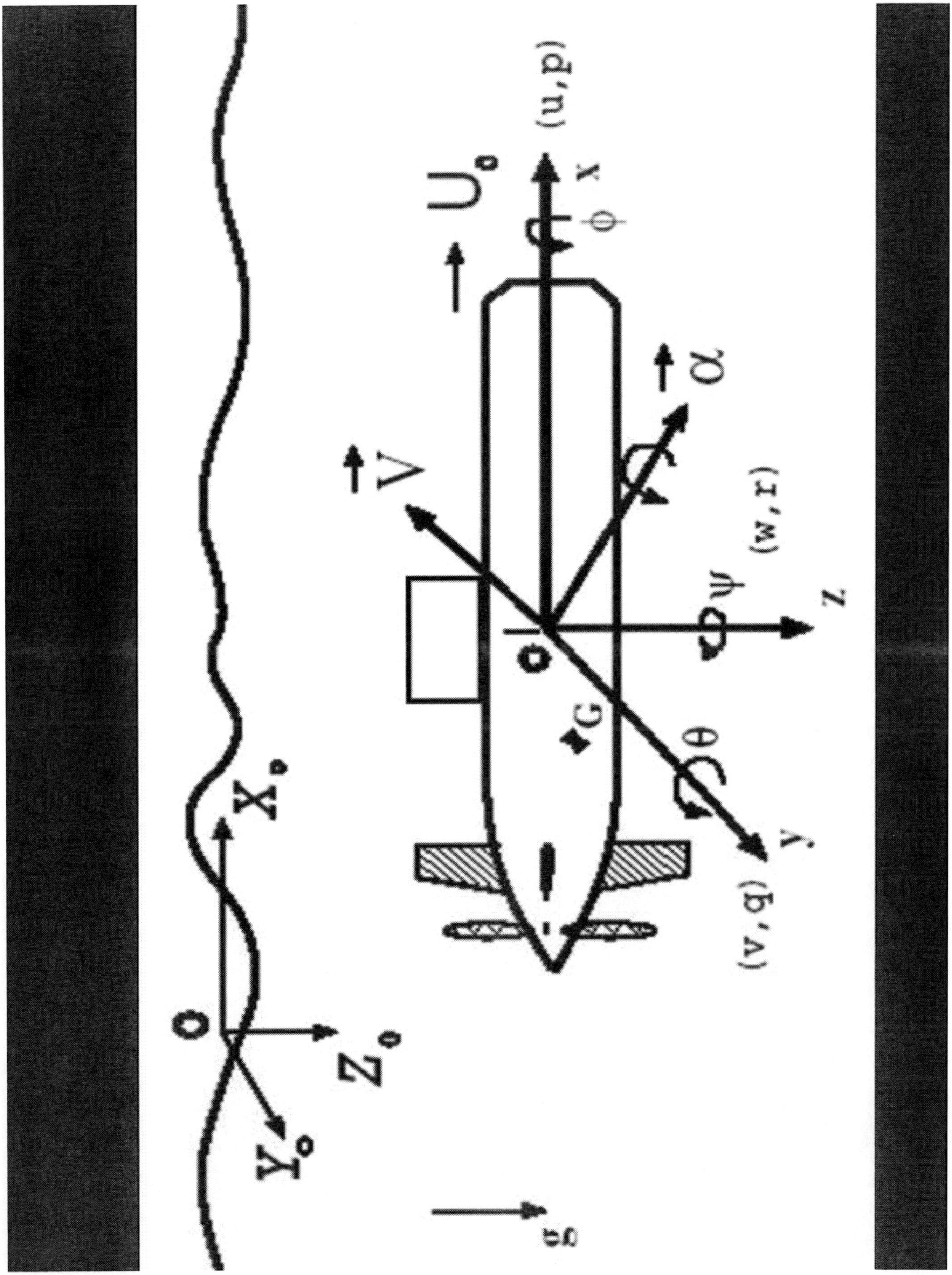
U₀
x
(u,p)
φ
α
V
(w,r)
ψ
z
o
G
θ
y
(v,q)
X₀
O
Z₀
Y₀
g

EQUATIONS OF MOTION

$$[A]\{\dot{V}\} - \{F_I\} = \{F_E\}$$

where,

- mass matrix: $[A] = [\mathrm{M}] + [\mathrm{M}']$
- acceleration: $\{\dot{V}\} = \{\dot{u} \quad \dot{v} \quad \dot{w} \quad \dot{p} \quad \dot{q} \quad \dot{r}\}^T$
- external force: $\{F_E\} = \{X, Y, Z, K, M, N\}^T$

Subscript notation: $X_{\dot{u}} = \partial X / \partial \dot{u}$

VARYING THE MATHEMATICAL MODEL

- In order to generalise the various mathematical models available for expressing $\{F_E\}$ in terms of HDCs, consider $\{F_E\}$ as the product of various HDCs (as matrix [B] of size 6 x *n*) with respective motion variables (as matrix [C] of size *n* x 6), where *n* depends on the particular model (typically 6 to 15).
- Thus,

$$\{F_E\} = diag\ ([B].[C])$$

- This also includes terms corresponding to hydrostatic forces and control forces.

VARIANTS CONSIDERED

- First Mathematical Model
  - Linear and nonlinear terms
- Second Mathematical Model
  - Partly linearised version, dropping squared velocity terms and retaining only those second order terms which are cross-products of velocity components
- Third Mathematical Model
  - Only strictly linear terms

FIRST MODEL

$[B_1]=$

$$
\begin{bmatrix}
X_{uu} & X_{vv} & X_{ww} & Z_{\dot{w}} & Z_{\dot{q}} & -Y_{\dot{r}} & -Y_{\dot{v}} & X_{\delta s \delta s} & X_{\delta r \delta r} \\
X_{\dot{u}} & Y_{v} & Y_{v|v|} & -Z_{\dot{w}} & -Z_{\dot{q}} & Y_{r} & Y_{\delta r} & 0 & 0 \\
Z_{w} & Z_{w|w|} & Y_{\dot{v}} & -X_{\dot{u}} & Z_{q} & Y_{\dot{r}} & Z_{\delta s} & 0 & 0 \\
K_{vw} & K_{p} & K_{v\dot{q}} & K_{\dot{q}r} & K_{w\dot{r}} & K_{\delta s} & K_{\delta r} & z_{G}W & 0 \\
M_{u\dot{w}} & M_{w} & M_{w|w|} & -N_{\dot{r}} & -Y_{\dot{r}} & -Z_{\dot{q}} & M_{q} & M_{\delta s} & -z_{G}W \\
N_{uv} & N_{v} & N_{v|v|} & Z_{\dot{q}} & M_{\dot{q}} & Y_{\dot{r}} & N_{r} & N_{\delta r} & 0
\end{bmatrix}
$$

FIRST MODEL

$$[C_1]=\begin{bmatrix} u^2 & ur & w & vw & uw & uv \\ v^2 & v & w|w| & p & w & v \\ w^2 & v|v| & vp & vq & w|w| & v|v| \\ wq & wp & uq & qr & rp & wp \\ q^2 & pq & q & wr & vp & pq \\ r^2 & r & pr & \delta s & uq & ur \\ vr & \delta r & \delta s & \delta r & q & r \\ (\delta s)^2 & & & c\theta.s\phi & \delta s & \delta r \\ (\delta r)^2 & & & & s\theta & \end{bmatrix}$$

SECOND MODEL

$$[B_2]=\begin{bmatrix} 0 & 0 & 0 & Z_{\dot{w}} & 0 & 0 & -Y_{\dot{v}} & 0 & 0 \\ X_{\dot{u}} & Y_v & 0 & -Z_{\dot{w}} & -Z_{\dot{q}} & Y_r & Y_{\delta r} & 0 & 0 \\ Z_w & 0 & Y_{\dot{v}} & -X_{\dot{u}} & Z_q & Y_{\dot{r}} & Z_{\delta s} & 0 & 0 \\ K_{vw} & K_p & K_{vq} & K_{qr} & K_{wr} & K_{\delta s} & K_{\delta r} & z_G W & 0 \\ M_{uw} & M_w & 0 & -N_{\dot{r}} & -Y_{\dot{r}} & -Z_{\dot{q}} & M_q & M_{\delta s} & -z_G W \\ N_{uv} & N_v & 0 & Z_{\dot{q}} & M_{\dot{q}} & Y_{\dot{r}} & N_r & N_{\delta r} & 0 \end{bmatrix}$$

SECOND MODEL

$$[C_2] = \begin{bmatrix} 0 & ur & w & vw & uw & uv \\ 0 & v & 0 & p & w & v \\ 0 & 0 & vp & vq & 0 & 0 \\ wq & wp & uq & qr & rp & wp \\ 0 & pq & q & wr & vp & pq \\ 0 & r & pr & \delta s & uq & ur \\ vr & \delta r & \delta s & \delta r & q & r \\ 0 & 0 & 0 & c\theta . s\phi & \delta s & \delta r \\ 0 & 0 & 0 & 0 & s\theta & 0 \end{bmatrix}$$

THIRD MODEL

$[B_3]=$

$$\begin{bmatrix} 0 & 0 & 0 & 0 & 0 & 0 & 0 & 0 & 0 \\ 0 & Y_v & 0 & 0 & 0 & Y_r & Y_{\delta r} & 0 & 0 \\ Z_w & 0 & 0 & 0 & Z_q & 0 & Z_{\delta s} & 0 & 0 \\ 0 & K_p & 0 & 0 & 0 & K_{\delta s} & K_{\delta r} & z_G W & 0 \\ 0 & M_w & 0 & 0 & 0 & 0 & M_q & M_{\delta s} & -z_G W \\ 0 & N_v & 0 & 0 & 0 & 0 & N_r & N_{\delta r} & 0 \end{bmatrix}$$

THIRD MODEL

$$[C_3]=\begin{bmatrix} 0 & 0 & w & 0 & 0 & 0 \\ 0 & v & 0 & p & w & v \\ 0 & 0 & 0 & 0 & 0 & 0 \\ 0 & 0 & 0 & 0 & 0 & 0 \\ 0 & 0 & q & 0 & 0 & 0 \\ 0 & r & 0 & \delta s & 0 & 0 \\ 0 & \delta r & \delta s & \delta r & q & r \\ 0 & 0 & 0 & c\theta . s\phi & \delta s & \delta r \\ 0 & 0 & 0 & 0 & s\theta & 0 \end{bmatrix}$$

TRAJECTORY SIMULATION

Approach

- The dynamic equations can thus be recast as:

$$\{\dot{u} \;\; \dot{v} \;\; \dot{w} \;\; \dot{p} \;\; \dot{q} \;\; \dot{r}\}^T = [A]^{-1}\{\{F_E\} + \{F_I\}\}$$

$$= [A]^{-1}\{\{diag[B].[C] + \{F_I\}\}$$

- Transform the velocities in the body system to the inertial (global) system
- Thus, we have 12 ordinary linear differential equation whose integration in time will determine the values of vehicle position X_O, Y_O, Z_O and orientation ϕ, θ, ψ with time.

Approach

- Inputs in the above equations of motion:
  - HDC values
  - Control inputs (plane deflection)
  - Speed
- A flexible trajectory simulation program was created, with scope for modifying the external force matrices [B] and [C], thus enabling the mathematical model to be readily changed

15/5 Zigzag at 5 knots – Axisymmetric Body

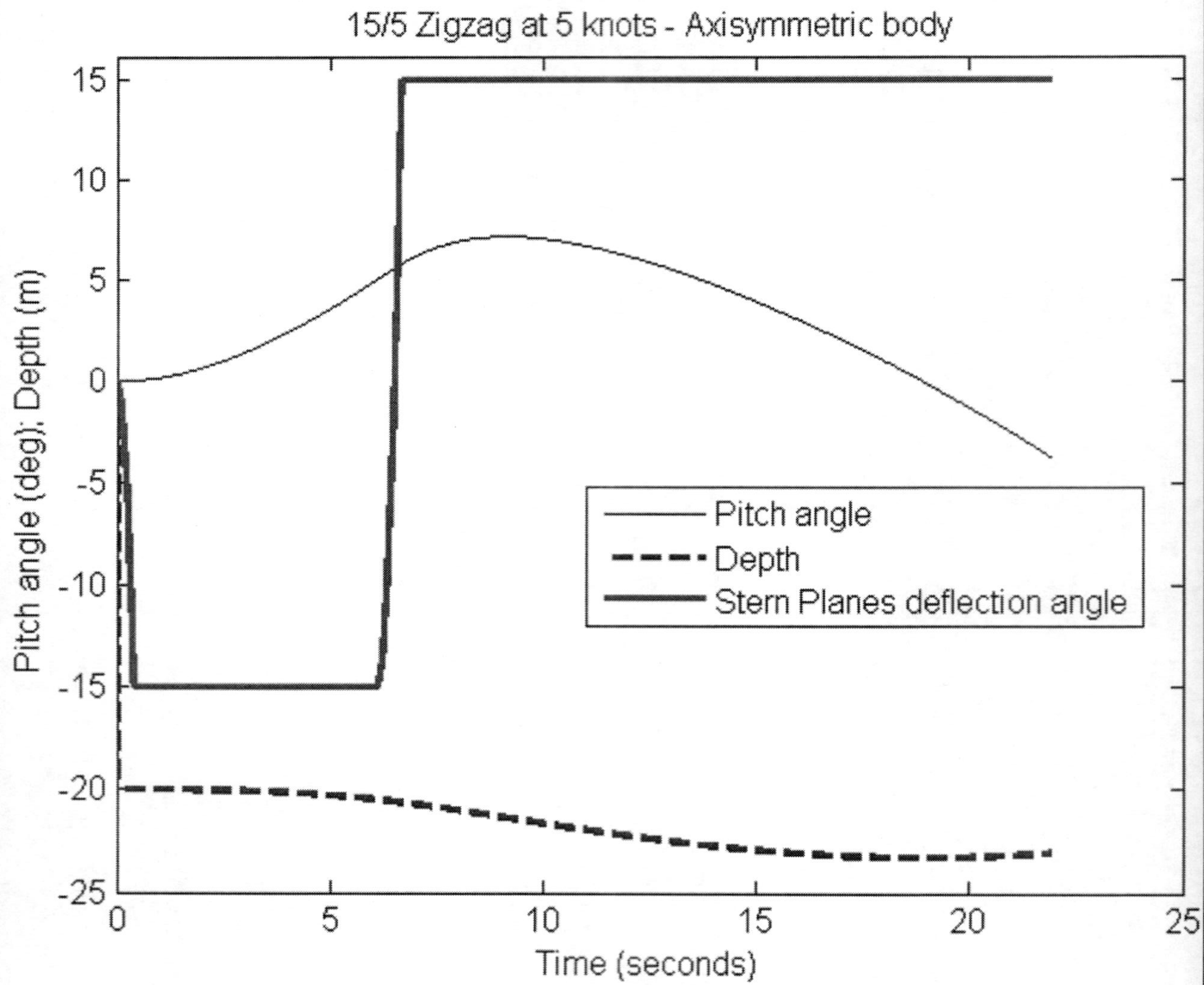

15/5 Zigzag at 5 knots - SUBOFF body

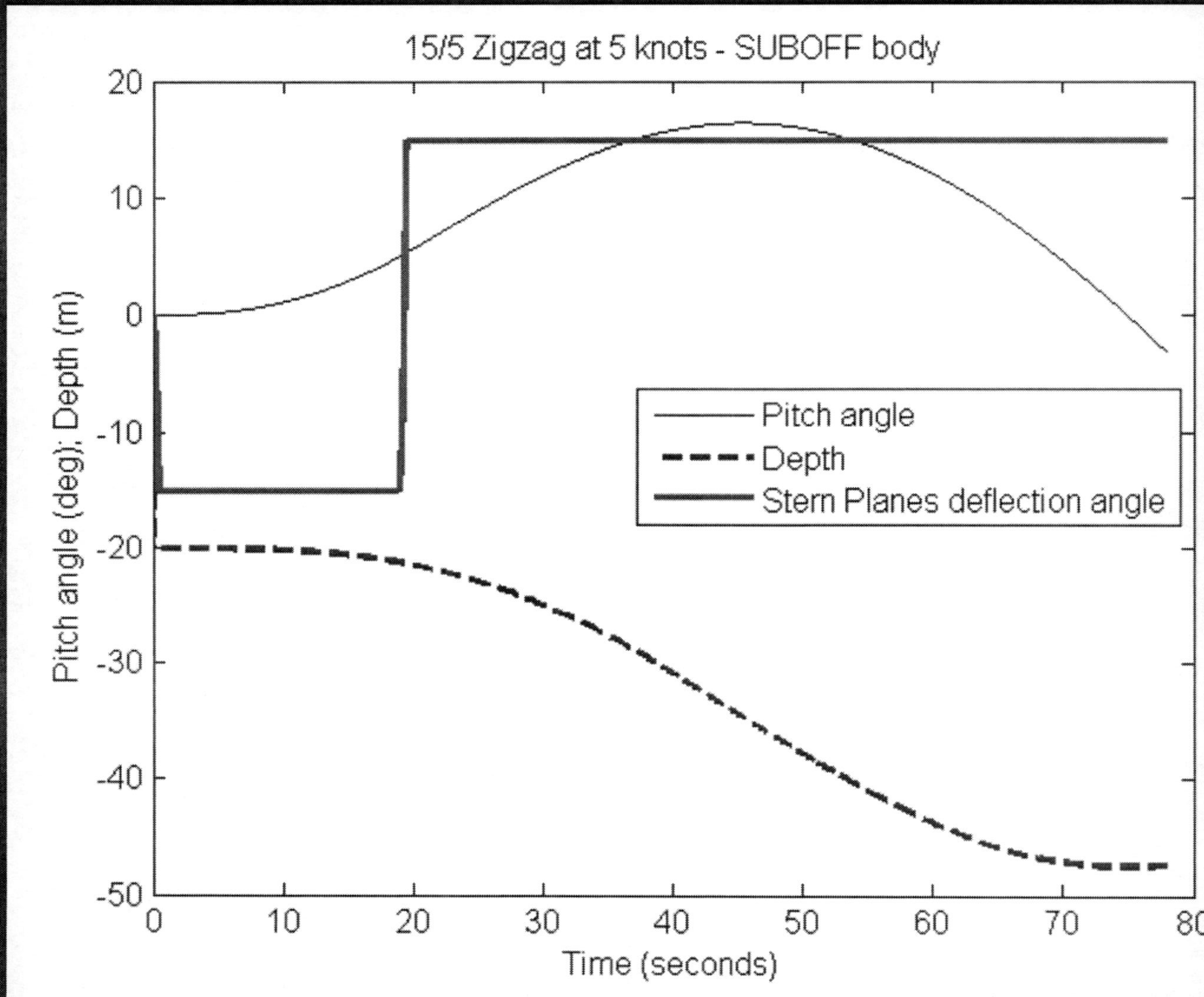

RESULTS – Fully Linear Model

Parameter	Axisym-metric body	SUBOFF body
Overshoot pitch angle (θ_O); deg	2.16	11.43
Overshoot change of depth (z_O); metres	2.81	26.34
Time to reach execute (t_e); seconds	6.12	18.96
Time to check pitch (t_c); seconds	9.16	45.48
Time to check depth (t_d); seconds	18.94	74.94

Effect of mathematical model

Parameters of Zigzag →	θ_O *(deg)*	z_O (m)	t_e (sec)	t_c (sec)	t_d (sec)
	(Difference from Model 3 values given in %)				
Model 1	2.40	2.91	6.28	9.44	19.14
	11.1%	*3.6%*	*2.6%*	*3.1%*	*1.1%*
Model 2	2.95	3.74	5.76	9.66	20.86
	36.6%	*33.1%*	*5.9%*	*5.5%*	*10.1%*
Model 3 (fully linear)	2.16	2.81	6.12	9.16	18.94

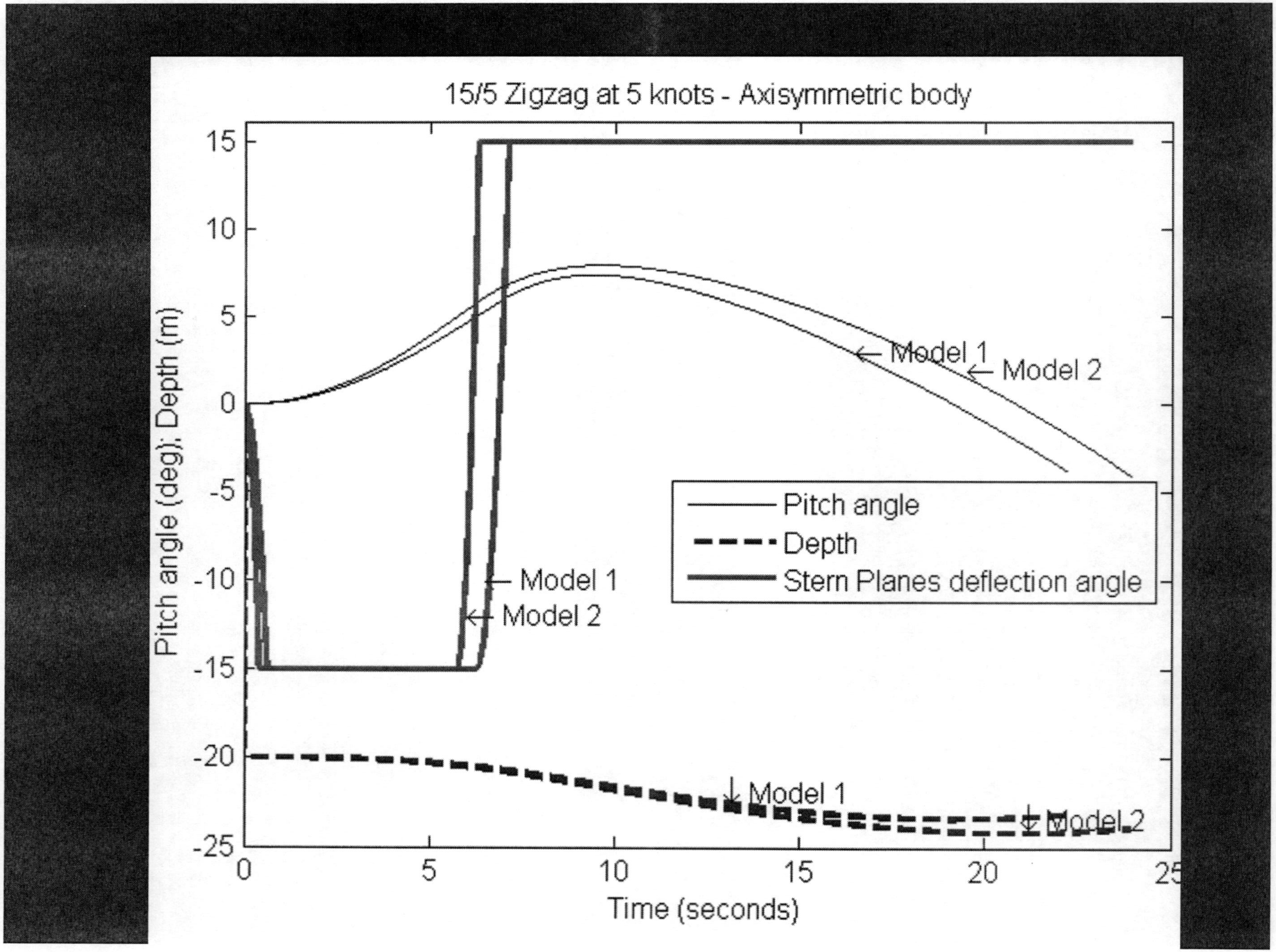

15/5 Zigzag at 5 knots - Axisymmetric body
Pitch angle (deg); Depth (m)
Time (seconds)
Pitch angle
Depth
Stern Planes deflection angle
Model 1
Model 2
Model 1
Model 2
Model 1
Model 2

HYDRODYNAMIC COEFFICIENTS

How Many HDCs?

- Mathematical model chosen dictates which (and how many) HDCs are to be evaluated.
- Maximum 36 linear HDCs.
- The number of second-order HDCs and various correction terms may vary from model to model.
- Model 1 uses total 55 HDCs; Models 2 & 3 use 24 HDCs.

Typical Uncertainty in Model Testing Results

HDC	Estimated Uncertainty
Static HDCs: Z_w', M_w', Y_v', N_v'	4 – 5 %
Rotary HDCs: Z_q', M_q', Y_r', N_r'	10%
Control HDCs	6 – 10%
Added mass HDCs: $Z_{\dot{w}}'$, $M_{\dot{q}}'$, $Y_{\dot{v}}'$, $N_{\dot{r}}'$	7%

Source: Roddy, 1990

Sensitivity Studies

- Sensitivity has been defined as the ratio of the relative change in some output variable, and the relative change in an input parameter, each compared to nominal values.

- Error level in one coefficient may be masked by errors in others.

Sensitivity Studies

- Most significant coefficients for submarine-like body:
  $M_q, M_{\delta s}, Z_{\delta s}, Z_q, Z_{\dot{w}}, Z_w, Z_{\dot{q}}, M_{\dot{w}}, M_{\dot{q}}$
- Most significant coefficients for axisymmetric body :
  $M_{\delta s}, Z_{\dot{w}}, Z_{\delta s}, M_{\dot{w}}, M_{\dot{q}}, M_q, Z_{\dot{q}}$
- These HDCs were varied, one at a time, by ±25 to 50% of their initial values and trajectory simulation was repeated using Model 3.
- As extreme case, all the HDC values were simultaneously changed by ±100% from initial values and trajectory simulation repeated.

Effect of change in all HDCs by ±50% for Axisymmetric Body

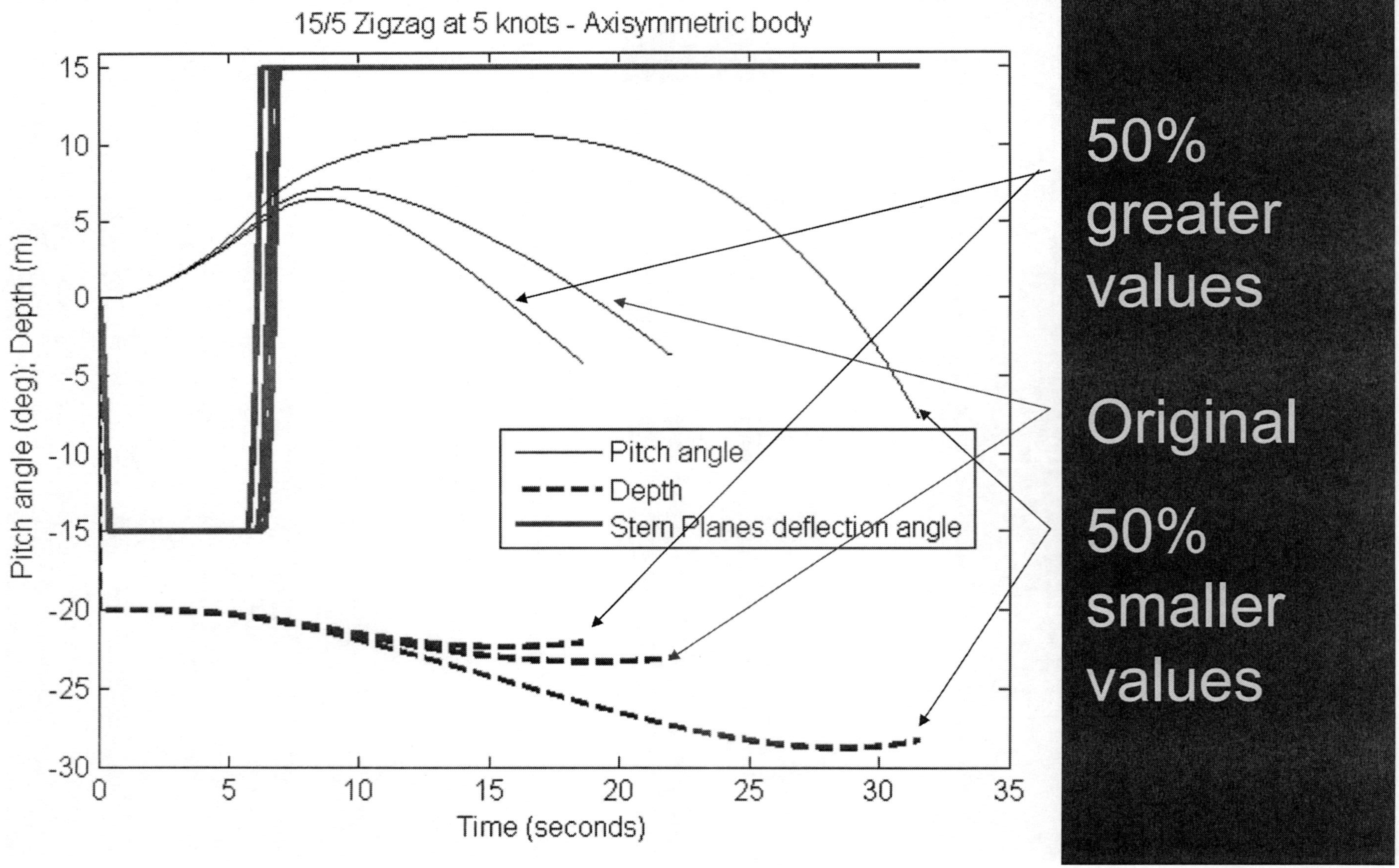

Results- Axisymmetric Body

Change in HDC by ↓		% Change in values of Zigzag Parameters				
		θ_o	z_o	t_e	t_c	t_d
$M_{\delta s}$	+50%	56.3	3.2	-23.9	-13.5	-18.0
	-25%	-48.5	-26.1	29.1	10.5	2.2
$Z_{\dot{w}}$	+50%	8.0	-4.6	-2.9	-1.5	-6.5
	-50%	-18.2	21.8	7.2	3.9	28.5
$Z_{\delta s}$	+50%	-50.6	-45.2	11.1	-6.6	-20.1
	-50%	68.7	55.5	-7.2	14.6	12.0
$Z_{\dot{q}}$	+50%	-2.9	-1.7	0.7	-0.2	-0.2
	-50%	3.0	1.8	-0.7	0.2	0.3

Results- Axisymmetric Body

Change in HDC by ↓		% Change in values of Zigzag Parameters				
		θ_O	z_O	t_e	t_c	t_d
$M_{\dot{w}}$	+50%	-2.1	-2.2	-0.7	-1.1	-1.3
	-50%	0.5	1.4	0.3	0.7	1.0
$M_{\dot{q}}$	+50%	-6.9	15.3	21.2	19.7	18.9
	-50%	0.5	-26.5	-25.8	-26.9	-25.9
M_q	+50%	-38.3	-28.7	9.8	-3.1	-8.7
	-50%	78.2	110.8	-8.2	18.3	44.2
M_w	+50%	-26.5	-6.5	7.5	-0.7	10.6
	-50%	17.6	-5.5	-5.6	-2.4	-10.7
All	+50%	-31.9	-34.9	3.9	-7.0	-17.8
	-50%	163.6	197.5	-6.5	69.0	50.4

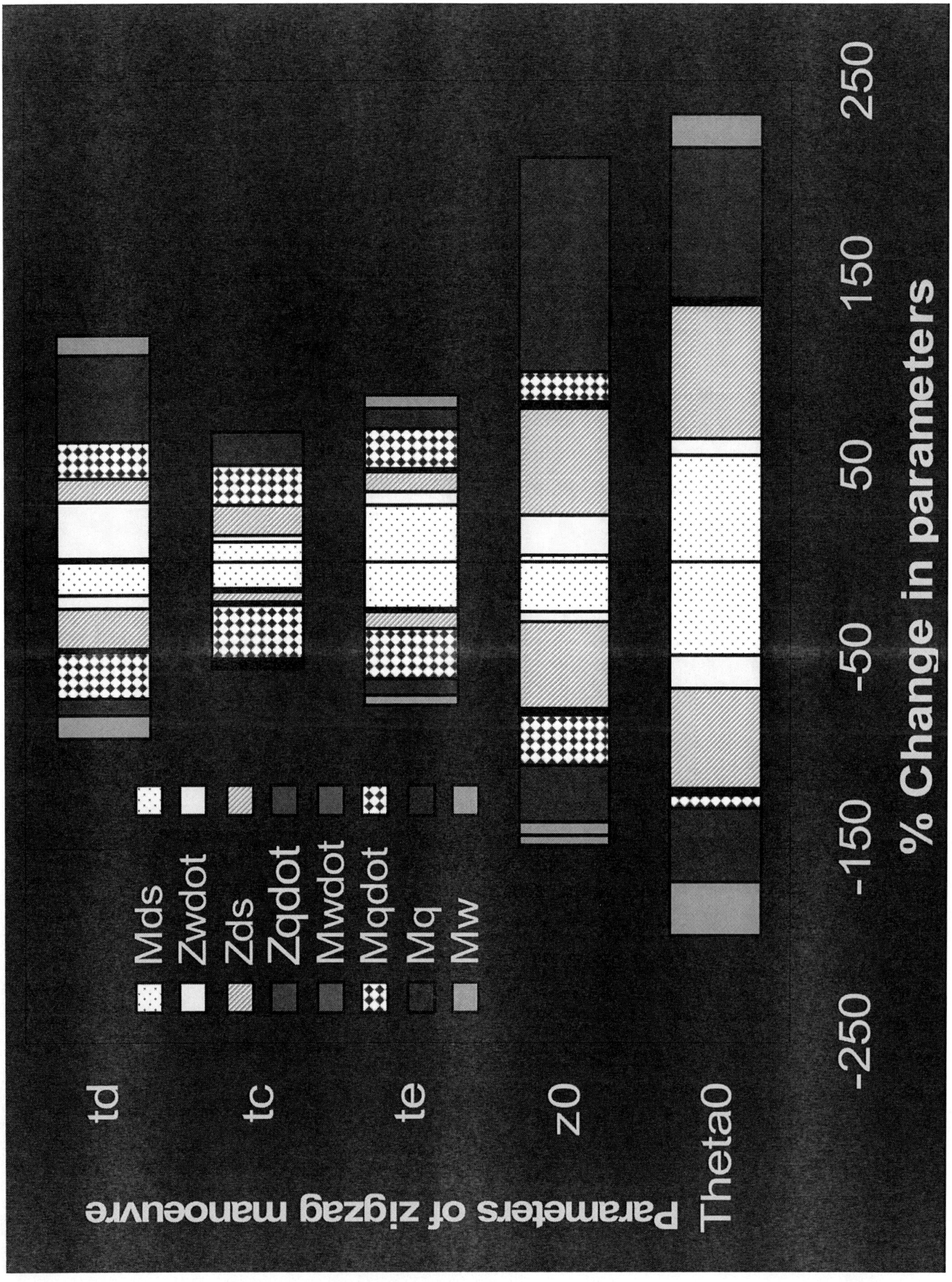
Parameters of zigzag manoeuvre
td
tc
te
z0
Theta0
Mds
Zwdot
Zds
Zqdot
Mwdot
Mqdot
Mq
Mw
-250
-150
-50
50
150
250
% Change in parameters

Analysis

- The three most important HDCs for the case considered are $M_{\delta S}$, $Z_{\delta S}$ and M_q.
- Change in value of these HDCs by ±50% causes change in at least one of the vertical plane zigzag manoeuvre parameters by more than 50%.
- The parameters (t_e), (t_c) and (t_d) are relatively less influenced (about ±100%) by change in HDC values, as compared to the parameters θ_O and z_O.

Effect of change in all HDCs by ±50% for SUBOFF Body

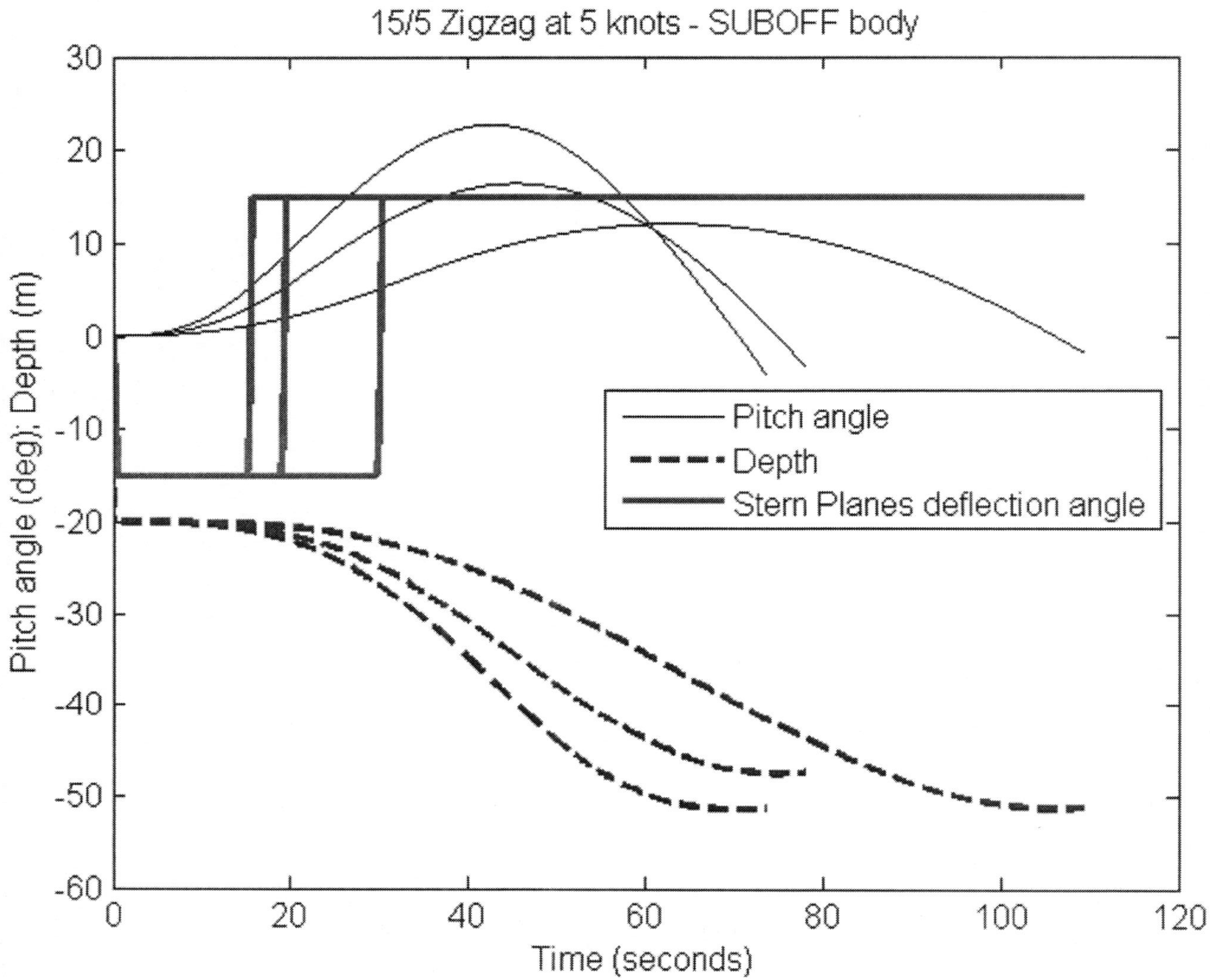

Results- SUBOFF Body

Change in HDC by ↓		% Change in Zigzag Parameters				
		θ_o	z_o	t_e	t_c	t_d
$M_{\delta s}$	+50%	-3.6	-7.7	-5.6	-5.6	-5.3
	-50%	4.4	9.4	6.6	6.5	6.1
$Z_{\dot{w}}$	+50%	15.7	19.1	5.3	11.5	11.4
	-50%	-17.4	-19.4	-6.4	-12.0	-11.4
$Z_{\delta s}$	+50%	17.3	0.3	-11.4	-6.6	-6.6
	-50%	-20.8	1.3	20.5	13.1	13.8
M_w	+50%	13.6	-2.4	-11.9	-9.5	-10.3
	-50%	-18.4	3.9	21.6	16.6	18.5
Z_q	+50%	-2.33	-2.58	0.11	-1.01	-0.93
	-50%	2.43	2.72	-0.11	1.01	0.96
All	+50%	54.9	16.0	-20.0	-5.9	-6.0
	-50%	-37.9	11.0	57.1	39.0	41.7

Analysis

- Only five HDCs for which variation by ±50% had an effect of more than 2% change in zigzag parameters.
- Most significant HDCs for the SUBOFF body simulation were $Z_{\dot{w}}, Z_{\delta s}, M_{w}$
- Thus, even for the same mathematical model, the body geometry has a significant influence on the relative importance of the HDCs.

Analysis

- Change in zigzag manoeuvre parameters due to HDC value variations does not exceed 21% when HDCs are varied one at a time.
- The % change is much lesser than the effect of variations in case of axisymmetric body.
- Even when all relevant HDCs are varied simultaneously by ±50% from original values, the variation in zigzag parameters is up to 55% only.

UNCERTAINTY IN FULL-SCALE TRIALS

Motivation

- Ultimate check for the manoeuvrability prediction is to compare with the measured values during full-scale trials.
- Such comparisons are not only rare, but are also fraught with uncertainties due to noise in the full-scale data.
- Prediction process needs to be only as accurate as measurements for validation.

Sources of Bias Error

- Errors in measuring instruments for depth, pitch angle and speed of the vehicle
- Set in inertial navigation or other motion / position-recording instruments
- Hydrostatic imbalance of the vehicle
- Changes in temperature of water with depth (affecting density and viscosity)
- Ocean currents at different depths

Sources of Precision Error

- Errors in recording and processing data, depending on frequency of measurements
- Unknown and random sea water disturbances
- Errors due to variation in time for applying control forces

Typical Instrumentation Least Count

- Depth: 1 metre
- Pitch angle: 1 degree
- Time: 0.5 seconds
- Plane angle: 1 degree
- Speed: 0.5 knots

Typical Instrumentation Bias – Case 1

Parameter	Typical Value	Instrument Uncertainty
Overshoot pitch angle (θ_O)	2 degrees	0.50
Overshoot change of depth (z_O)	5 metres	0.20
Time to reach execute (t_e)	12 seconds	0.017
Time to check pitch (t_c)	30 seconds	0.008
Time to check depth (t_d)	60 seconds	0.006

Precision Errors in Data Recording

- Estimated values are of the order of the least count of the instruments for depth, pitch angle, time and speed, leading to additional uncertainty values similar to those listed earlier.

Sea Water Density Variation

- Density of sea water (tropical waters) may vary from 1.015 to 1.030 tons/m^3.
- Change of depth in a region with sharp temperature gradient may cause imbalance between weight and buoyancy, up to 0.25% of the vehicle displacement.
- Uncertainty due to this source of error is estimated as 0.003 for θ_O and 0.005 for z_O

Error in Control Force Application

- Time required to apply plane angle when ordered may vary from operator to operator and may not exactly coincide with the moment the execute pitch angle is reached.
- Typical variations (low speeds) are of the order of 2 seconds, leading to uncertainty:
  - In all time parameters by 2 seconds
  - In the overshoot angle by 1-2 degrees
  - In overshoot change of depth by 2-5 metres.

Cumulative Uncertainty – Case 1

Parameter	Assumed Value	Total Uncertainty
Overshoot pitch angle (θ_O)	2 degrees	1.503
Overshoot change of depth (z_O)	5 metres	0.805
Time to reach execute (t_e)	30 seconds	0.250
Time to check pitch (t_c)	60 seconds	0.133
Time to check depth (t_d)	90 seconds	0.050

Effect of Parameter Values

- For different speeds, and certainly for different vehicles, the parameter values for the same manoeuvre would vary and accordingly the uncertainty levels would also change.
- For larger bodies and / or greater speeds (or plane deflections), the values of parameters measured are likely to be greater. In such a scenario, the associated uncertainties are likely to be lower.

Cumulative Uncertainty – Case 2

Parameter	Assumed Value	Total Uncertainty
Overshoot pitch angle (θ_O)	8 degrees	0.378
Overshoot change of depth (z_O)	15 metres	0.272
Time to reach execute (t_e)	15 seconds	0.200
Time to check pitch (t_c)	40 seconds	0.100
Time to check depth (t_d)	70 seconds	0.043

CONCLUSIONS

- Overall perspective is necessary to ensure that estimates at every stage of manoeuvrability prediction are reasonably accurate, while also optimising the time and effort required in the process.

- Choice of mathematical model (linear / non-linear) can influence the zigzag parameters by 1% to 37%, with the greater effect on the pitch overshoot and depth overshoot

- Experimental uncertainty in estimation of HDC values has been reported in the range of 5 to 10%.
- If all relevant HDCs are varied by ±50% simultaneously, the zigzag parameters change by up to 55% for SUBOFF and up to almost 200% for the axisymmetric body.
- Sensitivity of HDCs depends not only on the mathematical model adopted, but also on the manoeuvre and the body geometry

Uncertainty Estimates

- Summing up uncertainty for all stages of manoeuvrability prediction:

  60% to 250%

- Error magnitude of various parameters during full-scale measurements:

  5% to 150%

Lessons

- Predictions need to be only as accurate as can be measured and verified.
- The quantification of uncertainties at various stages can offer insight into margins for improvement at various stages of the prediction process.
- Further studies may consider other definitive manoeuvres, other bodies and other mathematical models.

THANK YOU

NUMERICAL INVESTIGATION OF AIR SUCTION IN TO A LOUVERED FUNNEL

D P Mishra, Institute of Technical Education and Research, Bhubaneswar, India
and S K Dash, Indian Institute of Technology, Kharagpur, India

SUMMARY

The entrainment of air into a louvered cylindrical funnel (cylindrical funnel with lateral openings used in naval and merchant ships) has been computed numerically. It was also found that there exists an optimum funnel diameter and optimum nozzle protrusion length (irrespective of the nozzle fluid temperature) for which the mass suction of air into the funnel can be the maximum. It was found that an inverted frustum with a value of $r_1/r_2 = 0.8$ could suck the maximum amount of air compared to a cylindrical funnel of same volume. It was found from the numerical experiments that for $Gr/Re^2 \leq 0.4$ the funnel inclination had no effect on the rate of mass suction while for $0.4 < Gr/Re^2 < 1$ the funnel inclination had marginal influence. As the value of Gr/Re^2 increased beyond 1 the influence of funnel inclination on rate of mass suction was found to be significant.

NOMENCLATURE

A_L	Louvers opening area (m^2)
D_{cd}	Diameter of computational domain (m)
D_f	Diameter of the funnel (m)
D_n	Nozzle diameter (m)
g	Acceleration due to gravity (m/s^2)
G_k	Production of turbulent kinetic energy (m^2/s^3)
Gr	Grashof Number $g\beta(T_{inlet} - T_{amb})H_f^3 / v^2$
H_{cd}	Height of the computational domain (m)
H_f	Height of the funnel (m)
k	Turbulent kinetic energy (m^2/s^2)
L_p	Protruding length of nozzle (m)
L_e	Entrance length (m)
$\dot{m}_{inlet}$	Mass flow rate through the nozzle (kg/s)
$\dot{m}_{suction}$	Air suction rate through the louvers (kg/s)
p	Pressure (Pa)
Pr	ν/α
Pr_t	Turbulent Prandtl number
Re_n	$\rho V_n D_n/\mu$
T	Temperature (K)
$\vec{v}$	Velocity vector
V_{inlet}	Outlet velocity at nozzle exit (m/s)

Greek symbols

υ	Kinematic viscosity (m^2/s)
α	Thermal diffusivity (m^2/s)
ρ	Density (kg/m^3)
μ	Shear viscosity (Pa.s)
μ_t	Turbulent viscosity (Pa.s)
μ^{eff}	Effective viscosity (Pa.s)
$\bar{\bar{\tau}}$	Stress tensor (Pa)
ε	Rate of dissipation (m^2/s^3)
σ_k	Turbulent Prandtl number for k
σ_ε	Turbulent Prandtl number for ε

Subscripts

n	Nozzle
∞	Free stream

1. INTRODUCTION

The funnel of naval or merchant ships is used to expel the combustion products of the gas turbine in to the atmosphere. Normally the combustion products are very hot (200 °C to 500 °C) when they reach the outlet of the nozzles protruding in to the funnel from the base of the ship (weather deck). The combustion products come out of the nozzle at a very high speed (typically at about 80 to 150 m/s) and there are lots of louvers (simply openings) around the periphery of the funnel through which the atmospheric air from the surroundings rushes in to the funnel due to the jet action created by the high speed jet coming out of the nozzle placed inside the funnel (the funnel and the louvers on it can be seen in Fig. 1). The fresh air that enters the funnel thus mixes with the hot gas and the mixture travels upward along the axis of the funnel at the expense of its own kinetic energy of the high speed jet and gets discharged into the atmosphere at the height of the funnel which is typically 7 to 10 m from the weather deck of the ship. By the time the mixture reaches the top of the funnel the temperature of the gas drops considerably due to the mixing with the cold air from the ambient. This is very much desired in a war ship because the hot gas from the nozzles if could have been thrown to the atmosphere without mixing with the fresh air then the enemy's ship could have captured the IR signal of the hot gas and thus could have spotted the war ship in the ocean (which is absolutely not desired). So it is desired that the gas turbine products of the ship should be released to the atmosphere through the funnel as low a temperature as possible.

Making the combustion product to attain a low temperature is only possible (in a very practical way) by sucking fresh air from the atmosphere through the louvers of the funnel. No extra power through the use of a pump is needed to do this. The jet action created by the high speed jet coming out from the nozzle is utilized to create low pressure zones in the funnel which will cause atmospheric air to be sucked into the funnel and mix with the hot gas thus making it to cool down. So to suck maximum air from the atmosphere is the objective of a best funnel design.

2. LITERATURE RIVIEW

Considerable research work has already been done by different researchers to avoid the interaction of exhaust emissions from the funnels (stacks) with other superstructures present on the ship. Baham and McCallum [1] presented guidelines for the selection of casing location, shape, height, and exit velocity for naval ship and they have also developed empirical mathematical expressions to predict plume temperature decay. Kulkarni et al. [2] reviewed the studies performed by various researchers and identified the smoke nuisance problem on ships depend on the turbulence zone over ship superstructure, downwash of stack, exhaust velocity of smoke and relative velocity between the wind and ship and angle of yaw. Seshadri et al. [3] on the basis of results from the experiments suggested several modifications in the geometry of the funnel to minimize the smoke ingress into GT intakes in the naval ships. Kulkarni et al. [4] made parametric studies of exhaust smoke-superstructure interaction on a naval ship using CFD tool.

In order to enhance entrainment and mixing, various techniques have been developed which include active and passive flow control. In active flow control, mixing can be enhanced by forcing the jet at a frequency close to its natural frequencies with the help of acoustic or electromagnetic devices; Zaman and Hussain [5]. However, passive flow control is much simpler and economical and can be achieved in several ways: placing of an obstruction–like blob body in the near field of the jet, introduction of tabs or notches at the nozzle exit (Tong and Warhaft [6] and Rajgopalan and Antonia [7]) and changing the nozzle cross section from circular to noncircular shape (Krothapalli et al [8]).

The entrainment of a turbulent jet in co-flow, with and without reaction has been studied by Han and Mungal [9]. From the experimental investigation they concluded that the co flow speed reduces the entrainment where as the buoyancy effectively improves the entrainment rate. Mixing and entrainment characteristics of circular and non circular confined jets have been investigated by Singh et al. [10]. They revealed that for a smaller mixing tube diameter and jet located at the inlet of the mixing tube, the circular jet entrains more than the noncircular jet. For a larger mixing tube or shifted jet locations, the noncircular jets entrain more ambient fluid. Mishra and Dash [11] determined the suction rate into a louvered funnel using CFD tools. They concluded that there exists optimum funnel diameter where the suction rate of air becomes the highest with all other parameters remaining fixed. Computational fluid dynamics analysis has been carried out by Kandakure et al. [12] to study the jet expansion and hydrodynamics in presence of confinement (draft tube). They have observed that with an increase in the draft tube diameter, the entrainment of the surrounding fluid into the draft tube increases while the average kinetic energy in the jet flow in draft tube decreases. The work found in literature so far does not site the use of CFD methods to determine the mass ingress of atmospheric air in to the funnel through the louvers and hence to arrive at optimum design of the funnel so as to obtain highest suction rate in to it.

3. MATHEMATICAL FORMULATION

The computational investigation is carried out for a cylindrical funnel (of diameter, D_f and height, H_f as shown in Fig. 1). The funnel is closed at the bottom and opened at the top to the surrounding atmosphere. At the bottom of the funnel a nozzle of diameter, D_n having a protrusion length of L_p (in to the funnel) is placed at the center to supply high velocity air jet. Another cylindrical computational domain of diameter, D_{cd}, (more than 6 times of D_f) and height, H_{cd} (2 to 3 times of H_f) is placed around the funnel so that boundary condition can be applied on the boundary of the computational domain for suction to take place near the louvers cut on the funnel. The louvers are kept rectangular in shape (L_H = 0.8 m

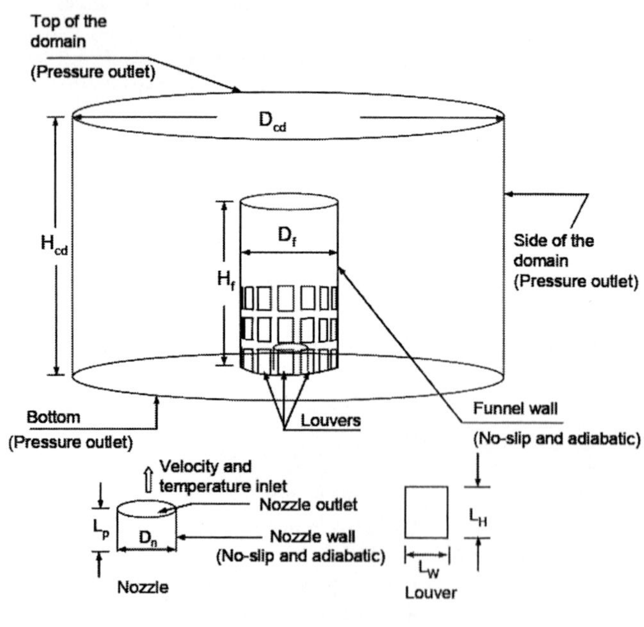

Figure 1: Schematic diagram of computational domain and a pictorial view of boundary conditions applied to it

and L_w= 0.6 m) and 12 louvers are placed in one stack with a distance of 0.5 m between each stack. The flow field in the domain would be computed by using three-dimensional incompressible Navier-Stokes equations with a two equation k-ε turbulence model along with the energy equation. The fluid used in the simulation is air, at temperatures of 300 to 700K, and is treated to be incompressible up to an injection velocity of 120m/s.

3.1 GOVERNING EQUATIONS

The governing equations for the above analysis can be written as:

Continuity

$$\nabla.(\rho\vec{v}) = 0 \quad (1)$$

Momentum

$$\nabla.(\rho\vec{v}\vec{v}) = -\nabla p + \nabla.(\bar{\bar{\tau}}) - (\rho_\infty - \rho)g \quad (2)$$

The density ρ is taken to be a function of temperature according to ideal gas law while laminar viscosity μ and thermal conductivity are kept constant. The Boussenesq approximation is not adopted in the model since the variation of ρ with temperature is tremendous in the range of operating parameters.

Energy

$$\frac{D(\rho T)}{Dt} = \frac{\partial}{\partial x_i}\left[\left(\frac{\mu}{\Pr} + \frac{\mu_t}{\Pr_t}\right)\frac{\partial T}{\partial x_i}\right] \quad (3)$$

The stress tensor $\bar{\bar{\tau}}$ is given by

$$\bar{\bar{\tau}} = \mu^{eff}\left(\nabla\vec{v} + \nabla\vec{v}^T\right) \quad (4)$$

effective viscosity $\mu^{eff} = (\mu + \mu_t)$

Turbulence kinetic energy - k

$$\nabla.(\rho\vec{v}k) = \nabla.\left(\frac{\mu_t}{\sigma_k}\nabla k\right) + G_k - \rho\varepsilon \quad (5)$$

Rate of dissipation of k

$$\nabla.(\rho\vec{v}\varepsilon) = \nabla.\left(\frac{\mu_t}{\sigma_\varepsilon}\nabla\varepsilon\right) + \frac{\varepsilon}{k}(C_{1\varepsilon}G_k - C_{2\varepsilon}\rho\varepsilon) \quad (6)$$

The turbulent viscosity μ_t is computed from

$$\mu_t = \rho C_\mu \frac{k^2}{\varepsilon} \quad (7)$$

The term G_k representing the production of turbulent kinetic energy is computed from

$$G_k = \mu_t\left(\nabla\vec{v} + (\nabla\vec{v})^T\right) : \nabla\vec{v} \quad (8)$$

σ_k and σ_ε are the Prandtl numbers for k and ε

The constants used in the above k-ε equations are the followings, although these constants are normally used for internal flow but for the present case we have not changed the constants (there is a provision in Fluent to change the constants).

$C_{1\varepsilon} = 1.44; C_{2\varepsilon} = 1.92; C_\mu = 0.09; \sigma_k = 1.0; \sigma_\varepsilon = 1.3$, $\Pr_t = 1$

3.2 BOUNDARY CONDITIONS

The boundary conditions can be seen from Fig. 1. The funnel wall and nozzle wall are solid and have been given a no-slip boundary condition. Pressure outlet boundary conditions has been imposed at the outer periphery, at the top surface and at the bottom of the annular computational domain and velocity inlet boundary condition has been employed at the nozzle outlet which supplies high velocity air in to the funnel.

At the pressure outlet boundary, the velocity will be computed from the local pressure field so as to satisfy the continuity but all other scalar variables such as, T, k and ε are computed from the zero gradient condition, Dash [13].

The turbulent quantities, k and ε, on the first near wall cell have been set from the equilibrium log law wall function as has been described by Jha and Dash [14, 15] and Jha et al. [16]. The turbulent intensity at the inlet of the nozzle has been set to 2% with the inlet velocity being known and the back flow turbulent intensity at all the pressure outlet boundaries have been set to 5%. If there is no back flow at a pressure outlet boundary then the values of k and ε are computed from the zero gradient condition at that location.

4. NUMERICAL SOLUTION PROCEDURE

Three-dimensional equations of mass, momentum, energy and turbulence have been integrated over the control volume and the subsequent equations have been discretized over the control volume using the finite volume technique to yield a set of algebraic equations which could be solved by the algebraic multi grid solver of **Fluent 6.3** in an iterative manner by imposing the above boundary conditions. First order upwind scheme (for convective variables) was considered for momentum as well as for the turbulent discretized equations. After a first-hand converged solution could be obtained the scheme was changed over to second order upwind so as to get little better accuracy (mass suction rate was changed by only 0.5%). A detail description of the types and number of grids used and the effect of domain size have been discussed separately in the results and discussions.

SIMPLE algorithm with a PRESTO scheme for the pressure velocity coupling was used for the pressure correction equation. Under relaxation factors of 0.3 for pressure, 0.7 for momentum and 0.5 for k and ε and 1 for temperature were used for the convergence of all the variables. Tetrahedral cells were used for the entire computational domain because it was the only choice in such a complicated geometry. Convergence of the discretized equations were said to have been achieved when the whole field residual for all the variables fell below 10^{-3} for u, v, w, p, k and ε whereas for energy the residual level was kept at 10^{-6}.

5. RESULTS AND DISCUSSIONS

5.1 DOMAIN AND GRID INDEPEDENCE TEST

Fig. 1 shows the louvered cylindrical funnel where the louver size is L_H by L_W. We did not vary the size of the louver to see its effect on the mass suction, rather we varied the louver opening area to see the effect on mass suction. The louver opening area was varied by putting many rows of louvers on the cylindrical funnel. For an actual ship operation the funnel may not be exactly cylindrical but from academic point of view it is kept cylindrical and the present study can always be extended to accommodate a particular shape of funnel for a real ship operation. To compute the mass suction into the funnel we put a bigger computational domain around the funnel and the boundary conditions were put to the bigger domain so that suction can take place in the funnel. The boundary conditions can be seen pictorially from Fig. 1. The height of the computational domain was kept to three times that of the funnel height and the diameter of the domain was kept to six times that of the diameter of the funnel.

Before proceeding to compute the mass suction rate in to the funnel we first did a grid independence test for a sample case. Tetrahedral cells were used to discretize the domain because of the lateral openings created on the cylindrical funnel. The number of meshes on the nozzle

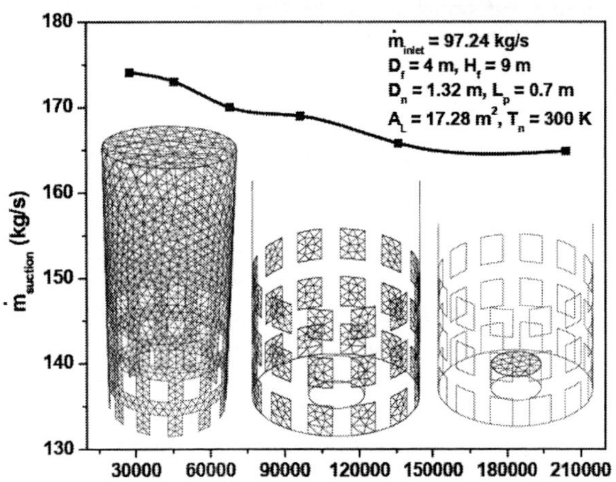

Figure 2: Effect of number of cells on air entrainment into the funnel

and on the louver openings is varied to mesh the entire region from a total cell of 30,000 to 90,000. Then from 90,000 onwards we adopted the regions around the louver, around the exit of the funnel and around the cylindrical surface where suction takes place to get finer meshes. The suction rate of air is shown in Fig. 2 versus the number of cells. It is apparent that after a number of cells of 130,000 the mass suction does not change at all. We finally adopted the finer meshes which give about 210,000 cells to perform all computations. However, when the funnel height and diameter changed we again performed the grid independence test before reporting any result

5.2 EFFECT OF DOMAIN

Fig. 3 shows the suction rate in to the funnel when the domain size is changing. We have taken three domain sizes as has been shown in Fig. 3. In each domain the diameter is kept to 24m. The height of the domain is varying from 9 to 27m. It is clear from Fig. 3 that the domain height does not have any appreciable effect on mass suction. We used a pressure outlet boundary on the top surface of the domain which will allow mass to go out of the funnel while at other locations it will allow mass from the outside atmosphere to come into the domain. Pressure outlet boundary is the only choice of the boundary condition for this kind of problem and we do not get any effect of the domain on mass suction when we keep the domain diameter to be 24m. Probably a much shorter domain diameter could have affected the mass suction but they were not tried in the computation.

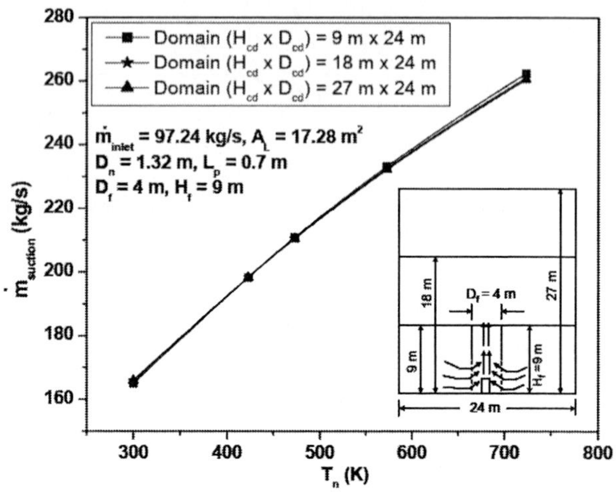

Figure 3: Effect of size of the domain on mass ingress of air into the funnel as a function of nozzle fluid temperature

5.3 EFFECT OF LOUVERS OPENING AREA ON ENTRAINMENT RATE

Fig. 4 shows the mass entrainment of air in to the funnel with different louvers opening area for a funnel of 4 m diameter and 9 m height (other operating parameters for the numerical simulation can be seen from the plot itself). The louvers opening area was increased by putting many rows of louvers on the cylindrical funnel with constant louver size of L_H by L_W. The mass suction of air in to the funnel is increasing with the increase in the louvers opening area and by increasing the louvers opening area to 4 times it is seen that the entrainment rate increases monotonically to 1.7 times. It is seen from the plot when louvers opening area is increased more and more the rise in mass suction is less and less. It is also quite clear from the Fig. 4 putting more louvers after certain louvers opening area (23 m^2 in the present case) does not pay more in terms of mass entrainment. It means the louvers present towards top of the funnel are not very effective as the surrounding atmosphere prevails at the top of the funnel.

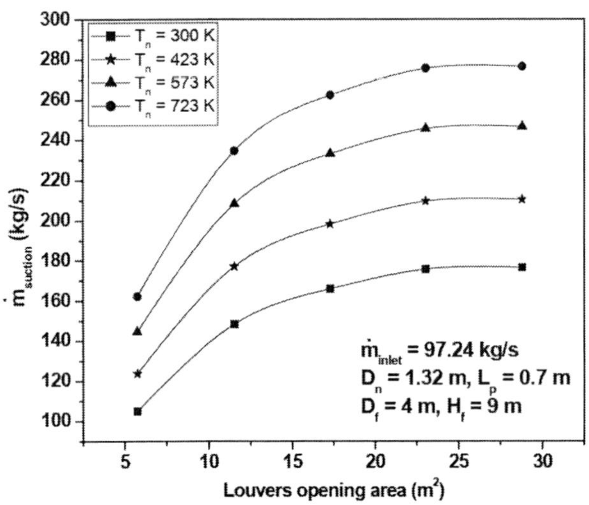

Figure 4: Mass ingress of air into the funnel as a function of louvers opening area and nozzle fluid temperature

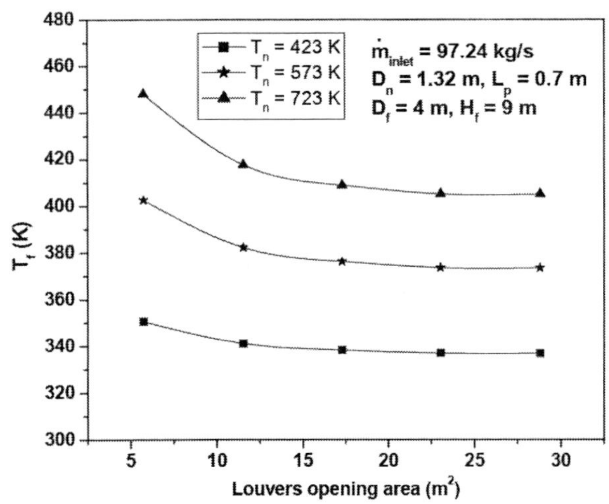

Figure 5: The funnel outlet temperature as a function of louvers opening area and nozzle fluid temperature

Fig. 5 shows the funnel outlet temperature (T_f) as a function of louvers opening area. It can be seen from the plot the T_f is decreasing continuously with the increase in louvers opening area. It is also seen from the plot that after a certain louvers opening area (23 m^2) T_f does not vary any more. As the louvers opening area increases the funnel sucks more ambient fluid in to it. So the low temperature ambient fluid mixes with the hot plume in the funnel and as a result the funnel exit temperature falls. When the louvers area still increases the mass suction in to the funnel does not increase significantly so the funnel exit temperature remains constant.

5.4 OPTIMUM NOZZLE PROTRUSION

The effect of mass suction due to the protruding length of the nozzle in to the funnel has been shown in Fig. 6. It can be seen that when the nozzle protrusion is increasing from 0 to 0.7 m the mass suction is increasing for all nozzle fluid temperature and the mass suction decreases slowly after that when the nozzle protrusion still increases. At 0.7 m of nozzle protrusion there is a maximum in the mass suction curve and this is the optimum protrusion length for the funnel where mass suction is the maximum. It is interesting to note that the optimum protrusion of the nozzle is not depending on the nozzle fluid temperature. Rather for all nozzle fluid temperatures the optimum is a unique value. When the nozzle protrudes too much into the funnel then it can cross the louvers opening and hence will induce less suction near the louver opening, so the mass suction will fall. When the nozzle will attain a position where the suction created due to the emerging jet is able to cover all the louver openings then the mass suction into the funnel becomes the highest.

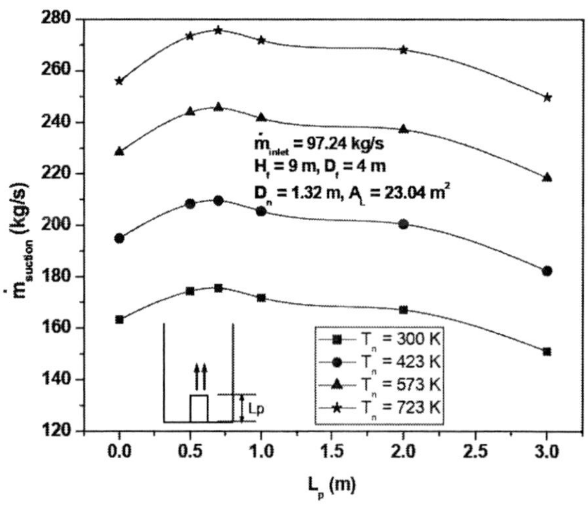

Figure 6: Mass ingress into the funnel as a function of protrusion length of the nozzle into the funnel (optimum protrusion length = 0.7 m)

5.5 OPTIMUM FUNNEL DIAMETER

Fig. 7 shows the existence of optimum funnel diameter for maximum air suction into it. The optimum diameter is a unique value around 4.5 m for all the nozzle fluid temperature. As the nozzle fluid temperature increases the mass suction into the funnel increases but maximum suction occurs at a constant diameter of the funnel which shows that there is an optimum diameter for which the mass suction can be the highest into the funnel. If the funnel diameter is too low it will offer more resistance to the air flow inside it and hence the mass suction will fall.

If the diameter is too large then the suction created by the nozzle jet will not be felt at the louver openings very strongly for which the mass suction in to the funnel will again fall. So this shows that there is an existence of optimum funnel diameter for maximum mass suction which is about 4.5 m for a funnel height of 9 m.

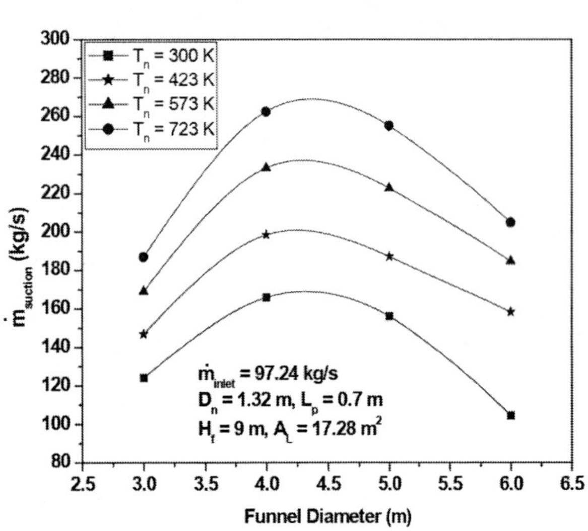

Figure 7: Entrainment of atmospheric air into the funnel as a function of diameter of the funnel (optimum funnel diameter = 4.5 m)

5.6 OPTIMUM SHAPE OF THE FUNNEL

So far we have been doing air suction in to a cylindrical louvered funnel. Keeping the volume of the cylinder same we could convert the cylinder into a frustum where the diameter at the bottom is $2r_1$ and at the top is $2r_2$. Keeping the height of the frustum again to 9 m we carried out the mass suction computation into the frustum while varying the ratio r_1/r_2 and keeping the height still at 9 m but keeping the volume to a constant. The mass suction of air in to the frustum shape funnel is shown in Fig. 8. It can be seen that there exists an optimum shape of the funnel, an inverted frustum with $r_1/r_2 = 0.8$, where the mass suction is highest. When the value of $r_1/r_2 = 0.8$ the frustum opens upward and causes the suction of air through the louvers in a perpendicular direction to the slant face of the nozzle. Due to this perpendicular entry the flow suffers less resistance for which the mass suction becomes highest at $r_1/r_2 = 0.8$. At other values of r_1/r_2 the flow near the louvers is not nicely streamlined. If r_1/r_2 is more than 1 then the frustum opens downward, or closes to the top, causing the air flow at the louvers to enter the funnel downward which is against the flow of the jet fluid. So the suction in to the funnel falls drastically which can be seen from Fig. 8. So there is the existence of an optimum funnel shape for maximum air suction in to it.

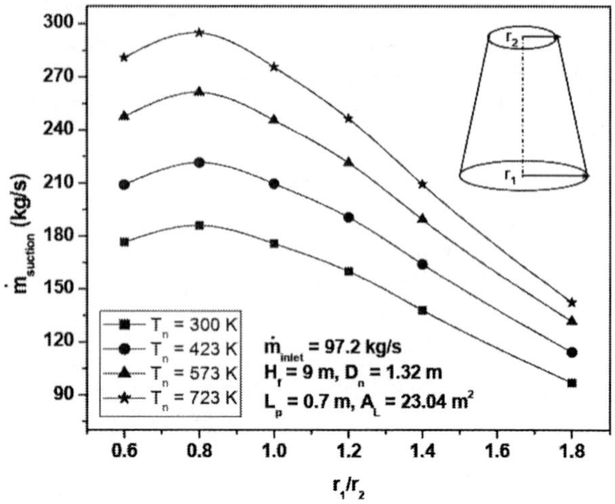

Figure 8: Mass ingress into the funnel as a function of shape of the funnel (optimum r_1/r_2 can be seen to be at 0.8 m)

5.7 INCLINED FUNNEL

We did CFD computation with hot fluid and at different Gr/Re^2 and also we have shown the mass suction in to the funnel as a function of its inclination (Fig. 9). It is clear that below $Gr/Re^2 = 0.176$ there is no effect of inclination on the mass suction which can be seen from Fig. 9. Fig. 10 shows a very large variation in Gr/Re^2 (from 0.4 to 102) where we could compute the mass ingress into the inclined funnel. When $Gr/Re^2 = 0.4$ there is no effect of funnel inclination on mass suction. As the value of Gr/Re^2 increases to 0.71 very minor effect on

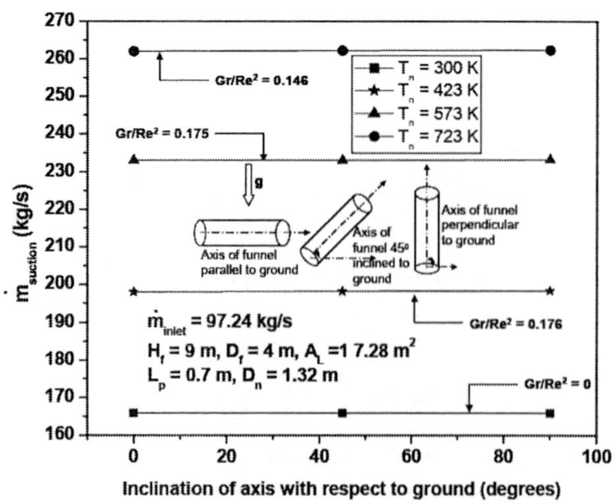

Figure 9: Air suction into the funnel as a function of nozzle fluid temperature and inclination of the funnel

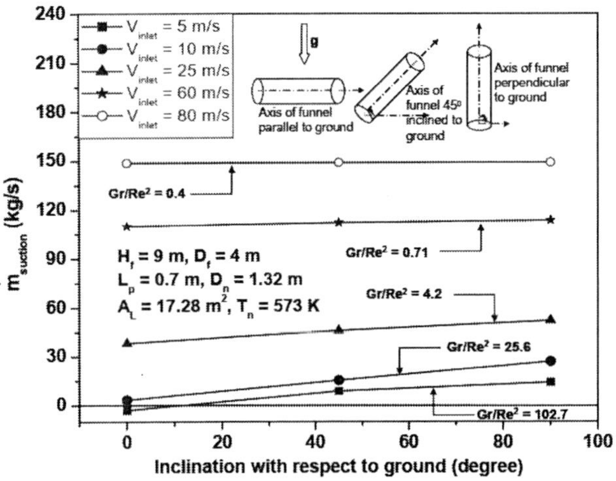

Figure 10 Effect of inclination of the funnel on mass ingress of air in to it when the nozzle ejects hot air at 573 K

mass suction could be seen due to the funnel inclination and the effect becomes much stronger as the value of Gr/Re^2 crosses 1. At Gr/Re^2 more than 25 the mass suction very much depends on the inclination of the funnel because the jet becomes too buoyant and for a horizontal funnel the jet instead of escaping out through the funnel exit tries to escape through the upper louvers on the funnel.

6. CONCLUSIONS

Mass suction rate in to a louvered cylindrical funnel has been computed by solving the conservation equations for mass, momentum and energy. The mass suction rate for a louvered funnel has shown that there exists optimum nozzle protrusion which brings highest suction in to the funnel, while the protrusion length is about 0.7m. There also exists optimum funnel diameter where the mass suction in to the funnel becomes highest irrespective of the nozzle fluid temperature. The optimum diameter was found to be 4.5m where the mass suction could rise by 50% by changing the diameter from 3 to 4.5m. There also exists optimum funnel height where the mass suction in to the funnel becomes the highest. Compared to a cylindrical funnel a frustum shape funnel could suck more air in to it and the optimum diameter ratio of the frustum base to top was found to be 0.8. It was seen that for $Gr/Re^2 \leq 0.4$ the funnel inclination had no effect on the rate of mass suction. When Gr/Re^2 increases beyond 0.4 then the inclination of the funnel has slight effect on the mass suction rate. When Gr/Re^2 still increases beyond 1 then the funnel inclination has significant effect on mass suction.

7. REFERENCES

1. BAHAM, G. J. and Mc CALLUM, D., 'Stack Design Technology for Naval and Merchant Ships', Trans. SNAME, Vol. 85, pp. 324-349, 1977.

2. KULKARNI, P. R., SINGH, S. N. and SESHADRI, V., 'Study of Smoke Nuisance Problem on Ships- a Review', Int. J Maritime Eng, Vol. 147, Part A2, pp.27-50, 2005.

3. SESHADRI, V., SINGH, S. N. and KULKARNI, P. R., 'Study of problem of exhaust smoke ingress into GT intakes of a naval ship, Journal of ship Technology', Vol. 2, No.1, pp. 22-35, 2006.

4. KULKARNI, P. R., SINGH, S. N. and SESHADRI, V., 'Parametric Studies of Exhaust Smoke-Superstructure Interaction on a Naval Ship using CFD', Computers and Fluids, Vol. 36 pp. 794-816, 2007.

5. ZAMAN, K. B. M. and HUSSAIN, A. K. M. F., 'Vertex Pairing in a Circular Jet Under Controlled Excitation', J. Fluid Mech., Vol. 101, pp. 449-491, 1980.

6. TONG, C. and WARHAFT, Z., 'Turbulence Suppression in a Jet by means of a fine ring', Phys. Fluids, Vol. 6, pp. 328-334,1994.

7. RAJGOPALAN, S. and ANTONIA, R. A., 'Turbulence Reduction in a Mixing Layer of a Plane jet using small cylinder', Exp. Fluids, Vol. 25, pp. 96-103, 1998.

8. KROTHAPALLI, A., BAGANOFF, D. and KARAMCHETI, K., 'On the Mixing of Rectangular Jet', J. Fluid Mech., Vol. 107, pp. 201-220, 1981.

9. HAN, D. and MUNGAL, M.G., 'Direct Measurement of Entrainment in Reacting/Nonreacting Turbulent Jets', Combustion and Flame, Vol. 124, pp. 370-386. 2001.

10. SINGH, G. S., SUNDARAJAN, T. and BHASKARAN, K. A., 'Mixing and Entrainment Characteristics of Circular and Noncircular Confined Jets', J. Fluid Engg., ASME, Vol. 125, pp. 835-842, 2003.

11. MISHRA, D. P. and DASH, S. K., 'Theoretical Estimation of Suction Rate due to high velocity Air Jet', Proceedings of ICTACEM 2007, IIT Kharagpur, India, 27-29 December 2007, IIT, Kharagpur, page number 611-613, 2007.

12. KANDAKURE, M. T., PATKAR, V. C. and PATWARDHAN, A. W., 'Characteristics of

Turbulent Confined Jets', Chemical Engineering and Processing, Vol. 47, pp. 1234-1245, 2008.

13. DASH, S. K., 'Heatline visualization in turbulent flow', Int. J. of Num. Methods for Heat and Fluid Flow, Vol. 6, No.4, pp. 37-46, 1996.

14. JHA, P. K. and DASH, S. K., 'Effect of outlet positions and various turbulence models on mixing in a single and multi strand tundish', Int. J. Num. Method for Heat and Fluid Flow, Vol. 12, No. 5, pp. 560-584, 2002.

15. JHA, P. K. and DASH, S. K., 'Employment of different turbulence models to the design of optimum steel flows in a tundish', Int. J. Numerical Methods for Heat and Fluid Flow, Vol. 14, No. 8, pp. 953-979,2004.

16. JHA, P. K., RAJEEV, R., MONDAL, S. S. and DASH, S. K., 'Mixing in a tundish and a choice of turbulence model for its prediction', Int. J. Numerical Methods for Heat and Fluid Flow, Vol. 13, No. 8, pp. 964-996, 2003.

8. AUTHORS BIOGRAPHY

Mr. Dipti Prasad Mishra, holds the current position of Asst. Professor, Department of Mechanical Engineering at Institute of Technical Education and Research, Bhubaneswar, Orissa, India – 751030. He has completed his Ph.d. program and submitted his thesis at IIT Kharagpur under the supervision of Prof. S.K.Dash. His area of interest is computational fluid dynamics and heat transfer. He has done experimental and CFD work to maximize suction in a cylindrical louvered funnel, specifically used in naval and merchant ships.

E mail: dipti2k12002@yahoo.co.in.

Prof. Sukanta K. Dash, holds the current position of Professor, Department of Mechanical Engineering at I.I.T. Kharagpur, India – 721302. He has over 24 years of experience in teaching and R&D. He has published more than 50 research papers. He has guided 8 Ph. D. projects and 54 M tech. projects. He is a consultant to a number of industrial organizations (like Tata steel, Bhaba atomic research center, Mazagon dock ltd, G.R.S.E. Ltd., I. F.G.L. Refractories lTd etc.). He has won Sukumar Rakshit award by MGMI (2001) and Sanak Nanavti Gold Medal from TATA Steel for the Best Application Paper in TATA SEARCH (2000). His areas of interest are fluid mechanics, heat transfer, two phase flow, computational fluid dynamics and their application to different industrial problems.

E mail: sdash@mech.iitkgp.ernet.in

Numerical Investigation of Air Suction into a Louvered Funnel

By

Dipti Prasad Mishra

Department of Mechanical Engineering
Institute of Technical Education & Research
SIKSHA 'O' ANUSANDHAN UNIVERSITY
Bhuabaneswar

INTRODUCTION

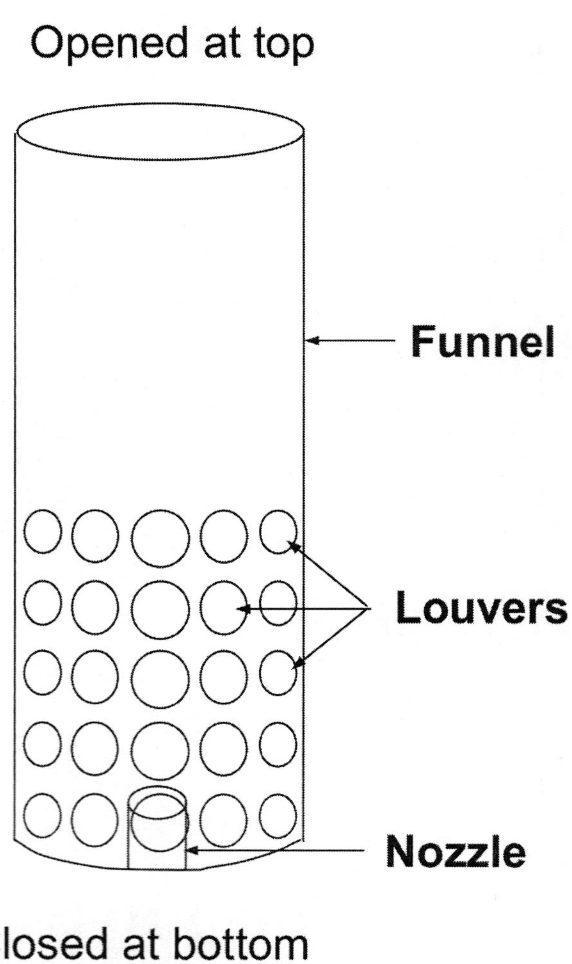

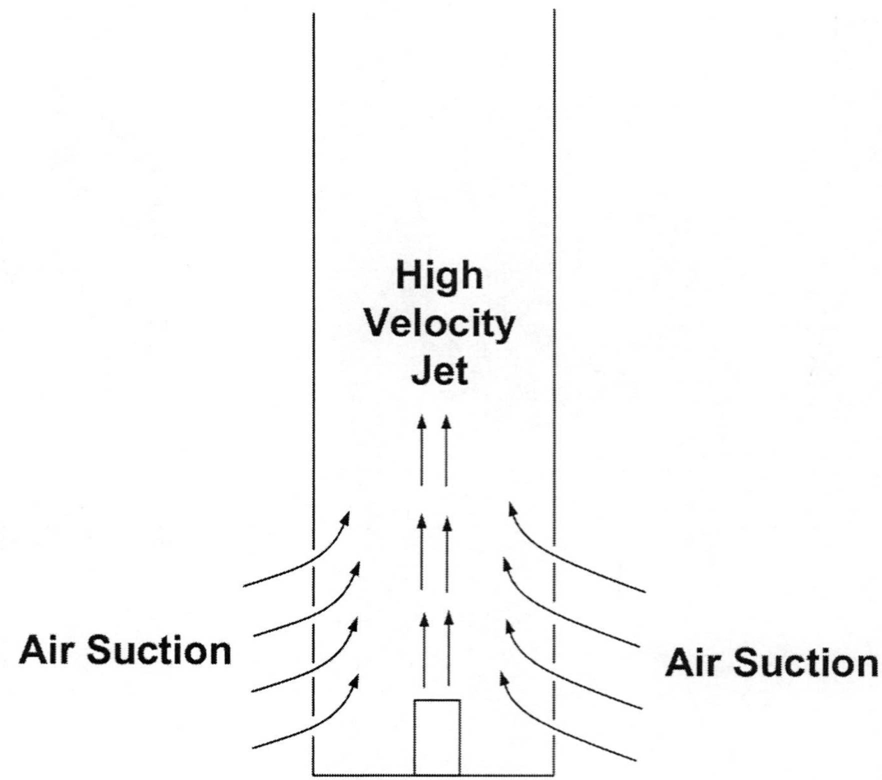

APPLICATIONS

Naval and Merchant ships

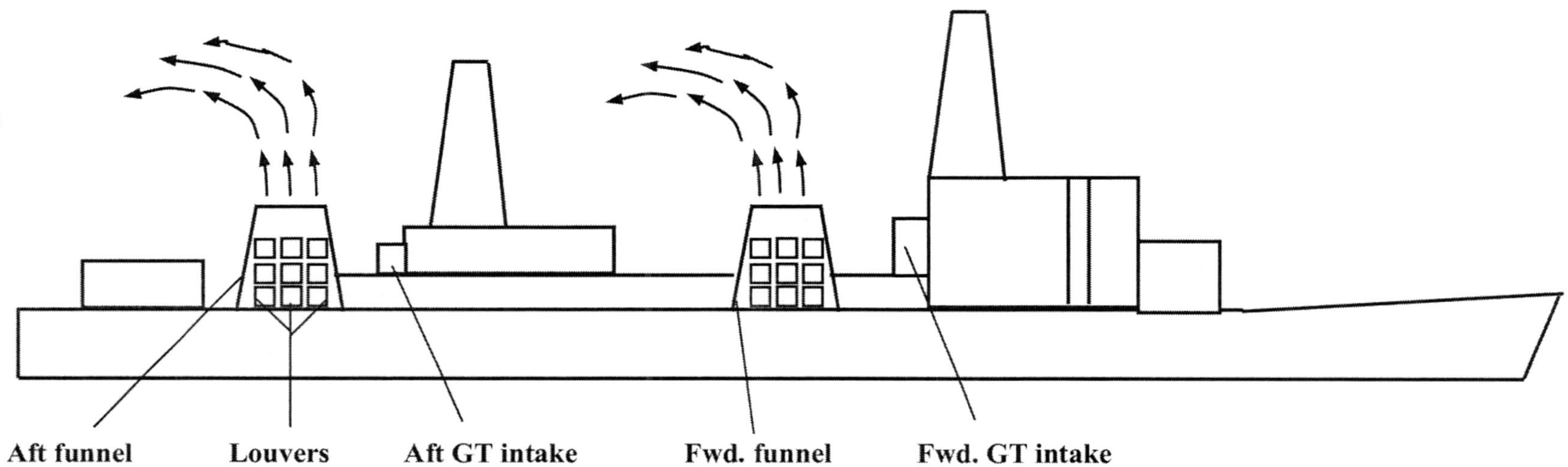

Contributions of Previous Researchers

Interaction of Exhaust Emissions with other Superstructure

Nolan (1946).	Based on wind tunnel experiments and concluded the stack gas velocities should be more than 150ft/s and height of the stack should be above the turbulent zone.
Acker (1952).	Based on wind tunnel experiments and concluded better performance can be achieved by increasing stack gas velocities.
Baham and McCallum (1977).	Provide guidelines for casing location, shape, height and exit velocity.
Seshadri et al. (2006).	Recommended for modification to the top side configuration of the stack to over come undesirable problem of ingress of exhaust smoke into GT intake.
Kulkarni et al. (2007).	Made a parametric study of exhaust smoke super structure interaction and predicted the flow path of exhaust plume.

Enhancement of Entrainment and Mixing

Zaman and Hussain (1980, 1981).	Active Flow Control: Forcing the jet at a frequency close to its natural frequencies with the help of acoustic or electromagnetic devices.
Tong and Warhaft (1994), Rajgopalan and Antonia (1998).	Passive flow control: Placing of an obstruction–like bluff body in the near field of the jet, introduction of tabs or notches at the nozzle exit.
Krothapalli et al. (1981), Ho and Gutmark (1987).	Passive flow control: Changing the nozzle cross section from circular to noncircular shape.
Singh et al. (2003).	Smaller mixing tube diameter and jet located at the inlet of the mixing tube, the circular jet entrains more than the noncircular jet and For a larger mixing tube or shifted jet locations, the noncircular jets entrain more ambient fluid.
Kandakure et al. (2008).	Presence of confinement (draft tube): With an increase in the draft tube diameter, the entrainment of the surrounding fluid into the draft tube increases

Objective

- Using CFD methodology to predict maximum air ingress into a real funnel of a naval or merchant ship.

Mathematical Modeling

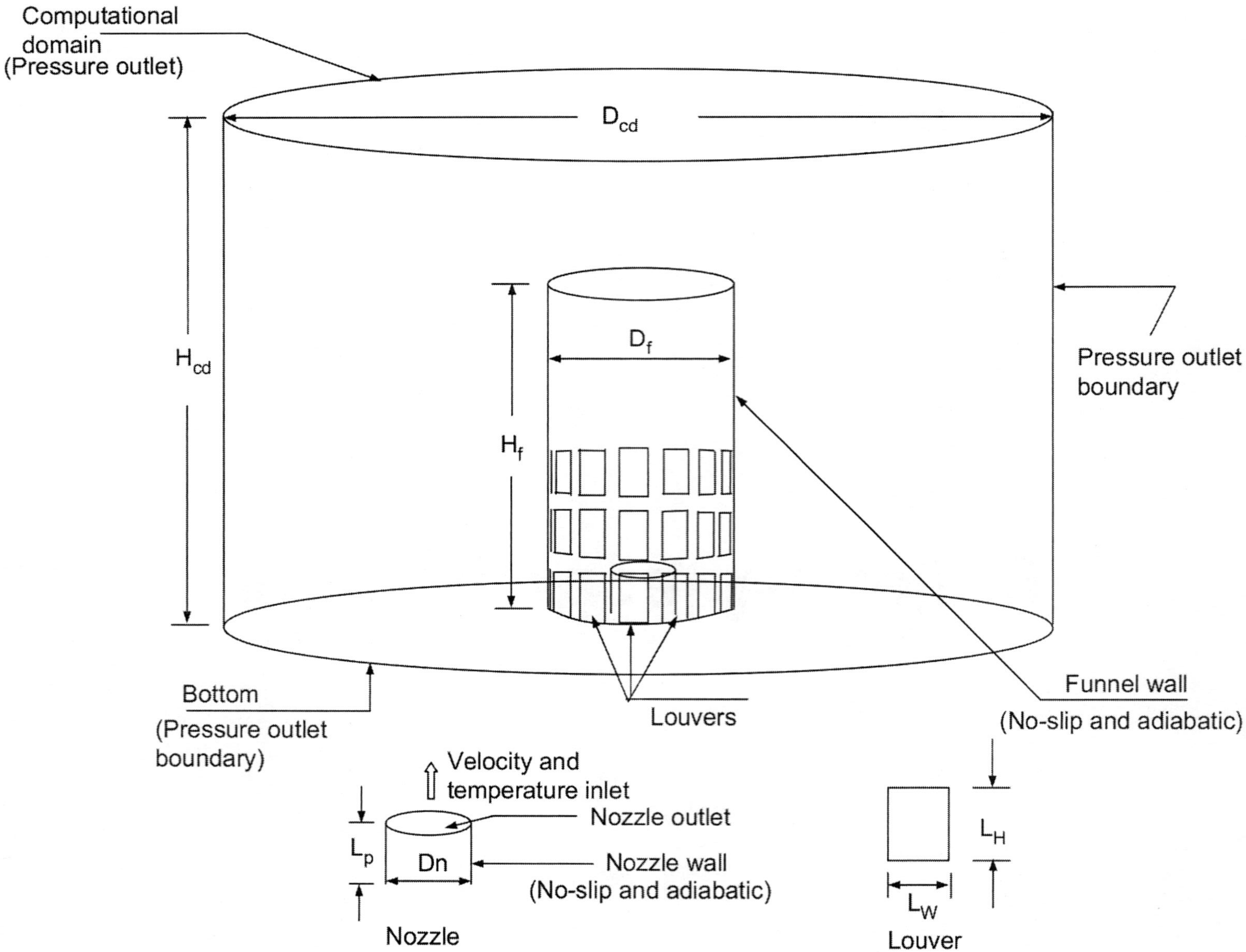

Schematic diagram of computational domain with the cylindrical funnel and a pictorial view of boundary conditions applied to it.

Governing Equations

1. Conservation of mass

$$\nabla.(\rho\vec{v}) = 0$$

2. Conservation of momentum

$$\nabla.\left(\rho\vec{v}\vec{v}\right) = -\nabla p + \nabla.\left(\mu^{eff}\nabla\vec{v}\right) - (\rho_{\infty} - \rho)g$$

3. Conservation of energy

$$\frac{D(\rho T)}{Dt} = \frac{\partial}{\partial x_i}\left[\left(\frac{\mu}{\Pr} + \frac{\mu_t}{\Pr_t}\right)\frac{\partial T}{\partial x_i}\right]$$

Turbulence modeling

Turbulence kinetic energy

$$\nabla.\left(\rho\vec{v}k\right) = \nabla.\left(\frac{\mu_t}{\sigma_k}\nabla k\right) + G_k - \rho\varepsilon$$

Turbulent rate of dissipation

$$\nabla.\left(\rho\vec{v}\varepsilon\right) = \nabla.\left(\frac{\mu_t}{\sigma_\varepsilon}\nabla\varepsilon\right) + \frac{\varepsilon}{k}\left(C_{1\varepsilon}G_k - C_{2\varepsilon}\rho\varepsilon\right)$$

Where G_k = Generation of turbulent kinetic energy due to mean velocity gradient and turbulent or eddy viscosity

$$\mu_t = \rho C_\mu \frac{k^2}{\varepsilon}$$

$$C_{1\varepsilon} = 1.44; C_{2\varepsilon} = 1.92; C_\mu = 0.09; \sigma_k = 1.0; \sigma_\varepsilon = 1.3 \quad ;\Pr_t = 1$$

Numerical Scheme

- Number of cells – 210000
- Cells – Unstructured tetrahedral cells
- Solver – Segregated solver
- Scheme – Implicit Scheme
- Pressure velocity coupling – SIMPLE Method
- Diffusion term – Second order Central difference
- Advection term – First order upwind
- Convergence Criteria (residuals) – 10^{-3} for all equations (Except Energy Equation)
- Convergence Criteria for Energy Equation (residual) - 10^{-6}

Results

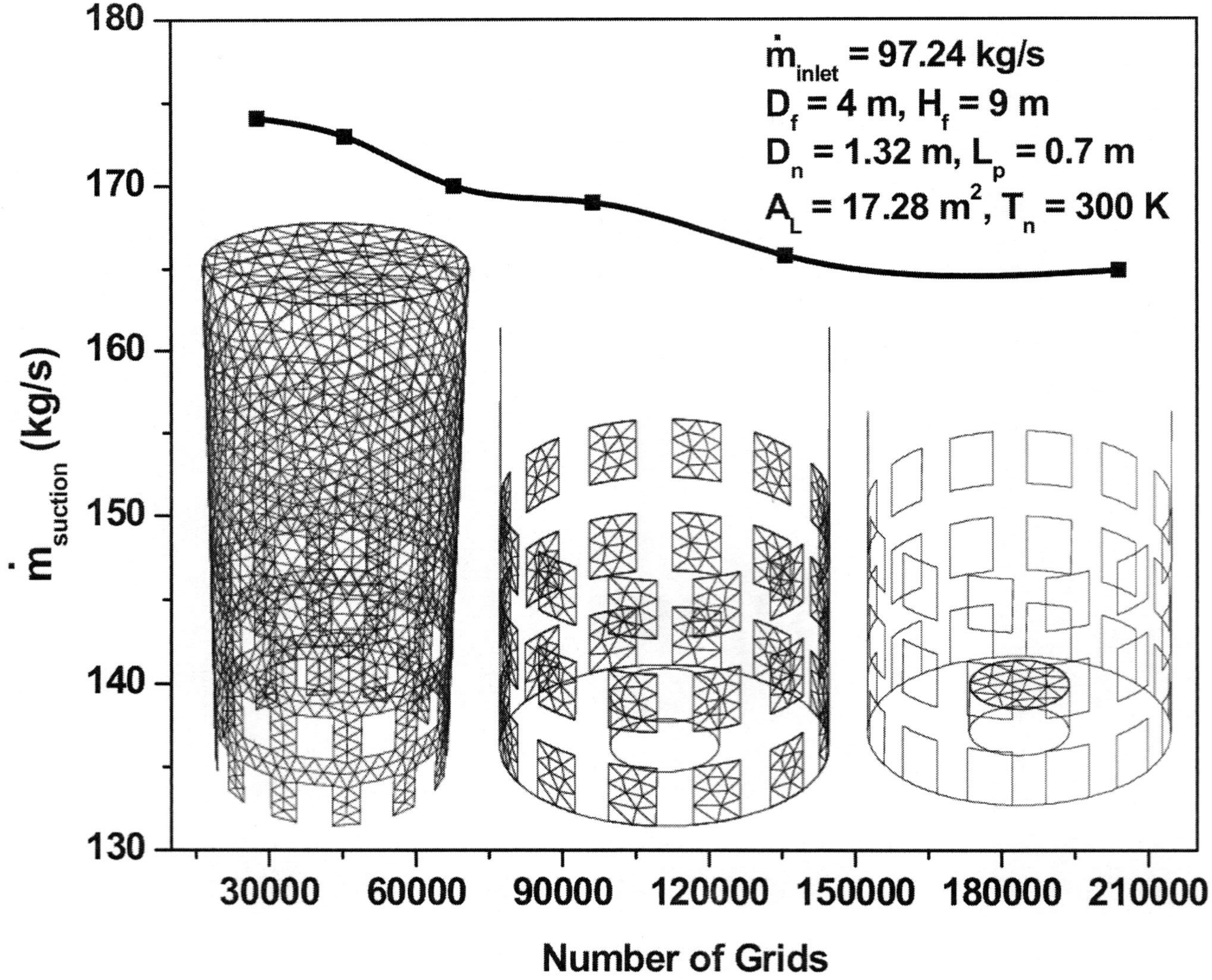

Effect of number of cells on air entrainment into the funnel

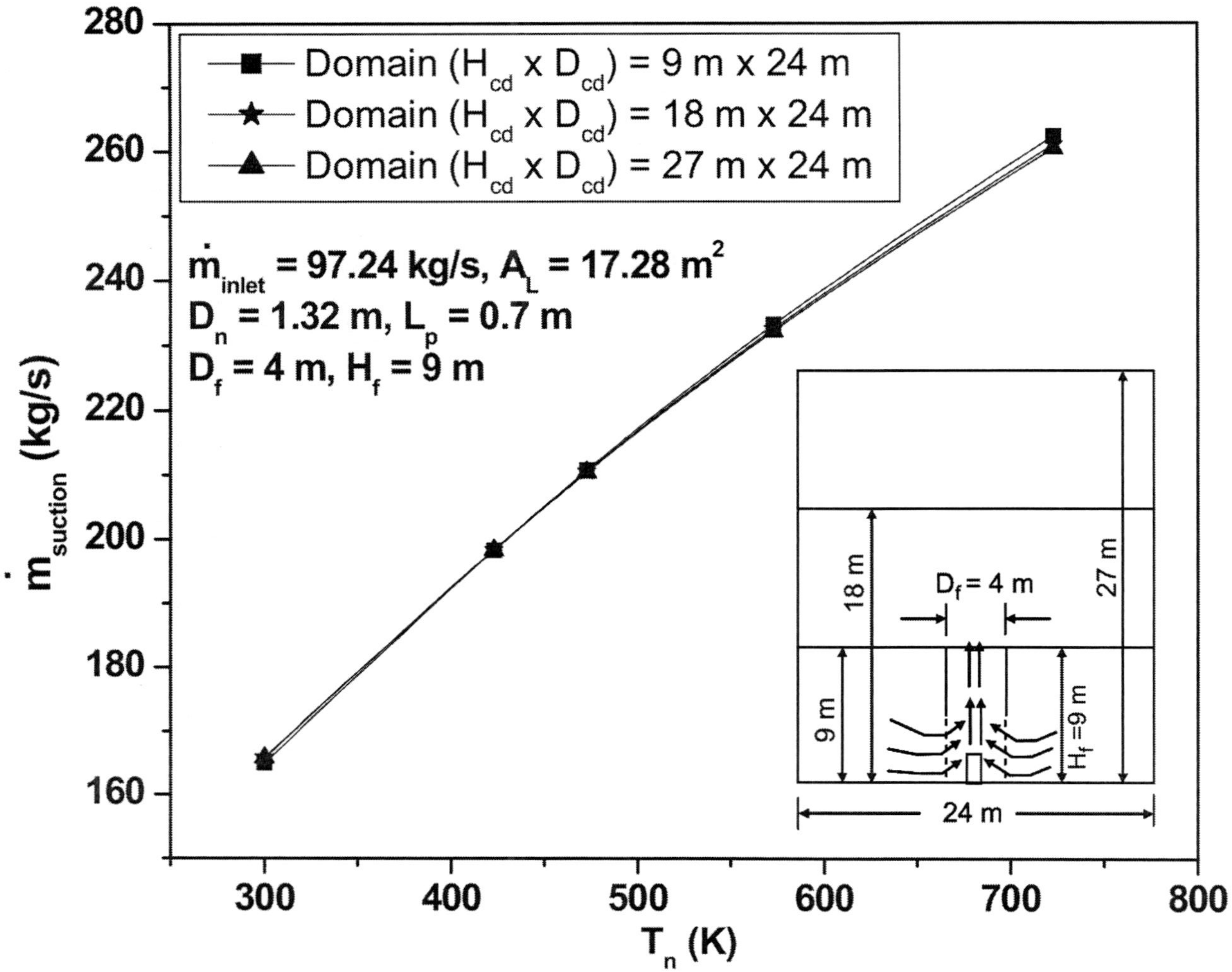

Effect of size of the domain on mass ingress of air into the funnel as a function of nozzle fluid temperature

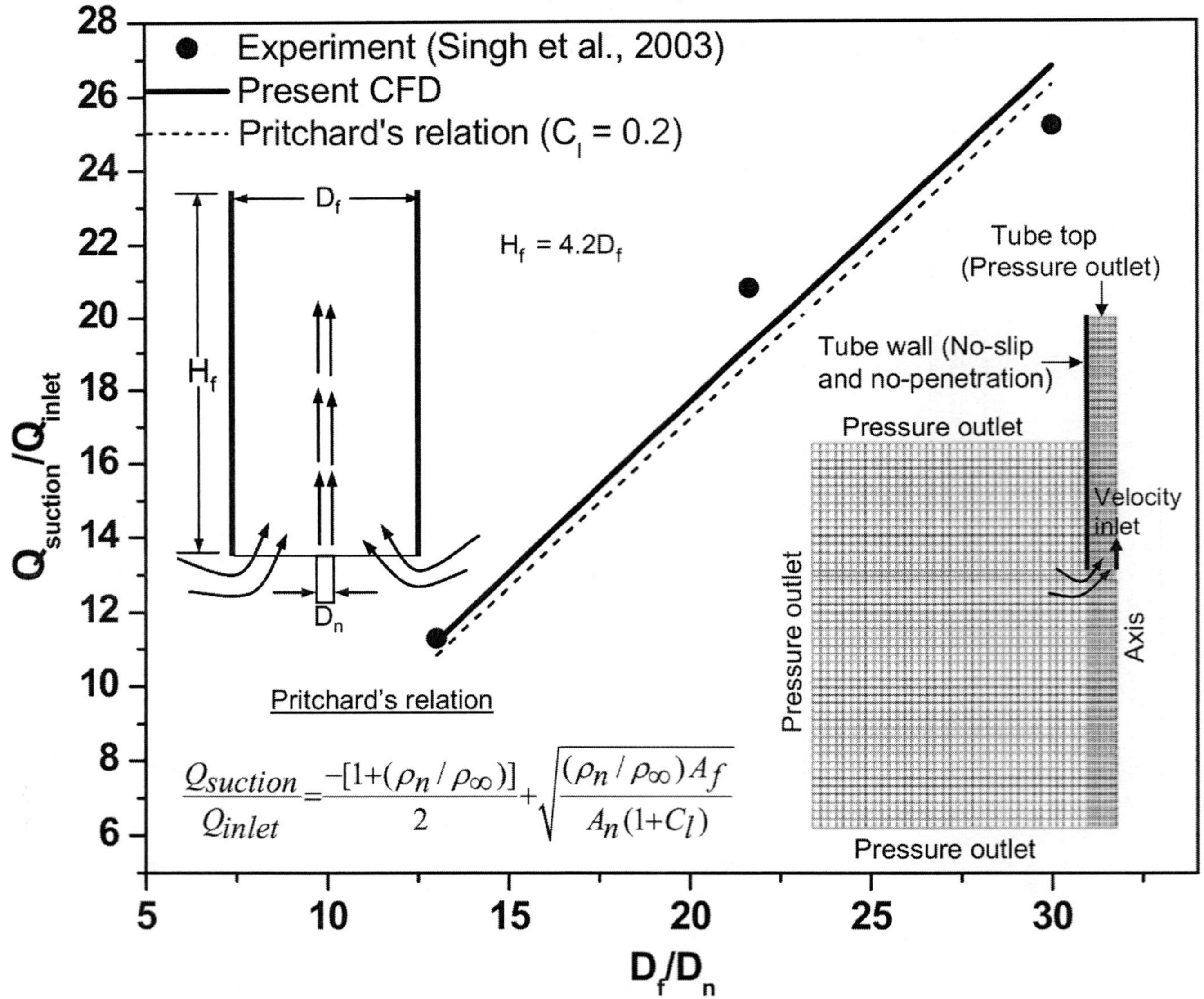

Entrainment ratio as a function of D_f/D_n for a wall constrained jet: A Comparison among the experiment, present CFD and Pritchard's relation

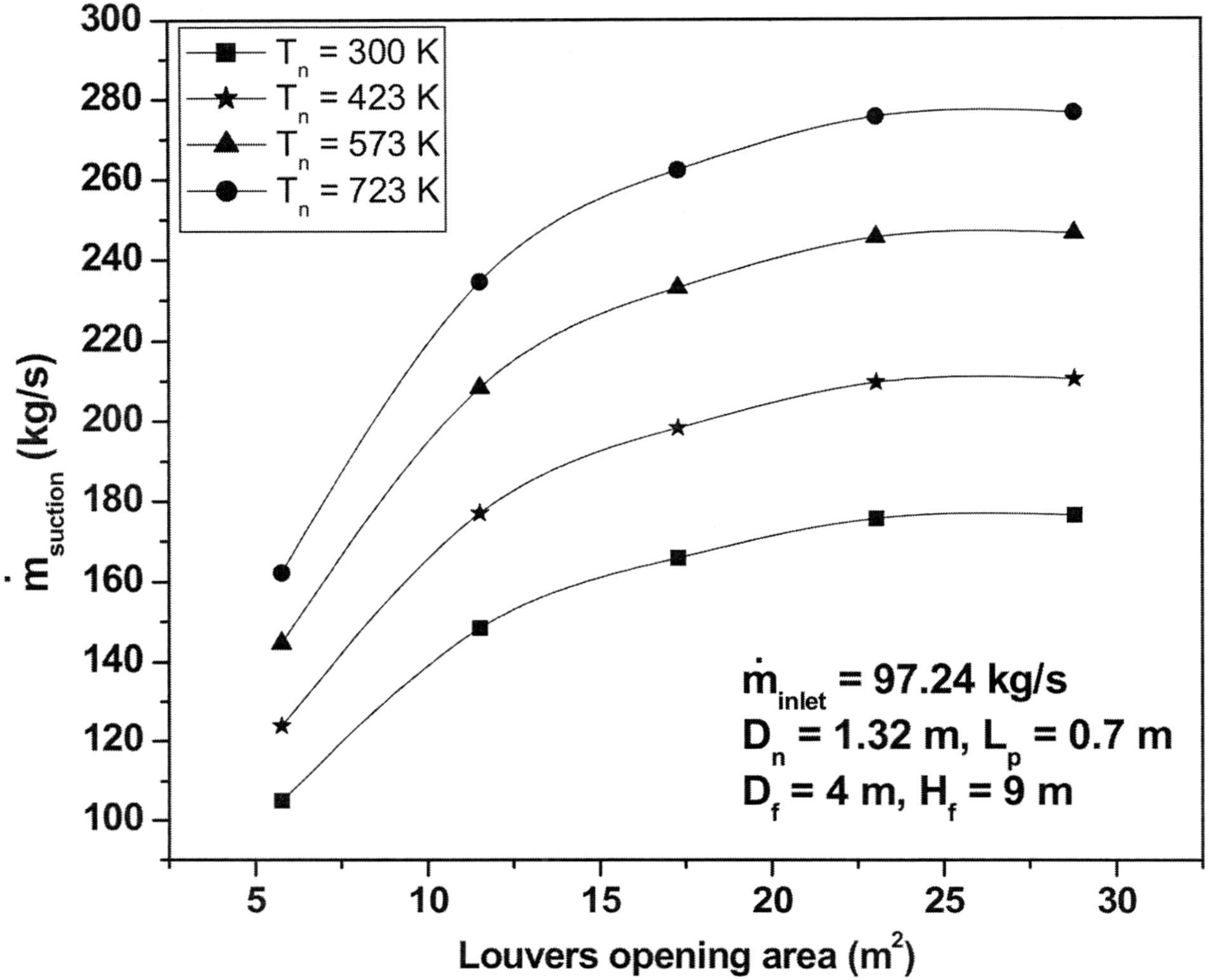

Mass ingress of air into the funnel as a function of louvers opening area and nozzle fluid temperature

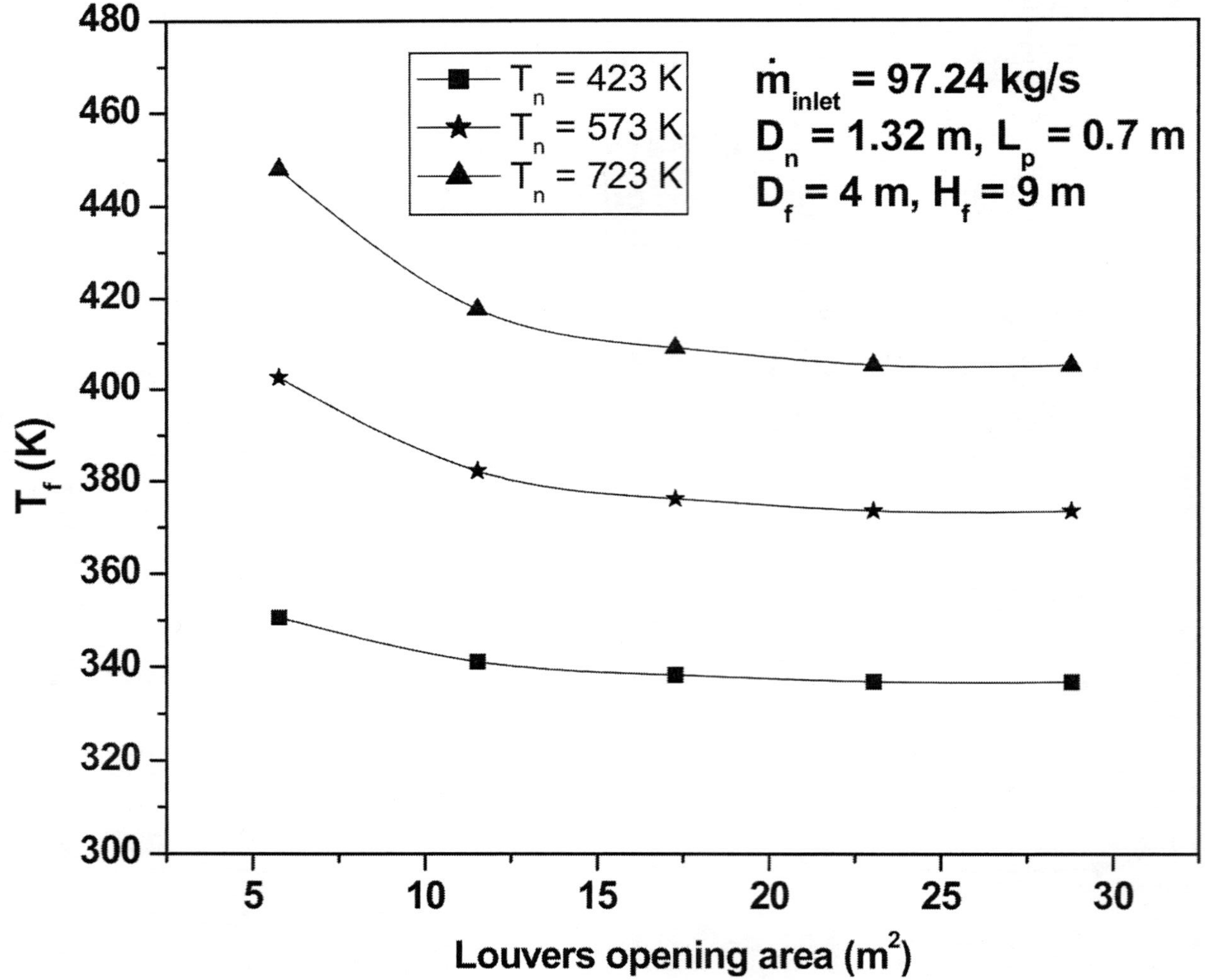

The funnel outlet temperature as a function of louvers opening area and nozzle fluid temperature

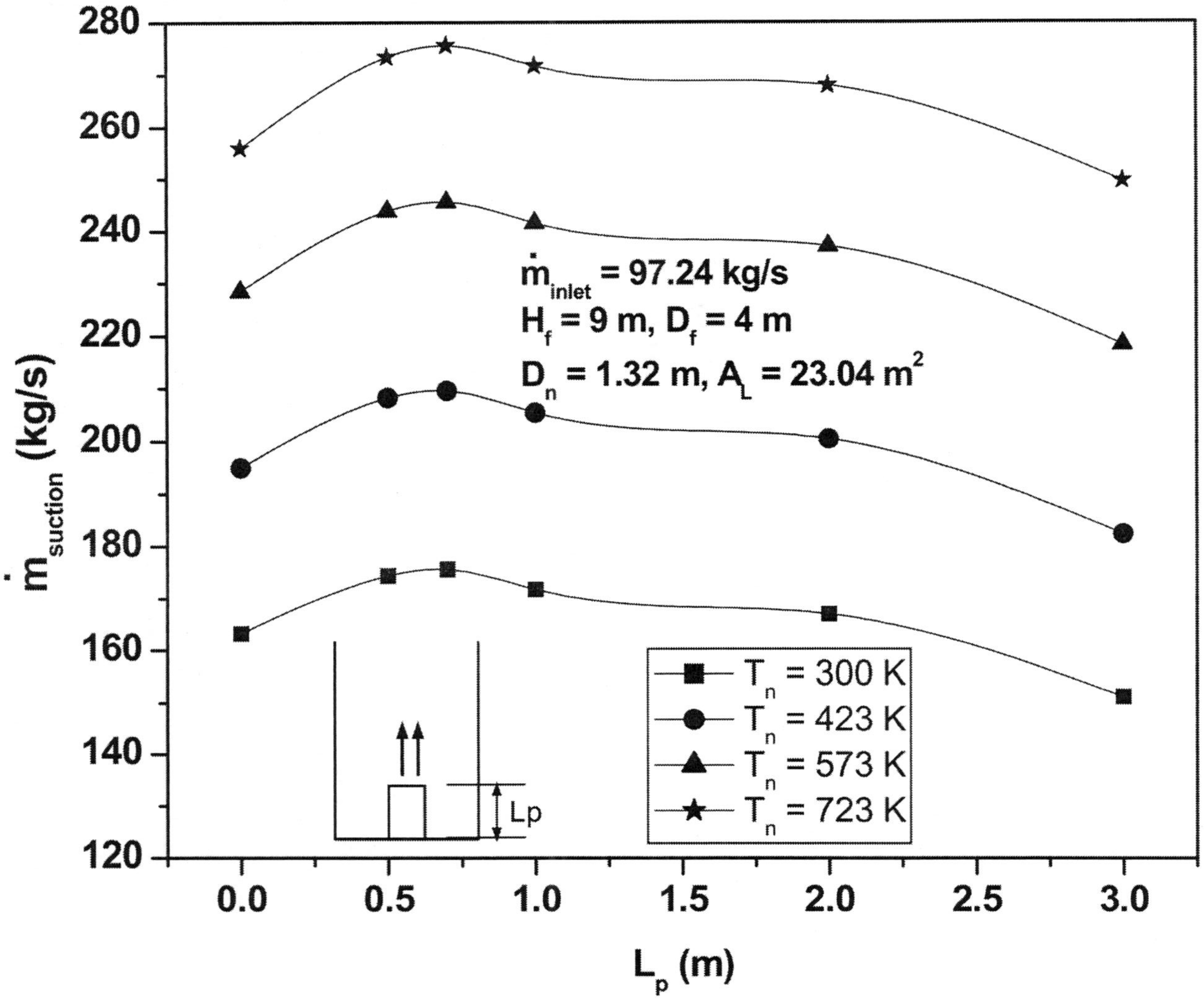

Mass ingress into the funnel as a function of protrusion length of the nozzle into the funnel (optimum protrusion length = 0.7 m)

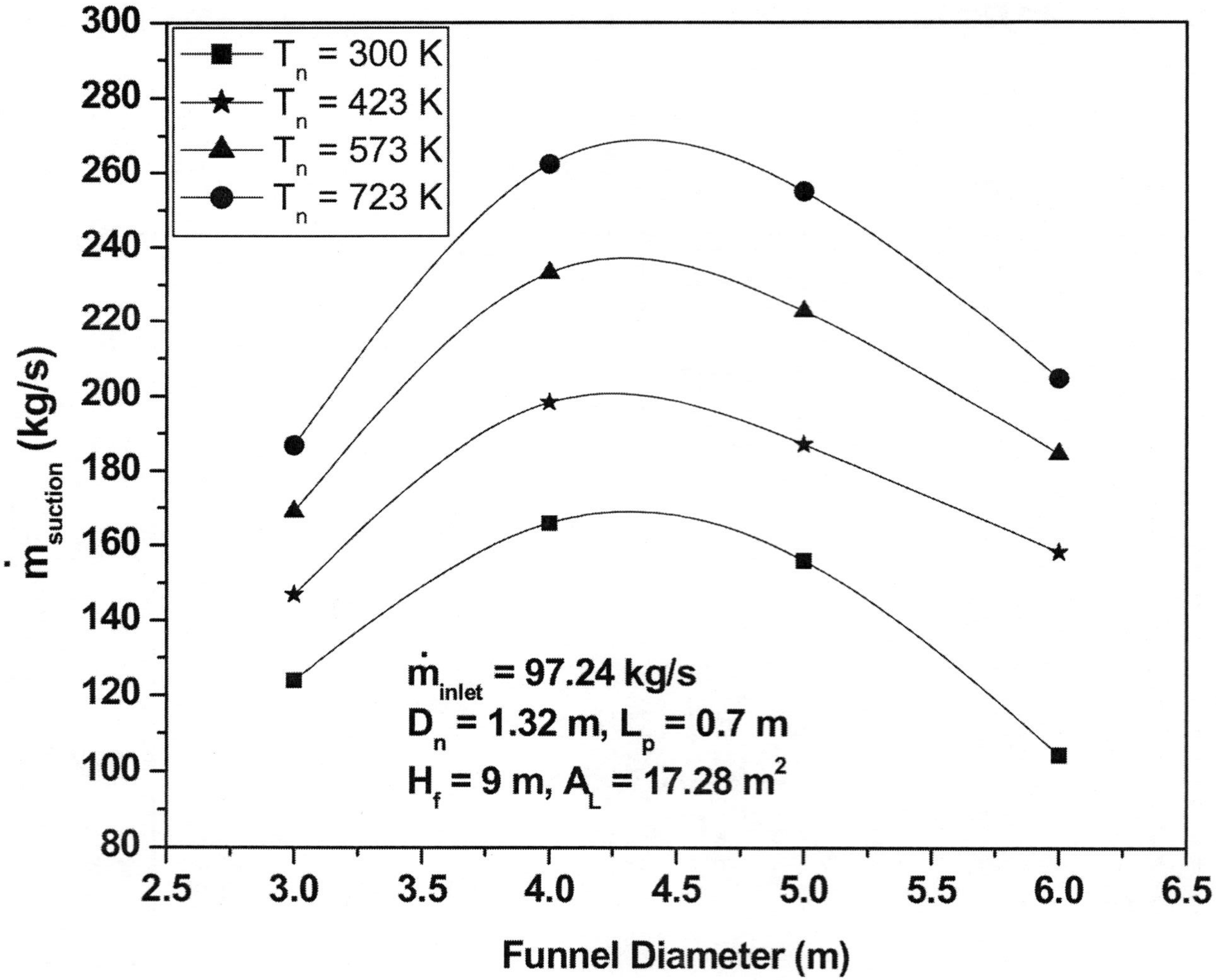

Entrainment of atmospheric air into the funnel as a function of diameter of the funnel (optimum funnel diameter = 4.5 m)

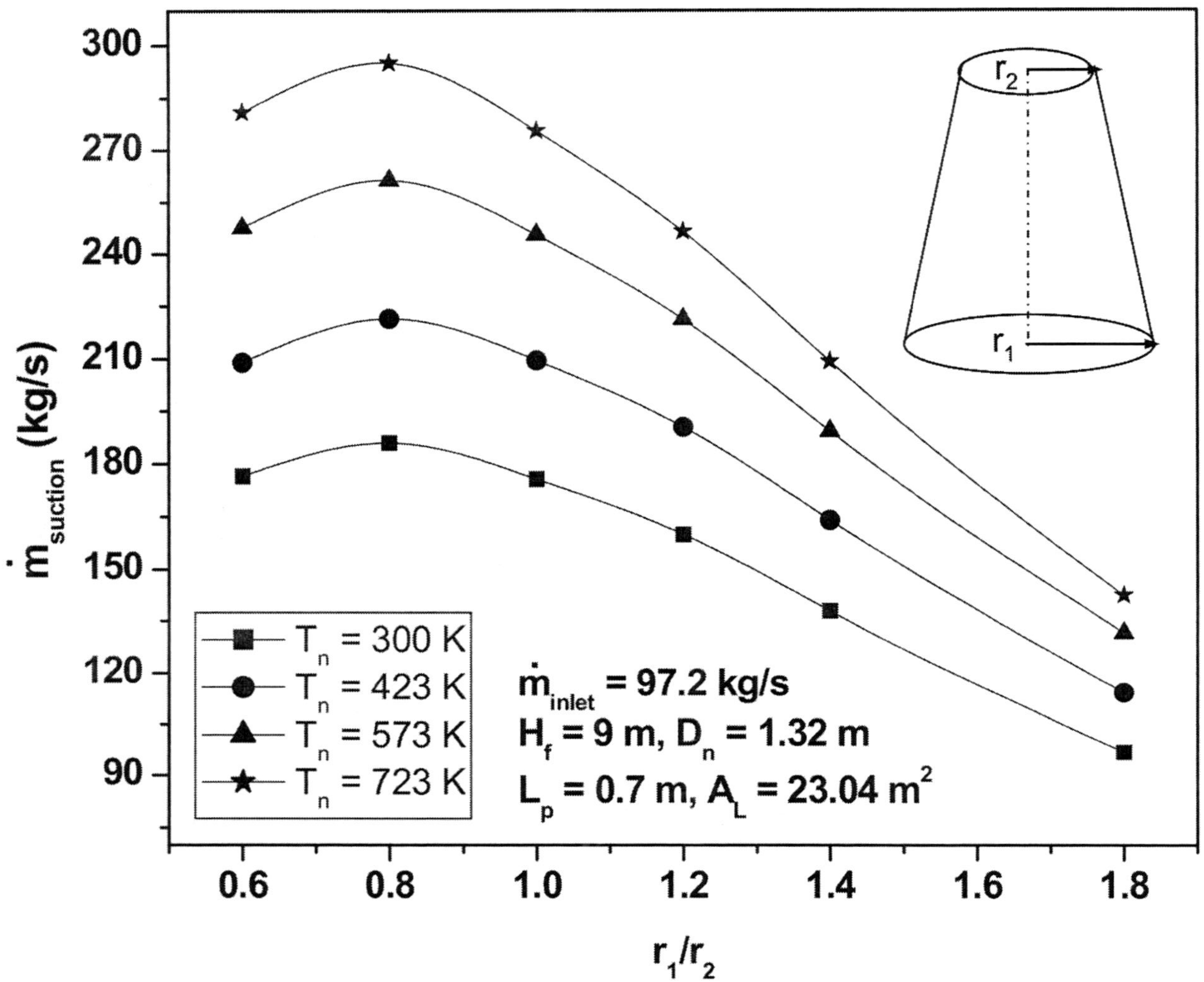

Mass ingress into the funnel as a function of shape of the funnel (optimum r_1/r_2 can be seen to be at 0.8 m)

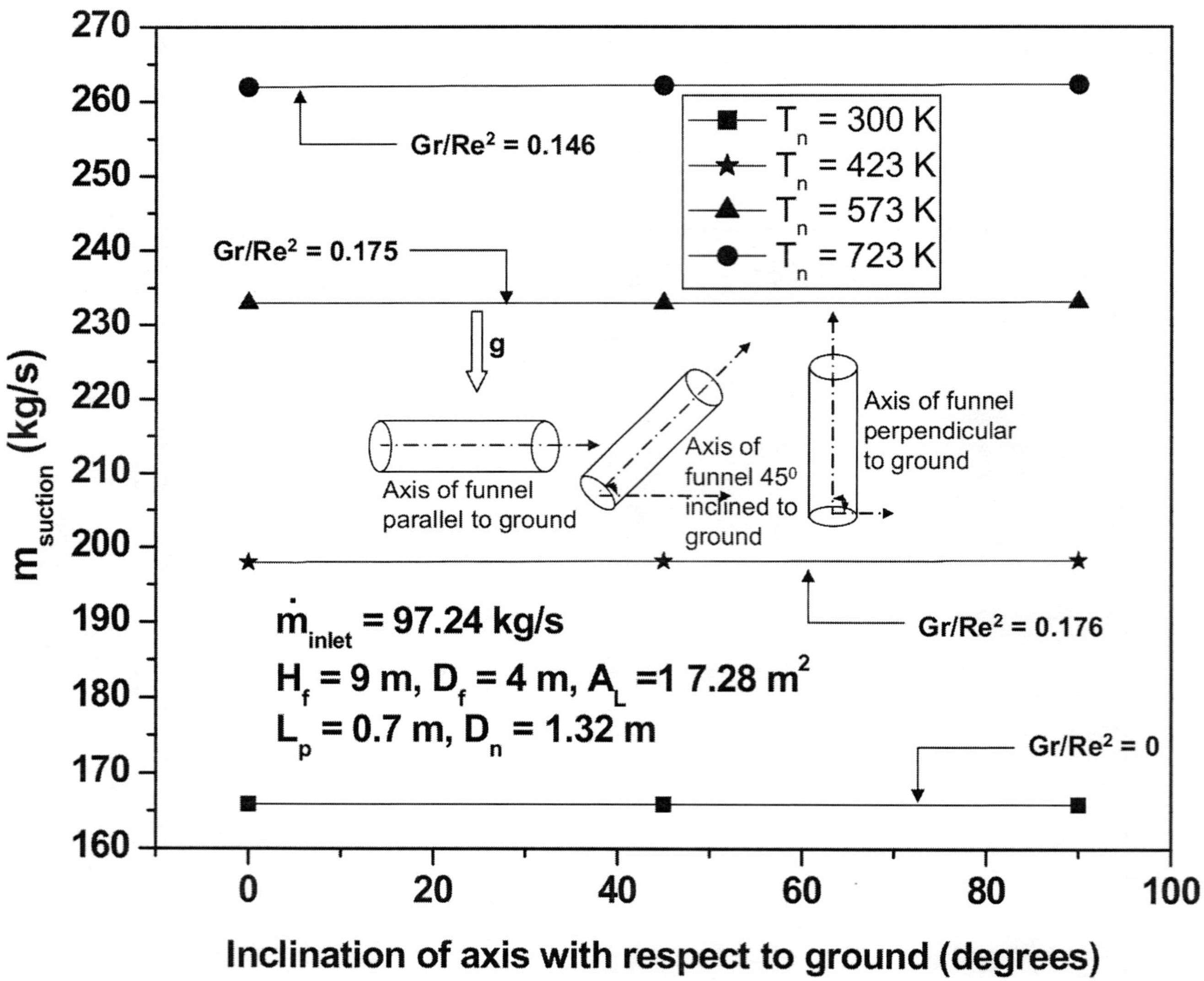

Air suction into the funnel as a function of nozzle fluid temperature and inclination of the funnel

Non-dimensional numbers for a particular funnel with different nozzle velocity which can decide the relative weightage of natural convection over forced convection.

Parameters are based on H_f = 9 m, D_f = 4 m, D_n = 1.32 m, T_n = 573 K and T_∞ = 300K

V_{inlet} (m/s)	5	10	20	30	40	50	60	80
Gr/Re^2	102.7	25.6	6.4	2.85	1.6	1.02	0.71	0.4
$Bo^{1/4}/Pe^{1/2}$	3.18	2.25	1.59	1.3	1.12	1.00	0.92	0.79

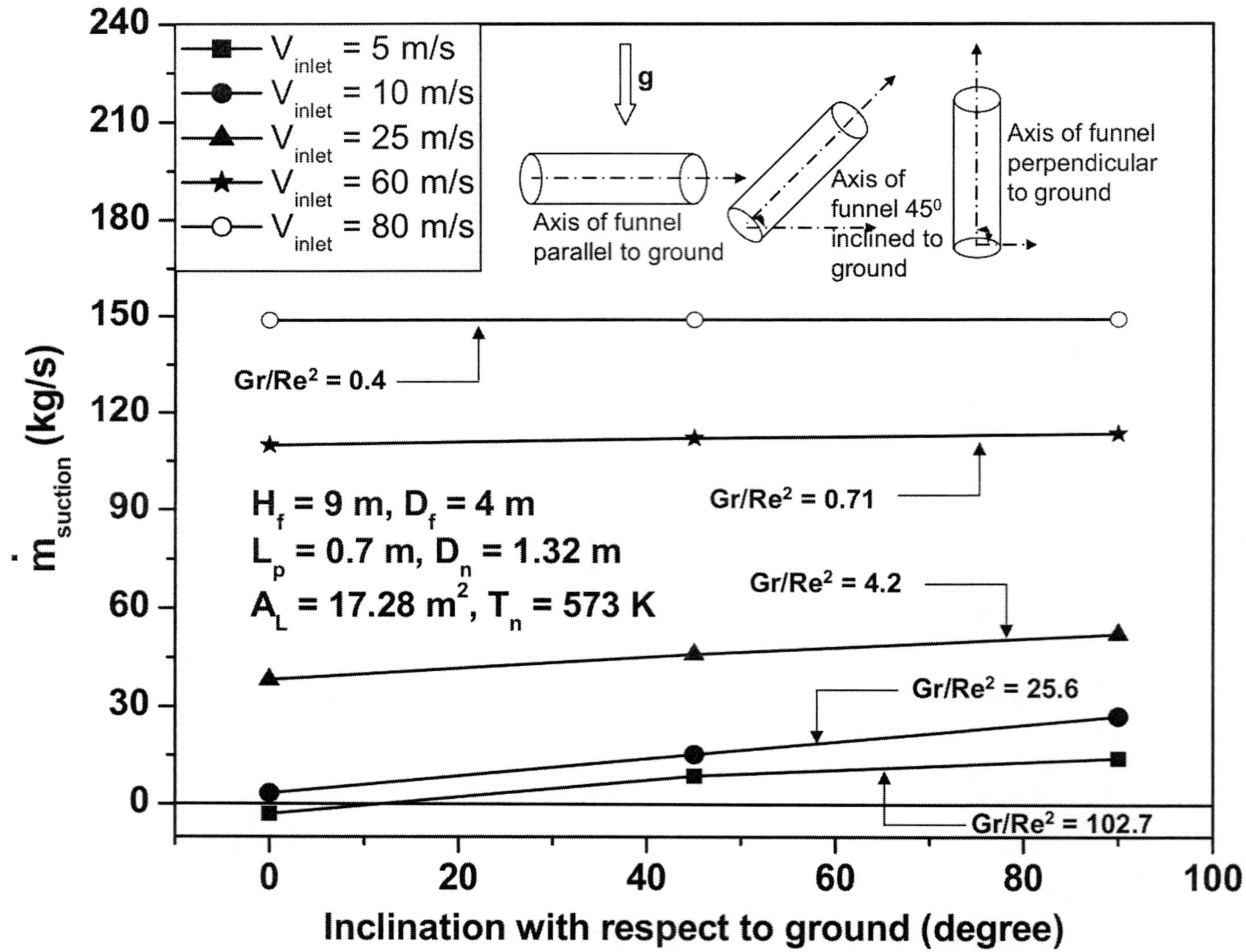

Effect of inclination of the funnel on mass ingress of air in to it when the nozzle ejects hot air at 573 K

$V_{Noz} = 5$ m/s, $T_{Noz} = 300$ K, $Gr/Re^2 = 0$

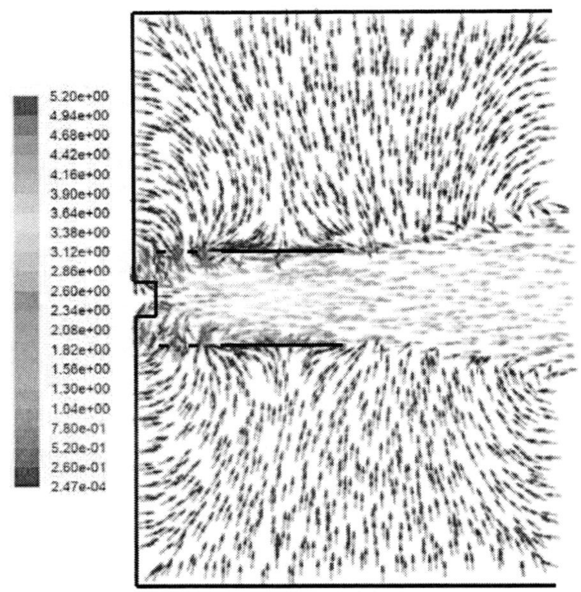

(a)

$V_{Noz} = 5$ m/s, $T_{Noz} = 573$ K, $Gr/Re^2 = 102.7$

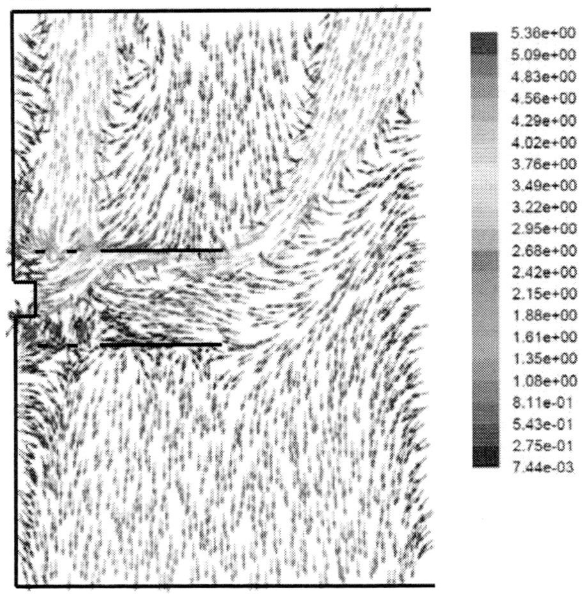

(b)

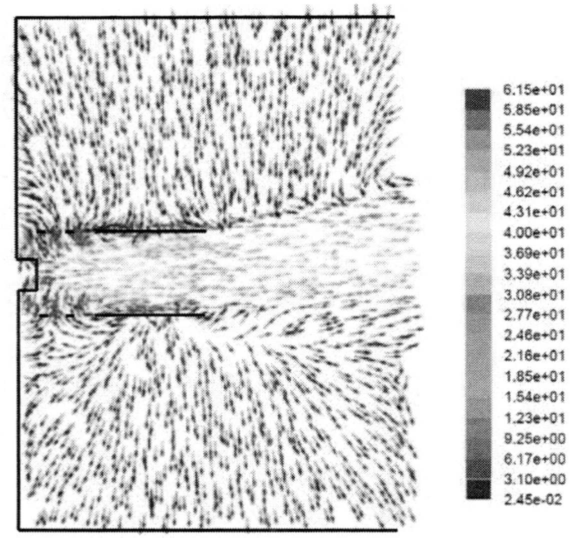

(c)

$V_{Noz} = 60$ m/s, $T_{Noz} = 573$ K, $Gr/Re^2 = 0.71$

Velocity field inside and outside the funnel for different nozzle inlet velocity and temperature when the funnel is horizontal.
$H_f = 9$ m, $D_f = 4$ m, $L_p = 0.7$ m, $D_n = 1.32$ m, $A_L = 17.28$ m^2

Conclusions

- The air entrainment rate is increased with increase in louvers opening area and nozzle flow temperature.
- There exists an optimum protrusion of the nozzle into the funnel, where the mass entrainment into it can be the maximum, and the optimum protrusion was found to be 0.7 m for the present set of computations irrespective of nozzle outlet temperature.
- There exists an optimum funnel diameter where the mass entrained in to it can be the maximum, and the optimum diameter was found to be 4.5 m, both the cases of hot and cold fluid flow conditions for a constant louvers opening area.

- Compared to a cylindrical funnel a frustum shape funnel could suck more air into it and the optimum diameter ratio of the frustum base to top was found to be 0.8.

- For hot nozzle jet the inclination of the funnel has effect on mass ingress of air into it, but after a certain critical value of the nozzle velocity, where $Gr/Re^2 = 1$, the inclination of the funnel has again no effect on the mass ingress.

References

Acker, H. G. (1952), Stack Design to Avoid Smoke Nuisance, Trans. SNAME, Vol. 60, pp. 566-593.

Baham, G. J. and Mc Callum, D. (1977), Stack Design Technology for Naval and Merchant Ships, Trans. SNAME, Vol. 85, pp. 324-349.

Becker, H. A., Hottel, H. C. and Williams, G. C. (1963), Mixing and Flow in Ducted Turbulent Jets, 9th Int. Symp. Combustion, The Combustion Institute, Pittsburg, PA, pp. 7-20.

Bejan, A. (2004), Convection heat transfer, John Wiley & Sons Inc., 3rd Edition, pp. 213.

Chassaing, P. (1979), Mélange turbulent de gaz inertes dans un jet de tube libre, Thése de doctorat d'état, institut National Polytechnique de Toulouse.

Cramers, P. H. M. R., Dierendonck, L. L. V. and Beenackers, A. A. C. M. (1992), Influence of Gas Density on the Gas Entrainment Rate and Gas Hold-up in Loop-Venturi Reactors, Chemical Engineering Science, Vol. 47, No. 9-11, pp. 2251-2256.

Das, S. K. and Biswas, M. N. (2006), Studies of Ejector-Venturi Fume Scrubber. Chemical Engineering Journal, Vol. 119, pp. 153-160.

Dash, S. K. (1996), Heatline visualization in turbulent flow. Int. J. of Num. Methods for Heat and Fluid Flow, Vol. 6, No.4, pp. 37-46.

Han, D. and Mungal, M.G. (2001), Direct Measurement of Entrainment in Reacting/Nonreacting Turbulent Jets, Combustion and Flame, Vol. 124, pp. 370-386.

Havelka, P., Linek, V., Sinkule, J., Zahradnik, J. and Fialova, M. (1997), Effect of the Ejector Configuration on the Gas Suction Rate and Gas Hold-up in Ejector Loop Reactors. Chemical Engineering Science, Vol. 52, No.11, pp. 1701-13.

Hill, B. J. (1972), Measurement of Local Entrainment Rate in the Initial Region of Axisymmetric Turbulent Air Jets, J. Fluid Mech., Vol. 51, pp. 773-779.

Ho, C. M. and Gutmark, E. (1987), Vertex induction and mass entrainment in a small-aspect-ratio elliptic jet, J. Fluid Mech., Vol. 179, pp. 383-405.

Holman, J. P. (1994), Experimental Methods for Engineers, McGraw-Hill International Editions, 6th Edition, pp. 49.

Hussain, F. and Husain, H. S. (1989), Elliptic jets Part 1. Characteristics of Unexcited and Excited Jets, J. Fluid Mech., Vol. 208, pp. 257-320.

Jha, P. K. and Dash, S. K. (2002), Effect of outlet positions and various turbulence models on mixing in a single and multi strand tundish. Int. J. Num. Method for Heat and Fluid Flow, Vol. 12, No. 5, pp. 560-584.

Jha, P. K. and Dash, S. K. (2004), Employment of different turbulence models to the design of optimum steel flows in a tundish. Int. J. Numerical Methods for Heat and Fluid Flow, Vol. 14, No. 8, pp. 953-979.

Mi, J., Nathan, G. J. and Luxton, R. E. (2000), Centerline Mixing Characteristics of Jets from Nine Differently Shaped Nozzles, Expt. Fluids, Vol. 28, pp. 93-94.

Prichard, R., Guy, J. J., and Conner, N. E. (1977), Industrial Gas Utilization, Bowker, New Providence, NJ.

Risso, F. and Fabre, J. (1997), Diffusive Turbulence in a Confined Jet Experiment, J. Fluid Mech., Vol. 337, pp. 233-261.

Singh, G. S., Sundararajan, T. and Bhaskaran, K. A. (2003), Mixing and Entrainment Characteristics of Circular and Noncircular Confined Jets, J. Fluid Engg., ASME, Vol. 125, pp. 835-842.

Tong, C. and Warhaft, Z. (1994), Turbulence Suppression in a Jet by means of a fine ring, Phys. Fluids, Vol. 6, pp. 328-334.

Weisgraber, T. H. and Liepmann, D. (1998), Turbulent Structure during Transition to Self-similarty in a round jet, Expt. Fluids, Vol. 24, pp. 210-224.

Zaman, K. B. M. and Hussain, A. K. M. F. (1980), Vertex Pairing in a Circular Jet Under Controlled Excitation, J. Fluid Mech., Vol. 101, pp. 449-491.

Nomenclature

A_L	:	Louvers opening area
Bo	:	Boussinesq number, $g\beta(T_{inlet} - T_{amb})H_f^3 / \alpha^2$
D_{cd}	:	Diameter of computational domain
D_f	:	Diameter of the funnel
D_L	:	Diameter of the louver
D_n	:	Nozzle diameter
g	:	Acceleration due to gravity
G_k	:	Production of turbulent kinetic energy
Gr	:	Grashof Number, $g\beta(T_{inlet} - T_{amb})H_f^3 / \nu^2$
H_{cd}	:	Height of the computational domain
H_f	:	Height of the funnel
K	:	Kelvin
k	:	Turbulent kinetic energy

L_p	:	Protruding length of nozzle
$\dot{m}_{inlet}$	:	Mass flow rate through the nozzle
$\dot{m}_{suction}$	:	Air suction rate through the louvers
p	:	Pressure
p_a	:	Atmospheric pressure
Pe	:	RePr
Pr	:	ν/α
Pr_t	:	Turbulent Prandtl number
Q_{inlet}	:	volume flow rate through the nozzle
Q_{suc}	:	volumetric air suction rate
Re	:	$\rho V_{inlet} D_n/\mu$
T	:	Temprature
V_n	:	Outlet velocity at nozzle exit

Greek symbols

ν	:	Kinematic viscosity
α	:	Thermal diffusivity
β	:	Volume expansion coefficient
ρ	:	Density
μ	:	Shear viscosity
μ_t	:	Turbulent viscosity
μ^{eff}	:	Effective viscosity
ε	:	Rate of dissipation
σ_k	:	Turbulent Prandtl number for *k*
σ_ε	:	Turbulent Prandtl number for ε

Thank you

A NUMERICAL AND EXPERIMENTAL INVESTIGATION INTO THE FREE STREAM HYDRODYNAMICS CHARACTERISTICS OF A SERIES OF FLAPPED RUDDERS USING SHIPFLOW^R

A. Bhattacharjee, Indian Institute of Technology, Kharagpur, India
O. P. Sha, Indian Institute of Technology, Kharagpur, India

SUMMARY

Manoeuvring of marine vehicles is defined as the controlled keeping and changing of course and speeds of the vessel, which is conventionally achieved by use of control surfaces, like rudders. In recent years, a number of high lift devices, like flap rudders have been developed primarily to increase lift at low speeds so as to achieve better low speed manoeuvring in ports and congested water ways. Flap rudders effectively work as a variable camber device and help in considerable increase of the lift force and also to delay stall. Such control surfaces in conjunction with both thrusters are able to achieve tighter turning circle diameter and better stopping ability at low speed. Kerwin et al. [1] carried out water tunnel experiments to determine free stream characteristics of a series of twelve rudders with systematic variation in the amount of flap area of the fixed part and movable part for a rudder with skeg. Initially, to calculate the effectiveness and robustness of the computational model, Kerwin's(1972) rudder is modelled and the lift, drag obtained from the computational analysis is compared with the water tunnel experiments of Kerwin(1972). The results are in good agreement and presented in the paper. Subsequently a series of flap rudder with varying percentage of flap areas (20%, 30%, 40%) and flap to rudder deflection are modelled and lift, drag and stall angle for varying angle of attack are numerically computed using RANS solver, SHIPFLOW^R. The results of this systematic study are presented. Finally for a few of the above variation of flap rudders physical models are constructed and tested into wind tunnel experimental facility of Naval Science and Technology Laboratory, (NSTL) Visakhapatnam. The experimental results are presented along with the computational results for these cases. It is expected that the current investigation will assist the naval architect, at the early design stage, to select the flap rudder geometry that will enhance the manoeuvring characteristic of the vessel and overcome the stringent present day IMO requirement.

NOMENCLATURE

α	Angle of attack (degrees)
υ	Kinematic viscosity (N s m^{-2})
ρ	Density of water (kg m^{-3})
Cl	Lift coefficient
Cd	Drag coefficient
S	Planform area(m^2)
V	Velocity (m/s)

1. INTRODUCTION

Manoeuvring of a marine vehicle can be viewed as the controlled change or retention of the direction of motion of a vessel and its speed in that direction. Whereas for a surface ship it includes all force, moment and motion vectors in any direction in the horizontal plane, for underwater vehicles, like submarine and ROV, manoeuvring encompasses the control of the path of the vehicle in three-dimensional space. The purpose of a control surface on a marine vehicle is to control the motion of the vessel. The control surface for a surface ship can consist of an all-moveable surface (commonly referred as the spade or under-hung rudder) or a combination of moveable and fixed surface (commonly called a horn, skeg, semi-balanced or gnomon rudder) to control the horizontal motion of the ship. In case of underwater vehicles, the control surface will additionally consist of a diving plane to control the vertical motion of such vehicles. The purpose of the control surface (rudder or diving plane) is to produce a control force resulting from its orientation and movement relative to the water. This in turn creates a moment on the vessel and causes it to reorient at an angle of attack to the flow.

Presently, bow thrusters in vessels are being used in vessels for easier berthing. These vessels can do away with the expenses of tugs for berthing at the dock. But to be effective the bow thrusters must work in tandem with high lift control surfaces such that the latter can generate large lateral force at low speeds. Special rudder types have been developed over the years by modification of the conventional rudders so as to obtain better performance. A number of high lift rudders and control surfaces have been proposed and employed primarily to enhance manoeuvring. The main objective of such devices is to delay the stall and to increase maximum lift achievable. The mission requirements for such design have been to obtain a tighter turning circle diameter, better stopping ability or higher rudder forces at low speed manoeuvres.

Kerwin et al. (1972) carried out water tunnel experiments to determine free stream characteristics of a series of twelve rudders with systematic variation in the amount of flap area of the movable part and the fixed part (skeg) for a rudder with skeg. Kerwin's (1972) rudder is modelled and the lift, drag obtained from the computational analysis is validated by comparison with published water tunnel experiments.

The robustness of the computational model is established as the SHIPFLOW^R results show a good agreement with experimental values Kerwin et al. (1972).

A series of flap rudder with varying percentage of flap areas (20%, 30%, 40%) and flap to rudder deflections (5°,10°,15° and 20°) are modelled and lift, drag and stall angle for varying angles of attack are numerically computed using RANS solver, SHIPFLOW^R. The results of this systematic study are presented for both ahead and astern condition. Finally for a few of the above variation of flap rudders physical models are constructed and tested in the wind tunnel experimental facility of Naval Science and Technology Laboratory, (NSTL) Visakhapatnam.

2. RANS SOLVER

SHIPFLOW$^{(R)}$ [2] is a special purpose computer code for predicting the flow around ship and around its component, developed at Chalmers University of Technology and FLOWTECH International, AB. To solve the flow around the hull two different approaches, i.e. global and zonal approaches are available in SHIPFLOW$^{(R)}$. A global approach means that the Navier-Stokes equations are solved in the whole flow domain. A zonal approach means that the flow domain is divided into different zones based on the flow characteristics inside, as shown in the Fig. 1. In Zone 1, which is characterised as inviscid and irrotational, a potential flow method is used. This method computes wave resistance and induced resistance, and provides input to a boundary layer method in Zone 2, where transition and boundary layer parameters on the forward half of the ship are predicted. In Zone 3 where the boundary layer gets very thick, a RANS code is used.

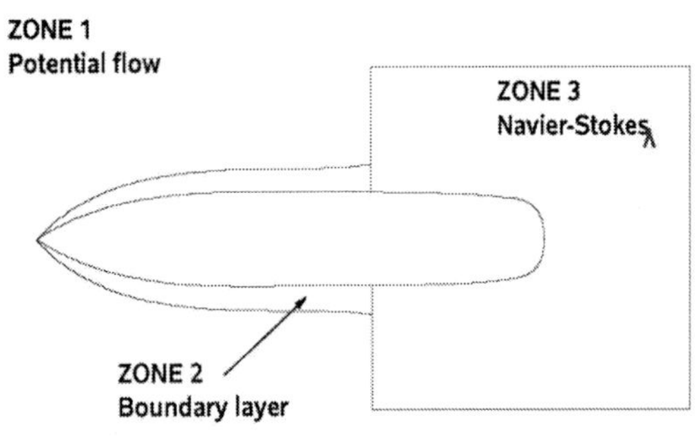

Figure 1: Different zonal approaches in SHIPFLOW®

XCHAP is based on a C++ class library CHAPMAN, which implements a finite volume method based code to solve the RANS equations. It uses several turbulence models (EASM, $k-\omega$ BSL, $k-\omega$ SST).The solver can be used with the zonal or global approach.

Equations for momentum, pressure and turbulence are fully coupled. Roe discretization is applied for the convective fluxes, and a central differencing is used for the diffusive fluxes. A flux correction with a min-mod limiter is used for increasing the accuracy to second order in regions of smooth flow. A local artificial time-step term is added to the equations and the discrete coupled equations are solved using an ADI-solver. Solving the RANS equations will give the time- averaged velocity, pressure and turbulent quantities.

XCHAP can use the grid provided by XGRID. Boundary values are generated from results provided by XPAN and XBOUND when the zonal approach is used. XCHAP can handle overlapping grids. Several parameterized models of appendages are available in the system, e.g. rudder and shaft. Grids can be imported from external gird generator too. There are also two actuator disk models available, a simple force model and a lifting line model. The flow related quantities such as velocity field, pressure field, local skin friction coefficient, friction and pressure resistance (drag) coefficients for the hull part covered by the grids are computed by XCHAP. The total resistance (drag) is calculated using the result from XPAN, XBOUND, and XCHAP.

3. VALIDATION OF KERWIN'S (1972) SERIES RUDDERS

Kerwin (1972) performed water tunnel experiments to determine the free stream characteristics of a series of rudders with systematic variations in the amount of flap area and flap balance. The results include plots of lift, drag, rudder moment and flap moment coefficients for a complete range of angles of attack and flap deflections. A comparison of these experimental results with the numerically simulated results using SHIPFLOW$^{(R)}$ for a number of cases is presented in this paper.

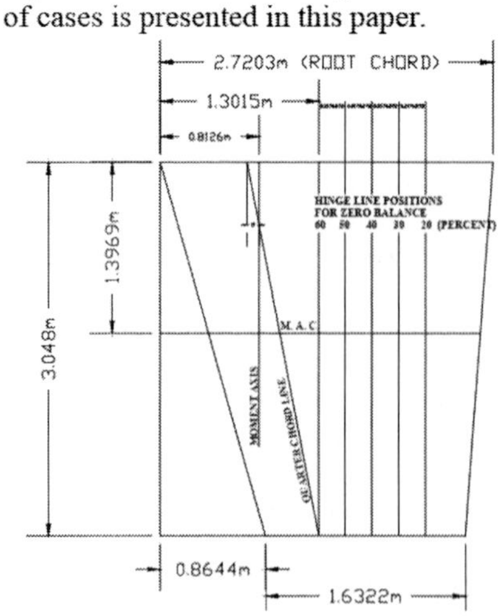

Figure 2: Basic outlines for "Kerwin (1972) Rudder" model series

Three different sections at three different span-wise positions of "Kerwin'(1972) Rudder" are shown in the Figs. 3 (a) to 3 (c). The figures are for a flap area of 30 % and flap deflection of 15 degrees.

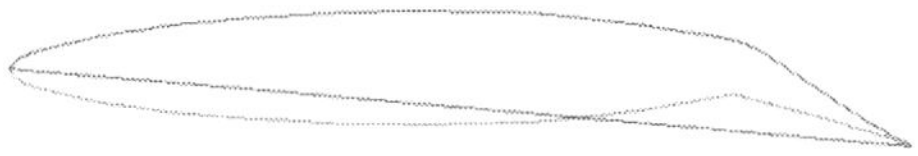

Figure 3(a) Section NACA66-010 used for the" Tip"

Figure 3(b) Section NACA66-015 used for the "Midsection"

Figure 3(c) Section NACA66-020 used for the "ROOT"

Fig. 4 (a) to (c) shows the root sections of the rudder models with different flap proportions (20%, 30%, and 40%) and the corresponding hinge points.

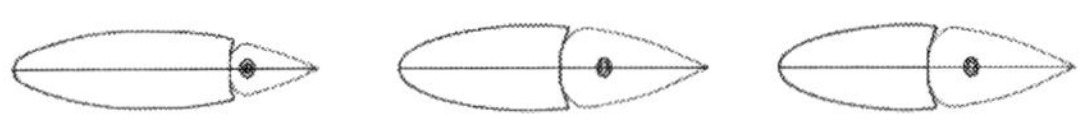

Fig. 4(a)20% Flap Fig. 4(b) 30%Flap Fig. 4(c)40% Flap

The validations of the Kerwin's (1972) experimental results are presented from Case 1 to Case 5. Each case represents a rudder configuration in Kerwin's (1972) experimental setup. Case 6 investigates the behaviour of the Kerwin's (1972) rudder in astern condition, simulated by SHIPFLOW$^{(R)}$. The results for this case cannot be validated as no experiments were done regarding astern condition of flap rudders.

Figure 5 shows the simulated model of the "Kerwin (1972) rudder" which has varying section thickness at root mid-depth and tip.

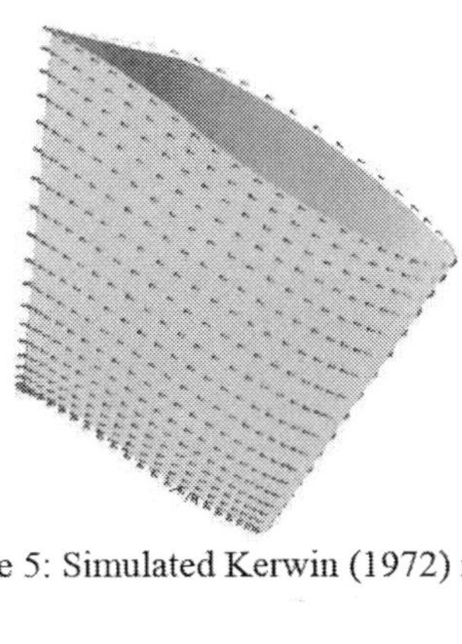

Figure 5: Simulated Kerwin (1972) rudder

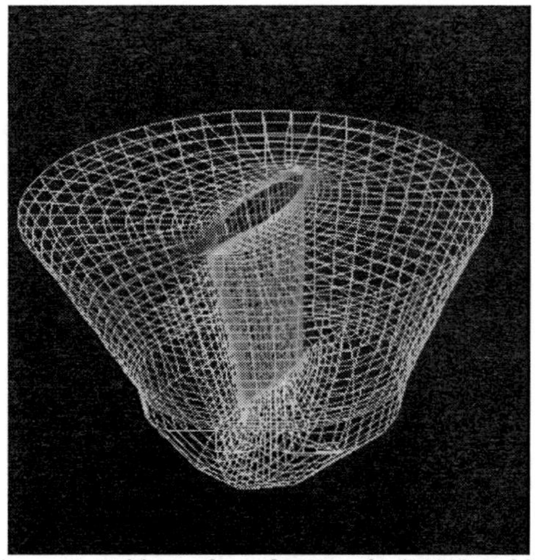

Figure 6: Grid envelope for Kerwin (1972) rudder

The considered for Kerwin (1972) rudder model was a wall mounted all-movable rudder built of a symmetric NACA66 series section, and equipped with a square tip, and tapered cross-section, as shown in Figure 6 and 7.

Rudder Specifications:

➢ Taper ratio	0.60
➢ Geometric aspect ratio	1.40
➢ Root section NACA66	0.20t/c ratio
➢ Tip section NACA66	0.10t/c ratio
➢ Span	10.0 inch
➢ Sweep of quarter chord	11.0 deg aft
➢ Mean Aerodynamic chord	7.289 inch
➢ Area, A	71.40 sq.in
➢ Reynolds Number	1.2×10^6

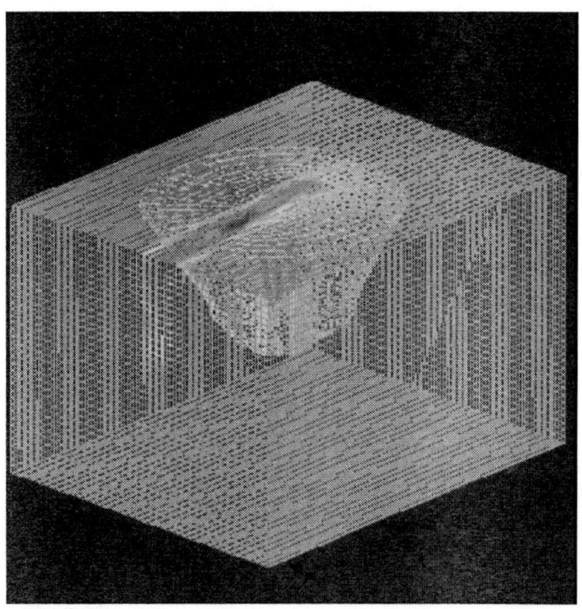

Figure 7: Kerwin (1972) rudder within the computational "BOX"

CASE 1

Table 1 shows different simulation parameters for Kerwin (1972) all-movable rudder. Figures 8 and 9 show the variation of lift coefficient and drag coefficient with angle of attack. The simulated all-movable rudder is wall mounted. It is attached to the floor and root gap between the rudder and the water tunnel floor is ignored due to modelling limitations. This makes the effective aspect ratio of the simulated rudder greater than the experimental rudder, which eventually has increased the slope of the Cl-α curve for the simulated rudder. The value of Cl_{max} is almost same but the stall angle has reduced due to higher effective aspect ratio of the simulated rudder. The Cd values at lower angle of attack show good agreement with experimental results, as shown in Figure 9. But Cd values are greater for higher angle of attack for simulated rudder when compared with kerwin's (1972) experimental results.

Table 1: Test matrix of Kerwin's (1972) all-movable rudder simulated in SHIPFLOW(R)

Foil Type	Span	Rudder angle of attack in degrees				
NACA66-020 for root and NACA66-010 for tip section.	Finite -3D flow	0	5	10	15	20

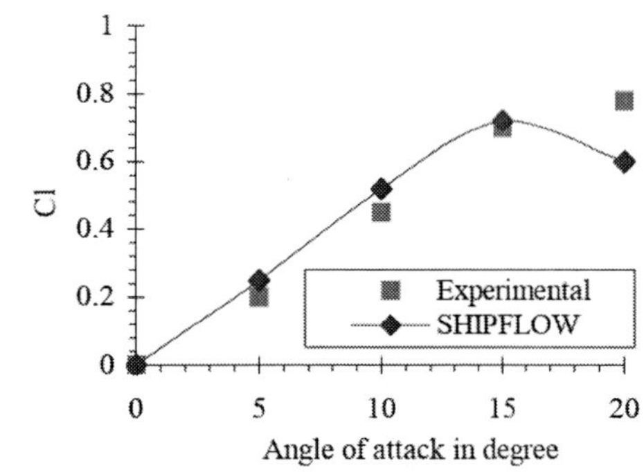

Figure 8: Plot of lift coefficient vs. angle of attack in degrees

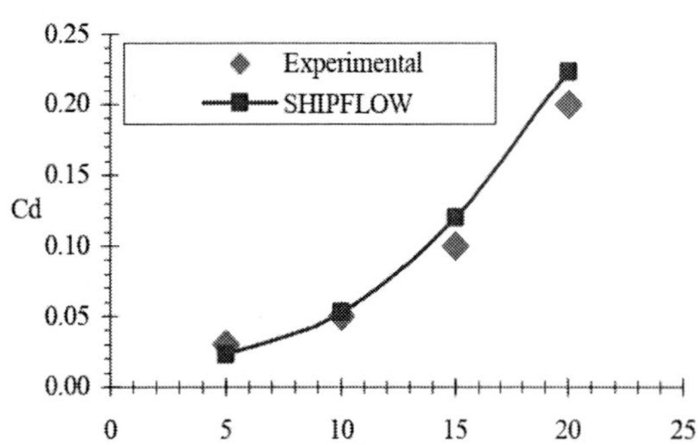

Figure 9: Plot of drag coefficient vs. angle of attack in degrees

CASE 2

Table 2 shows different simulation parameters for Kerwin's (1972) flap rudder, with 20% flap area and 10 degree flap deflection. Figure 10 and 12 show the variation of lift coefficient and drag coefficient with angle of attack. Simulated results of the flap rudders follow the similar trend as of the all-movable rudders, i.e. higher lift curve slope, lesser stall angle and same Cl_{max} values as compared to experimental values. Cd values also show good agreement with experimental results at lower angle of attack.

Table 2: Test matrix of Kerwin's (1972) flap rudder simulated in SHIPFLOW(R)
(Flap area of 20% and flap deflection of 10 degree)

Foil Type	Flap area	Flap deflection	Rudder angle in degrees				
NACA66-aerofoil section.	20%	10°	0	5	10	15	20

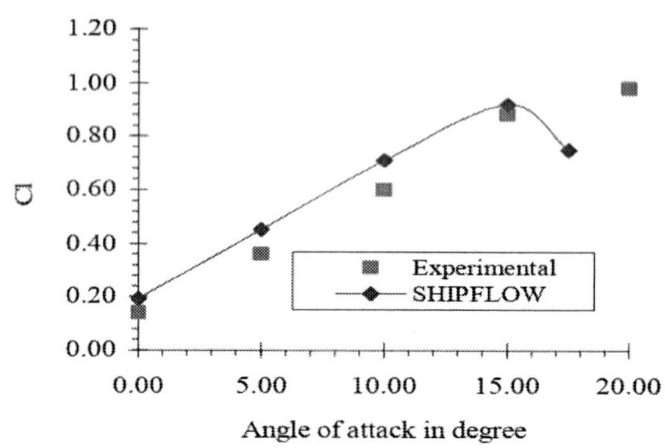

Figure 10: Plot of lift coefficient vs. angle of attack in degrees

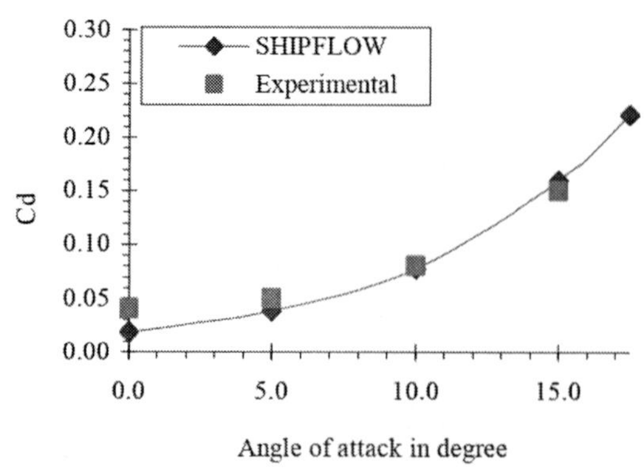

Figure 11: Plot of drag coefficient vs. angle of attack in degrees

CASE 3

Table 3 shows different simulation parameters for Kerwin's (1972) flap rudder, with 20% flap area and 15 degree flap deflection. Figures 12 and 13 show the variation of lift coefficient and drag coefficient with angle of attack. Simulated results of the flap rudders follow the similar trend as of the all-movable rudders, i.e. higher lift curve slope, lesser stall angle and same Cl_{max} values as compared to experimental values. Cd values also show good agreement with experimental results at lower angle of attack.

Table 3: Test matrix of Kerwin's (1972) flap rudder simulated in SHIPFLOW$^{(R)}$
(Flap area of 20% and flap deflection of 15 degree)

Foil Type	Flap area	Flap deflection	Rudder angle in degrees				
NACA66-aerofoil section.	20%	15°	0	5	10	15	20

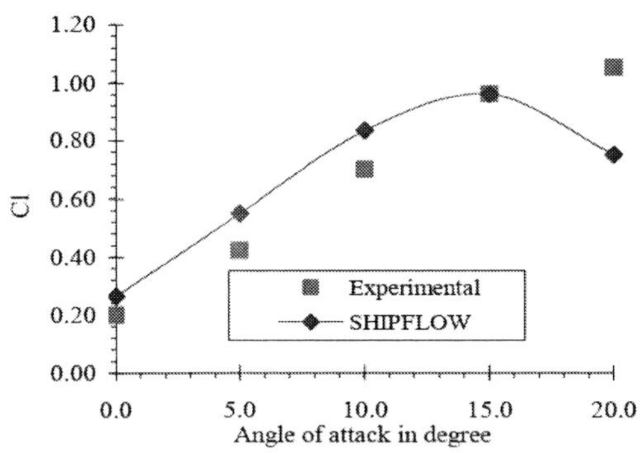

Figure 12: Plot of lift coefficient vs. angle of attack in degrees

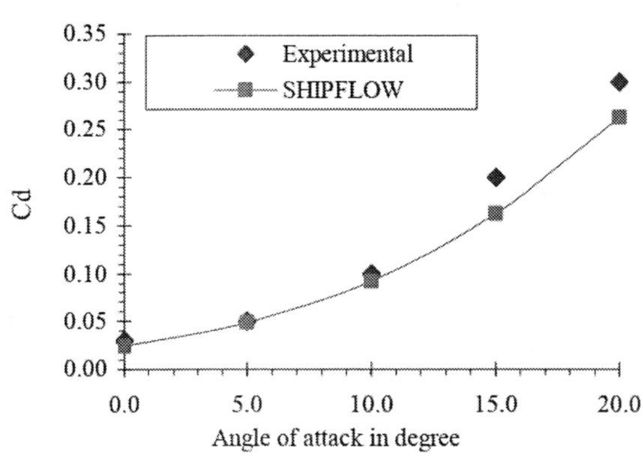

Figure 13: Plot of drag coefficient vs. angle of attack in degrees

CASE 4

Table 4 shows different simulation parameters for Kerwin's (1972) flap rudder, with 30% flap area and 10 degree flap deflection. Figures 14 and 15 show the variation of lift coefficient and drag coefficient with angle of attack. Similar conclusions can be drawn like Cases 2 and 3.

Table 4: Test matrix of Kerwin's (1972) flap rudder simulated in SHIPFLOW$^{(R)}$
(Flap area of 30% and flap deflection of 10 degree)

Foil Type	Flap area	Flap deflection	Rudder angle in degrees				
NACA66-aerofoil section.	30%	10°	0	5	10	15	20

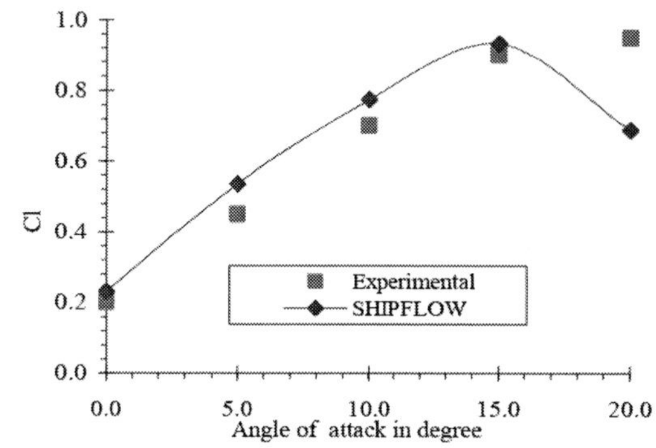

Figure 14: Plot of lift coefficient vs. angle of attack in degrees

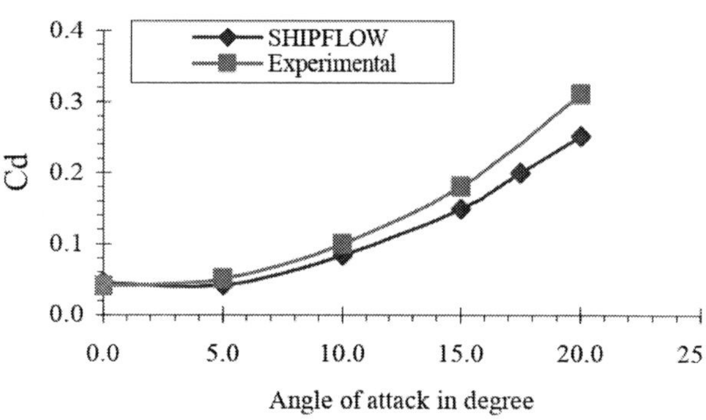

Figure 15: Plot of drag coefficient vs. angle of attack in degrees

CASE 5

Table 5 shows different simulation parameters for Kerwin's (1972) flap rudder with 30% flap area and 15 degree flap deflection. Figures 16 and 17 show the variation of lift coefficient and drag coefficient with angle of attack. Similar conclusions can be drawn like Cases 2, 3 and 4.

Table 5: Test matrix of Kerwin's (1972) flap rudder simulated in SHIPFLOW(R)
(Flap area of 30% and flap deflection of 15 degree)

Foil Type	Flap area	Flap deflection	Rudder angle in degrees				
NACA66-aerofoil section.	30%	15°	0	5	10	15	20

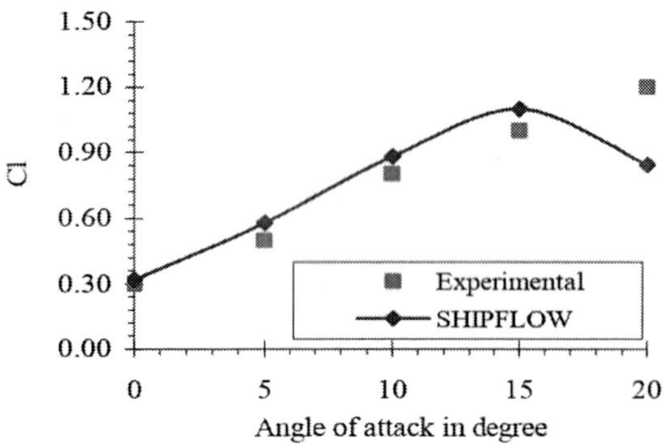

Figure 16: Plot of lift coefficient vs. angle of attack in degrees

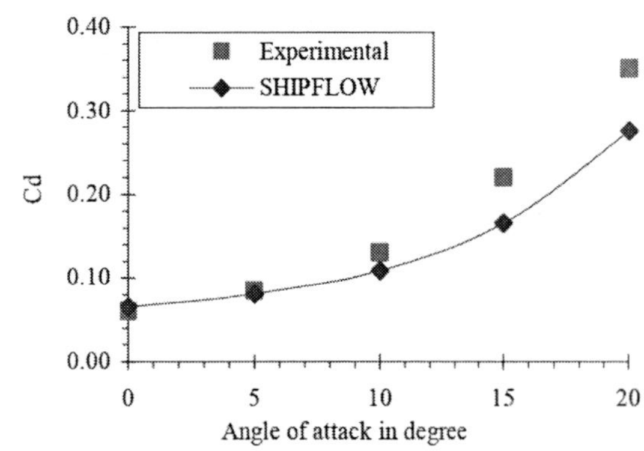

Figure 17: Plot of drag coefficient vs. angle of attack in degrees

CASE 6

Table 6 shows different simulation parameters for Kerwin's (1972) flap rudder with 20% flap area and 15 degree flap deflection. Figures 18 and 19 show the variation of lift coefficient (Cl) and drag coefficient (Cd) with angle of attack in astern condition. Flap area is 20% and the flap deflection is 15 degree. Due to absence of experimental data in astern condition, the results cannot be validated. The lift coefficient values at astern condition are much lower than ahead condition.

Table 6: Test matrix of Kerwin's (1972) flap rudder simulated in SHIPFLOW(R)
(Flap area of 20% and flap deflection of 15 degree)

ASTERN CONDITION:

Foil Type	Flap area	Flap deflection	Rudder angle in degrees				
NACA66-aerofoil section.	20%	15°	0	5	10	15	20

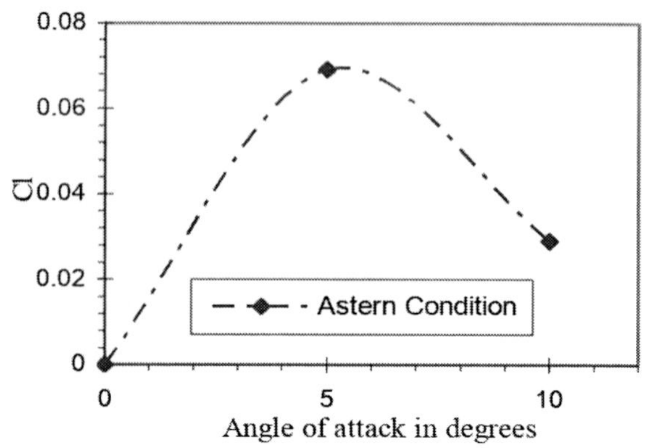

Figure 18: Plot of lift coefficient vs. angle of attack in degrees

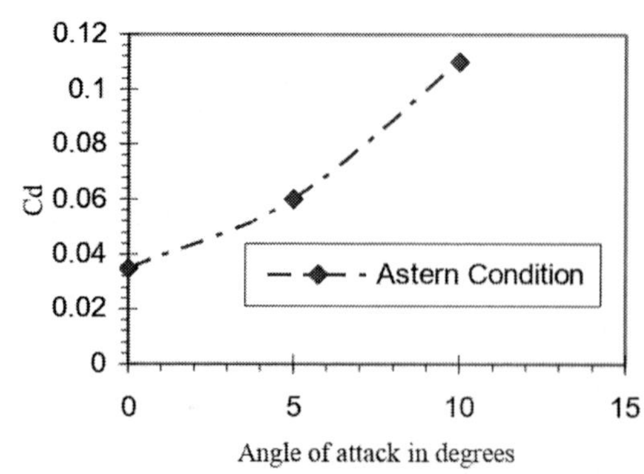

Figure 19: Plot of drag coefficient vs. angle of attack in degrees

3.1 DISCUSSIONS

The following major conclusions are drawn from the above investigations:

- For Kerwin's (1972) rudder, certain factors like root gap and flap gap have a significant effect on the hydrodynamic characteristics. These factors were assigned specific values in the experimental case, but were neglected in case of numerical computation due to some modelling limitations.

  This provides a possible explanation for the 5% to 15% variation of Cd, Cl and stall angle between the experimental and computational cases.

- In the attempt of modelling the rudder in SHIPFLOW$^{(R)}$ for higher flap area and flap deflection angle, cells of negative volume were formed. They were mainly concentrated in the zone surrounding the leading and trailing edges. The results thus obtained were erroneous. All this restricted the range of simulation of flap rudders to a flap area of 30% and a deflection angle of 15 degree.

4. FREESTREAM FLAP RUDDER CHARACTERISTICS

In the present work [3], investigation is carried out for understanding the behaviour of high lift control surfaces like flap rudder in freestream; ahead and astern condition. A series of flap rudder with varying percentage of flap areas (20%, 30%, 40%) and flap to rudder deflection are modelled and lift, drag and stall angle for varying angle of attack are numerically computed using RANS solver, SHIPFLOW^R. The results of this systematic study for both ahead and astern condition are presented. Finally, for a few of the above variation of flap rudders physical models are constructed and tested into wind tunnel experimental facility of Naval Science and Technology Laboratory, (NSTL) Visakhapatnam.

The following simulated cases are discussed below

- Free stream flap rudders, validated with NSTL experimental results.
- Free stream flap rudders, without validation.

4.1 EXPERIMENTAL SETUP OF WIND TUNNEL, NSTL

Figure 20 shows a typical NACA 0018 hydrofoil section of geometric aspect ratio 3(span 708mm, chord 236 mm) and maximum thickness (t/c) 0.15 mm, mounted in the wind tunnel test section. Figure 21 shows a foil incidence setup. Foil Incidence System (FIS) was developed for carrying out foil tests for 2-D flow, holding the foil horizontally and spanning full breadth of the tunnel. Angle of attack can be varied manually from -45 degree to +45 degree with a resolution of 0.1 degree. 5-component dynamometer was fitted on to the foil at the centre of pressure position with a securing clamp. The N1, N2, S1, S2 and axial outputs of the dynamometer were connected to an amplifier. The amplifier output voltages were acquired by multimeters. The simulated models, analysed in SHIPFLOW^R , are shown in Figures 22 and 23.

Figure 20: NACA 0018 in wind tunnel

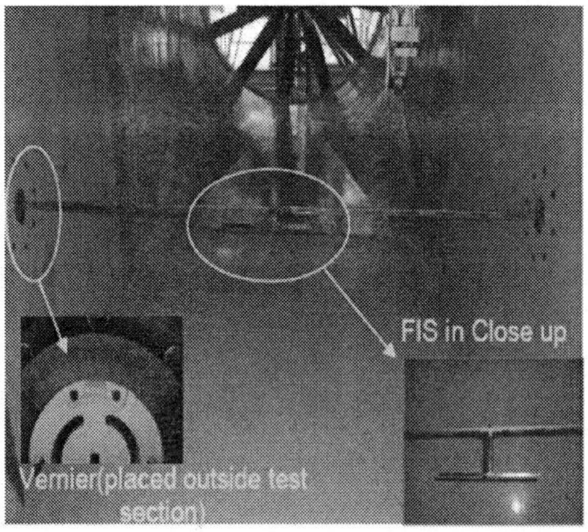

Figure 21: Foil incidence system

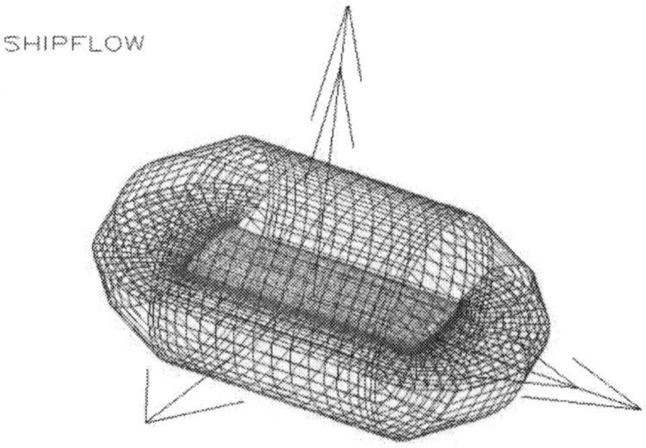

Figure 22: Rudder modelled with both ends closed Y axis (horizontal placed)

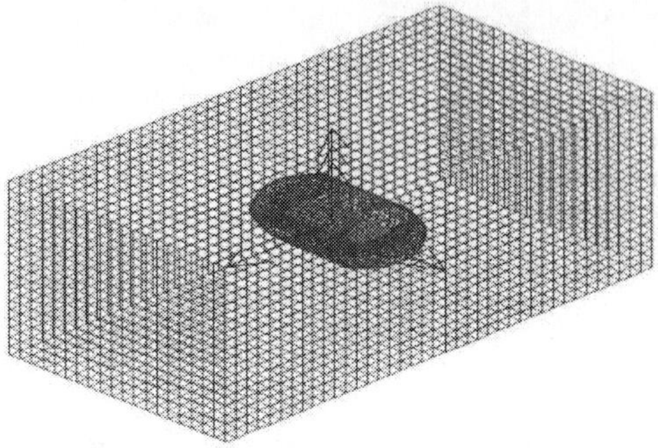

Figure 23: NSTL rudder within the computational "BOX"

CASE 7

Table 7 shows different simulation parameters in test matrix for NSTL flap rudder with flap area 40% and flap deflection 5 degree. Figures 24 and 25 show the variation of lift coefficient and drag coefficient with angle of attack at 15m/s. The results of SHIPFLOW(R) compared fairly well with those experimental results, as shown in Figure 24 and 25 except at higher angle of attack.

Table 7: Test matrix for NSTL flap rudder simulated in SHIPFLOW(R).

(Flap area of 40%, flap deflection of 5 degree and velocity=15m/s)

Foil Type	Flap area	Flap deflection	Rudder angle				
NACA 0018-section	40%	5°	0°	5°	10°	15°	20°

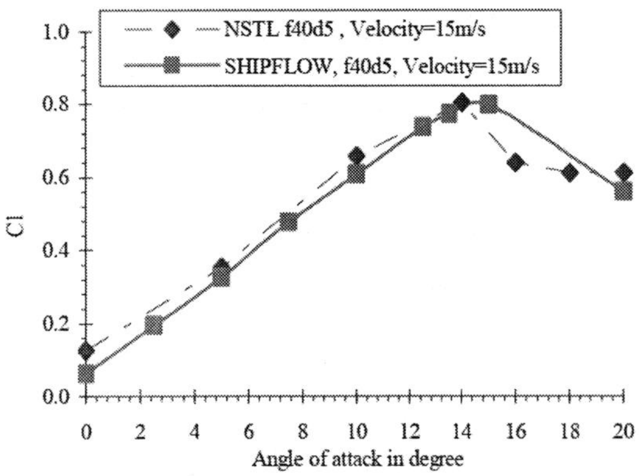

Figure 24: Plot of lift coefficient vs. angle of attack in degrees

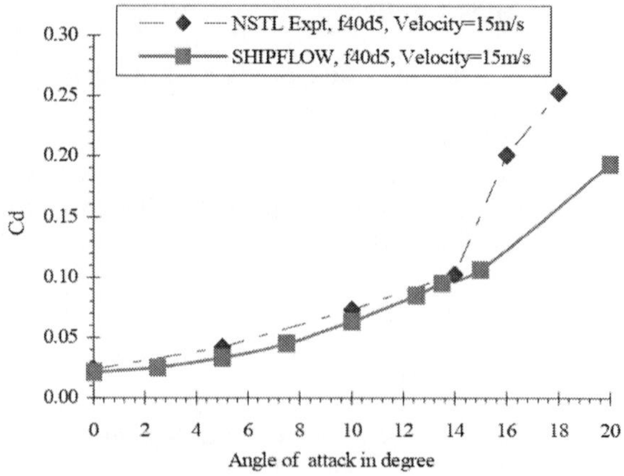

Figure 25: Plot of drag coefficient vs. angle of attack in degrees

4.2 VELOCITY PROFILE OF RUDDER IN AHEAD CONDITION

Figure 26 shows the pre-stall velocity profile at an angle of attack 10 degree and Figure 27 shows the velocity profile over the rudder after stall has happened. Pre-stall velocity profile shows uniform stream line pattern whereas fully reversed flow can be observed from Figure 27 at 20 degree angle of attack.

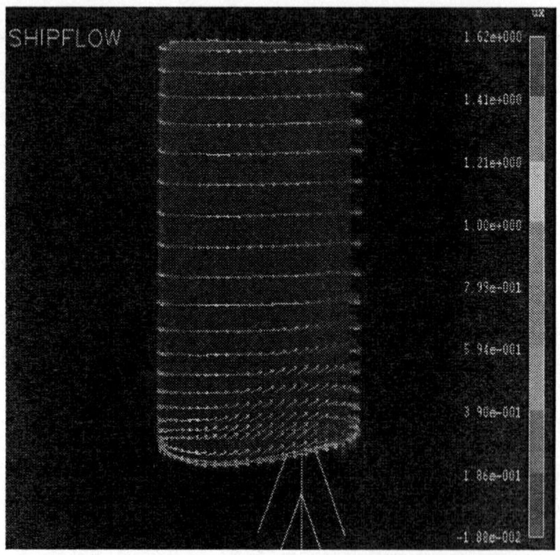

Figure 26: Velocity contour at pre-stall angle of attack (angle of attack =10 degree)

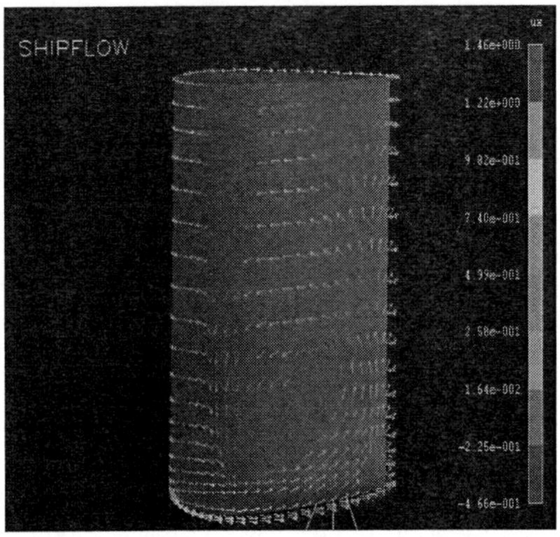

Figure 27: Velocity contour: Separation clearly visible at higher angle of attack, (angle of attack =20 degree)

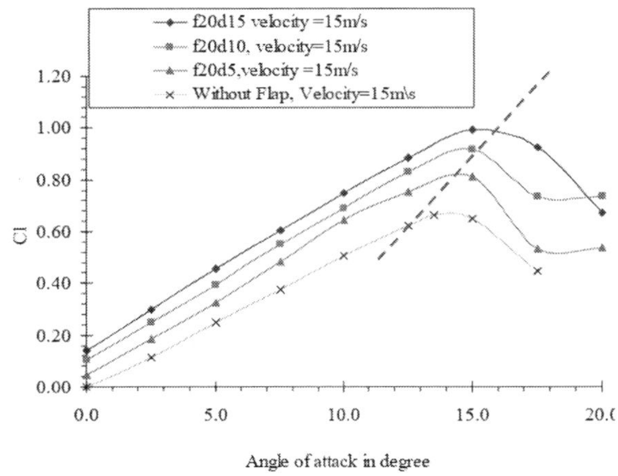

Figure 28: Plot of lift coefficient vs. angle of attack in degrees

Because of the encouraging results from SHIPFLOW(R) simulation, it was decided that a systematic simulated study can be done for flap rudders of varying flap area and flap deflection over a range of angles of attack. These data generated, will help the designer in selecting a more specific flap rudder configuration. Further investigation of flap rudders for varying flap area and flap deflections are shown from Case 8 to Case 10.

4.3 SHIPFLOWCOMPUTATIONAL RESULTS FOR VARYING FLAP PERCENTAGE AND FLAP DEFLECTION; WITHOUT EXPERIMENTAL VALIDATION

CASE 8

Table 8 shows different simulation parameters in test matrix for comparative studies of NSTL flap rudder with flap area 20% and flap deflection varying form 0 to 15 degree. Figures 28 and 29 show the variation of lift coefficient and drag coefficient with angle of attack at velocity 15m/s. Dash line in the Figure 28 shows the variation of stall angle with different flap percentage.

Table 8: Test matrix for comparative study of NSTL flap rudder by SHIPFLOW(R)

(Flap area of 20% and flap deflection of 0-15 degree, velocity =15m/s)

Foil Type	Flap area	Flap deflection	Rudder angle				
NACA 0018 section	20%	0-15°	0°	5°	10°	15°	20°

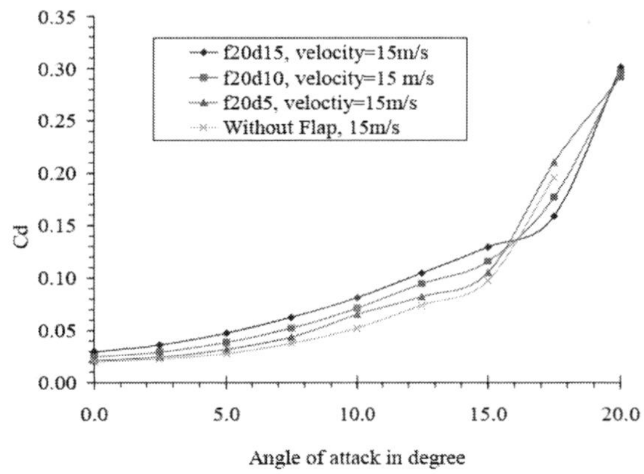

Figure 29: Plot of drag coefficient vs. angle of attack in degrees

CASE 9

Table 9 shows different simulation parameters in test matrix for comparative studies of NSTL flap rudders with flap area 40% and flap deflection varying form 5 to 15 degree. Figures 30 and 31 show the variation of lift coefficient and drag coefficient with angle of attack at velocity 30m/s. Dash line in the Figure 30 shows the variation of stall angle with different flap percentage.

Table 9: Test matrix for comparative study of NSTL flap rudder by SHIPFLOW(R)

(Flap area of 0-40% and flap deflection of 0-15 degree, velocity =30m/s)

Foil Type	Flap area	Flap deflection	Rudder angle				
NACA 0018 section	40%	0-15°	0°	5°	10°	15°	20°

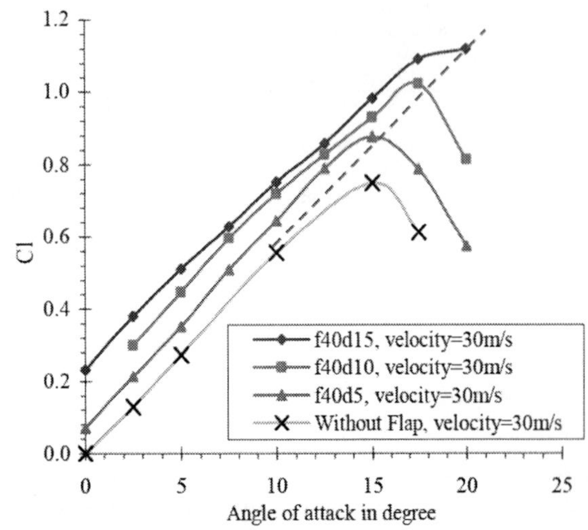

Figure 30: Plot of lift coefficient vs. angle of attack in degrees

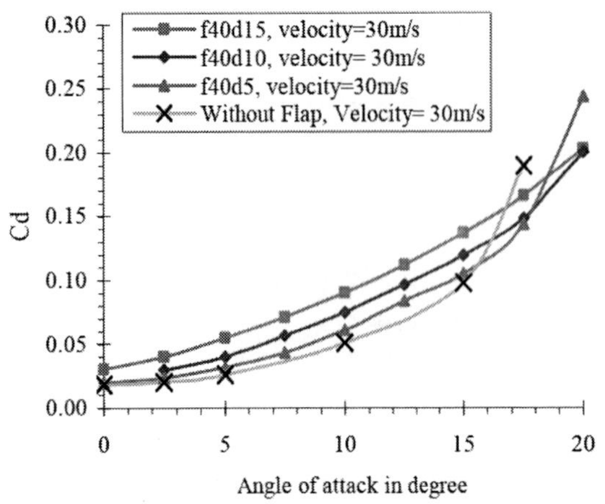

Figure 31: Plot of drag coefficient vs. angle of attack in degrees.

CASE 10 (Astern Condition)

Table 10 shows the test matrix of NSTL flap rudders with flap area 40% and flap deflection 15 degree. Both the ahead and astern condition results are presented for better comparison. Figures 32 and 33 show the variation of lift coefficient and drag coefficient with angle of attack at velocity 30m/s. Dash line in the Figure 32 shows the variation of stall angle with different flow condition.

Table 10: Test matrix for NSTL flap rudder simulated in SHIPFLOW(R)

(Flap area of 40% and flap deflection of 15 degree)

Foil Type	Flap area	Flap deflection	Rudder angle				
NACA 0018 section	40%	15°	0°	5°	10°	15°	20°

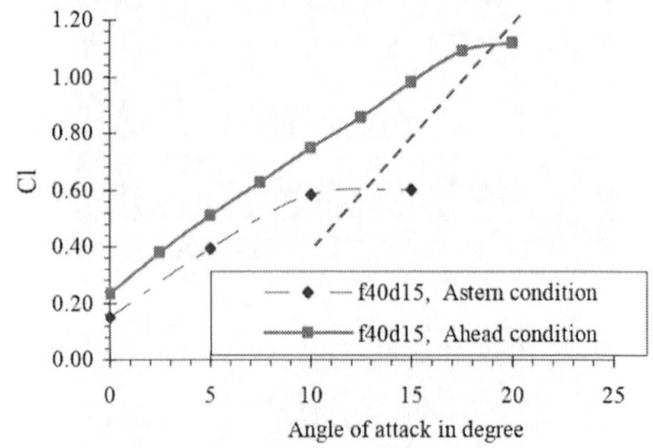

Figure 32: Plot of lift coefficient vs. angle of attack in degrees

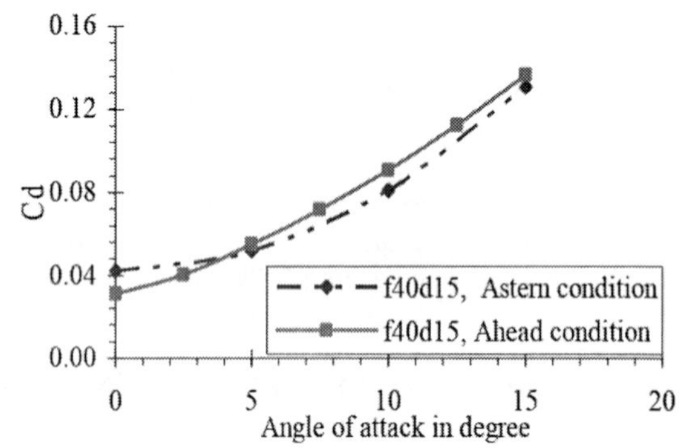

Figure 33: Plot of drag coefficient vs. angle of attack in degrees.

4.4 DISCUSSIONS.

- The results of SHIPFLOW(R) tallies fairly well with all-movable NSTL experimental results The results of the investigations clearly indicate an increase in lift coefficient and delay in stall with higher flap percentage and flap deflection. For a flap percentage of 20% and 15 degree flap deflection the lift coefficient increase by nearly 40% when compared with an all-moveable rudder. However, this is accompanied by 25 % in the drag coefficient value.

- A steep increase in the value of drag coefficient for larger angles of attack is observed for all-movable rudder. However, for flap rudder having higher flap deflection the increase in the value of drag coefficient for larger angles of attack is much more gradual because of delayed stall angles.

- There is a large reduction in the value of lift coefficient in astern condition when compared with ahead condition. The reduction is larger for flap rudders when compared with all moveable rudders. From the results of Case 10 one can observe that there is a 40% reduction in the value of lift coefficient for flap rudder (flap percentage 40% flap deflection 15 degree) when compared to 25% for all-moveable rudder). However, the values of lift coefficient in astern condition for flap rudder are still higher than that of all-moveable rudder.

- There is little change in the value of drag coefficient for flap rudder in ahead and astern condition. However, the increment in the value of drag coefficient is much larger for all-moveable rudders in astern condition. The lift to drag ratio in astern condition is higher for flap rudder than all-moveable rudder.

5. CONCLUSIONS

The prime objective of this work is to investigate the hydrodynamic characteristics of flap rudders in different operating conditions. The present work delineates an experimental investigation with a numerical one in which the "free flow" characteristics (lift, drag and centre of pressure) of flap rudders are calculated by using Reynolds-Averaged Navier-Stokes (RANS) solver, SHIPFLOW^R. The effectiveness and robustness of the computational results are established by comparing the numerical results with experimental results.

Published literature in this area suggests that a numerically simulated solution for flap rudders consistent with experimental results is not available for practical application. There is very limited literature (Whicker and Fehlner, 1958) available on the performance of all movable rudders in astern condition and no data on the behaviour of flap rudders in astern condition is available in public domain. The insufficiency of the available literature restricts the designer from selecting flap rudders at the initial design stage.

This will help the designer, at the early design stage, to select the flap rudder geometry (area, flap deflection, rate of flap deflection) that will enhance the manoeuvring characteristics of the vessel and satisfy the stringent International Maritime Organisation (IMO) manoeuvring requirement.

6. ACKNOWLEDGEMENTS

This work was carried out as a part of the sponsored research project on "An Investigation into the Hydrodynamic Characteristics of Foils with and without Flap" and funded by Naval Research Board, New Delhi. The authors would thankful to Naval Science and Technological Laboratory, Visakhapatnam, for extending their experimental facilities.

7. REFERENCES

[1]. Kerwin, J. E., Mandel, P. and Lewis, S. D., "An Experimental Study of Series of Flapped Rudders", Journal of Ship Research, The Society of Naval Architects and Marine Engineers, December 1972, 221 – 239.

[2]. Rengström, B., SHIPFLOW^R manual, FLOWTECH International AB, Sweden, 2009.

[3]. Bhattacharjee, A., "An Investigation into the Hydrodynamic Characteristics of All-Movable and Flap Rudders". M.S. Thesis, Indian Institute of Technology, Kharagpur, India, 2009.

8. AUTHORS BIOGRAPHY

Mr. Anandarup Bhattacharjee was born in December, 1979, at Berhampore, West Bengal. He graduated with Physics Honours from Calcutta University. He received his Bachelor's degree in Aeronautical Engineering from Aeronautical Society of India, New Delhi in the year 2005. He joined Master of Science (by research) program, in the Department of Ocean Engineering and Naval Architecture, Indian Institute of Technology, Kharagpur, in the year 2007 as a Junior Research Fellow. The author's interest lies in the field of Computational Fluid Dynamics and Hydrodynamics.

Dr. O. P. Sha received his B.Tech. (Hons.) in Naval Architecture from the Indian Institute of Technology, Kharagpur in 1980; M.Sc. in Ship Production Technology from the University of Strathclyde, Scotland, United Kingdom, in 1982; and Ph.D. in Engineering from the Indian Institute of Technology, Kharagpur in 1996. Presently, he is a professor in the Department of Ocean Engineering and Naval Architecture at the Indian Institute of Technology, Kharagpur, India. He teaches a wide variety of subjects at the Institute and his main areas of interest include ship design and production, CAD/CAM and high performance vehicles. He is involved in a number of sponsored research and industrial consultancy projects.

INTERNATIONAL CONFERENCE ON
SHIP & OCEAN TECHNOLOGY - ICSOT 2009

An Numerical and Experimental Investigation into the Freestream Hydrodynamic Characteristics of a Series of Flap Rudders using SHIPLOW^R

A Bhattacharjee[1], O P Sha[2]

[1] *Dept of Ocean Engg. & Naval Architecture, IIT Kharagpur, India*
[2] *Dept of Ocean Engg. & Naval Architecture, IIT Kharagpur, India*

BRIEF OVERVIEW OF PRESENTATION

- ❑ Introduction
- ❑ Background
- ❑ Validation & Verification of the SHIPFLOW Code
- ❑ Free Stream Rudder Characteristics Investigation
- ❑ Conclusion
- ❑ References

INTRODUCTION

- Manoeuvring – controlled change or retention of the direction of motion and its speed in that direction.
- Surface ships – it includes all force, moment and motion vectors in any direction in the horizontal plane.
- Submarine or AUV – manoeuvring encompasses the control of the path of the vehicle in three-dimensional space.
- Control surface – produce a control force resulting from its orientation and movement relative to the water. This in turn creates a moment on the vessel and causes it to reorient at an angle of attack to the flow.

INTRODUCTION

- Control Surface in ships – in form of rudders

- Conventional Ship rudders can be classified as
  - ✓ Types of rudder
    - ✓ Simplex type
    - ✓ Spade rudder
    - ✓ Horn type
  - ✓ Balance area forward of rudder stock
    - ✓ Balanced rudder – simplex, spade
    - ✓ Semi-balanced rudder – horn
    - ✓ Unbalanced rudder

Conventional Rudders used in Ocean Going Vessels

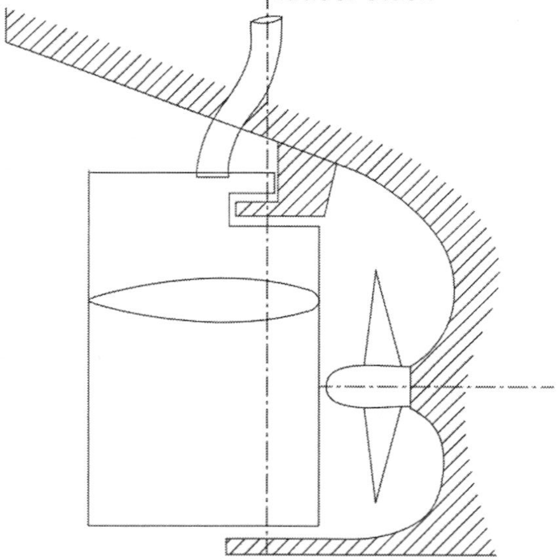

Simplex rudder

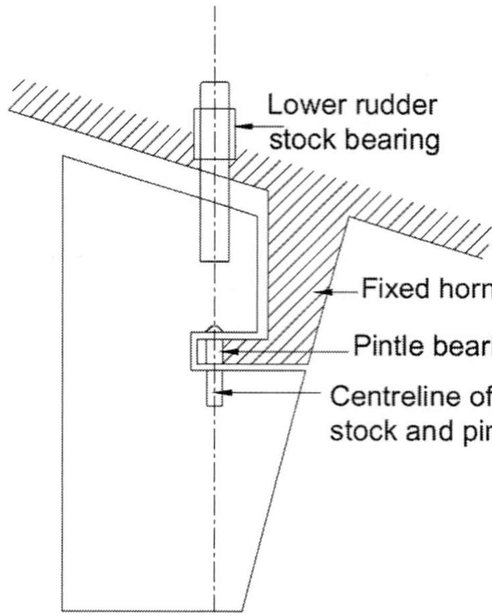

Horn, Skeg, Semi-balanced
or Gnomom rudder

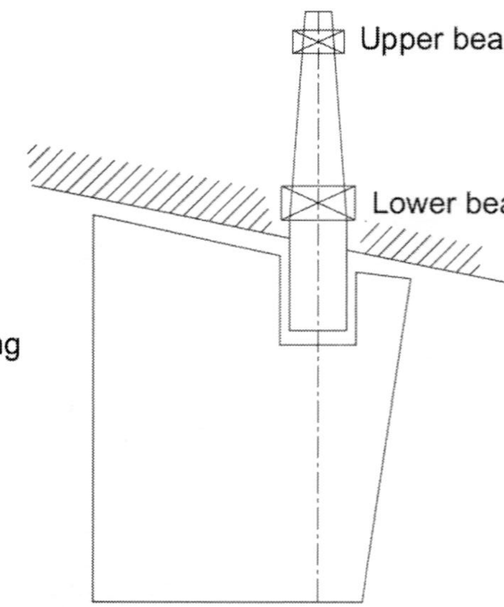

Spade or Underhung rudder

Background

- Now a days, bow thrusters in vessels are being used for better berthing purpose. These vessels can do away with the expenses of tugs for berthing at the dock.

- But to be effective the bow thrusters must work in tandem with high lift control surfaces such that the latter can generate large lateral force.

- A number of high lift rudders and control surfaces have been proposed and employed, primarily to enhance low-speed manoeuvring.

- The main objective of such devices is to delay stall and to increase maximum lift achievable. The mission requirements for such design have been to obtain a tighter turning circle diameter, better stopping ability or higher rudder forces at low speed manoeuvres.

Background

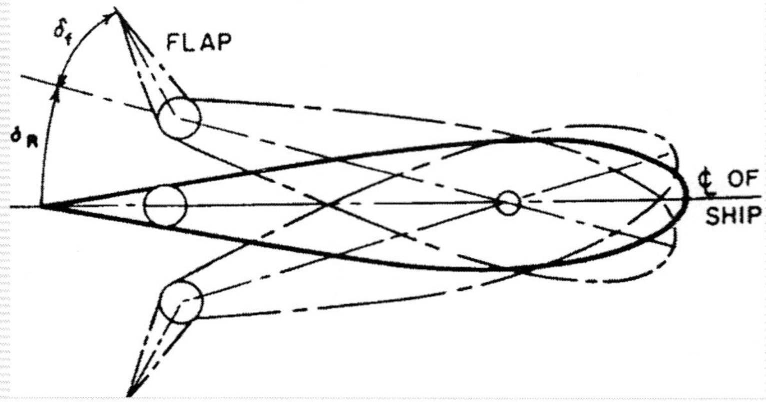

Flapped control surfaces, effectively work as a variable camber devices and for a given control surface incidence, can produce a significant increase in lift. Due to their facility to produce high lift, they tend to be applied in situations where space may be limited, such as for rudder required to produce high maneuvering forces.

Background

Flap Rudder

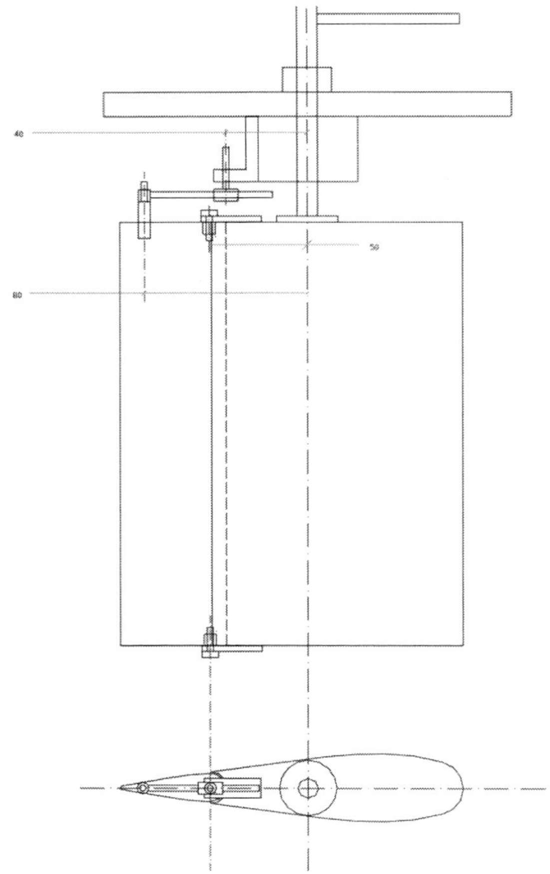

AUTOCAD Diagram of Flap Rudder

Background

- The significant change in the level of understanding of the behavior of control surfaces, with relatively low aspect ratio suitable for marine applications, occurred with the publication of the extensive free stream tests of Whicker and Fehlner (1958).

- Harrington (1981) presents an extensive review of rudder torque prediction.The paper made use of Whicker and Fehlner (1958) data for calculating the rudder force, centre of pressure and rudder torque.

- The results were compared with the Joessel (1873) method. The frictional losses in the rudder bearing were estimated and compared with full scale measurements.

Background

- It can be seen that the publication of the free stream characteristics of various rudder series, including all movable, flapped and semi balanced skeg rudders, has enabled a more rigorous analysis and physically correct approach to be used for the design of marine rudders and control surface.

- The free stream characteristics of the rudder represent its performance in the absence of the hull, propeller, or appendages.

- In order to take account of the presence of hull and/or propeller, corrections are applied to the rudder aspect ratio, the inflow velocity to the rudder and to the rudder angle of attack to yield their effective values.

Background

- Kato and Motora (1968) presented one of the first systematic sets of open water tests on flap rudder.

- Water tunnel experiments were performed by Kerwin, Mandel, and Lewis (1972) to determine the free stream characteristics of a series of twelve rudders with the systematic variation in the amount of flap area and flap balance.

SHIPFLOW

A special purpose computer code for predicting the flow around ship and around its component, developed at Chalmers university of Technology and Flowtech International AB.

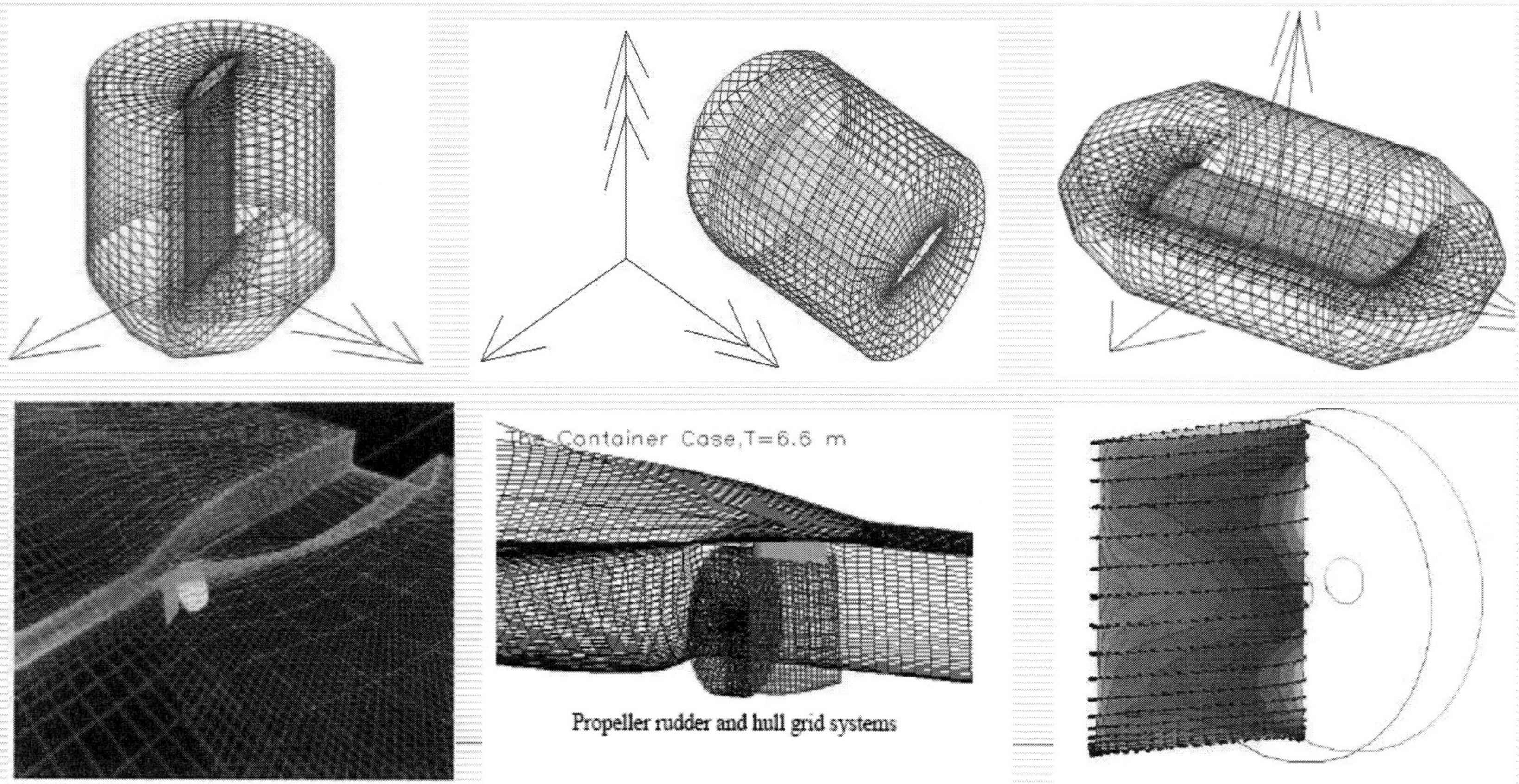

Modelling rudder in SHIPFLOW

SHIPFLOW

To solve the flow around the hull two different approaches, i.e. global and zonal approaches are available in SHIPFLOW. A global approach means the Navier-Stokes equation are solved in the whole flow domain. A zonal approach means that the flow domain is divided into different zones based on the flow characteristics.

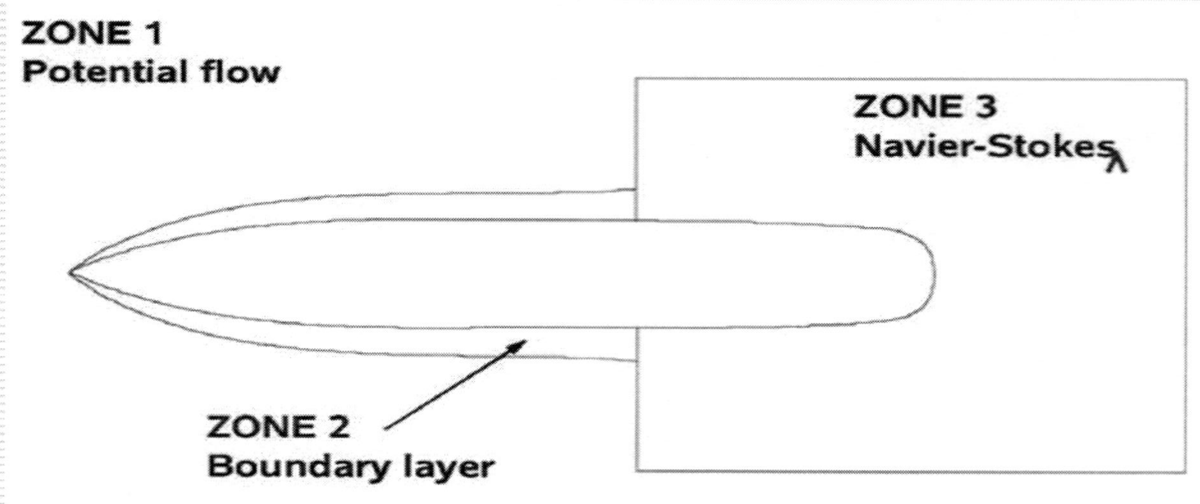

SHIPFLOW

PRE-PROCESSORS	FLOW PROGRAMS			POST-PROCESSOR
	POTENTIAL FLOW	*BOUNDARY LAYER*	*NAVIER-STOKES*	
XMESH XGRID	XPAN	XBOUND	XCHAP XVISC	1. SPOST 2. Tecplot

- XCHAP is based on a C++ class library CHAPMAN, which implements a Finite Volume method based code to solve the RANS code.
- It uses several turbulence model**s (**EASM, k-wBSL,K-wSST).The solver can be used for both Zonal and Global approach. XCHAP can handle overlapping grids.
- Equations for momentum, pressure and turbulence are fully coupled. Roe discretization is applied for the convective fluxes and a central differencing is used for the diffusive fluxes.
- The discrete coupled equations are solved using an ADI-solver

Verification & Validation of the SHIPFLOW Code.

- To establish the effectiveness and robustness of the computational model few rudder cases have been modeled in the SHIPFLOW^R and the numerical results are validated.

- Kerwin et al (1981). performed one of the first water tunnel, flap rudder experiment by varying systematically the flap percentage and flap deflection angle and investigated the free stream characteristics of flap rudder.

- To justify the computational code with respect to flap rudders, Kerwin's rudders are being modelled and simulated and the numerical results are validated with the experimental results.

Verification & Validation of the SHIPFLOW Code

- A series of flap rudder with varying percentage of flap areas (20%, 30%, 40%) and flap to rudder deflections (5°,10°,15° and 20°) are modelled and lift, drag and stall angle for varying angles of attack are numerically computed using RANS solver, SHIPFLOW^R.

- The results of this systematic study are presented for both ahead and astern condition.

- Finally for a few of the above variation of flap rudders physical models are constructed and tested in the wind tunnel experimental facility of Naval Science and Technology Laboratory, (NSTL) Visakhapatnam.

Kerwin(1972) Experiments - Rudder with Flap

- Kerwin et al. (1972) carried out water tunnel experiments to determine free stream characteristics of a series of twelve rudders with systematic variation in the amount of flap area of the fixed part and movable part for a rudder with skeg.

- Initially, to calculate the effectiveness and robustness of the computational model, Kerwin's rudder is modelled and the lift, drag obtained from the computational analysis is compared with the water tunnel experiments of Kerwin. The results are in good agreement and presented in the paper.

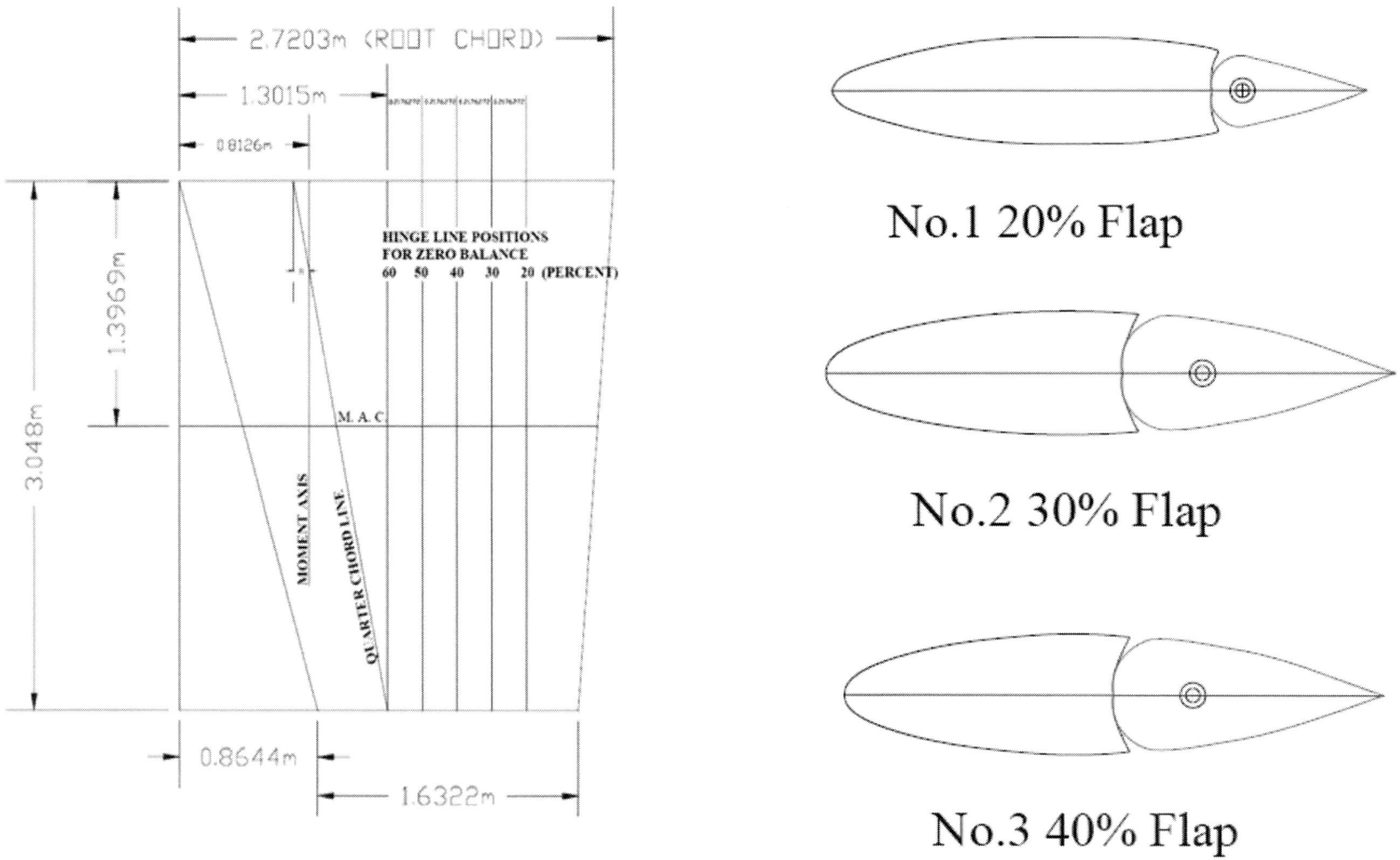

Schematic diagram of Kerwin rudder

Rudder Specification:

➢ Taper ratio	0.60
➢ Geometric aspect ratio	1.40
➢ Root section NACA66	0.20t/c ratio
➢ Tip section NACA66	0.10t/c ratio
➢ Span	10.0 inch
➢ Sweep of quarter chord	11.0 deg aft
➢ Mean Aerodynamic chord	7.289 inch
➢ Area, A	71.40 sq.in
➢ Reynolds Number	1.2×10^6

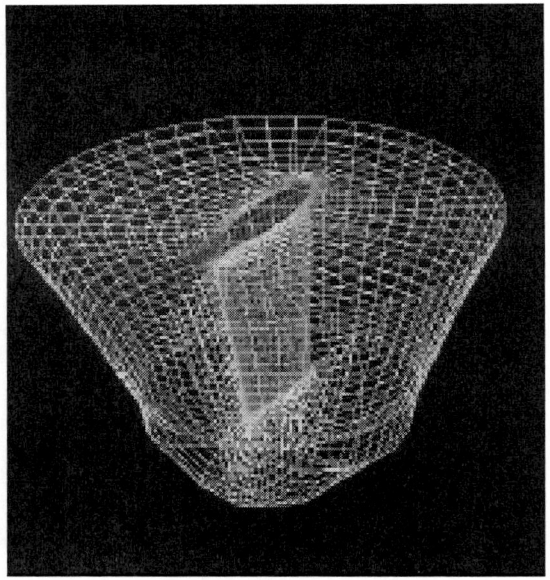

Grid envelope for Kerwin (1972) rudder

Kerwin (1972) rudder within the computational "BOX"

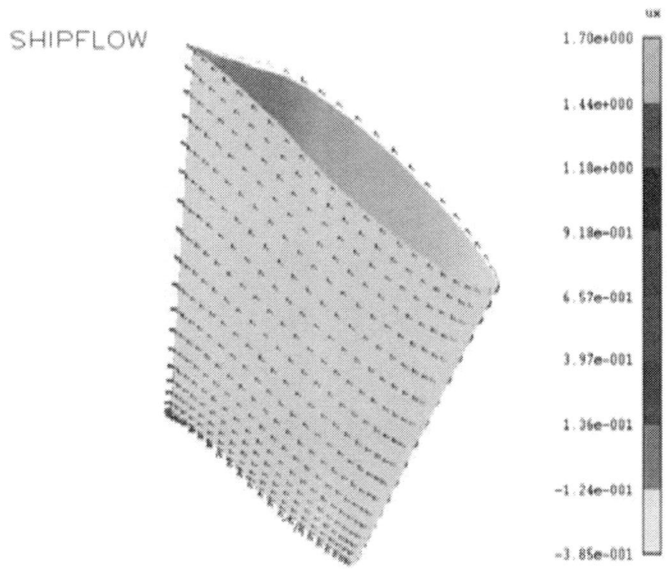

Simulated Kerwin (1972) rudder

CASE 1: All Movable

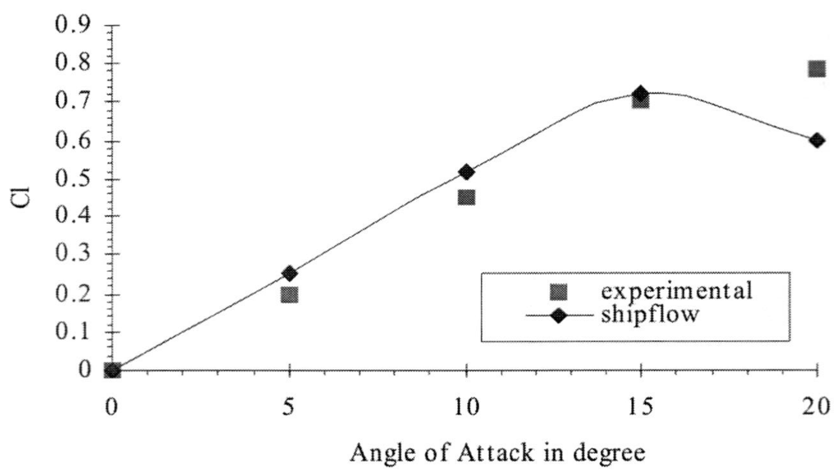

Plot of lift coefficient vs. angle of attack in degrees

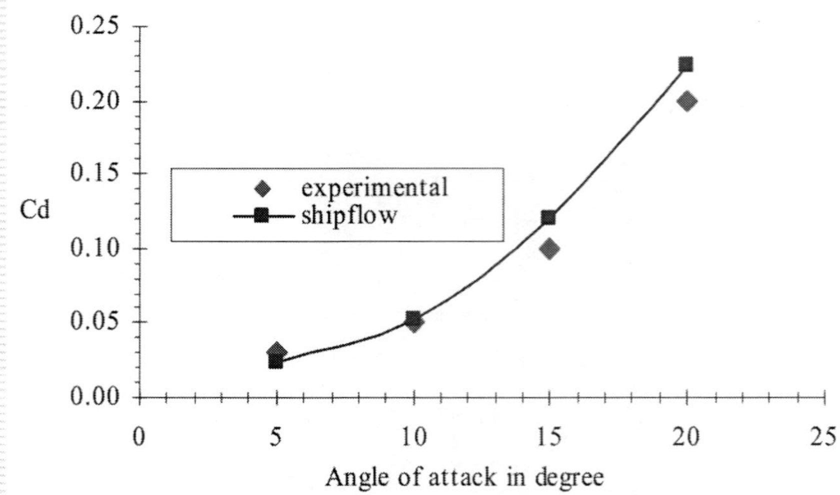

Plot of drag coefficient vs. angle of attack in degrees

CASE 2: Flap %=20; & flap deflection =10deg.

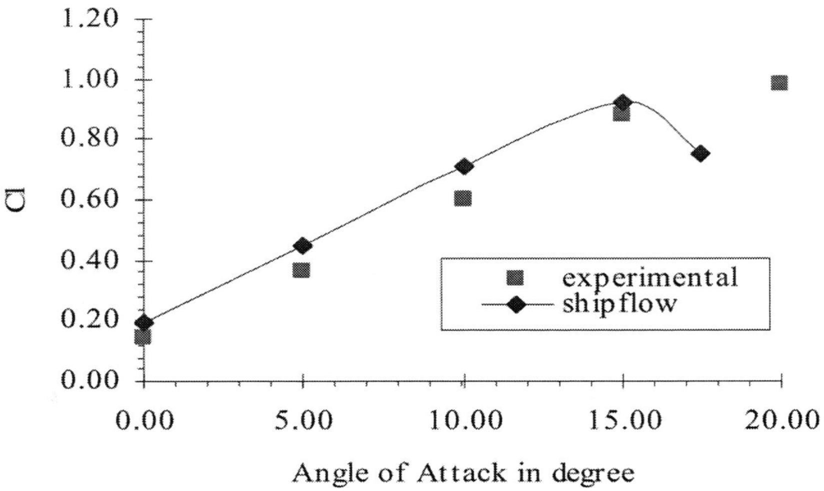

Plot of lift coefficient vs. angle of attack in degrees

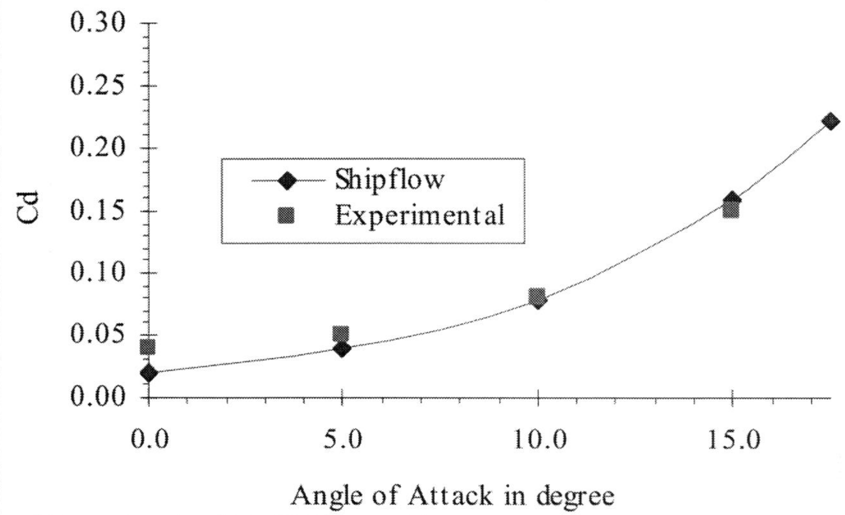

Plot of drag coefficient vs. angle of attack in degrees

CASE 3: Flap %=20; & flap deflection =15deg.

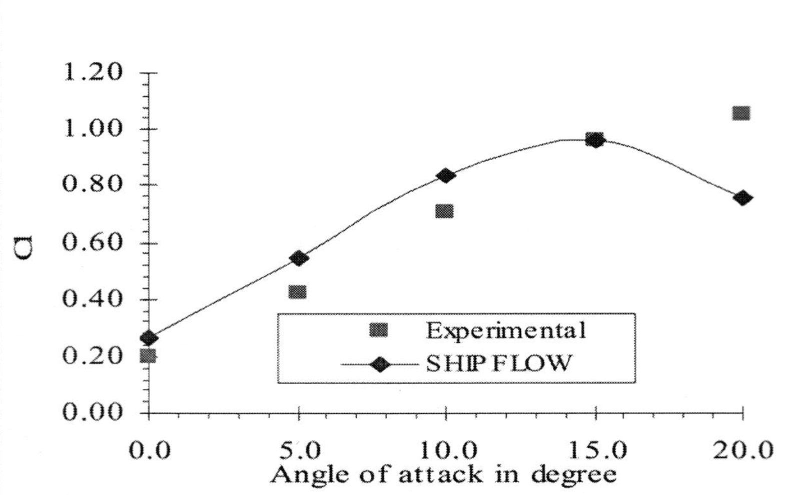

Plot of lift coefficient vs. angle of attack in degrees

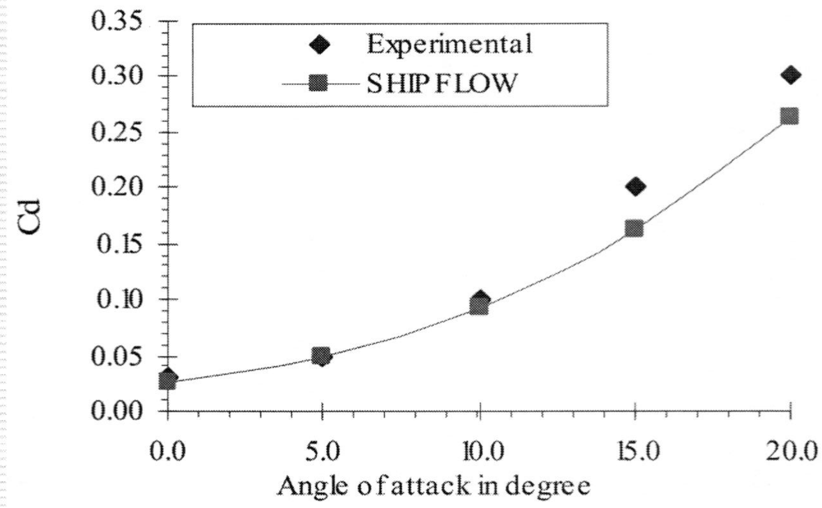

Plot of drag coefficient vs. angle of attack in degrees

CASE 4: Flap %=30; Flap deflection =15 deg.

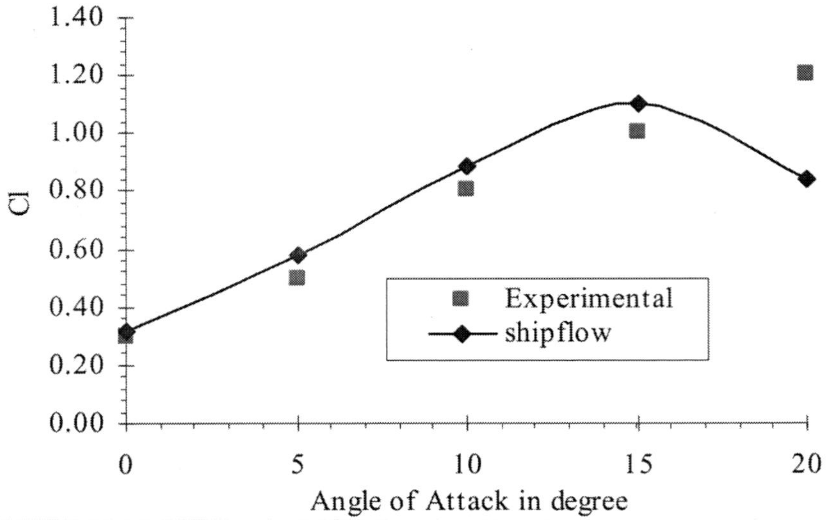

Plot of lift coefficient against. angle of attack in degrees

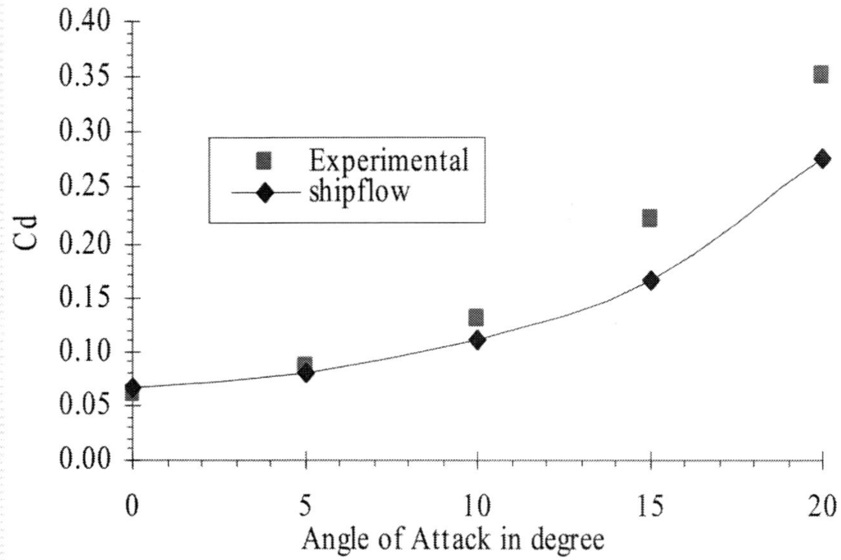

Plot of drag coefficient against. angle of attack in degrees

CASE 5: Flap %=30 & Flap deflection =10 deg.

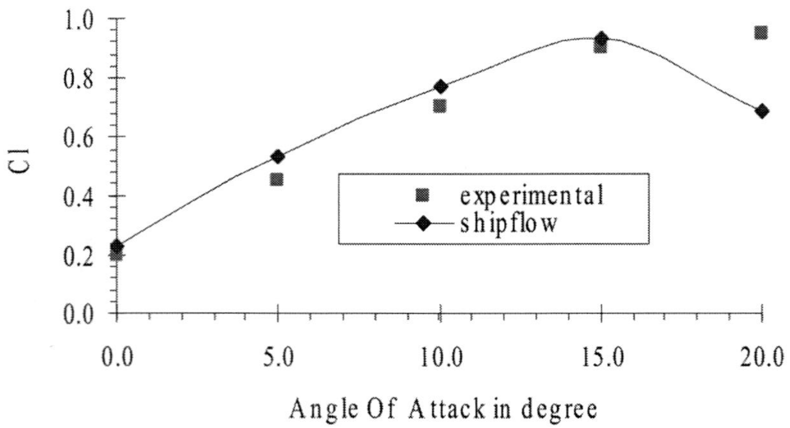

Plot of lift coefficient against. angle of attack in degrees

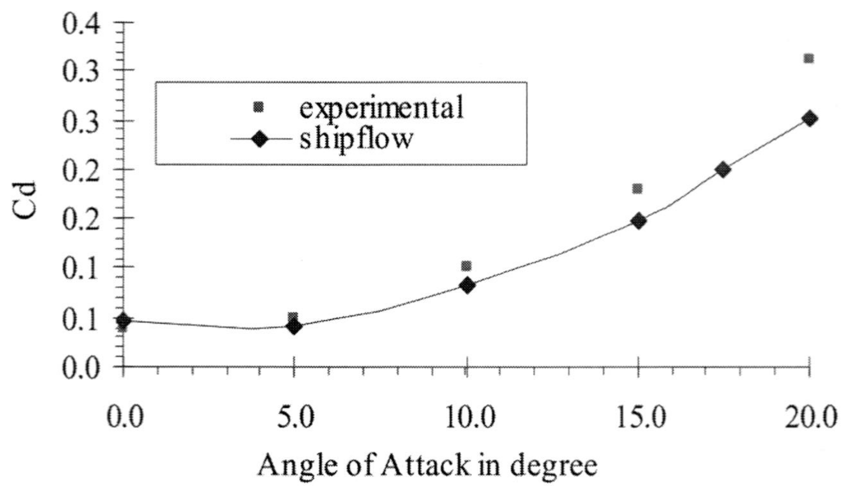

Plot of drag coefficient against. angle of attack in degrees

CASE 6: Flap %=20 & Flap deflection =15 deg.

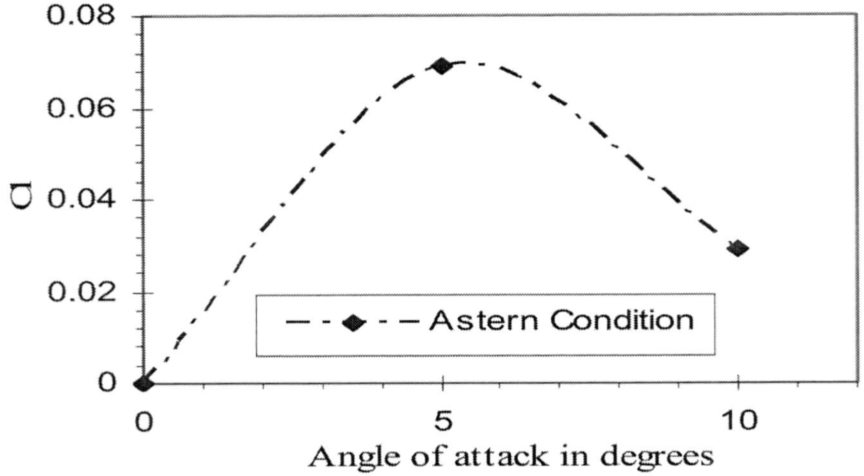

Due to absence of experimental data in astern condition, the results cannot be validated. The lift coefficient values at astern condition are much lower than ahead condition.

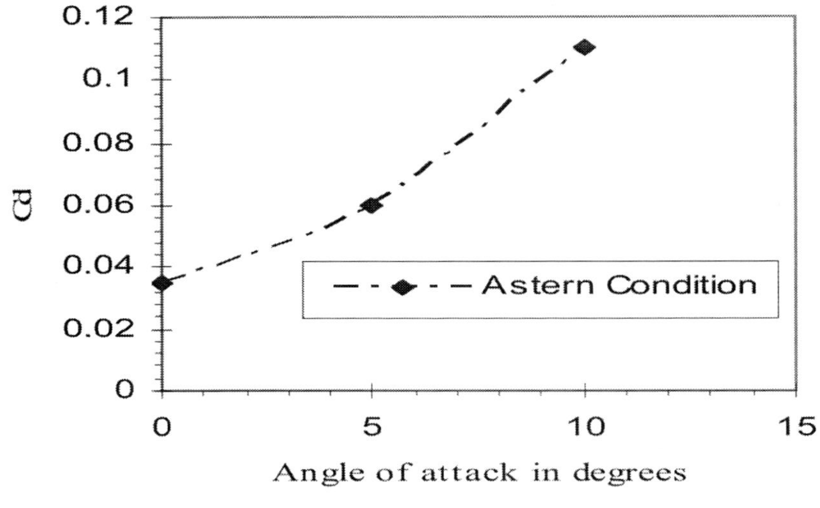

Conclusions from experimental validation with Shipflow

- The results obtained by numerical simulations are in well agreement with the experimental results. This is reflected in the range of variation (5% to 15%) of Cd and Cl.

- For Kerwin's (1972) rudder, certain factors like
  - ✓ root gap and flap gap have a significant effect on the hydrodynamic characteristics.
  - ✓ These factors were assigned specific values in the experimental case, but were neglected in case of numerical computation due to some modelling limitations.

Conclusions from experimental validation with Shipflow

- The simulated all-movable rudder is wall mounted.
  - ✓ It is attached to the floor and root gap between the rudder and the water tunnel floor is ignored due to modelling limitations.
  - ✓ This makes the effective aspect ratio of the simulated rudder greater than the experimental rudder, which eventually has increased the slope of the Cl- curve for the simulated rudder.
- This provides a possible explanation for the 5% to 15% variation of Cd, Cl and stall angle between the experimental and computational cases.

Conclusions from experimental validation with Shipflow

- In the attempt of modelling the rudder in SHIPFLOW for higher flap area and flap deflection angle, *cells of negative volume* were formed.

- They were mainly concentrated in the zone surrounding the *leading* and *trailing edges*. The results thus obtained were erroneous.

- All this restricted the range of simulation of flap rudders to a
  - ✓ flap area of 30%
  - ✓ flap deflection angle of 15 degree.

Free Stream Rudder Characteristics

- It can be seen that the publication of the free stream characteristics of various rudder series, including all movable, flapped and semi balanced skeg rudders, has enabled a more rigorous analysis and physically correct approach to be used for the design of marine rudders and control surface.

- The free stream characteristics of the rudder represent its performance in the absence of the hull, propeller, or appendages.

- In order to take account of the presence of hull and/or propeller, corrections are applied to the rudder aspect ratio, the inflow velocity to the rudder and to the rudder angle of attack to yield their effective values.

TESTING OF FOILS IN WIND TUNNEL, NSTL

Salient features of the Wind tunnel:

Test Section - 1.5 X 1.5 X 4.0 m Long

- Plenum Chamber - 4.3 X 4.3 X 4.0 m Long
- Contraction - Section varying from 4.3 X 4.3m square to 1.5 X 1.5 m square- 4.0m Long.
- Diffuser - Section varying from 1.5 X 1.5 m square to 3.048m Dia. Circle, 7.8 m Long
- DC Variable speed motor - 125 kW at 750 rpm
- Fan - 3.04m Diameter made of CFRP (12 bladed)
- Maximum design speed at test section of the tunnel - 55m/sec
- The height of the tunnel axis from floor level is 4570mm

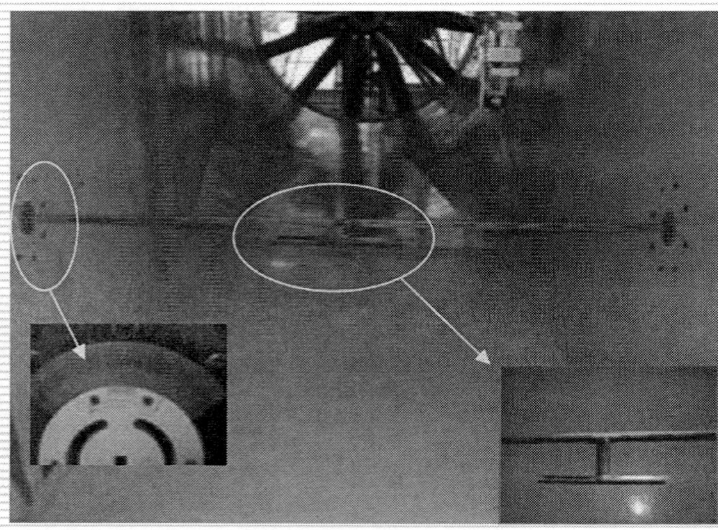

Modelled and Simulated NACA0018

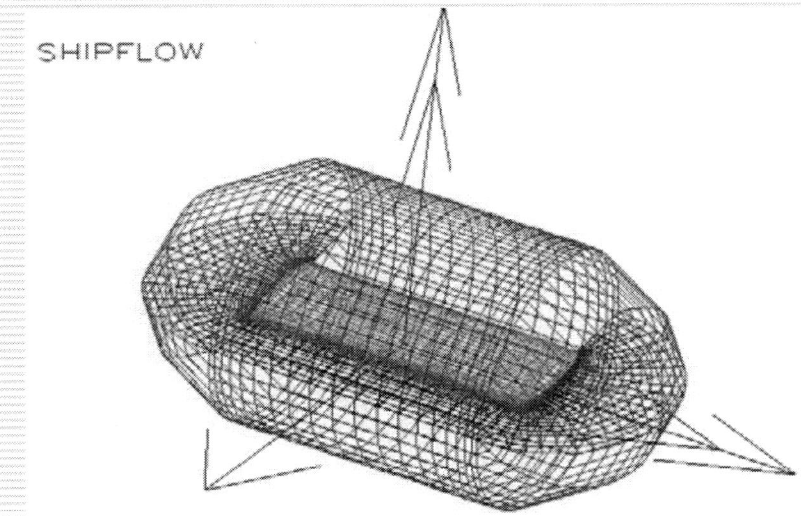

Rudder modelled with both ends closed Y axis (horizontal placed)

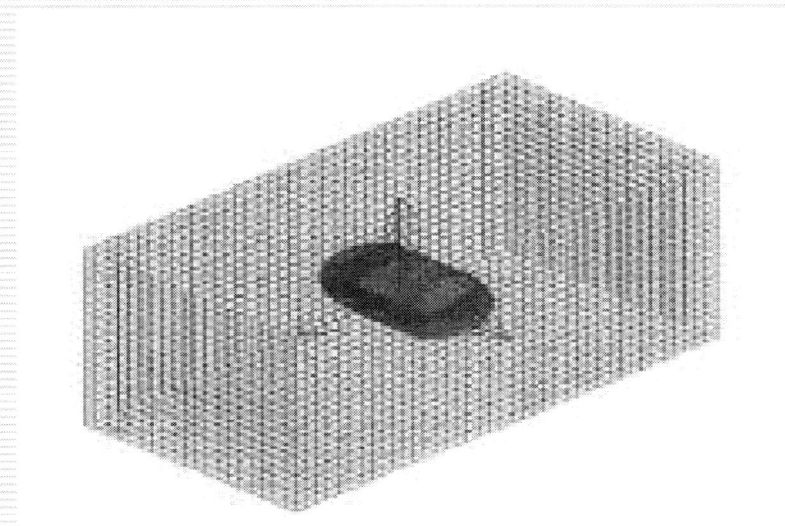

NSTL rudder within the computational "BOX"

TESTING OF FOILS IN WIND TUNNEL, NSTL

TESTING OF FOILS IN WIND TUNNEL, NSTL

Geometry : Span = 708 mm Chord = 236 mm

Foil Type	Condition		Flap deflection angle in degree				Rudder angle of attack in degree				
Symmetrical NACA0018	Without flap 3D flow						0	5	10	15	20
Symmetrical NACA0018	With flap 3D flow	20%flap	5	10	15	20	0	5	10	15	20
		30%flap	5	10	15		0	5	10	15	20
		40%flap	5	10	15		0	5	10	15	20

Tests done for velocity 15 m/s and 20 m/s

Case 1: All movable Rudders (without Flap)

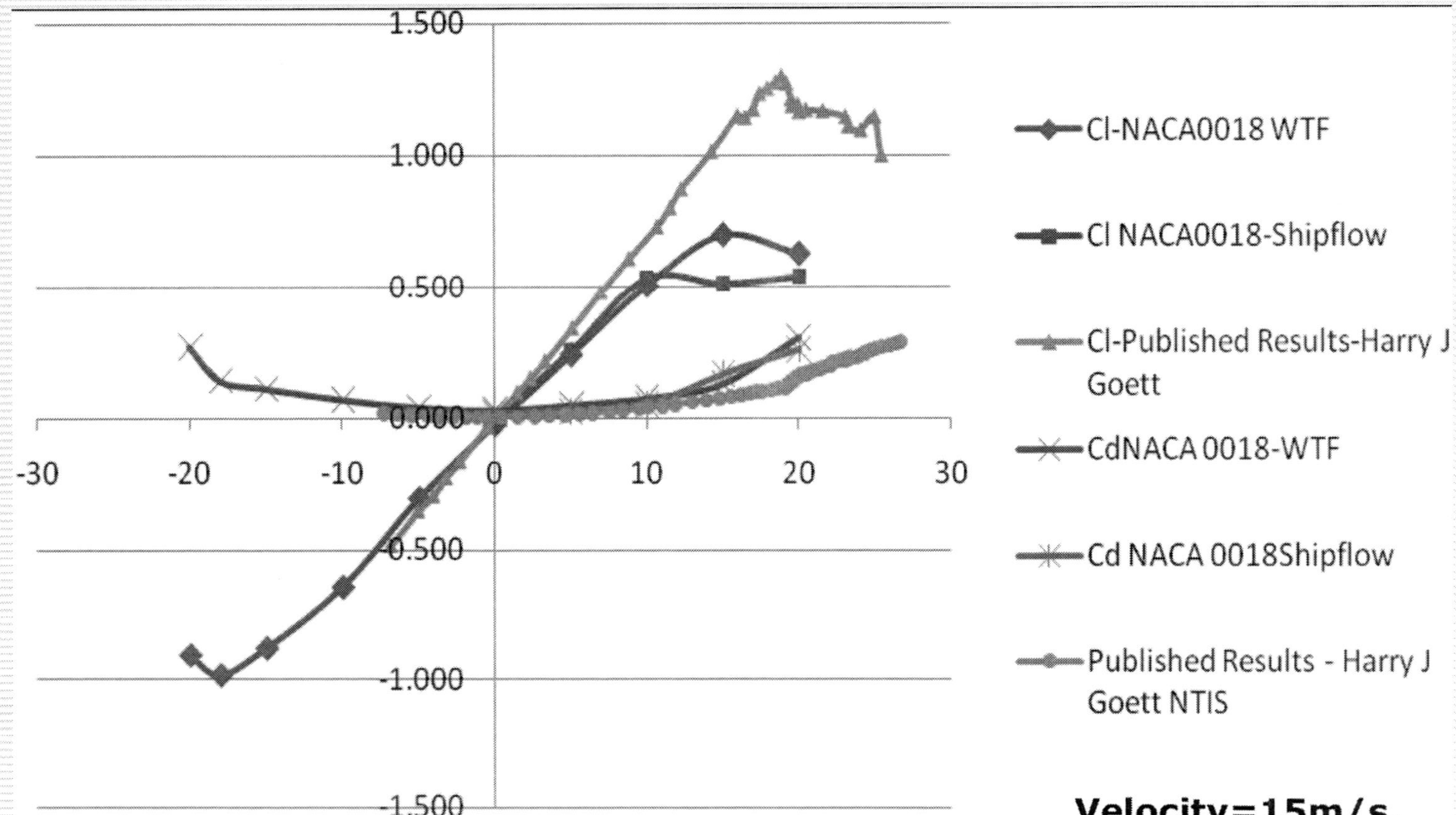

Plot of lift and drag coefficient with angle of attack

Velocity=15m/s,
Reynolds No=0.24e6,
Aspect Ratio=3.0

Case 2: Flap Rudder; Flap area=20%, deflection=10 & 5 deg

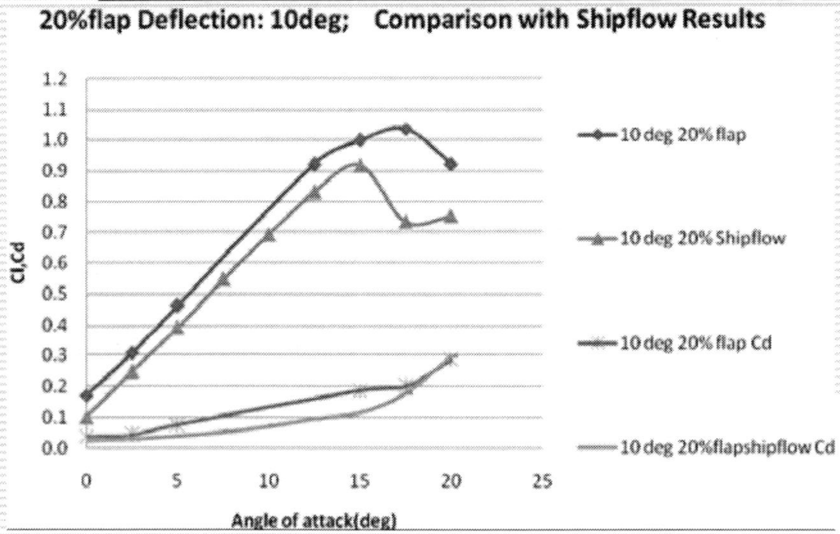

Velocity=15m/s,
Reynolds No=0.24e6,
Aspect Ratio=3.0

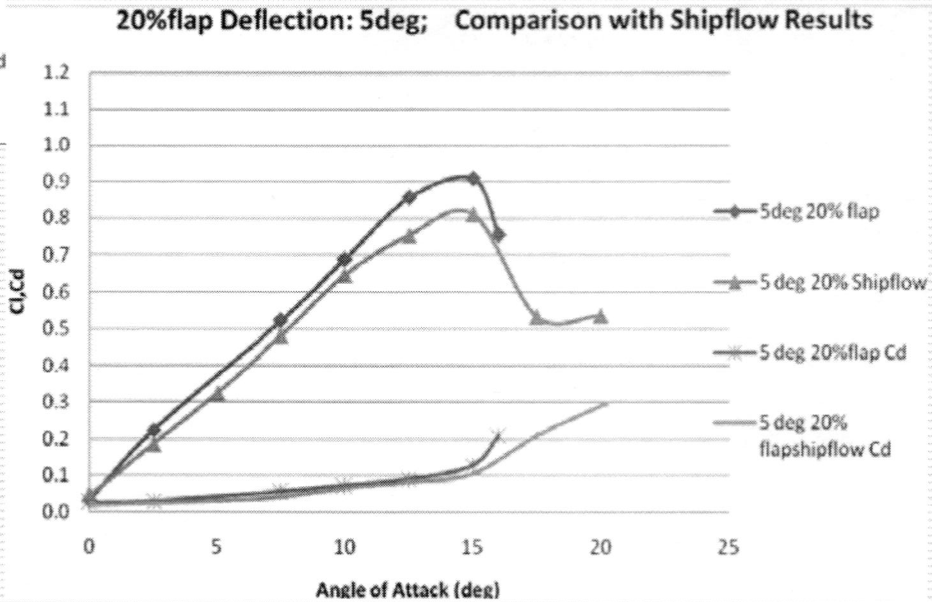

Case 3: Flap Rudder; Flap area=40%, deflection=10 & 5 deg

40%flap Deflection: 10deg; Comp with Shipflow Results

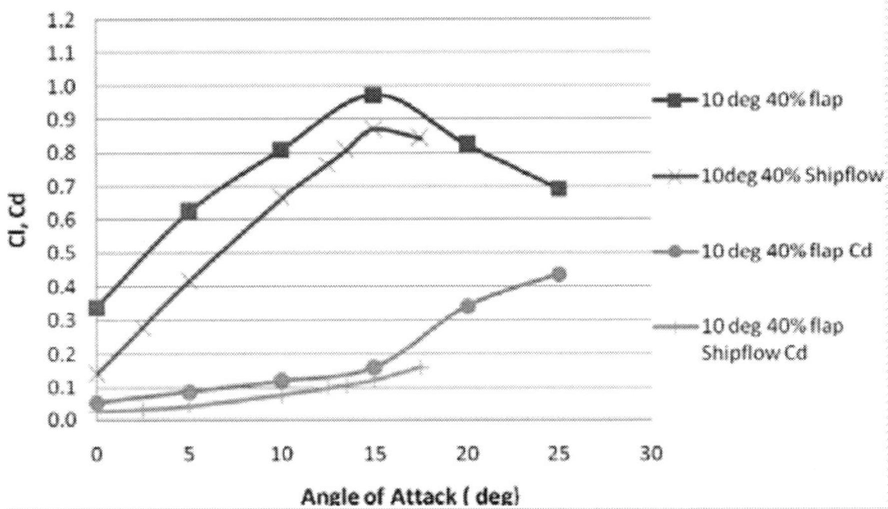

Velocity=15m/s,
Reynolds No=0.24e6,
Aspect Ratio=3.0

40%flap Deflection: 5deg; Comparison with Shipflow Results

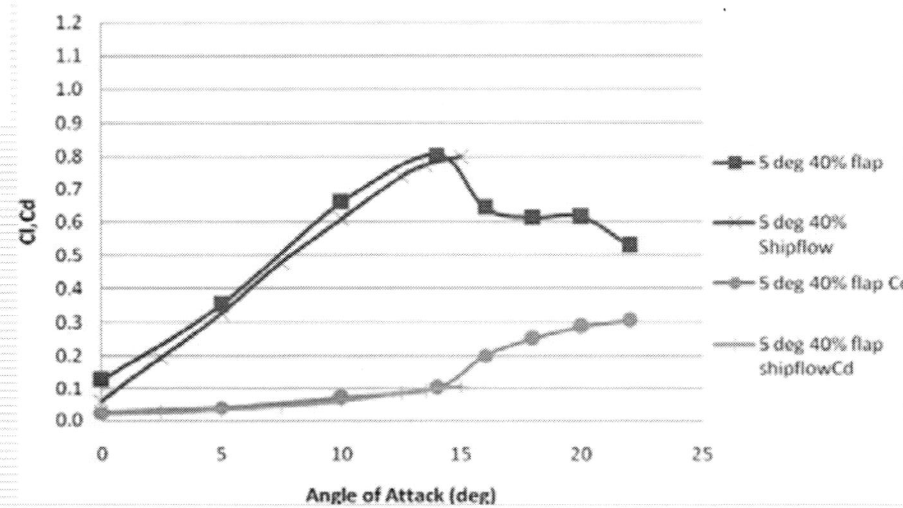

Case 4: Flap = 40% ,Velocity = 15m/s Reynolds No =0.24e6

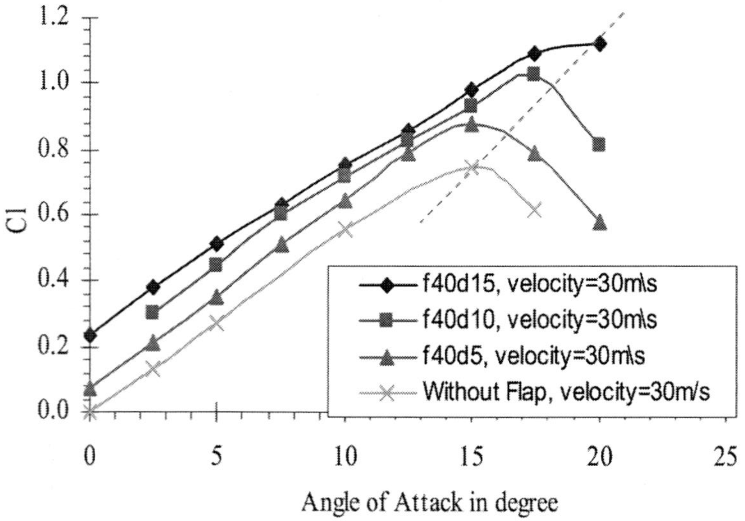

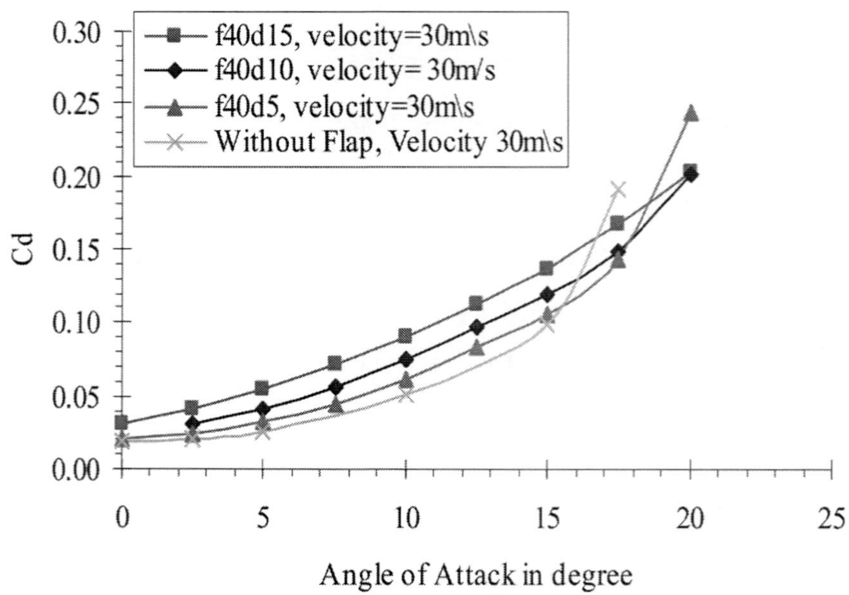

Case 5: Flap = 20% ,Velocity = 15m/s Reynolds No =0.24e6

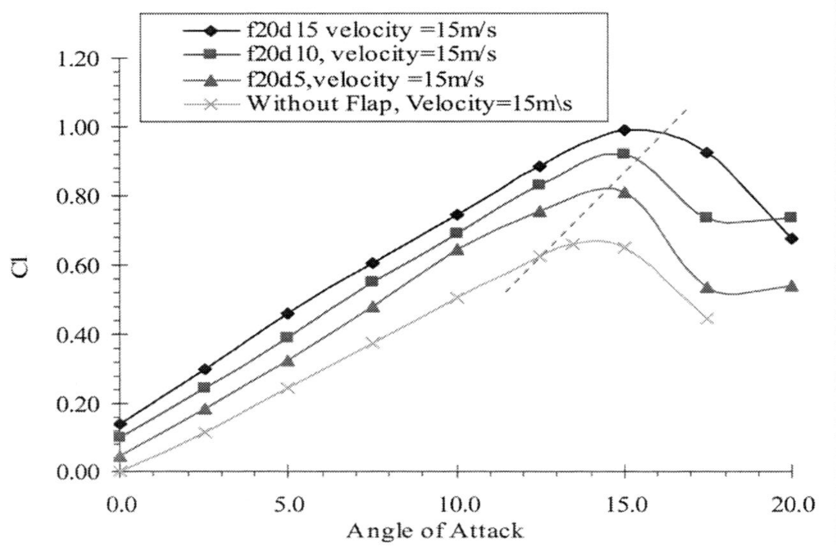

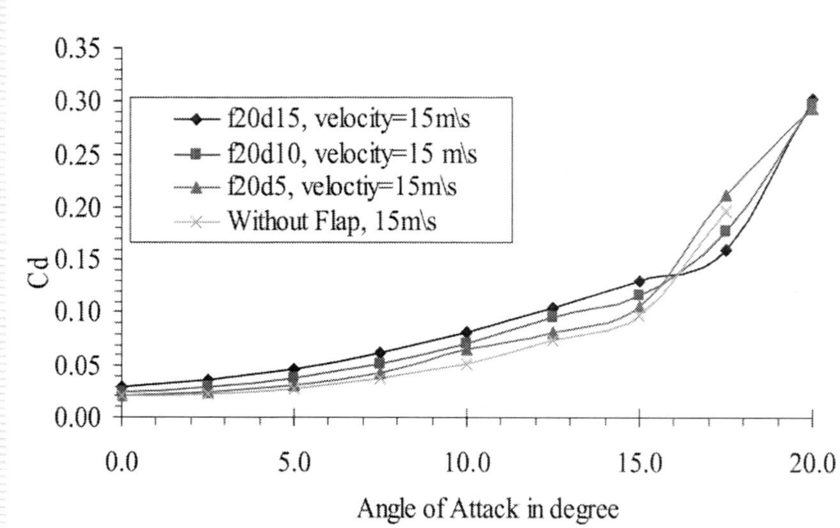

Astern Condition

Case 6: All Movable, Velocity = 30m/s Reynolds No =0.48e6

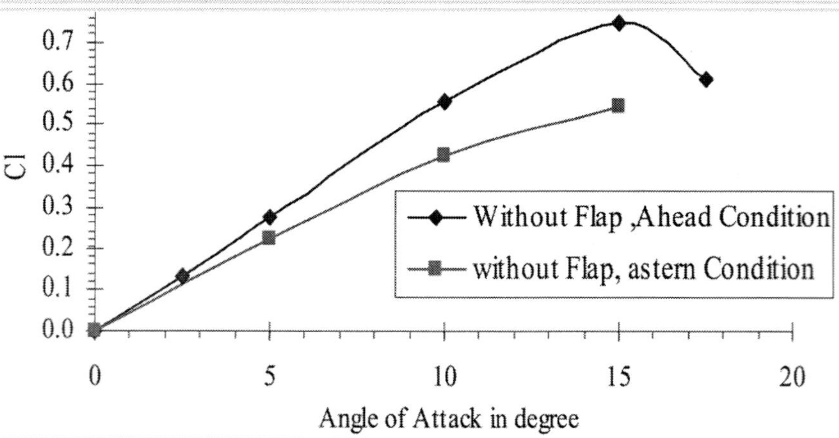

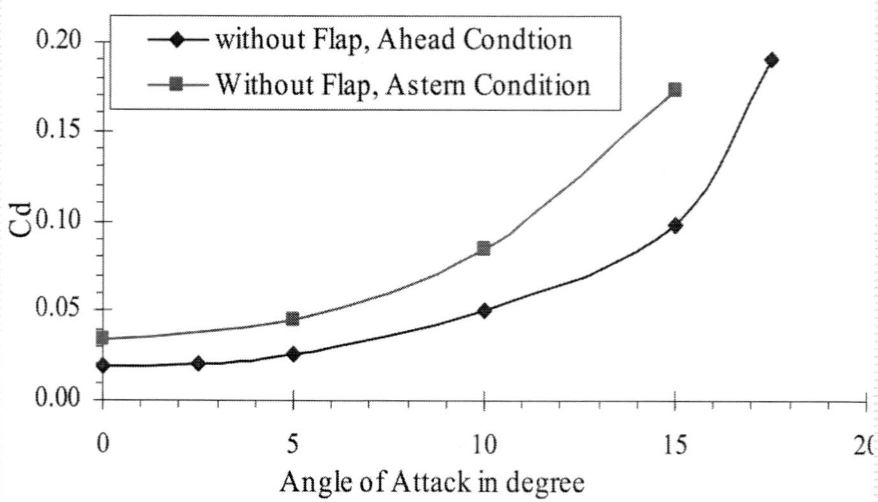

Case 7: Flap = 20%, Deflection=15 deg, Reynolds No =0.48e6 Velocity = 30m/s

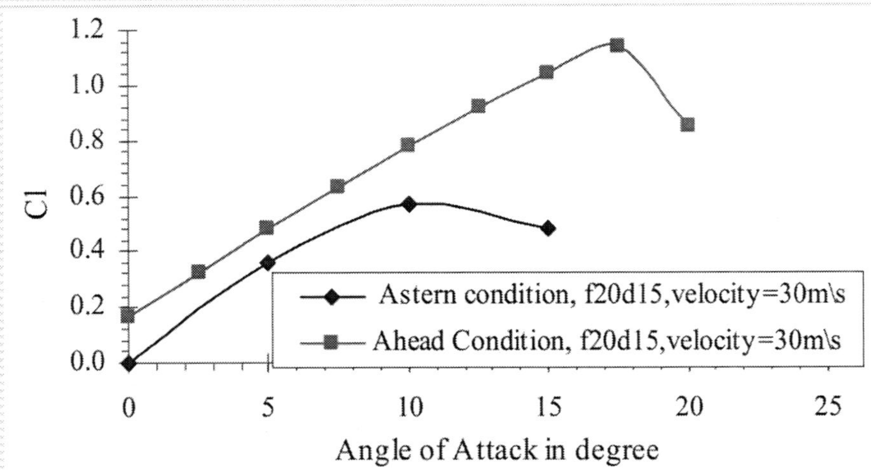

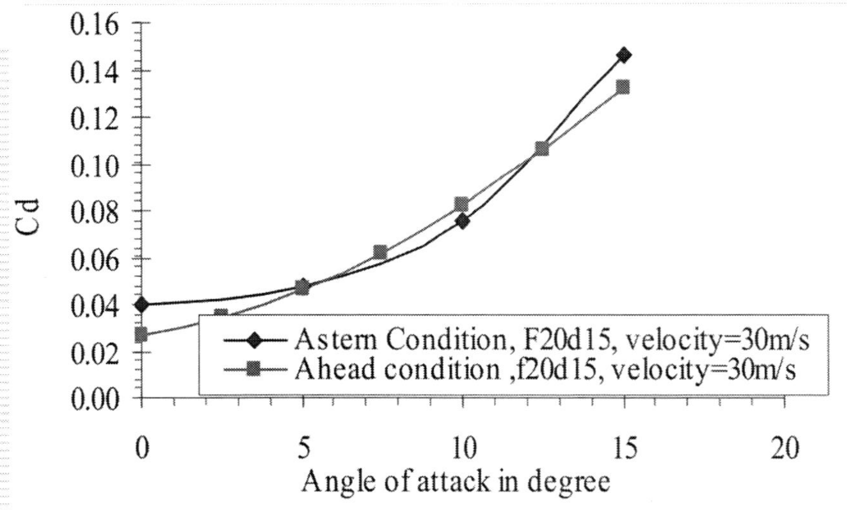

Case 8: Flap = 40%, deflection =15 deg, Reynolds No =0.48e6, Velocity = 30m/s

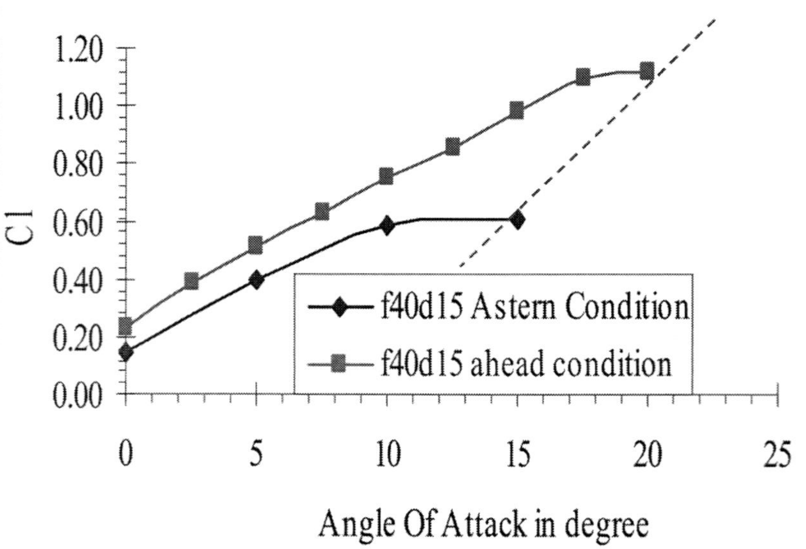

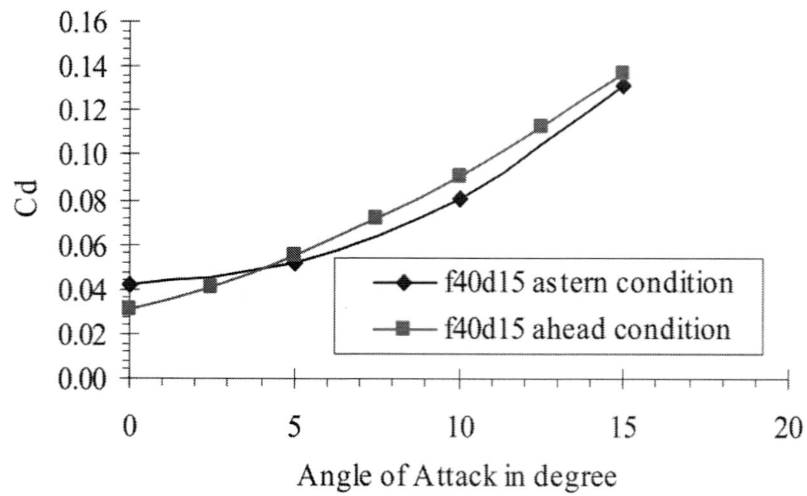

Case 9: Comparative study, Reynolds No =0.48e6, Velocity = 30m/s

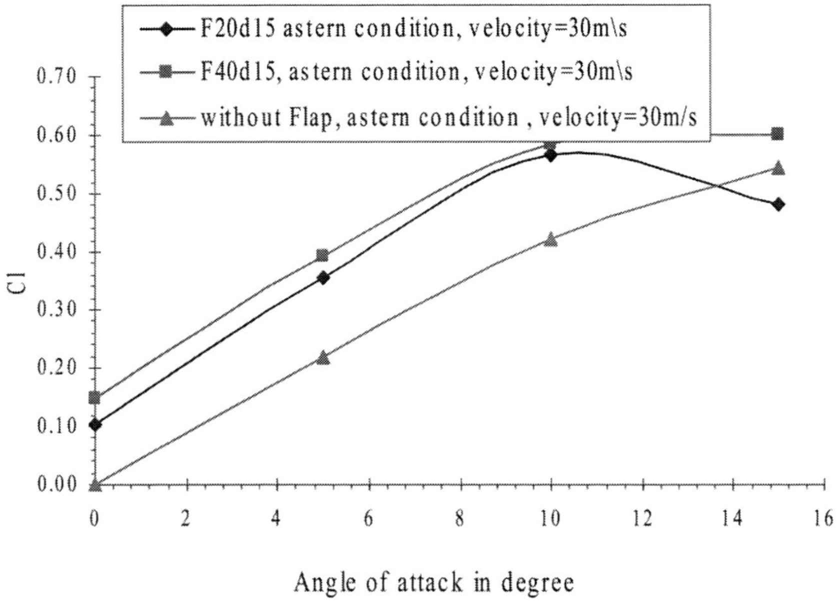

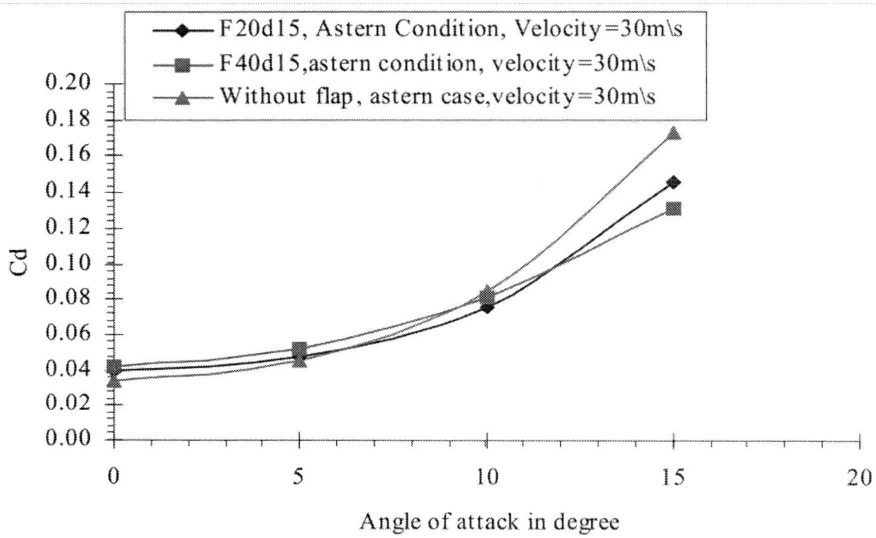

Velocity Profile of the rudder

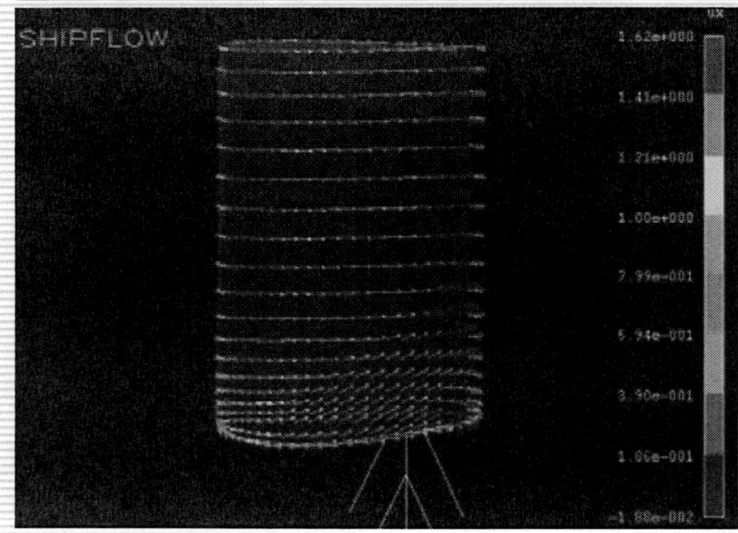

Pre-Stall Angle of attack=10 degrees

"Shipflow" results showing well agreement with the experimental results. After angle of attack 15 degrees it is showing the decrement in the lift coefficient value and predicting the stall characteristics which are well supported by the streamlines in the above figure.

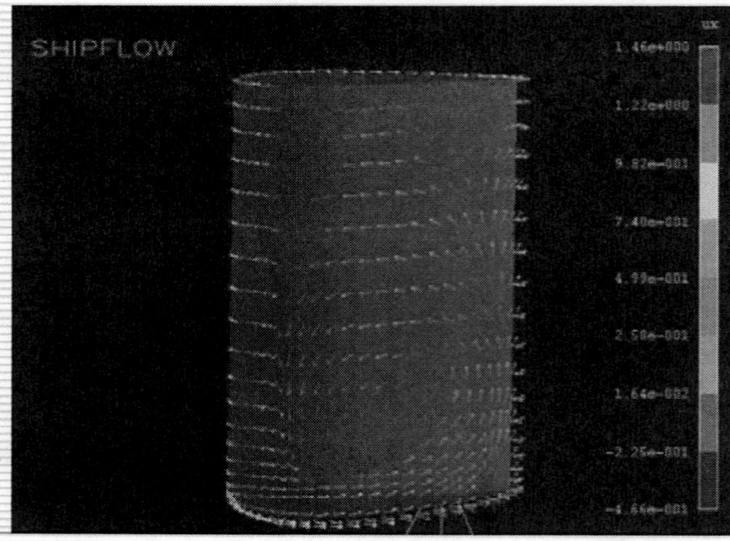

Separation Clearly visible at Higher Angle of Attack degrees

Observations from Free Stream Rudder Characteristics

- results clearly indicate an increase in
  - ✓ lift coefficient
  - ✓ and delay in stall

  with higher flap percentage and flap deflection.

- For a flap percentage of 20% and 15 degrees flap deflection the lift coefficient increase by nearly 40% when compared with an all-moveable rudder. However, this is accompanied by 25% increase in the drag coefficient value.

Observations from Free Stream Rudder Characteristics

- There is a large reduction in the value of lift coefficient in astern condition when compared with ahead condition.
- The reduction is larger for flap rudders when compared with all moveable rudders. From the results one can observe that there is a 40% reduction in the value of lift coeff. for flap rudder (flap % 40; flap deflection 15 deg) when compared to 25% for all-moveable rudder.
- However, the values of lift coefficient in astern condition for flap rudder are still higher than that of all-moveable rudder.

Observations from Free Stream Rudder Characteristics

- There is little change in the value of drag coeff. for flap rudder in ahead and astern condition. However, the increment in the value of drag coefficient is much larger for all-moveable rudders in astern condition.

- The lift to drag ratio in astern condition is higher for flap rudder than all-moveable rudder.

- A steep increase in the value of drag coefficient for larger angles of attack is observed for all-moveable rudder. However, for flap rudder having higher flap deflection the increase in the value of drag coefficient for larger angles of attack is much more gradual because of delayed stall angles.

CONCLUSION

- The prime objective of this work is to investigate the hydrodynamic characteristics of flap rudders in different operating conditions.

- The present work delineates an experimental investigation with a numerical one in which the "free flow" characteristics (lift and drag coeffcient) of flap rudders are calculated by using Reynolds-Averaged Navier-Stokes (RANS) solver, SHIPFLOW.

- The effectiveness and robustness of the computational results are established by comparing the numerical results with experimental results.

CONCLUSION

- Published literature in this area suggests that a numerically simulated solution for flap rudders consistent with experimental results is not available for practical application.

- There is very limited literature (Whicker and Fehlner, 1958) available on the performance of all movable rudders in astern condition and no data on the behaviour of flap rudders in astern condition is available in public domain.

- The insufficiency of the available literature restricts the designer from selecting flap rudders at the initial design stage.

CONCLUSION

- This will help the designer, at the early design stage, to select the flap rudder geometry (area, flap deflection, rate of flap deflection) that will enhance the manoeuvring characteristics of the vessel and satisfy the stringent International Maritime Organisation (IMO) manoeuvring requirement.

References

Bhattacharjee, A., "An Investigation into the Hydrodynamic Characteristics of All-Movable and Flap Rudders". M.S. Thesis, Indian Institute of Technology, Kharagpur, India, 2009.

Baker, G.S and Bottomley, G.H. (1921/22). "Maneuvering of ships. Part 1-Unbalanced rudders of single screw ships". Transactions of the Institute of the engineers and Shipbuilders in Scotland, vol.65, 1921/22, pp. 522-583.

Cole, A.P. (1938), "Destroyer turning circles". Transactions of The Royal Institute of Naval Architects, Vol. 80, 1938, pp. 32-57.

Felli, M., Greco, L., Salvatore, F., Di Felice, F., and Soave, M. (2006), Experimental and theoretical investigation of propeller rudder interaction phenomenon, 26th Symposium on Naval Hydrodynamics, Vol. III, pp.105-120, Rome, Italy.

Gawn. R.W.L. (1943), Steering experiments, Part I Transactions of the Royal Institute of Naval Architect.vol.85, 1943, pp 35-73.

Harrigton, R.L. (1981), Rudder torque prediction. Transactions of The Society of Naval Architects and Marine Engineers, Vol. 89. pp. 23-90.

Joessel, (1873), Rapport Sur des experience relatives aux gouvernalis. Memorial du Gennie Maritime, Rapport 9.

References

Kerwin, J. E., Mandel, P. and Lewis, S. Dean, (Dec. 1972.), "An Experimental Study of a Series of Flapped Rudders", Journal of Ship Research, SNAME.

Lewis, E. V. (Ed), (1998), "Principles of Naval Architecture", Volume III, SNAME.

Molland, A.F. and Turnock, S.R., (2002),Flow Straightening effects on a ship rudder due to upstream propeller and hull, International Shipbuilding Progress, Vol. 49, No3, pp.195-214.

Molland A.F. and Turnock S.R, (1991) Wind tunnel investigation of the influence of propeller loading on ship rudder performance, Ship Science report No.46, Scholl of Engineering Science, and University of Southampton.

Whicker, L.F and Fehlner, L.F. (1958), Free Stream characteristic of family of low aspect ratio control surface for application to ship design. DTMB Rep. 933.December.

THANK YOU

GENERATION OF LARGE COMPOUND CURVED SURFACES USED IN SHIP BY LINE HEATING USING OXY-ACETYLENE GAS FLAME

P Biswas, Department of Ocean Engg & Naval Architecture, Indian Institute of Technology, Kharagpur, India
N R Mandal, Department of Ocean Engg & Naval Architecture, Indian Institute of Technology, Kharagpur, India
O P Sha, Department of Ocean Engg & Naval Architecture, Indian Institute of Technology, Kharagpur, India

SUMMARY

One of the most tedious and human skill dependent operations is plate forming for producing compound curve plates in shipbuilding industry. For generation of the heating line pattern to achieve the target surface, the surface properties (i.e. grid point, gauss curvature, principle curvatures and their directions) are required. The strain distribution pattern and their magnitude of a curved surface need to be evaluated to determine the extent of line heating. In this present work a suitable program has been developed to determine the surface properties and the heating line pattern by using B-Spline surface theory. For getting the initial template (i.e. 2-D blank) and strain distribution over the plate, a numerical technique and an optimized strain field distribution approach based on minimization of total strain energy has been used. Finally this technique has been successfully applied to develop a large compound curved surface used in ship

KEYWORDS

B-spline surface, Surface fit, Strain field distribution, Principal curvature, Gaussian curvature, Blank generation, Heating line generation, oxy-acetylene gas flame

1. INTRODUCTION

Line heating is a method of plate forming by means of local heating in linear segments. It can be gainfully utilized in shipbuilding industry, and other relevant plate forming operations in allied industries. Now a days very few skilled shipwrights are there, who are capable of performing the process of line heating. The skill is purely based on experience and intuition. Hence it is all users essential to develop a proper science based for this art of line heating. Eventually this will pave the way for automation of the plate bending process. In this present investigation line heating by oxy-fuel gas torch has been presented.

Early surface development procedures were implemented in shipyards based on geodesic development during the last three decades, mainly for ship hull plates whose Gaussian curvature is very small. Letcher [1] presented a basic geometric theory for flattening and fabrication of doubly curved plates. The mapping from the curved surface to its planar development was modeled by adding in-plane strains to the curved surface. The strain field was obtained by solving a generalized Poisson's equation with the source term equal to the Gaussian curvature. However, since the problem is formulated as a boundary value problem, a good solution relies on a well specified boundary condition which was hard to know beforehand.

Ueda et al. [2] investigated the relation between the final shape of a plate and the inherent strain. They computed the strain caused by deformation from the initial configuration to the final one using large deformation elastic FEM analysis. Since the initial configuration is usually the projection of the doubly curved surface onto the x–y plane, their approach can only be applied to the cases when the 3D surface is relatively flat, i.e., the curvature is small.
Cho et al. [3] presented an algorithm to approximately develop a doubly curved surface by minimizing the mapping error function for locally isometric mapping between a given and developed surface net. The method was applied to construct an auxiliary planar domain of triangulation for tessellating trimmed parametric surface patches, which sufficiently preserved the shape of triangles when mapped into three-dimensional space. Again, the applicability of Cho's method for metal forming is unclear.

Jang et al [4] used a simplification where a circular, axisymmetric disc with springs represented the heated region. The spring constants were determined from the theory of an infinite plate with a hole. The disk was then distributed along the heating path to simulate the effect of actual heating and the resulting strains were converted into equivalent forces by integration. Zenkour [5] gave a solution for the bending of thin rectangular plates with uniform, linear, and quadratic thickness variation based on classical thin-plate theory.
Various mapping techniques emulate the actual forming process [6] so that the cut plates after necessary shaping fit one another perfectly at the erection stage. Unfortunately, this does not work very well, as the mappings are typically based on differential geometry (which assumes evenly distributed deformation) in contrast to line heating and other forming methods which deform the plates locally. Therefore a proper knowledge and better control of the forming process is important.

The works of the above investigators indicate that the thermal history and residual deformation in Line heating process [7] is strongly affected by many parameters and by their interactions. In this work a suitable surface generation and its surface properties prediction was carried out using b-spline surface theory. A Genetic algorithm (GA) based optimized strain field distribution approach [8] based on minimization of total strain energy and a numerical method was applied for strain field distribution and blank generation of a given compound curved surface.

2. SURFACE THEORY FOR GENERATION OF COMPOUND CURVED SURFACE

Surfaces can be broadly divided in to two categories a) developable and b) non-developable. Mathematically the developable surfaces are the ones that have zero Gaussian curvature. This means that the developable surface can be developed without stretching or tearing the surface into a flat surface. The developable surfaces are rolled only on one direction and so can also be called as rolled surfaces. Developable surfaces are highly favored in metal forming because they can be easily formed from metal flat sheets. There is least amount of residual stress left after the surface has been formed.

On the other hand a non-developable surface is defined mathematically as a surface that has non-zero Gaussian curvature in at least some parts of the surface. This means that there is stretching of the flat plate during the formation of the surface. Although developable surfaces are highly desirable for the ease of production but sometimes hydrodynamic, structural or aesthetic requirements may lead to non-developable surfaces. The flow diagram for generation of compound curved surfaces is shown in Fig.1.

3. RESULTS AND DISCUSSION

Theoretically a surface patches of a ship have been generated. The location of the surface patches which were taken from the ship are shown in Table 1. The control points of ship surface patch is shown in Tables 1 and. Patch was generated using B-Spline surface theory as shown in Figs. 2.

The overall strain field distribution pattern for patch-is shown in Fig.3. And the blanks based on the following strain field distribution for the patch is shown in Fig. 4. The heating line patterns to generate patch is shown in Fig.5.

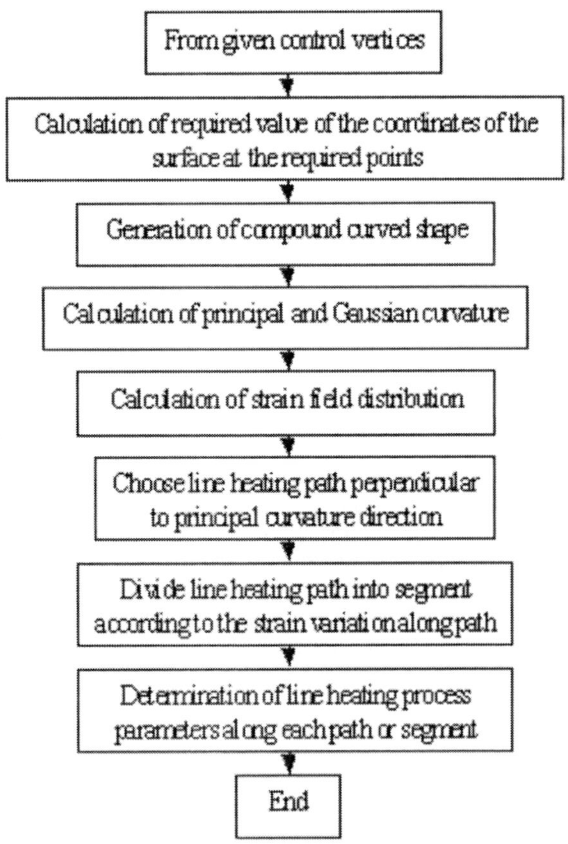

Fig. 1 Flow diagram for compound curved surface generation using line heating

Table 1: Offsets of surface-patch: Ship=asw(c), Block=bc10, Part=104

WATERLINE 500mm

FR NO	HALF-BREADTH	HEIGHT AB
9.5	712.45	500.00
9.0	755.28	500.00
8.5	793.09	500.00
8.0	836.47	500.00
7.5	880.81	500.00
7.0	928.08	500.00
6.5	978.59	500.00
6.0	1032.40	500.00

WATERLINE 750mm

FR NO	HALF-BREADTH	HEIGHT AB
9.5	749.30	750.00
9.0	756.08	750.00
8.5	770.52	750.00
8.0	788.07	750.00
7.5	809.42	750.00
7.0	842.35	750.00
6.5	881.68	750.00
6.0	927.44	750.00

WATERLINE 1000mm

FR NO	HALF-BREADTH	HEIGHT AB
9.5	857.94	1000.00
9.0	829.05	1000.00
8.5	802.20	1000.00
8.0	788.41	1000.00
7.5	781.17	1000.00
7.0	785.91	1000.00
6.5	806.20	1000.00
6.0	838.24	1000.00

WATERLINE 1250mm

FR NO	HALF-BREADTH	HEIGHT AB
9.5	1010.87	1250.00
9.0	957.06	1250.00
8.5	906.68	1250.00
8.0	862.70	1250.00
7.5	825.92	1250.00
7.0	800.19	1250.00
6.5	784.96	1250.00
6.0	786.82	1250.00

WATERLINE 1500mm

FR NO	HALF-BREADTH	HEIGHT AB
9.5	1168.01	1500.00
9.0	1098.76	1500.00
8.5	1037.07	1500.00
8.0	976.94	1500.00
7.5	920.38	1500.00
7.0	868.32	1500.00
6.5	823.66	1500.00
6.0	790.99	1500.00

WATERLINE 1750mm

FR NO	HALF-BREADTH	HEIGHT AB
9.5	1326.08	1750.00
9.0	1245.30	1750.00
8.5	1172.07	1750.00
8.0	1100.51	1750.00
7.5	1030.65	1750.00
7.0	965.45	1750.00
6.5	899.35	1750.00
6.0	836.52	1750.00

WATERLINE 2000mm

FR NO	HALF-BREADTH	HEIGHT AB
9.5	1484.98	2000.00
9.0	1395.17	2000.00
8.5	1310.39	2000.00
8.0	1227.56	2000.00
7.5	1146.37	2000.00
7.0	1069.83	2000.00
6.5	990.11	2000.00
6.0	908.95	2000.00

BOTTOM SEAM

FR NO	HALF-BREADTH	HEIGHT AB
9.5	728.49	650.00
9.0	751.29	650.00
8.5	779.11	650.00
8.0	806.12	650.00
7.5	835.93	650.00
7.0	875.36	650.00
6.5	919.34	650.00
6.0	968.85	650.00

TOP SEAM

FR NO	HALF-BREADTH	HEIGHT AB
9.5	1437.79	1925.87
9.0	1339.45	1907.53
8.5	1248.16	1888.23
8.0	1160.25	1868.52
7.5	1075.67	1848.39
7.0	997.56	1827.88
6.5	919.50	1806.38
6.0	844.95	1783.78

AFT SEAM

FR NO	HALF-BREADTH	HEIGHT AB
FR9.0-150	744.91	500.00
FR9.0-150	753.83	750.00
FR9.0-150	835.06	1000.00
FR9.0-150	969.81	1250.00
FR9.0-150	1115.32	1500.00
FR9.0-150	1264.95	1750.00
FR9.0-150	1417.36	2000.00

FORD SEAM

FR NO	HALF-BREADTH	HEIGHT AB
FR6.0-300	1005.71	500.00
FR6.0-300	904.23	750.00
FR6.0-300	821.09	1000.00
FR6.0-300	783.31	1250.00
FR6.0-300	805.21	1500.00
FR6.0-300	866.95	1750.00
FR6.0-300	949.65	2000.00

FRAME.NO	X
9.5	88600
9.0	89200
8.5	89800
8.0	90400
7.5	91000
7.0	91600
6.5	92200
6.0	92800

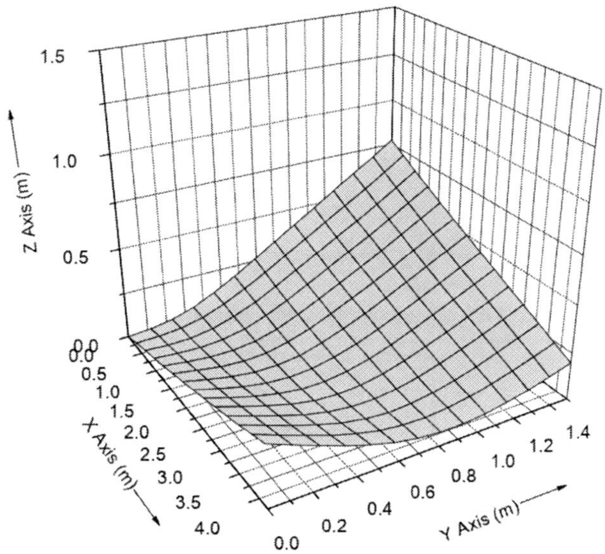

Fig. 2 Theoretically developed 3-D Shape of surface patch

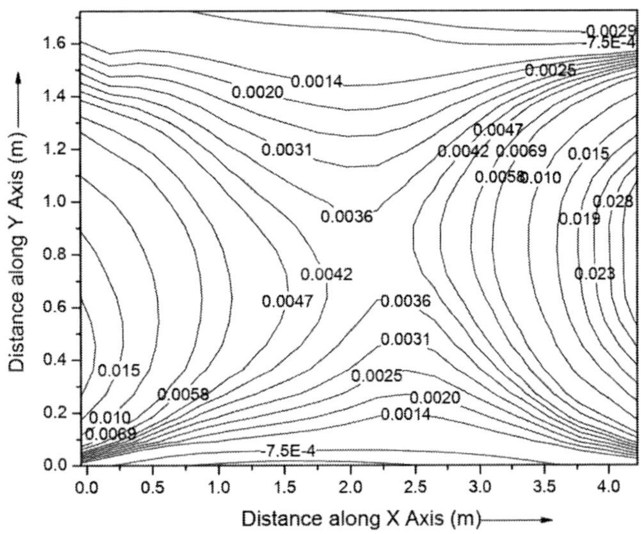

Fig. 3 Strain field distribution of surface patch

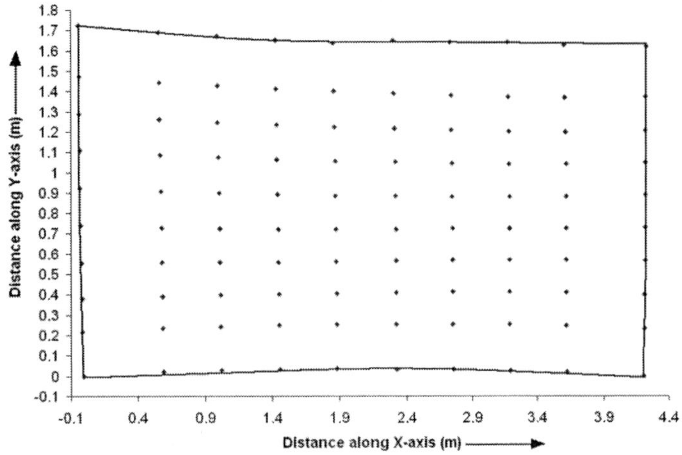

Fig. 4 2-D Blank of surface patch

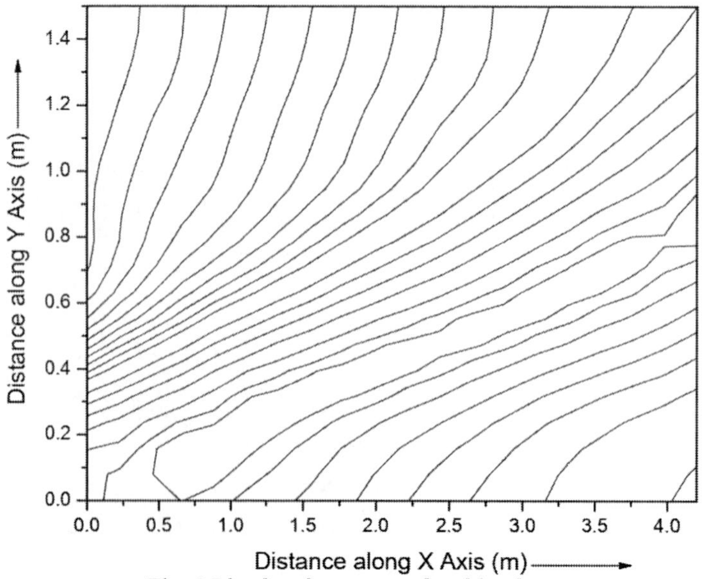

Fig. 5 Line heating pattern for ship plate

4. Experimental details

An experiment was conducted in GRSE Ltd, Kotkata, for developing the above surface patch. First of all the blank was cut in desired shape by an automatic plasma cutting machine from a plate of 10mm thick mild steel plate and the heating lines were drawn on the blank automatically by an numerical control machine as shown in Fig.6. Then line heating operation were carried out manually by oxy-acetylene gas flame though the heating line by the designed heating parameters [9]. The developed shape after first set of heating operation is shown in Fig. 7. The shape obtained from experiment is well comparable with original shape.

Fig.6 Heating line pattern on blank

5. Conclusions

The following conclusion are drawn from the about the investigations

1. B-spline surface theory has been successfully applied to generate the surface (i.e. grid points) and to determine its surface properties to determine the line heating pattern.
2. A GA based approach has been developed for generating the optimized strain field distribution of the 3-D patch.
3. The numerical method has been successfully developed by considering the boundary strain obtained from optimized strain field distribution for generating the overall strain field distribution and 2-D blank generation of the 3-D patch.
4. The results are encouraging for developing a system for automating plate forming operation using line heating.
5. Line heating technique can be effectively used to generate any shape those are encountered in ship hull forms.

Fig.7 Experimentally developed shape after first set of heating operation

6. References

1. Manning, J.R., Computerized pattern cutting: Methods based on an isometric tree, *Computer-Aided Design,* Vol.12 (1), pp. 43–47, 1980.

2. Hinds, B.K., McCartney, J. and Woods, G., Pattern development for 3D surfaces, *Computer- Aided Design*, Vol.23, No.8, pp. 583–592, 1991.

3. Azariadis, P. and Aspragathos, N., Design of plane developments of doubly curved surfaces, Computer-Aided Design, Vol. 29, No.10, pp.675–685, 1997.

4. Letcher, J.S. Jr., Lofting and fabrication of compound-curved plates, *J. Ship Research*, Vol.37, No.2, pp.166–175, 1993.

5. Ueda, K., Murakawa, H., Rashwan, A.M., Okumoto, Y. and Kamichika, R., Development of computer-aided process planning system for plate bending by line heating (report 1)—relation between final form of plate and inherent strain, *J. Ship Production*, Vol.10, No.1, 59–67, 1994.

6. Cho, W., Patrikalakis, N.M. and Peraire, J., Approximate development of trimmed patches for surface tessellation, *Computer-Aided Design*, Vol. 30, No.14, pp. 1077–1087, 199

7. Jang, C.D,S.I Seo, and D.E.Ko., A study on preditions of deformations of plates due to line heating using a simplified thermal elasto-plastic analysis, *Journal of Ship Production*, Vol.13, No.1, pp.P 22-27, February 1997.

8. Zenkour, A. M., An Exact Solution for the Bending of Thin Rectangular Plates with Uniform, Linear, and Quadratic Thickness Variations, *Int. J. Mech. Sci.*, Vol.**45**, pp. 295–315, 2003.

9. Biswas Pankaj, Mandal N. R., Sha O. P., Prediction of Process Parameters of Line Heating Process by Three Dimensional Finite Element Method, *Proc. IMechE, J. Engineering for the Maritime Environment,* Vol. 221 Part M, 2007, pp 17-30.

9. AUTHORS BIOGRAPHY

Pankaj Biswas holds the current position of Research Associate in the department of Ocean Engineering and Naval Architecture, Indian Institute of Technology, Kharagpur, India.
Nisith Ranjan Mandal holds the current position of Professor, Head in the department of Ocean Engineering and Naval Architecture, Indian Institute of Technology, Kharagpur, India.
Om Prakash Sha holds the current position of Professor in the department of Ocean Engineering and Naval Architecture, Indian Institute of Technology, Kharagpur, India.

Generation of large compound curved surfaces used in ship by line heating using oxy-acetylene gas flame

by

Dr. Pankaj Biswas[1], Prof. N.R. Mandal[2] and Prof. O.P. Sha[3]

DEPARTMENT OF OCEAN ENGINEERING AND NAVAL ARCHITECTURE
INDIAN INSTITUTE OF TECHNOLOGY
11TH DECEMBER 2009

Introduction

- Methods of plate forming

  - ***Mechanical forming & Thermo-mechanical forming***

  - Line heating

    Line heating is a method of forming curved plates by means of localized spot or linear heating

  - Used mainly in shipbuilding industry, also being actively considered in automotive industry and other plate forming requirements

- Kinematics of line heating

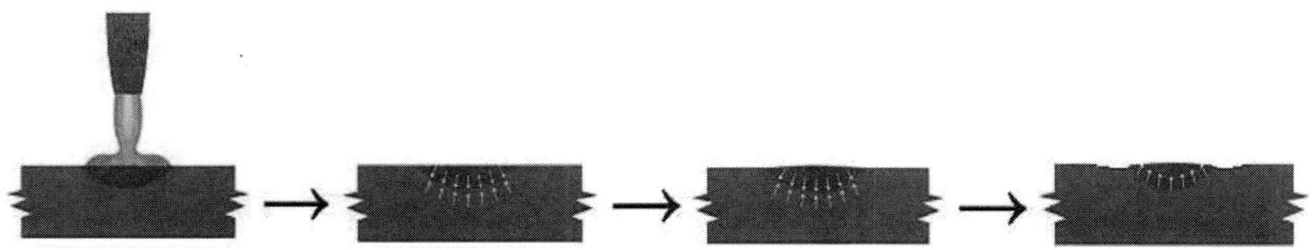

Objective

❖ The objective is theoretical determination of heating parameters and heating pattern to achieve a given 3-D large curved surface.

❑ The following steps will be followed to achieve this objective:

- ✓ Nonlinear elasto-plastic thermo-mechanical analysis of line heating induced stress, strains and resulting residual deformations.
- ✓ Estimation of strain field distribution for a given curved surface.
- ✓ Blank generation from the given patch.
- ✓ Determination of heating path and suitable heating parameters required to produce the patch.
- ✓ Experimental validation of the proposed scheme for generation of compound surfaces.

Outline of Prediction of process parameters

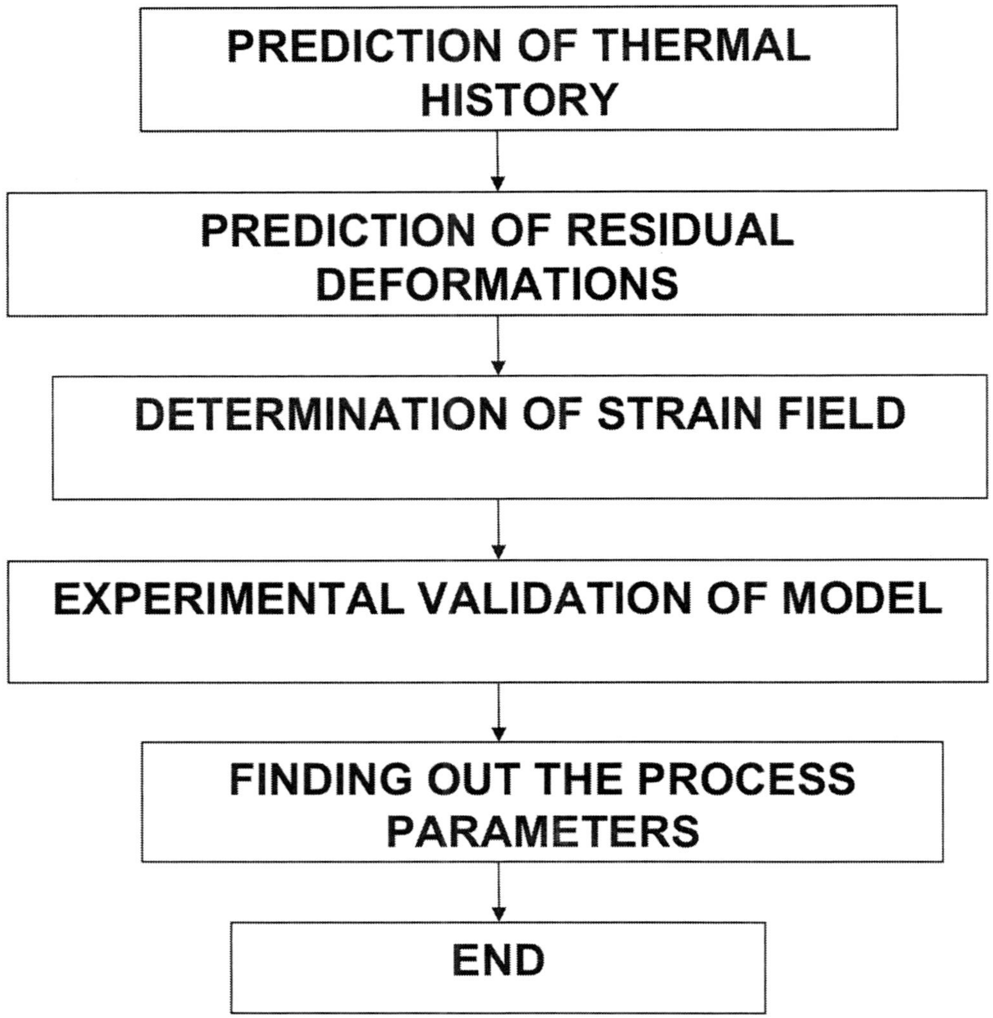

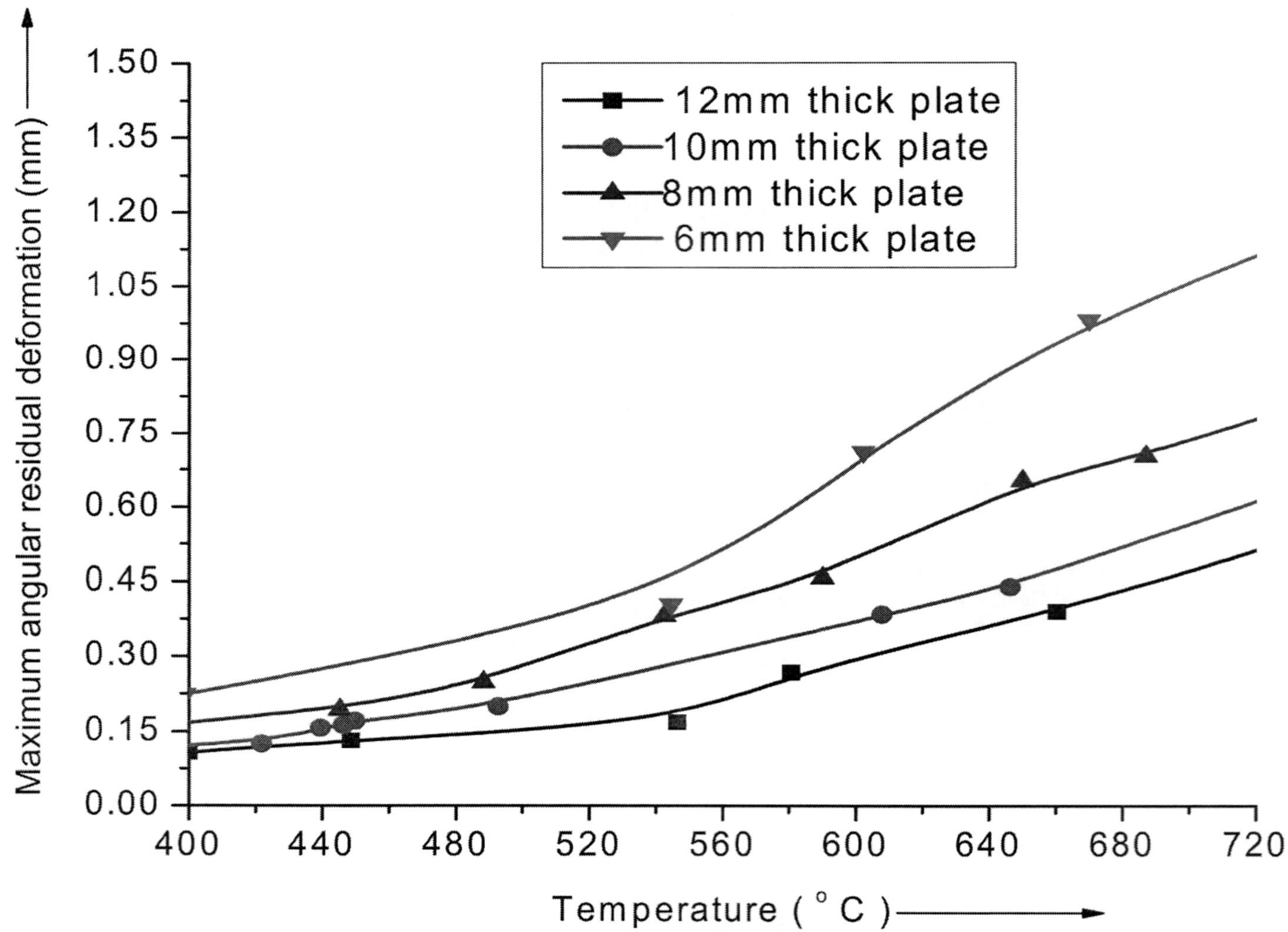

Variation of residual deformation with peak temperature for different plate thicknesses

Strain Distributions for Single Pass and Double Pass Line Heating

Parameters for line heating operation

Sl. No.	Plate thickness (mm)	Heat input (J/s)	Torch speed (mm/s)
1	10	5550	10

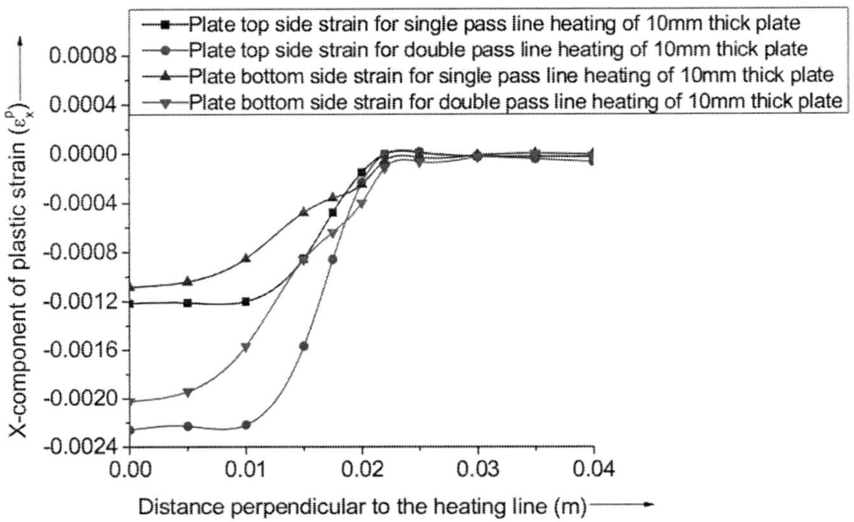

X-component of strain perpendicular to the heating line

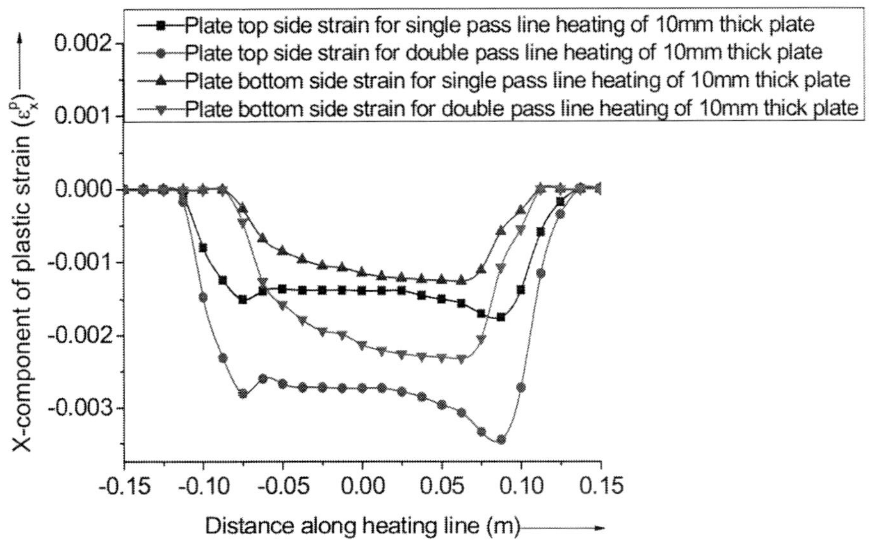

X-component of strain along heating line

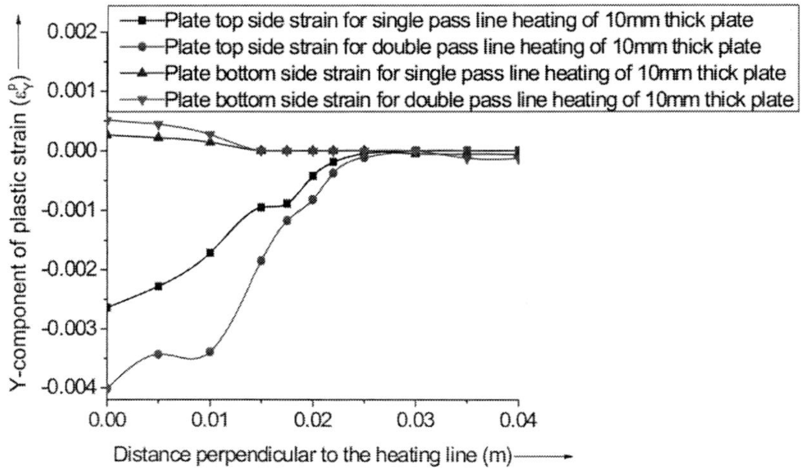

Y-component of strain perpendicular to the heating line

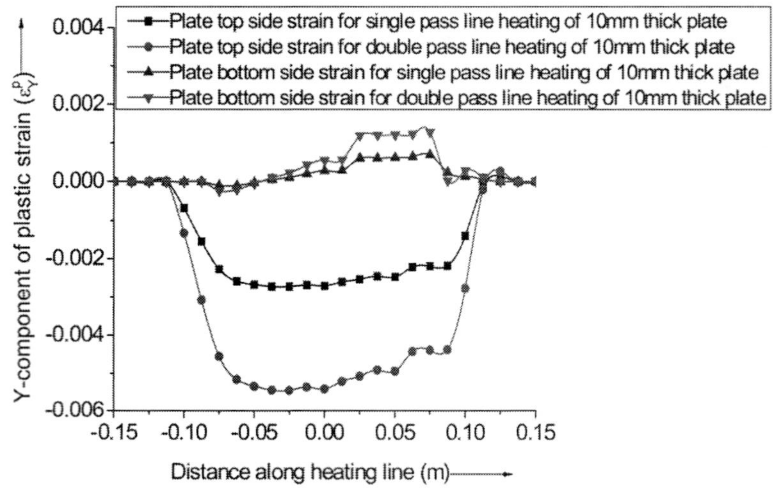

Y-component of strain along heating line

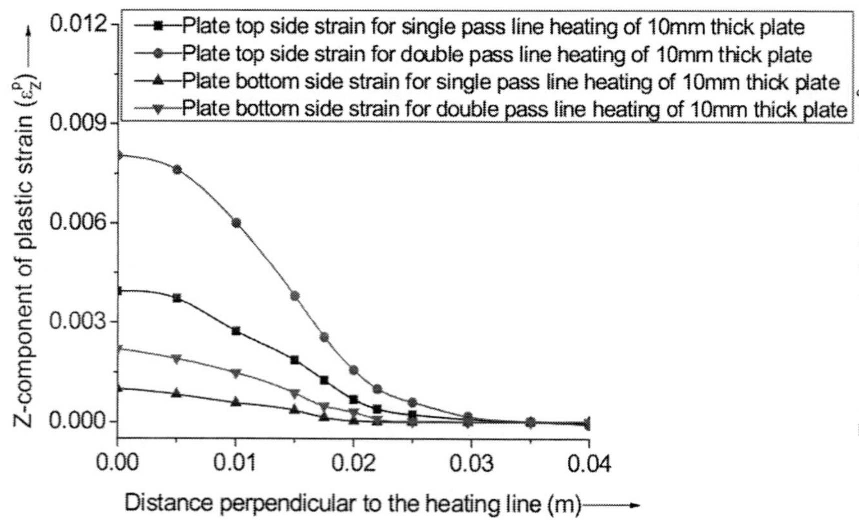

Z-component of strain perpendicular to the heating line

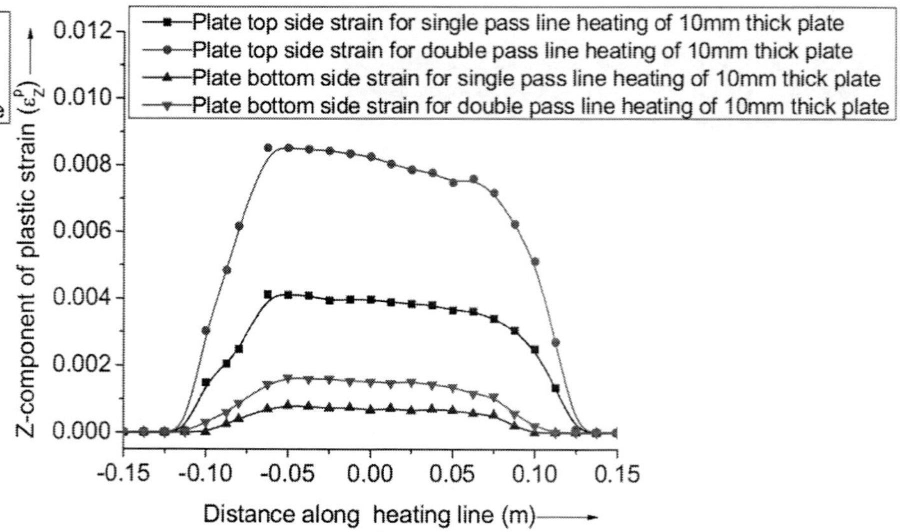

Z-component of strain along heating line

Average Plastic Strains vs. Peak temperature:

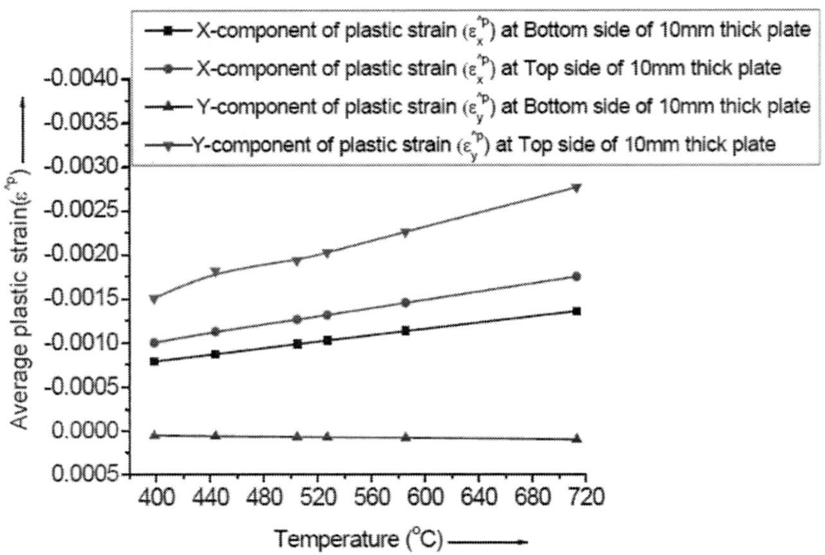

Fig. X and Y-component of average plastic strain on top and bottom side of 10mm thick plate

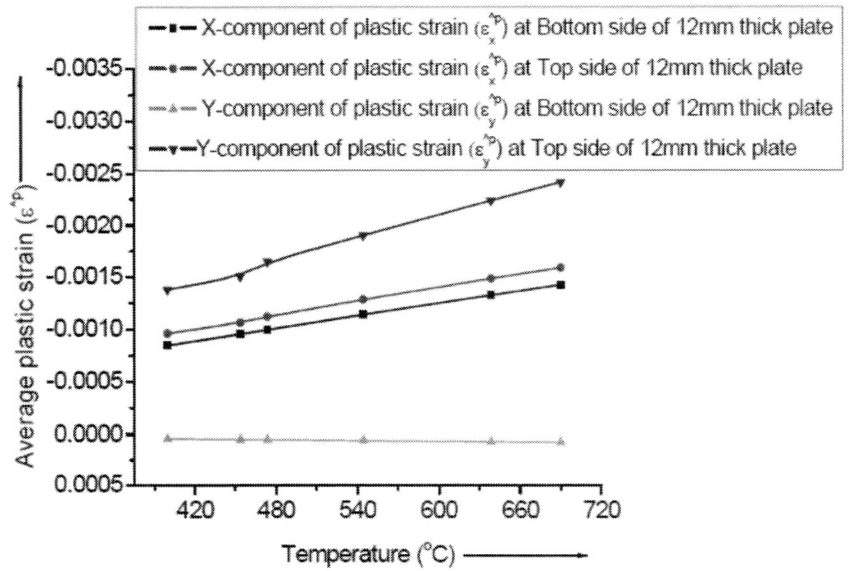

Fig. X and Y-component of average plastic strain on top and bottom side of 12mm thick plate

Development of Compound Curvature Surface using Line Heating

- A curved surface can be divided into two categories

  - ✓ Developable surfaces – more desirable
  - ✓ Non-developable surface – required for structural, hydrodynamic and aesthetic purposes

- Objective

  - ➢ Theoretically find the 2D shape of the blank
  - ➢ Determine the heating path and the heat input to get the desired curved surface

Curved surface and its planer development:

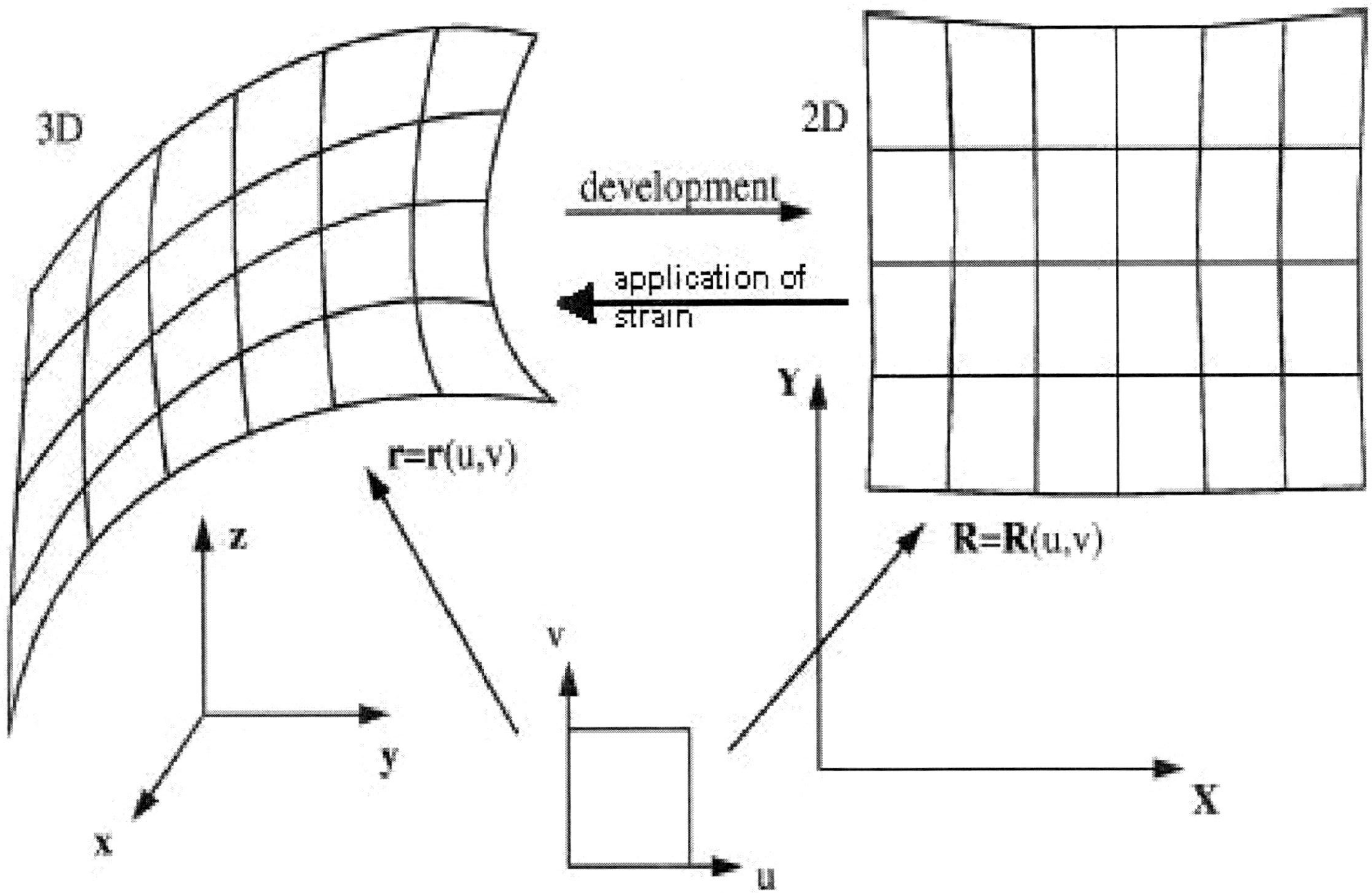

Fig. Curved surface and its planar development.

Outline of process planning for generation of compound curved surface

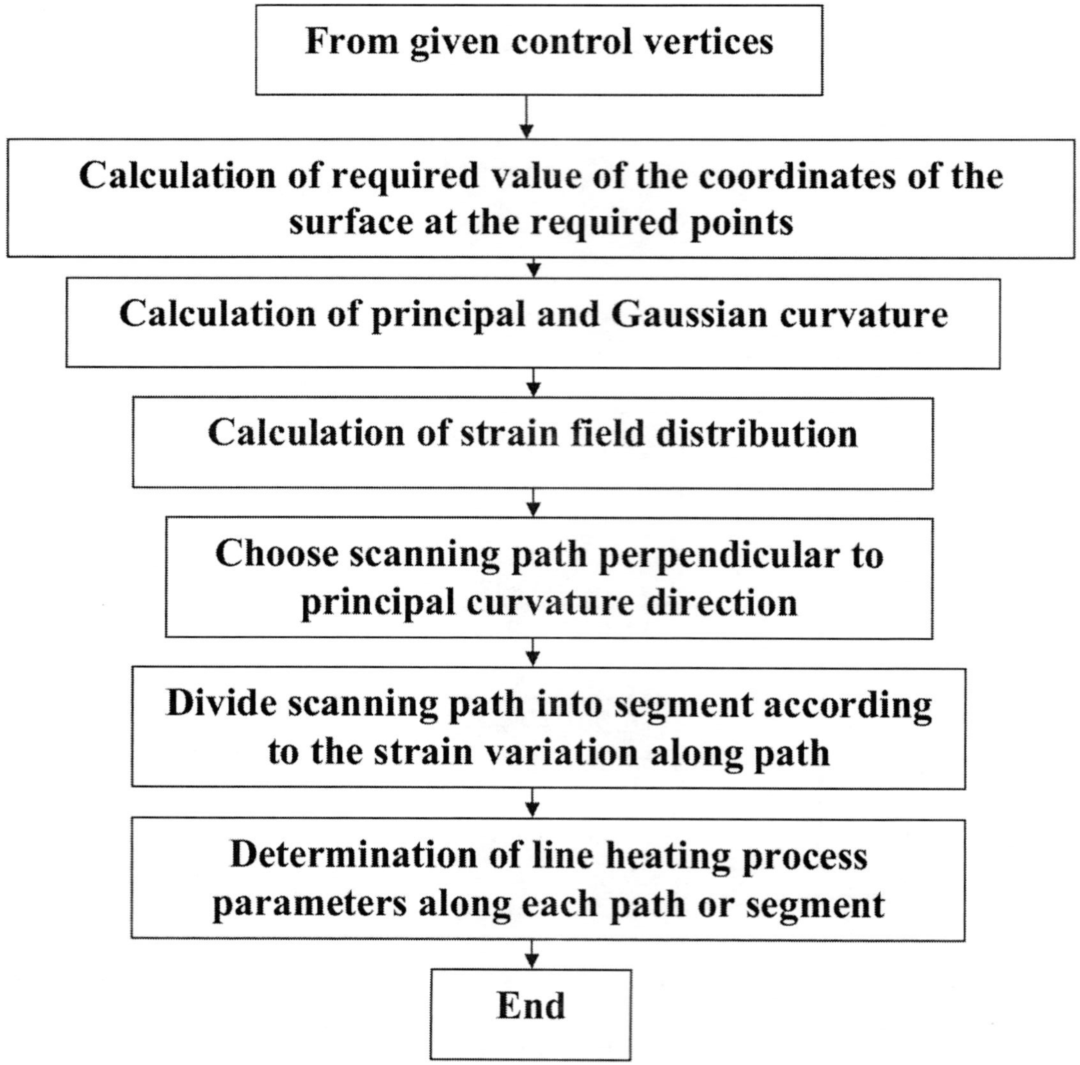

Constructing the desired surface from the given control points

- Input
  - Control vertices
  - Degree of curve along each parametric line
  - Number of output grid points along each direction

- Output
  - Required value of the co-ordinates of the surface at the required points.

Example of a pillow shape surface obtained from the given control points:

Inputs:
- **Control vertices =16**
- **Degree of curve along each parametric line=3**
- **Number of output grid points along each direction=10**

X (m)	Y (m)	Z (m)
0	0	0
0.11	0	0.024
0.34	0	0.024
0.45	0	0
0	0.11	0.014
0.11	0.11	0.046
0.34	0.11	0.046
0.45	0.11	0.014
0	0.34	0.014
0.11	0.34	0.046
0.34	0.34	0.046
0.45	0.34	0.014
0	0.45	0
0.11	0.45	0.024
0.34	0.45	0.024
0.45	0.45	0

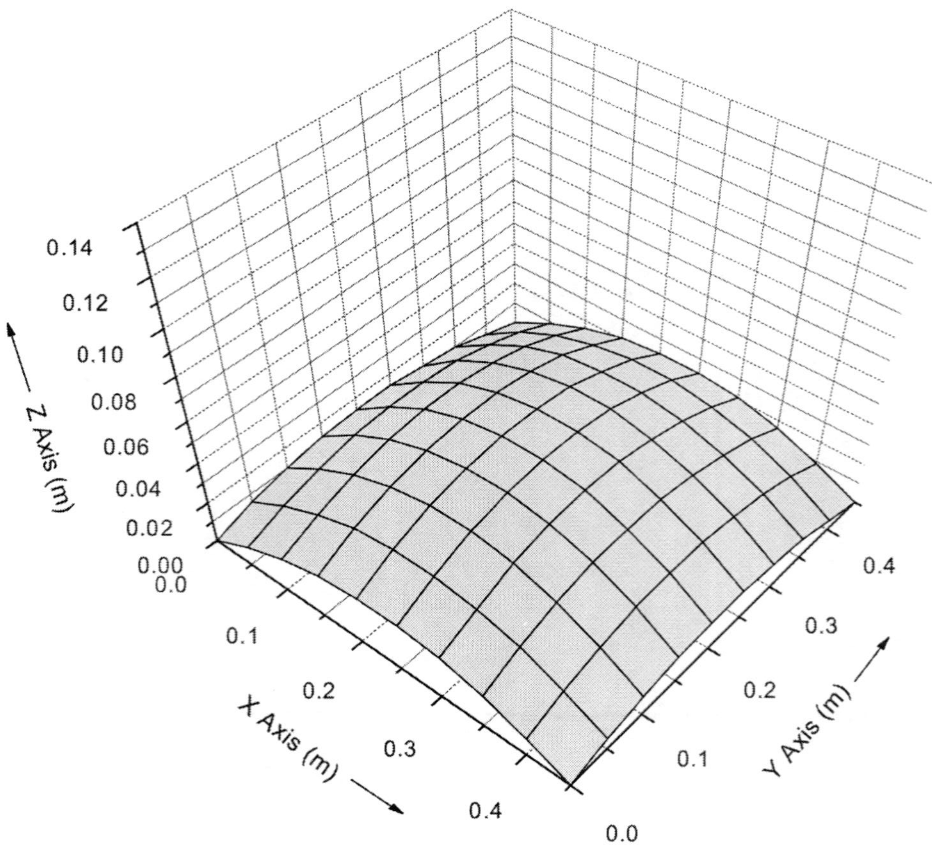

Strain Field Generation

✓ There can be multiple blanks that give the same final curved surface.

Here strain energy has been minimized, so that

- Less material requirement
- Work minimized

❑ Strain Field for Planar Development:

An approach based on minimization of the total strain energy after adding the strain field to the given doubly curved surface, which maps to a planar shape.

❖ Mathematically :

$$\min \iint_D \left\{ \left(\varepsilon^s\right)^2 + \left(\varepsilon^t\right)^2 \right\} \left| r_u \times r_v \right| du dv = \min \iint_D \left\{ \left(\varepsilon^s\right)^2 + \left(\varepsilon^t\right)^2 \right\} \sqrt{EG - F^2}\, du dv$$

Solution obtained using Optimized GA.

Strain field distribution along u and v-direction for pillow patch

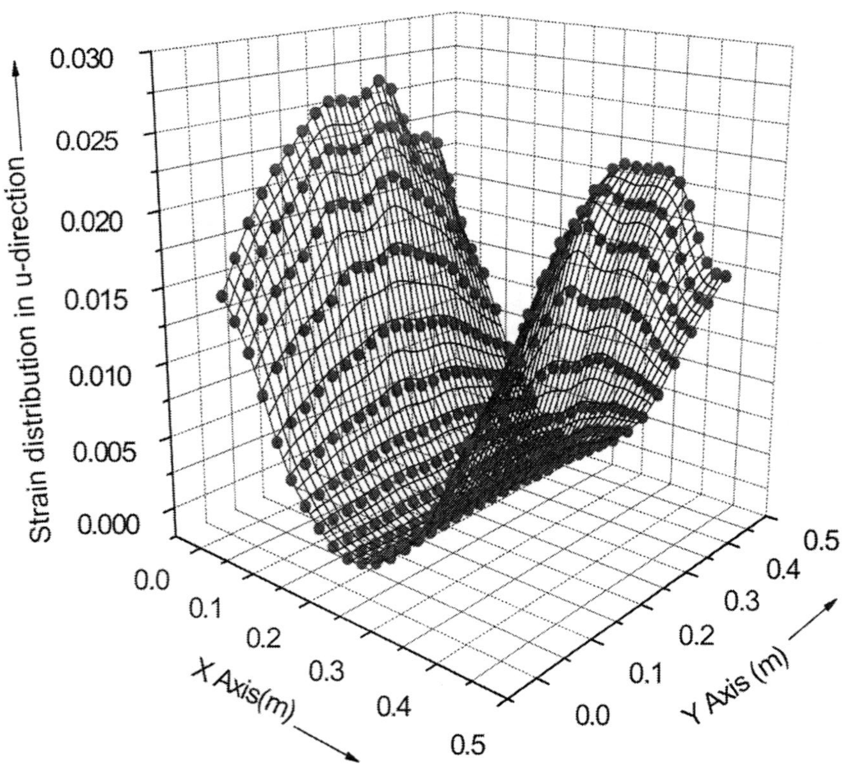

Strain field distribution along u-direction for pillow patch

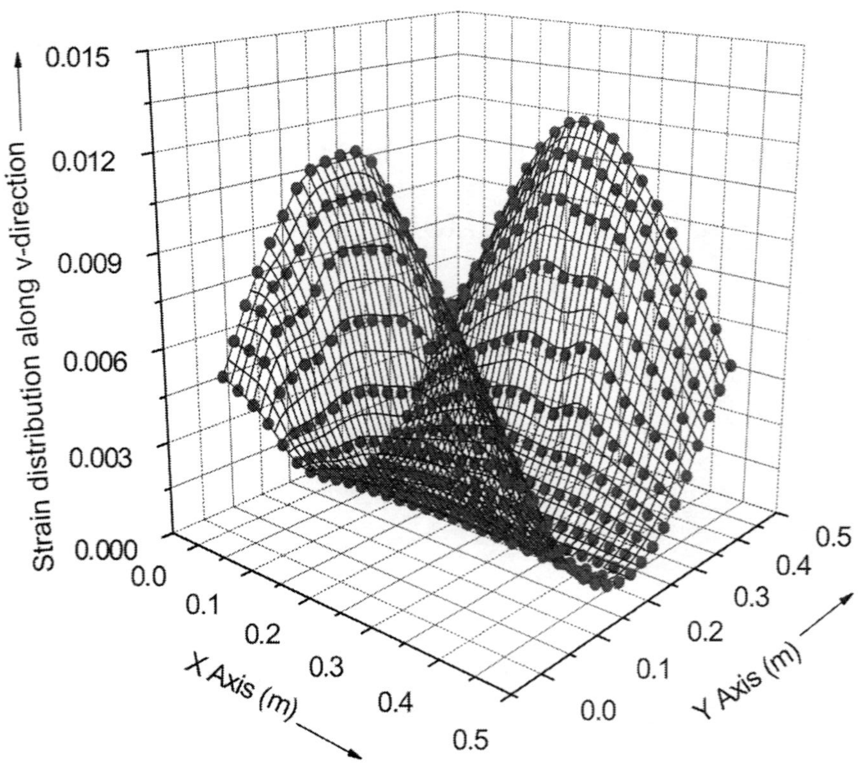

Strain field distribution along v-direction for pillow patch

Overall Strain Field distribution and Blank Generation

❖ The expression of link equation

$$(X_i - X_j)^2 + (Y_i - Y_j)^2 = \left[1 + 0.5(\varepsilon_i + \varepsilon_j)(\cos^2\theta + A\sin^2\theta)\right]^2 d^2$$

❖ If there is ‘n’ number of nodes then there is 3n number of unknown.

❖ It was seen that $3n$ is greater than L.

➢ Assumptions made:

❖ Specification of all boundary strain (M in number).

❖ Removing rigid-body degrees of freedom by, specifying X and Y for one node.

❖ **The solution of the system of nonlinear equations was obtained by well-known Newton-Raphson iteration method.**

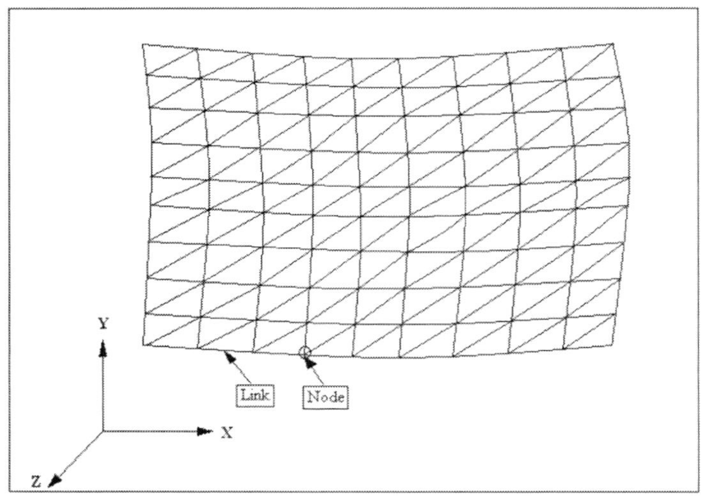

Over all strain field distribution, 2-D blank and heating line pattern for pillow patch

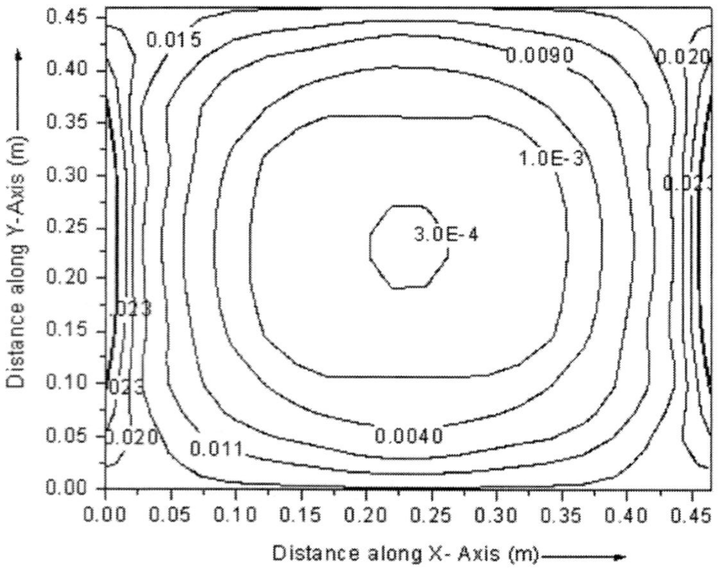

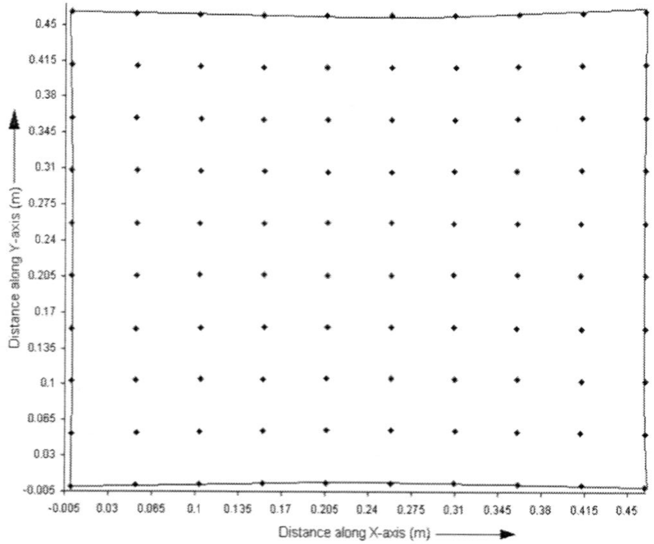

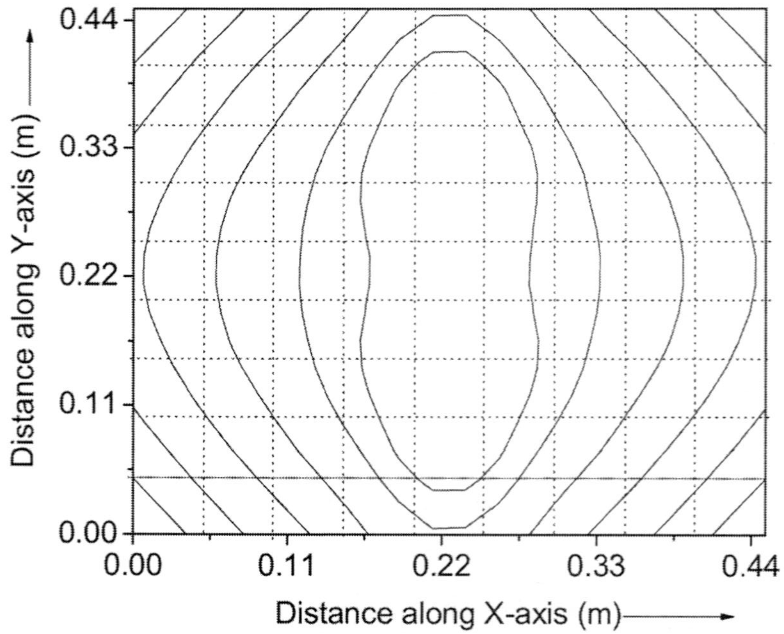

Experimentally measured and developed pillow patch shape

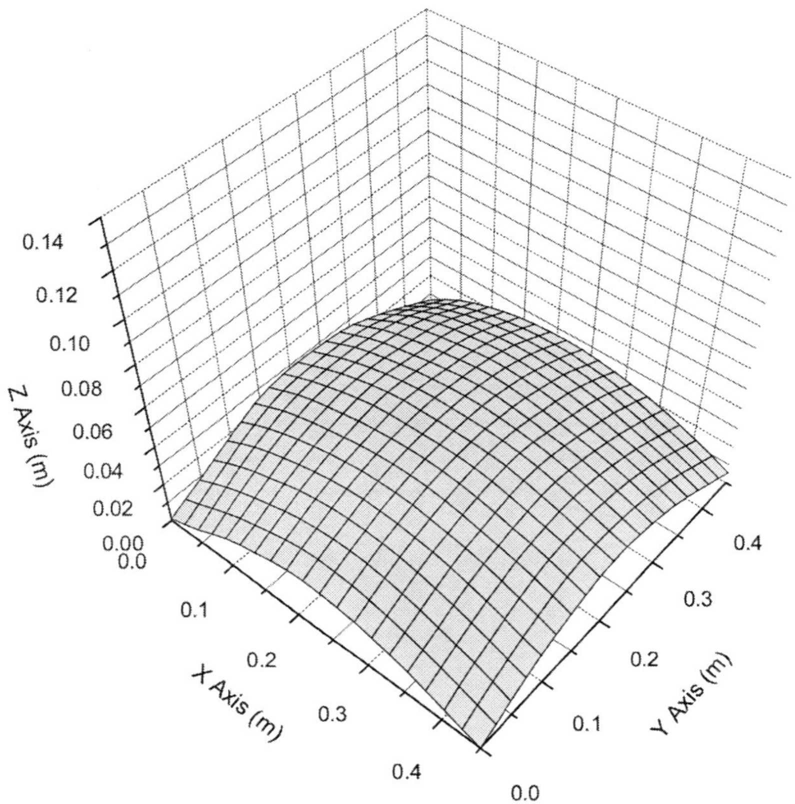

Position of surface patches (SP1 and SP2) in ship

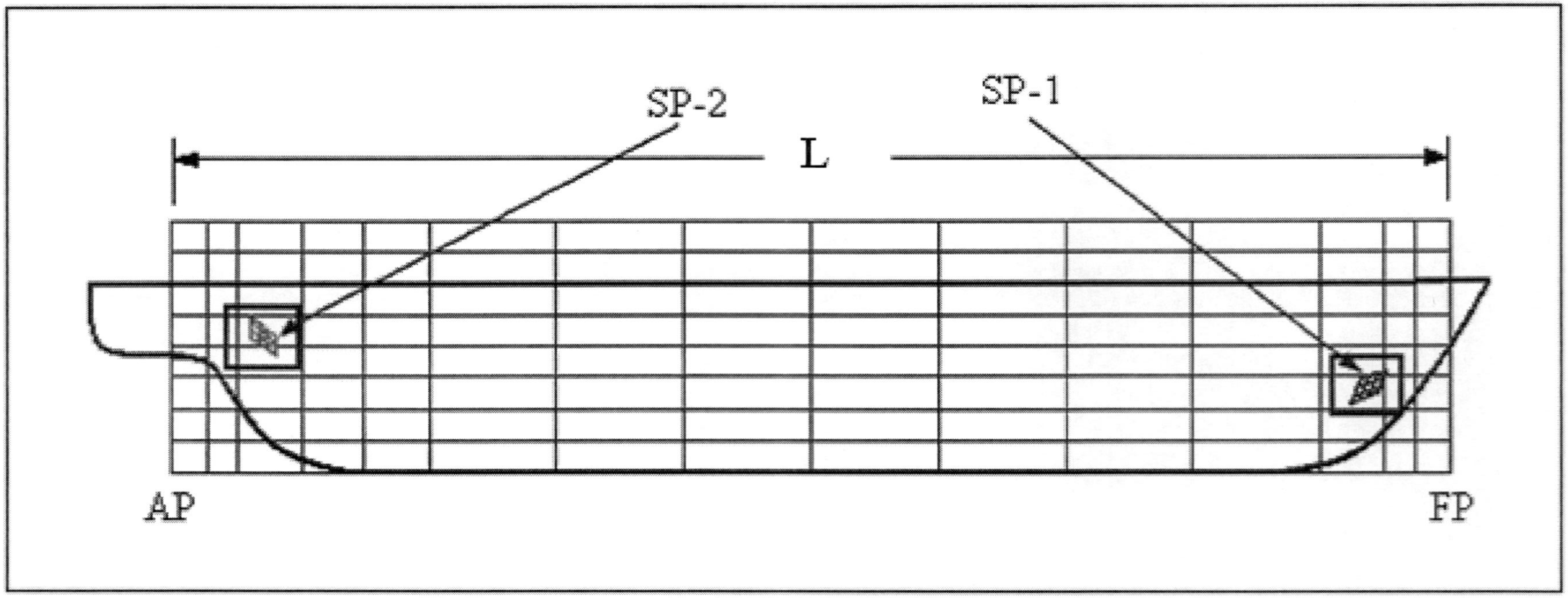

Co-ordinate for patch SP-1 and generated surface patch SP-1

X (m)	Y (m)	Z (m)
0	0	0
0.15	0	-0.004
0.3	0	-0.008
0.45	0	-0.0093
0	0.15	0.0278
0.15	0.15	0.0264
0.3	0.15	0.0223
0.45	0.15	0.0261
0	0.3	0.0175
0.15	0.3	0.0209
0.3	0.3	0.0287
0.45	0.3	0.0419
0	0.45	-0.004
0.15	0.45	0.0071
0.3	0.45	0.021
0.45	0.45	0.0371

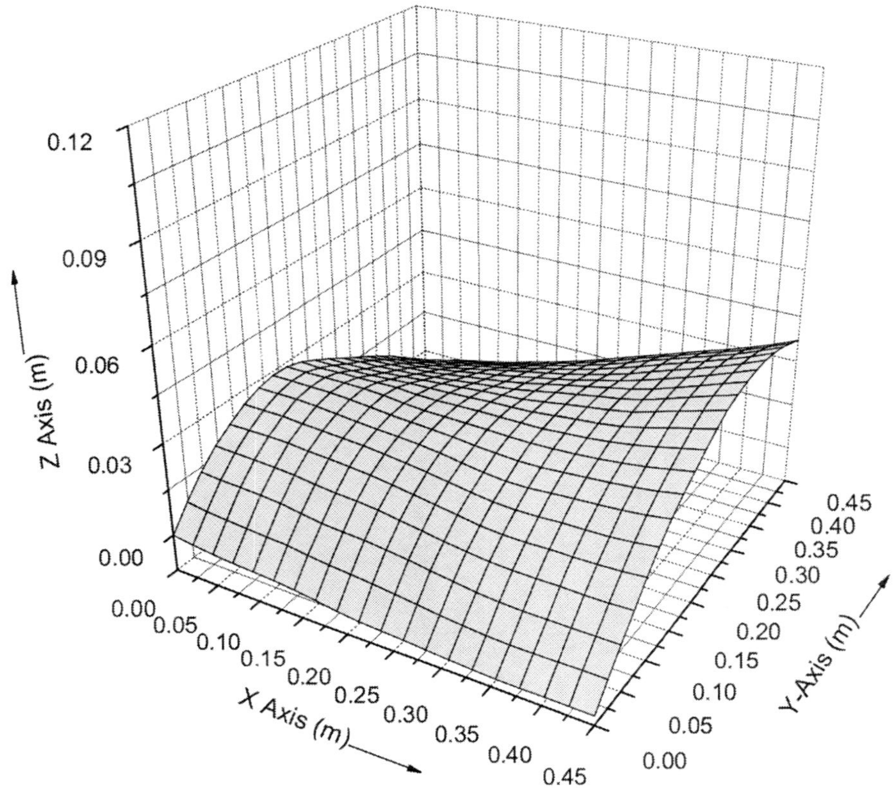

Over all strain field distribution, 2-D blank and heating pattern for pillow patch surface patch 1 (SP1)

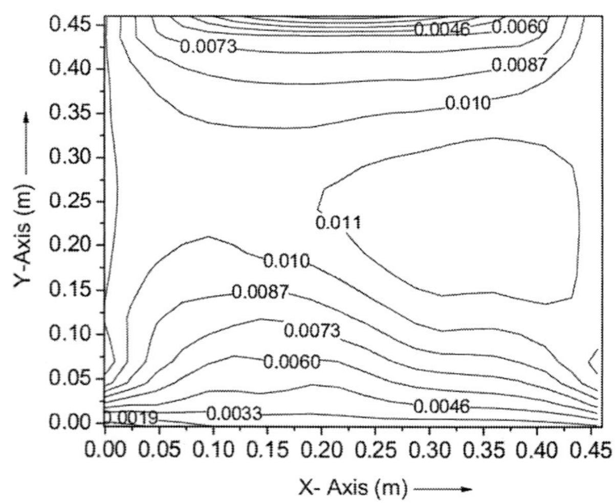

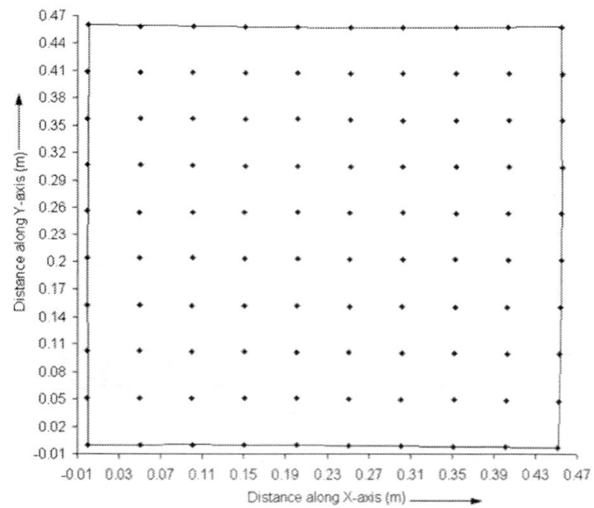

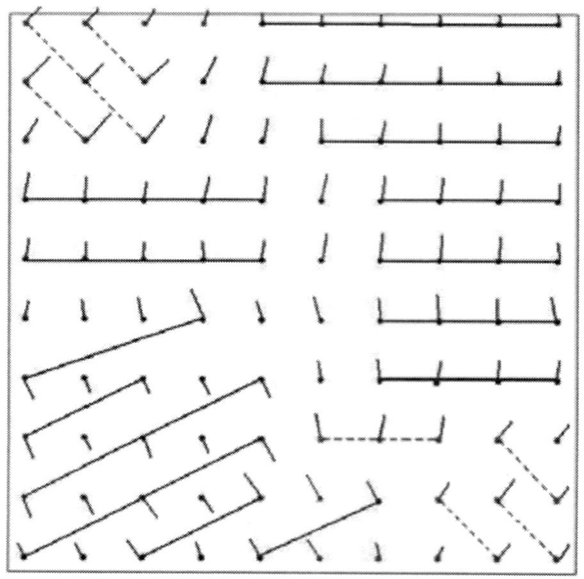

Experimentally measured and developed of surface patch 1

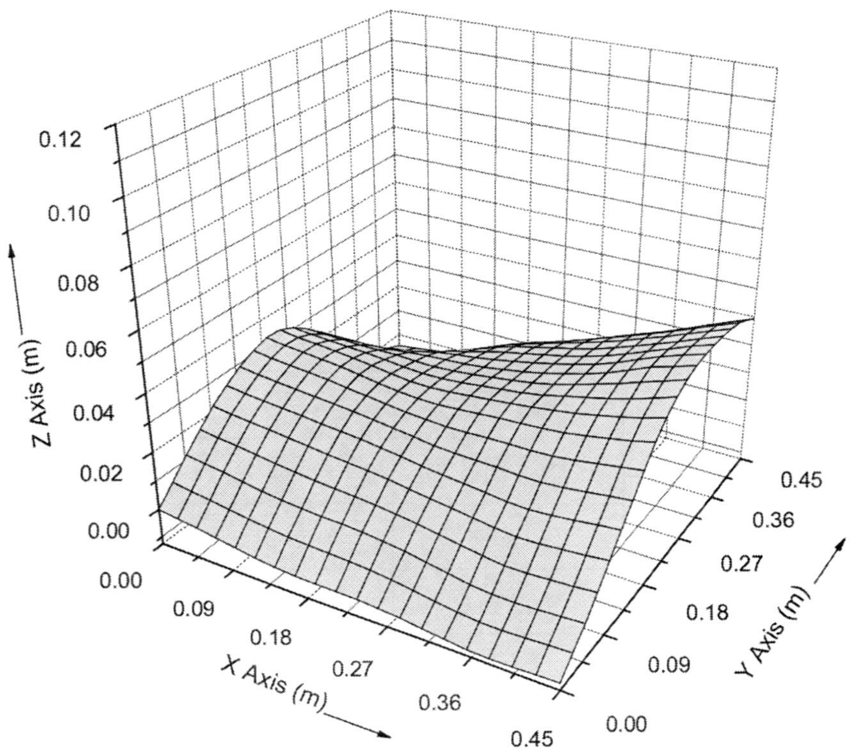

Co-ordinate for patch 2 (SP-2) and generated surface patch (SP-2)

X (m)	Y (m)	Z (m)
0	0	0
0.15	0	0.0176
0.3	0	0.00214
0.45	0	-0.0124
0	0.15	-0.0104
0.15	0.15	0.0153
0.3	0.15	0.0041
0.45	0.15	-0.0108
0	0.3	-0.0027
0.15	0.3	0.0031
0.3	0.3	-0.0082
0.45	0.3	-0.00205
0	0.45	0.00053
0.15	0.45	0
0.3	0.45	-0.0072
0.45	0.45	0

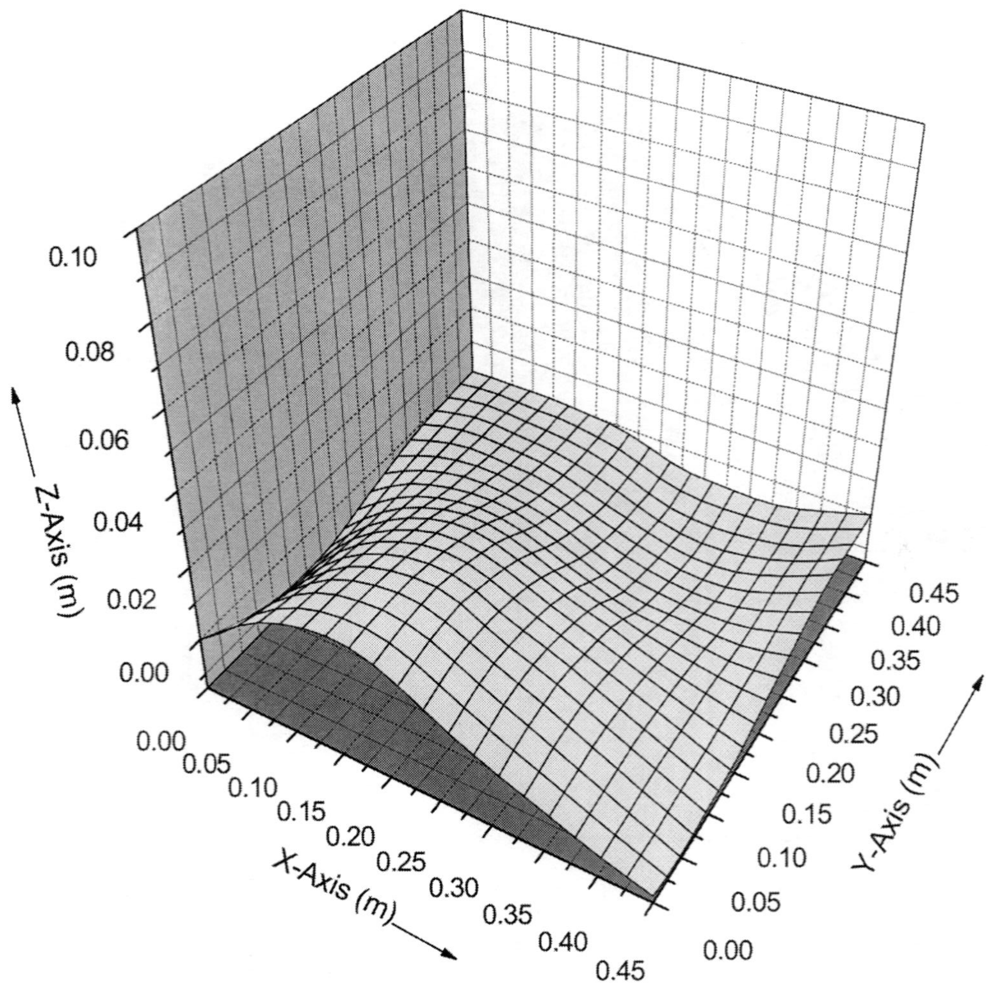

Over all strain field distribution, 2-D blank and heating pattern for pillow patch surface patch 2 (SP2)

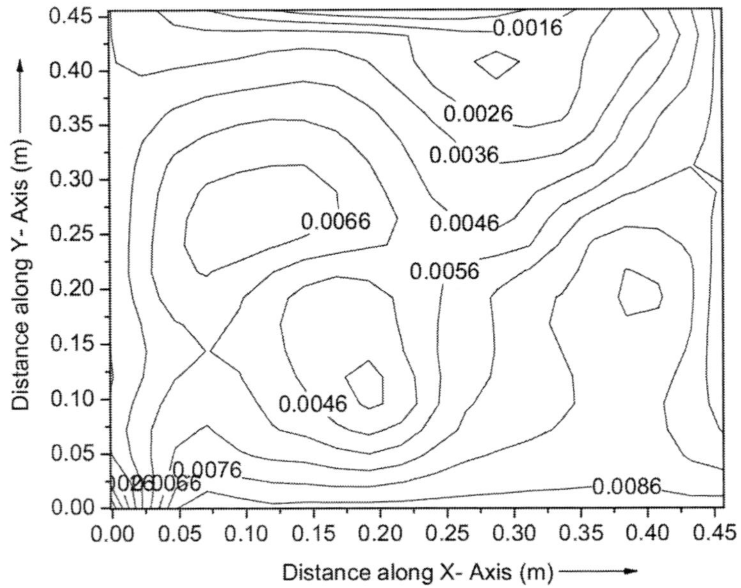

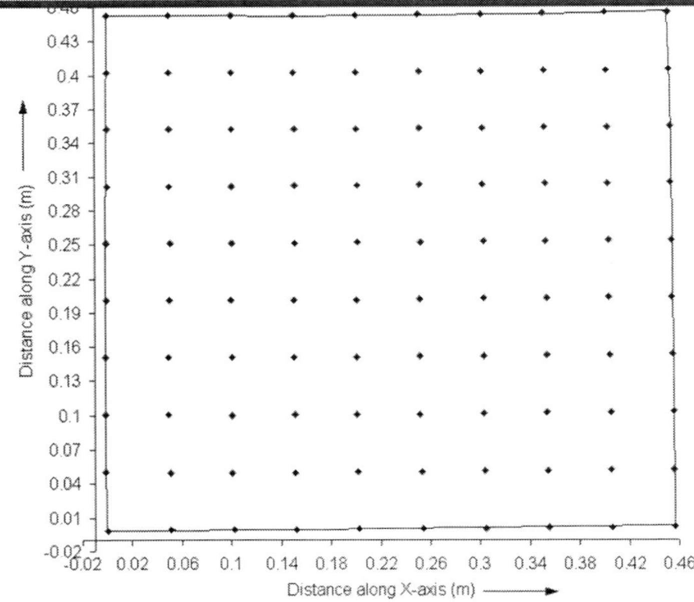

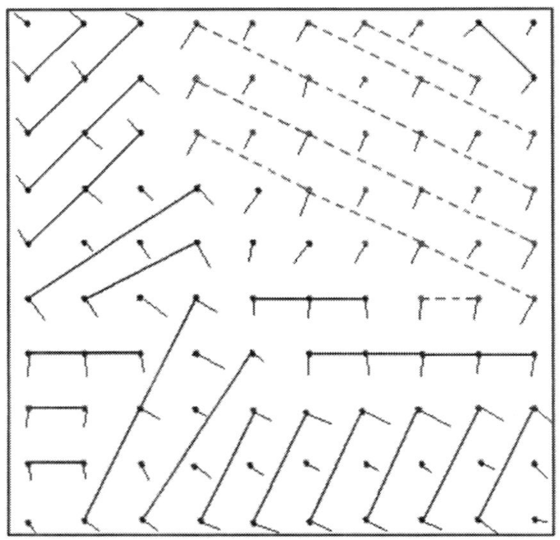

Experimentally measured and developed of surface patch 2

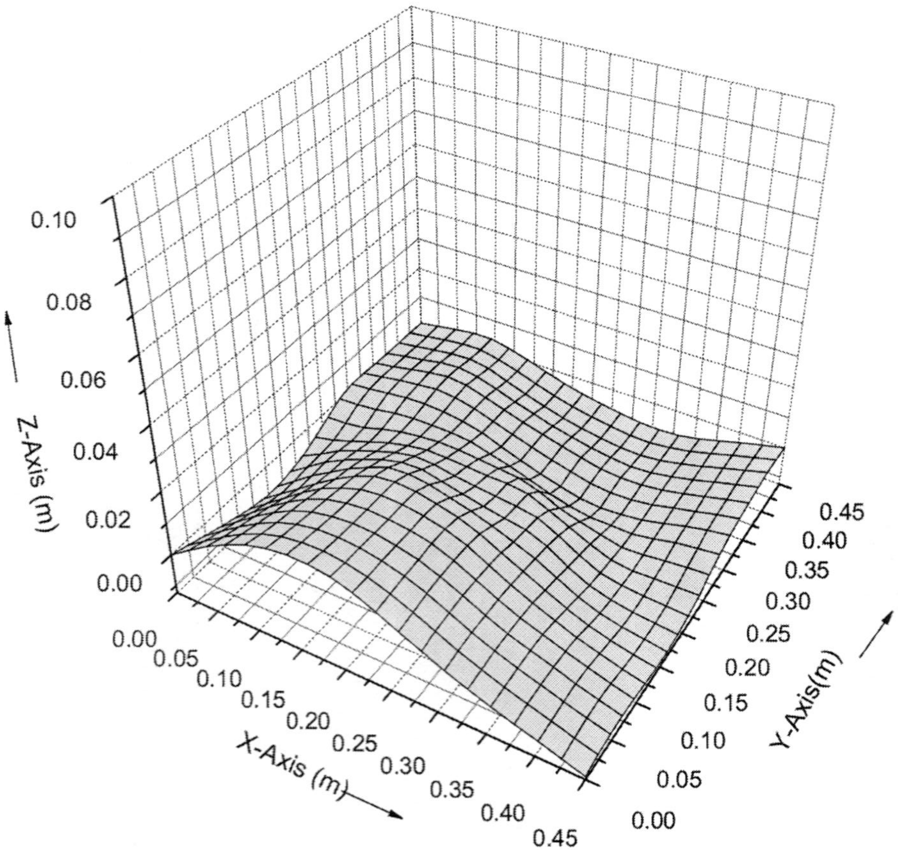

Generation of large compound curved surfaces:

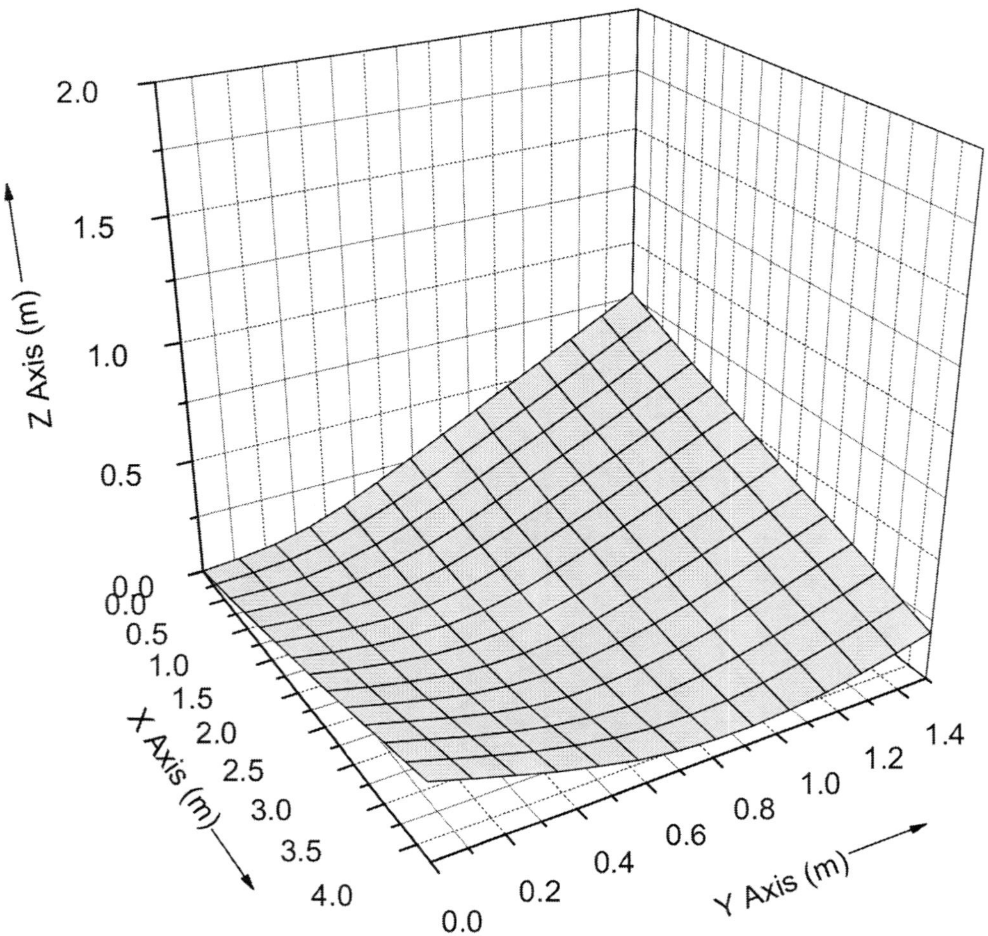

Fig. Theoretically developed 3-D Shape of surface patch

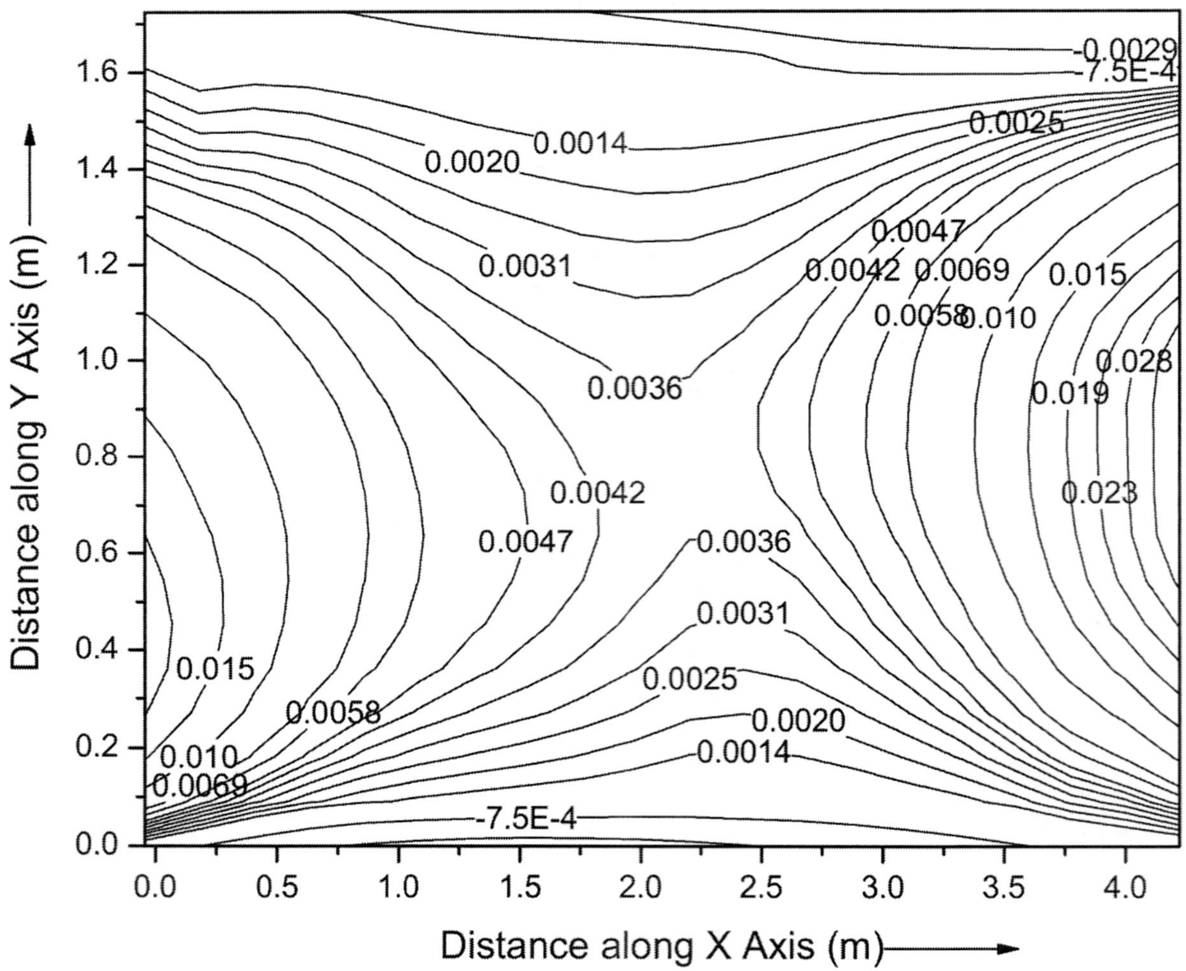

Fig. Strain field distribution of surface patch

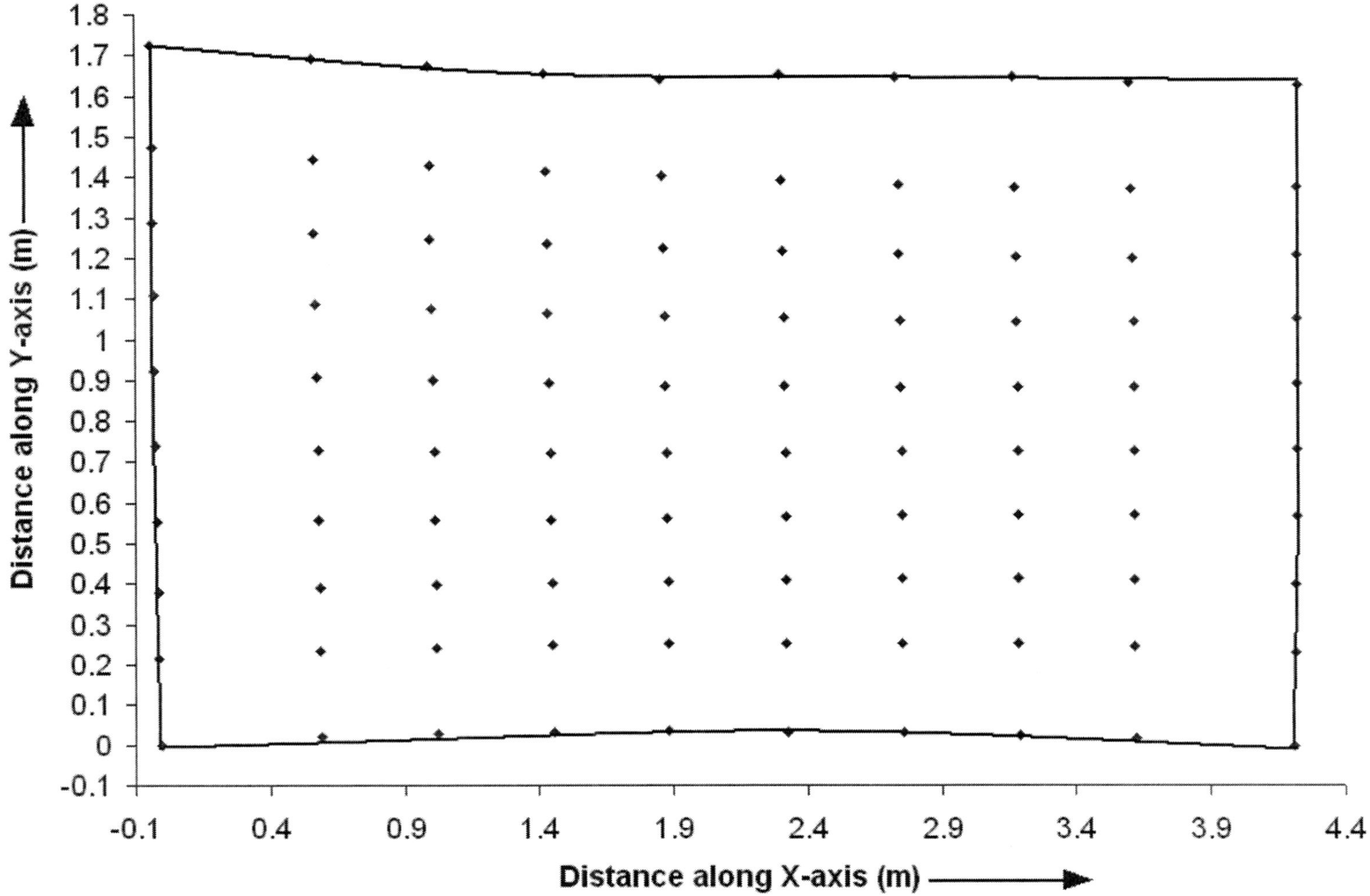

Fig.2-D Blank of surface patch

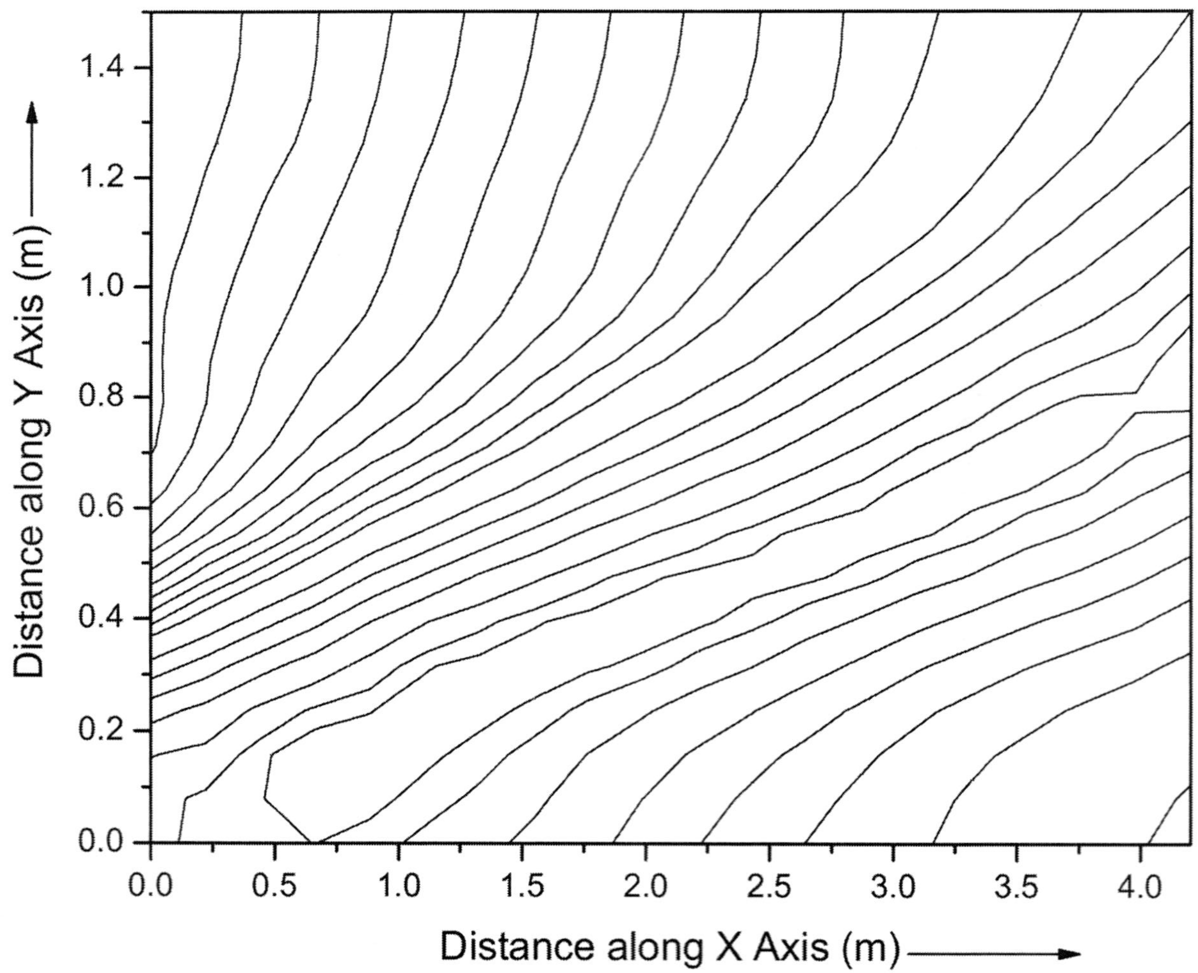

Fig. Line heating pattern for ship plate

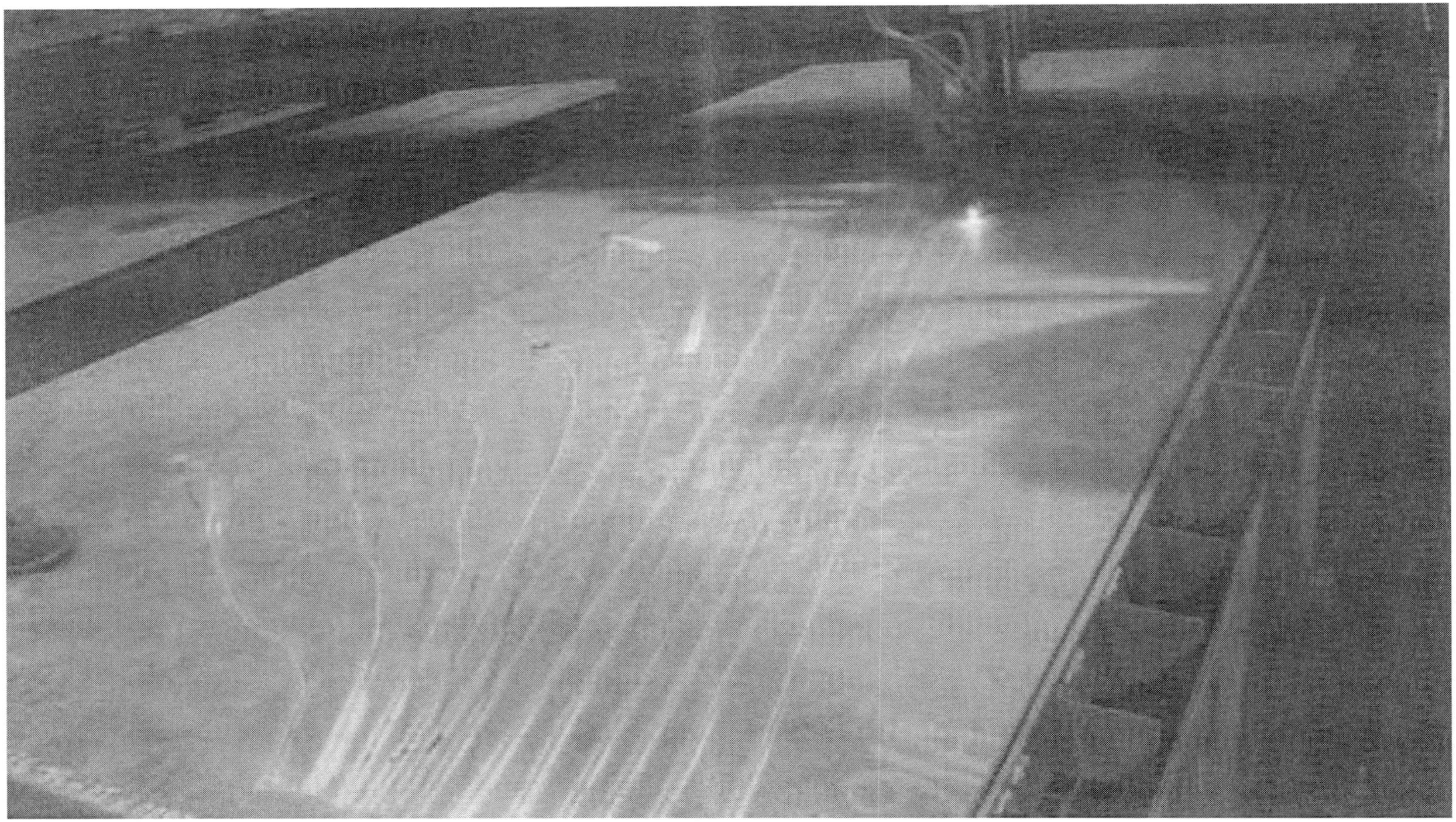

Fig. Heating line pattern on blank

Fig. Experimentally developed shape after first set of heating operation

Fig. Experimentally developed shape after 2nd set of heating operation

Conclusions

- For development of any type of compound curved surfaces B-Spline surface theory was successfully applied.

- The surface properties i.e. Gauss and principle curvatures and direction of principle curvatures of a surface patch were derived for generating the pattern of heating lines.

- The numerical method was successfully developed by considering the boundary strain obtained from optimized strain field distribution for generating the overall strain field distribution and development of the 2-D blank of a given 3-D patch.

- Experiments were conducted to produce the desired compound curved surfaces. The plate shapes thus generated compared well as the plate shape is concerned, however with respect to the magnitude of deformation the deviation was within 5% to 15%.

- The computation methodology developed in this study for production of large compound curved surfaces opens up a very promising possibility towards automating the production process of compound curved panels for ship hulls.

NAMASKAR

RELIABILITY BASED SHIP DEISGN

S V Khair, Indian Institute of Technology, & Cybermarine Knowledge System Pvt. Ltd., and
Dr. A Bhar Indian Institute of Technology, Kharagpur

SUMMARY

The main objective of ship structural design is to ensure the safety, functional and performance requirement of the structural element of ship for specified time period under specified environment. And this must be accomplished under the condition of uncertainty related to the loads especially wave loads, and strength of the structure. The reliability based design methods allows for the uncertainties in the design variables (both in loading and structural resistance) to be explicitly taken into account and so provide an explicit frame work for establishing safety levels. The main objective of this project is to provide an overview of the reliability based design method and how it can be used for ship design. During the course of the study the different failure modes of the important structural elements of ship; hull girder, stiffened panel and unstiffened panel are studied and limit state equation is written for each failure mode. The results are expressed in partial safety factor format (LRFD) for future use in design.

NOMENCLATURE

a	=	Length or span of plate; the length or span of the panel between transverse webs
b	=	Width of plate; distance between longitudinal stiffeners
a/b	=	Aspect ratio of plate
l	=	Stiffener span
s	=	Stiffener spacing
t	=	Plate thickness
t_p	=	Thickness of attached plating
t_w	=	Web thickness
A_f	=	Faceplate cross-sectional area
A_s	=	Cross-sectional area of the stiffener without attached plate
E	=	Young's modulus
M_{sw}	=	Still-water bending moment
M_{vw}	=	Vertical wave bending moment
M_u	=	Ultimate moment capacity
M_p	=	Plastic moment capacity
P_{st}	=	Still-water pressure
P_w	=	Wave-induced pressure
Z	=	Section modulus of mid-ship section
Z_s	=	Section modulus of the stiffener with attached plating
β_o	=	Target Reliability index
$\gamma_{M_{sw}}$	=	Load factor for still-water bending moment
$\gamma_{M_{vw}}$	=	Load factor for vertical wave-induced bending moment
$\gamma_{P_{sw}}$	=	Load factor for still-water pressure
$\gamma_{P_{vw}}$	=	Load factor for wave-induced water pressure
σ_E	=	Euler Buckling stress
σ_Y	=	Yield stress of the material
σ_{cr}	=	Critical buckling stress

1. INTRODUCTION

A ship structure is a complex, thin stiffened shell structure subjected to random loads resulting from the action of winds, waves, current, ice etc. The design of ship structure depends upon these random loads and structure capacity to resist them. There are many uncertainties to be dealt with, in design of marine structure. First there is uncertainty of the loads, especially those arising from waves. Second, there are uncertainties regarding material properties, such as yield stress, fatigue strength, notch toughness, and corrosion rate. Third, there will be always some degree of uncertainty in the analysis of the structure as complex as ship, as analysis and mathematical modeling of the physical environment necessarily involve assumptions, approximations, and idealizations. Fourth, there can be variations and hence uncertainties in the quality of construction and these factors also have a significant influence on the strength of a structure. Finally, there are uncertainties of operations, such as operating errors resulting from human action. Since there are always uncertainties, and hence some risk of failure, and it is impossible to make structure absolutely safe. Instead it can be made only "sufficiently safe" which means risk can be brought down to the acceptable level. Historically theses uncertainties are compensated by safety factor in deterministic design methods. This safety factor is based on experience and subjective judgment. The main shortcoming of these deterministic methods is that it fails to give the indication of level of safety or probability of failure. In recent times, with increased emphasis on safety of structure at sea, it's become necessary to account uncertainties more rationally. The reliability based design methods are essential tools for considering the uncertainties more rationally. The reliability based design methods account for the uncertainties with statistical basis.

This paper aims at giving an overview of how the reliability based design methods can be used for deigning of main structural elements of the ship; girder, stiffener and unstiffened panel. The limit states for different failure modes of these structural elements of ship are studied and design equation for each failure mode is produced in Load and Resistance Factor Design (LRFD) format. The Load and resistance factors which are also known as partial safety factors (PSF) to be used for designing of structural elements are computed for range of Target reliability index β_O

2. STRUCTURAL RELIABILITY THEORY

The most common definition of the reliability of structure is, "*the probability of a structure performing its intended function over specified period of time and under specified operating conditions.*" [1]

Thus the reliability of any structural component with respect to a given mode of failure is defined as the probability that it will not fail:

Reliability (β) = 1 – Failure Probability
(1)

There are various methods mentioned in different literatures [1, 5] for evaluation of the Reliability and Failure probability of the structure. These methods are broadly classified based on the amount of information required and difficulty level as, Level- I, Level- II and Level- III methods.

The Level-I methods are the simplest of all the reliability methods and also known as *First Moment Methods*. In these methods the uncertainties are accounted by means of safety factors, whose values are predetermined probabilistically using Level – II methods. Load and Resistance Factor Design (LRFD), and Partial Safety Method comes under Level-I category. The Level – II methods are bit more complex as compared to Level – I methods, and used simplified probabilistic theory to evaluate the probability of failure and reliability of structure. In these methods, the various random variables are represented by their first and second moments and hence theses methods are also known as, *Second Moment Methods*. These methods are used for calculation of partial safety factors (PSF) to be used in level-I methods, and for evaluation of reliability of structure in direct reliability based design. In both cases the acceptable level of risk (Target reliability index) and correlation between different basic variables must be defined beforehand. The different Level- II methods are as follows;

- Cornell Safety Index Method (FOSM)
- Hasofer and Lind Safety Index Method (FORM)
- Second Order Reliability Method (SORM)

Level-III methods are the most complex and rigorous methods, which utilized the complete probability distribution, function of all the random variables to calculate the probability of failure. Because of the complexity and economical constraints these methods are not normally used in design of ship. The more about these methods can be read in references [1, 5, and 15]

The basics steps involved in the direct reliability based design of structure are as follows;

1. Identification of all the possible modes of failure and selection of a failure criterion for each mode of failure.
2. Definition of the limit state functions associated with the identified failure modes.
3. Calculation of the various loads applied on the structure. (e.g. static loads, transient loads, low and high frequency loads, vibrational loads, impact loads, and residual stress etc.)
4. Structural response analysis for determination of load effects. (e.g. stresses and deformations)
5. Selection of load and load effect combinations.
6. Identification and statistical modeling of uncertainties related to the random variables associated with strength and load effects.
7. Selection of target reliability index β.
8. Structural reliability analysis (e.g. calculation of the safety index β or probability of failure P_f) and comparison with the target safety index β_o.
9. Calculation of partial safety factors (PSF)
10. Sensitivity analysis.

3. SHIP STRUCTURAL RELIABILITY ANALYSIS

For reliability based design of ship components like hull girder, stiffened panel between bulkheads or web frames and unstiffened plates between stiffeners generally three types of failure modes are considered; primary, secondary and tertiary. The limit states associated with these three failure modes are classified as the serviceability limit states and ultimate limit states depending upon the type of failure mode.

3.1 HULL GIRDER

Hull girder failure modes consist of the fully plastic mode, the initial yielding mode and collapse modes. The collapse mode includes buckling and post-buckling strength of the hull and is always the governing mode of failure. The fully plastic mode provides an upper bound on the ultimate strength, which is never attained in a hull of normal configurations. The initial yield mode assumes that buckling does not occur prior to yielding, which is a function of the standard elastic section modulus of hull girder and yield strength of the material, normally used in current design practice. Their limit states can be expressed as following,

3.1.1 Initial Yielding of Hull Girder

The limit state for initial yielding failure of hull girder is given as;

$$M = g(X) = ZX_1 - M_{sw,0}X_2 - M_{vw,0}X_3 \qquad (2)$$

Where X_1, X_2 and X_3 represent the model uncertainty in predicting the section modulus of hull girder, SWBM and VBWM respectively. The values of SWBM $M_{sw,0}$ and VBWM $M_{vw,0}$ are calculated as per the unified requirements of Longitudinal Strength of Classification societies [16]. And the model uncertainties associated with random variables are accounted as per the information taken from Table 2 and 3. Then using the FORM the partial safety factors are evaluated for the range of the Target reliability index as mention in Table 4, 5 and 6 for this type failure

3.1.2 Fully Plastic Failure Mode of Hull Girder

The limit state for fully plastic failure of hull girder is given as;

$$M = g(X) = M_p X_4 - M_{sw,0}X_2 - M_{vw,0}X_3 \qquad (3)$$

Where X_4, X_2 and X_3 represent the model uncertainty in predicting the plastic moment capacity of hull girder, SWBM and VBWM respectively. The value of Plastic moment capacity of hull girder can be evaluated using the formula given in [12]

3.1.3 Collapse Failure of Hull Girder

The limit state for collapse or ultimate failure of hull girder is given as;

$$M = g(X) = M_u X_5 - M_{sw,0}X_2 - M_{vw,0}X_3 \qquad (4)$$

Where X_5, X_2 and X_3 represent the model uncertainty in predicting the ultimate moment capacity of hull girder, SWBM and VBWM respectively. The value of ultimate moment capacity of hull girder can be evaluated using the formula given in [12]

3.2 STIFFENED PANEL

Stiffened panels are the most commonly used structural element in a ship; appearing in decks, bottoms, bulkheads, and side shell. The primary purpose of the panel is to absorb out of plane (or lateral) loads and distribute those loads to the ship's primary structure. It also serves to carry part of the longitudinal bending stress because of the orientation of the stiffeners. The amount of in-plane compression or tension experienced depends primarily on the location of the panel. Deck panels tend to experience large in-plane compression and small lateral pressures. Bottom panels experience large in-plane tension and compression, but usually with very significant lateral pressures.

The stiffened panels can be further subdivided as the gross panel and stiffened sub-panel. Gross panel is a panel of plating which has stiffeners running in two orthogonal directions. This panel is bounded by other structure, which have significantly greater stiffness in the planes of the loads when compared to the panel and its stiffeners. These boundaries would be provided by structure, such as transverse bulkheads, longitudinal bulkheads, side shell, or large longitudinal girders. The stiffened sub-panel is panel having stiffeners running in one direction and bounded on each end by a transverse structure which has significantly greater stiffness in the plane of the lateral load. The sides of the panel are defined by the presence of a large structural member which has greater stiffness in bending and much greater stiffness in axial loading. Structural members such as keels, bottom girders, longitudinal bulkheads, deck girders, etc., can act as the side boundaries of the panel.

The collapse of a gross panel can be prevented by choosing the size of the transverse stiffeners so that they provide sufficient flexural rigidity and act as support to the smaller longitudinal stiffener to enforce nodes at the location of the transverse stiffeners. If the transverse stiffeners act as nodes, then the collapse of the stiffened panel is controlled by the strength of the longitudinally stiffened sub-panel. For stiffened sub-panel three important failure modes are there [8];

Mode – I : Compressive collapse of the stiffener flange
Mode – II ; Compressive collapse of plate flange
Mode – III ; Combined compressive collapse of the plate flange and tensile failure of the stiffener flange.

In case of the ship Mode – I failure occurs in longitudinally stiffened sub-panel and Mode – II failure occurs in deck, Bottom and side shell plating sub-panel. The limit state equation for Mode – I failure is given as [15];

$$g(x) = \sigma_Y' - \sigma_n - \frac{M_{max}}{Z_s} \qquad (5)$$

Where σ_Y' and σ_n represents the equivalent yield stress and hull girder stress respectively. And M_{max} is the induced bending moment under the action of lateral pressure is given as;

$$M_{max} = \phi \frac{psl^2}{m} \qquad (6)$$

Substituting the values σ_Y' and σ_n in above limit state equation,

$$M = g(X) = Z_s X_1' - \frac{Z_s}{Z}\left(M_{sw,0}X_2 + M_{vw,0}X_3\right) - X_2' \frac{\left(P_{st,0}X_4 + P_{w,0}X_5\right)sl^2}{m} \qquad (7)$$

Table 1. For axially and laterally loaded stiffener

	Axial Compressive Stress	
	Simply Supported Stiffener	Fixed Stiffener
Pressure	Acting on the stiffener side	Acting on the plating side
σ_Y'	σ_Y	$\sigma_Y\sqrt{1-3(\tau/X_1)^2}$
τ	0	$\tau=\frac{psl}{2A_w}$
m	8	12
ϕ	$\frac{1}{1-\sigma_n/\sigma_E}$	$\frac{1}{1-0.18\sigma_n/\sigma_E}$

Where X_1, X_2, X_3, X_4, and X_5 represent the model uncertainty in predicting the section modulus of stiffener, SWBM, VBWM and still water and wave pressure respectively.

3.3 UNSTIFFENED PANEL

Unstiffened panel or plates are important components in ship structure, and the global failure of stiffened panel can be partially controlled by designing the strength of plate elements between stiffeners.

The mainly two types of loads, lateral load (water pressure) and the in-plane loads (axial compressive loads and shear loads), affect the strength of the plate. The effect of lateral pressure on the strength of plates subjected to in-plane is so complex that there are no simple models available to predict the strength of plates under these types of loading. Researchers has demonstrated that, the lateral pressure has negligible effect on the both the uniaxial and biaxial compressive strength when b/t is less than 50 (Becker et al. 1970). However, for values of the ratio b/t greater than 50, the lateral pressure can have a negative impact on the biaxial strength. Therefore, it is recommended to treat lateral pressure as an uncoupled load from other in-plane loads, and to design for them individually and separately [13]. For plates there are four limits states, serviceability and ultimate limit state for each type of load

3.3.1 Limit states for plate under the action of lateral pressure

The Serviceability Limit state equation for initial yielding of simply supported plate under action of lateral pressure is given as;

$$g(x)=\sigma_Y-\sqrt{k_1^2P^2\left(\frac{b}{t}\right)^4+k_2^2P^2\left(\frac{b}{t}\right)^4-k_1k_2P^2\left(\frac{b}{t}\right)^4} \quad (8)$$

$$g(x)=X_1-\left(P_{st,0}X_2+P_{w,0}X_3\right)\left(\frac{b}{t}\right)^2\sqrt{k_1^2+k_2^2-k_1k_2} \quad (9)$$

Where X_1, X_2, and X_3 represent the model uncertainty in predicting the yield stress of material, still water pressure and wave pressure respectively.

The Ultimate strength Limit state equation for collapse of simply supported plate under action of lateral pressure is given as ;

$$g(x)=\sigma_u-(P_{st}+P_w) \quad (10)$$

Where σ_u = ultimate strength of plate and it is computed using empirical relations given by Mansour (1986) and Hughe (1988). For reference relation given by Mansour is mentioned here;

$$\sigma_u=k_4\frac{\sigma_Y t^2}{ab} \quad (11)$$

k_4 = a coefficient that depends on the aspect ratio of plate (a/b, such that a>b) and its boundary conditions.

3.3.2 Limit states for plate under the action of In-plane loads

The different in- plane loads acting on the plate are; uniaxial compression, biaxial compression, shear and combined biaxial compression and shear. The buckling of plate under action of uniaxial compression is an failure mode for ship plating .

The serviceability limit state for buckling failure of plate under uni-axial compression is given as;

$$g(x)=\sigma_{cr}-\sigma_n \quad (12)$$

Where σ_{cr} and σ_n represent the critical buckling stress (Elastic) and hull girder stresses. And the critical buckling stress σ_{cr} is computed as;

$\sigma_{cr}=\sigma_E$ for $\sigma_E\le\sigma_Y/2$

$\sigma_{cr}=\sigma_Y(1-\sigma_Y/4\sigma_E)$ for $\sigma_E>\sigma_Y/2$

Where σ_E = Euler critical buckling stress and is given as;

$$\sigma_E=k_1\frac{\pi^2E}{12\left(1-\upsilon^2\right)}\left(\frac{b}{t}\right)^2 \quad (13)$$

The ultimate strength limit state for buckling failure of plate under uni-axial compression is given as;

$$g(x)=F_u-\sigma_n \quad (14)$$

Where F_u and σ_n represent the ultimate buckling strength of plate and hull girder stresses. And the ultimate buckling strength of plate is computed as given in [8, 13]

4. RESULTS & DISCUSSION

In preceding section the limit state equation for different failure modes of ship structural elements are mentioned. For some of these limit state equations direct reliability analysis is carried using FORM for example ship. An diving support vessel (*DSV*) is considered as the example ship, whose principal particulars and midship section are as follows;

L = 91.00 M,
B = 22.00 M,
D = 6.95 M,
T = 5.00 M,
Disp = 5780 T
and C_b = 0.61.

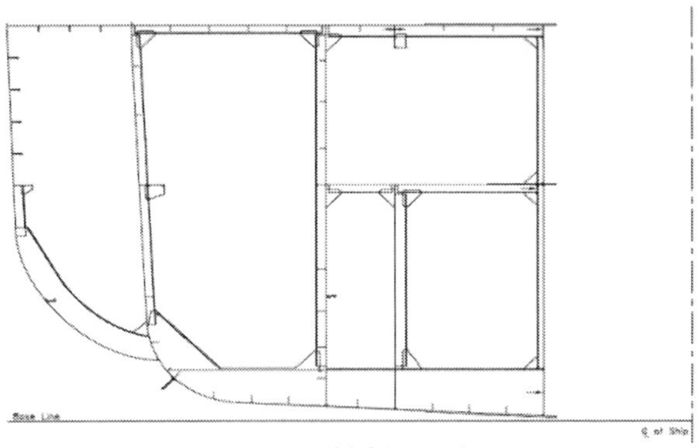

Figure. 1 Midship section

In the direct reliability analysis the uncertainty associated with load and strength variables are accounted using the information given in Table 1 and Table 2. And the range of the Target reliability for the particular failure mode is selected from information given in Tables 4, 5, 6 and 7. Form the direct reliability analysis the partial safety factors (PSF) are computed for each failure mode and range of target reliability index selected.

Table 2. Recommended Probabilistic characteristic of strength basic random variables (Assakkaf 1998 and Atua 1998) (From [12])

Variable	Nominal Value	Statistical Information		
		Mean	COV	Distribution Type
OS F_y	F_y	1.11 F_y	0.07	lognormal
HS F_y	F_y	1.22 F_y	0.09	lognormal
F_u	F_u	1.05 F_u	0.05	normal
E	E	1.04 E	0.02	normal
Z	Z_r	1.04 Z_r	0.05	lognormal
υ	0.3	0.3	0	---
M_y	F_yZ	$\bar{F}_y\bar{Z}$	0.15	lognormal
M_P	F_yZ_P	$\bar{F}_y\bar{Z}_P$	0.18	lognormal

Table 3. Recommended Probabilistic characteristic of Load random variables (Atua 1998) (From [12])

Variable	Distribution type	Mean to Nominal Ratio	COV
Still-water bending moment M_{sw}	Normal	0.4 to 0.6 for commercial ships, and 0.7 for naval vessels	0.3 to 0.9 for commercial ships, and 0.15 for naval vessels
Life-time Extreme wave-induced bending moment M_W	Largest extreme value (type I)	1.0	0.1 to 0.2
Whipping Bending Moment M_D	Extreme value (type I) Exponential	Mean value can be determined using formulae based on spectral analysis	0.2 to 0.3
Springing bending moment M_{SP}	Extreme value (type I)	1.0	0.3
Hydrostatic pressure due to Stillwater P_{sw}	Normal	0.4 to 0.6 for commercial ships, and 0.7 for naval vessels	0.15
Hydrostatic pressure due to waves P_W	Largest extreme value (type I)	1.0	0.15
Hydrostatic pressure due to dynamic effects P_D	Largest extreme value (type I)	1.0	0.25
Hydrostatic pressure due to combined wave and dynamic loads P_{WD}	Weiibull	1.0	0.25

Table 4. Target Safety Indices – Mansour's Proposal (From [15])

Failure Mode	Commercial Ships		Naval Ships	
	P_f	β_0	P_f	β_0
Primary (Initial yield)	2.9×10^{-7}	5.0	1.0×10^{-9}	6.0
Primary (Ultimate strength)	2.3×10^{-4}	3.5	3.2×10^{-5}	4.0
Secondary (Ultimate strength)	6.2×10^{-3}	2.5	1.4×10^{-3}	3.0
Tertiary (Ultimate strength)	2.3×10^{-2}	2.0	6.2×10^{-3}	2.5
Fatigue Very serious Serious Not Serious	 1.4×10^{-3} 6.2×10^{-3} 1.6×10^{-1}	 3.1 2.5 1.0	 1.3×10^{-4} 1.4×10^{-3} 6.7×10^{-2}	 3.5 3.0 1.5

Table 5. Target Safety Indices – Based on Seriousness of Failure (From [15])

Seriousness of failure	Safety		Serviceability	
	P_f	β_0	P_f	β_0
Extreme	1.0×10^{-3}	3.1	5.1×10^{-6}	4.5
Severe	6.2×10^{-3}	2.5	3.1×10^{-5}	4.1
Moderate	2.3×10^{-2}	2.0	1.2×10^{-4}	3.7

Table 6. Recommendations for range target reliability index for Hull girder failure (Atua 1998) (From [12])

Range	Reference
4.0 to 6.0 (Sagging)	Mansour et al (1995)
5.0 to 6.0 (Hogging)	Mansour et al (1995)

Table 7. Recommendations for range target reliability index for Stiffened panel Limit states (From [13])

Range	Reference
3.5 to 4.5	Mansour et al (1995)

4.1 HULL GIRDER

The design equation for initial yielding failure of hull girder, in LRFD format is;

$$Z \geq \frac{\gamma_{M_{SW}} M_{SW,0} + \gamma_{M_{VW}} M_{VW,0}}{\Phi_R (\sigma_Y)_{\min}} \qquad (15)$$

Where the values of partial safety factors corresponding to target reliability index can be selected from following graph;

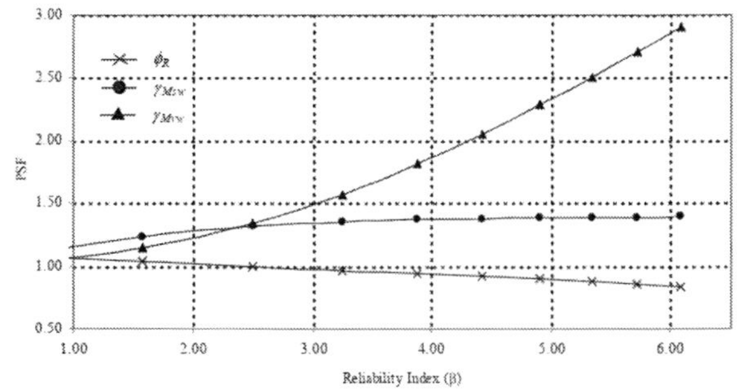

Figure 2. Partial Safety Factors for Initial yielding Failure of Hull Girder.

And the variation of the reliability index with section modulus for example ship is as follows;

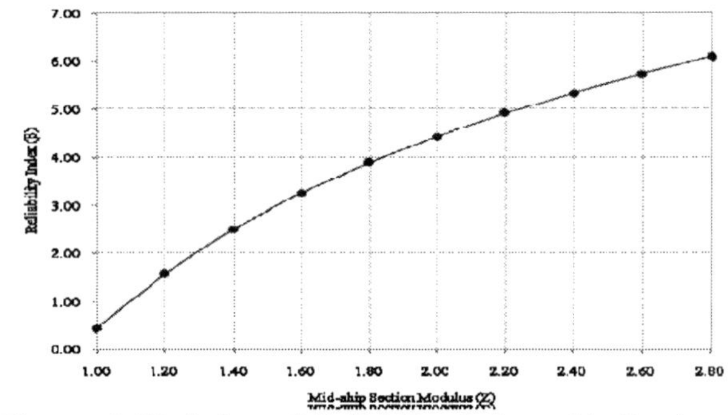

Figure. 3 Variation of Reliability Index with Section Modulus (Z)

4.2 STIFFENED PANEL

The design equation for initial yielding failure of stiffener flange (Mode – I), in LRFD format is;

$$Z_s \geq \frac{\frac{Z_S}{Z}\left(\gamma_{M_{SW}} M_{SW,0} + \gamma_{M_{VW}} M_{VW,0}\right) + \left(\frac{1}{1 - 0.18 \frac{\gamma_{M_{SW}} M_{SW,0} + \gamma_{M_{VW}} M_{VW,0}}{Z \sigma_E}}\right) \frac{\left(\gamma_{P_{SW}} P_{St,0} + \gamma_{P_{VW}} P_{W,0}\right) s l^2}{m}}{\Phi_R (\sigma_Y)_{\min}} \qquad (16)$$

Where the values of partial safety factors corresponding to target reliability index can be selected from following graph;

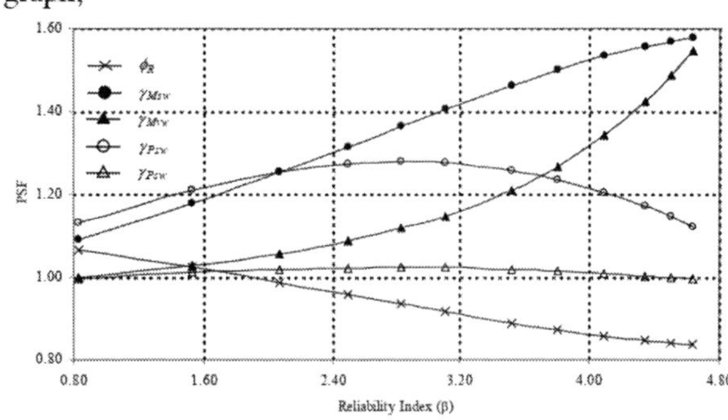

Figure 4. Partial Safety Factors for Initial Yielding Failure of Stiffener flange (Mode – I).

4.2 UNSTIFFENED PANEL

The design equation for initial yielding failure of unstiffened panel under action of lateral pressure, in LRFD format is;

$$\left(\frac{t}{b}\right)^2 \geq \frac{\left(\gamma_{P_{sw}} P_{st,0} + \gamma_{P_{vw}} P_{w,0}\right)\sqrt{k_1^2 + k_2^2 - k_1 k_2}}{\Phi_R(\sigma_Y)_{\min}} \qquad (17)$$

Where the values of partial safety factors corresponding to target reliability index can be selected from following graph

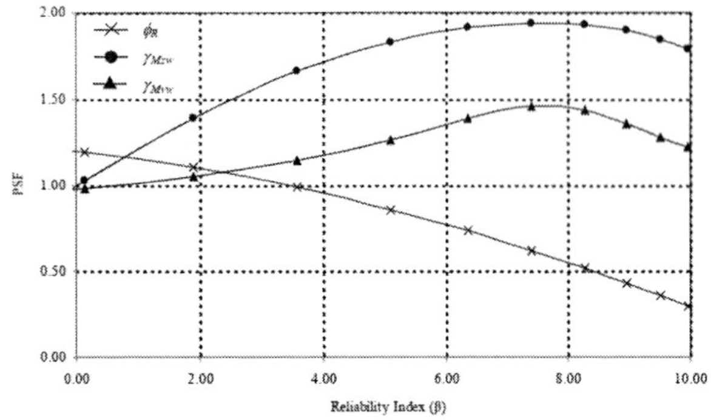

Figure 5. Partial Safety Factors for Initial Yielding Failure of Unstiffened Panel

The design equation buckling failure of unstiffened panel under action uniaxial compression, in LRFD format is;

$$\Phi_R(\sigma_Y)_{\min}\left(1 - \Phi_R(\sigma_Y)_{\min}/4\sigma_E\right) \geq \frac{\gamma_{M_{sw}} M_{sw,0} + \gamma_{M_{vw}} M_{vw,0}}{Z} \qquad (18)$$

Where the values of partial safety factors corresponding to target reliability index can be selected from following graph

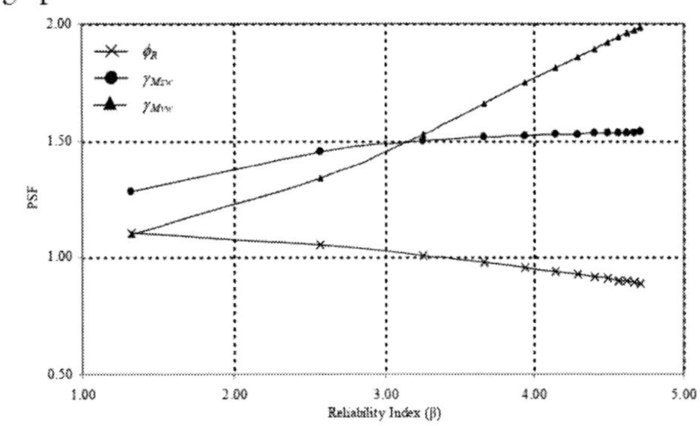

Figure 6. Partial Safety Factors for Buckling Failure of Unstiffened Panel

5. REFERENCES

1. Ranganathan, R. "Structural reliability Analysis and Design", *Jaico Publication House, Mumbai.* (1999)
2. Hughes, O.F., "Ship Structural Design, A Rationally-Based, Computer-Aided Optimization Approach" *The Society of Naval Architects and Marine Engineers, Jersey City, New Jersey.* (1988)
3. Timoshenko, S.P., and Goodier, J.N., "Theory of Elasticity", *McGraw- Hill.* (1951)
4. Timoshenko, S.P., and krieger, S.W.,"Theory of Plates and Shells", *McGraw- Hill.* (1951)
5. Mansor, A. "An Introduction to Structural Reliability Theory" *SSC-351, Ship Structure Committee, NTIS, Springfield, Virginia.* (1990)
6. Mansor, A., Lin, M., Hovem, L., and Thayamballi, A., "Probability- Based ship Design Procedures: A Demonstration" *SSC-368, Ship Structure Committee, NTIS, Springfield, Virginia.* (1993)
7. Mansor, A., and Thayamballi, A., "Probability Based ship Design: Loads and Load Combinations" *SSC-373, Ship Structure Committee, NTIS, Springfield, Virginia.* (1994)
8. Mansor, A., Wirsching, P.H., White, G.J., and Ayyub, B.M., "Probability Based ship Design: Implementation of Design Guidelines" *SSC-392, Ship Structure Committee, NTIS, Springfield, Virginia.* (1995)
9. Ayyub, B.M., Beach, J., and Packard, T., "Methodology for the Development of Reliability-Based Design Criteria for Surface Ship Structures," *Naval Engineers Journal, ASNE, 107(1), Jan. 1995, 45-61.*
10. Ayyub, B.M., Assakkaf., I.A., and Auta, K.I., " Reliability- Based Load and Resistance Factor Design (LRFD) of Hull Girders for Surface Ships " *Journal of Naval Engineers, Vol. 112, No.4, July* (2000).
11. Ayyub, B.M., Assakkaf., I.A., Beach, J., Metlton, W., Natale, N., and Colnley, J.A., "Methodology for the Development of Reliability- Based Load and Resistance Factor Design (LRFD) guidelines for Ship Structures," *Naval Engineers Journal, ASNE, 114(2).* (2002)
12. Ayyub, B.M., Assakkaf., I.A., Sikora, J., Adamchack, J., and Auta, K., " Reliability- Based Load and Resistance Factor Design (LRFD) guidelines for Hull Girder Bending," *Naval Engineers Journal, ASNE, 114(2).* (2002)
13. Assakkaf., I.A., Ayyub, B.M., Hess, P.E., and Atua, K.A., " Reliability- Based Load and Resistance Factor Design (LRFD) guidelines for Stiffened Panels and Grillages of Ship Structure," (2004)
14. Assakkaf., I.A., Ayyub, B.M., Hess, P.E., and Knight, D.E., " Reliability- Based Load and Resistance Factor Design (LRFD) guidelines for Unstiffened Panels of Ship Structure," (2004)
15. Dr. Dominique Beghin "Reliability- Based Structural Design", *Technical and Research Bulletin 2-32, SNAME.* ,(2007)
16. *RINA rules for the classification of ships-2008.*

RELIABILITY BASED SHIP DESIGN

Prepared By :- Shridhar Vinayak Khair (07NA6013)

Under guidance of :- Prof. Ashoke Bhar

Introduction

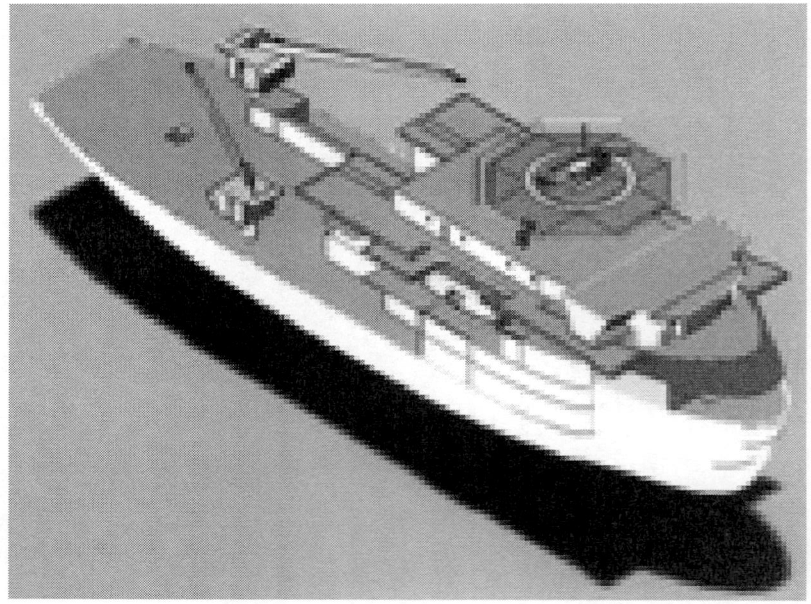

River Ro-Ro Ferry

LOA = 88.44 M

LBP = 86.00 M

B = 18.44 M

D = 6.95 M

Multi Support Vessel

LOA = 91.00 M

LBP = 86.00 M

B = 22.00 M

D = 5.00 M

General arrangement of converted MSV

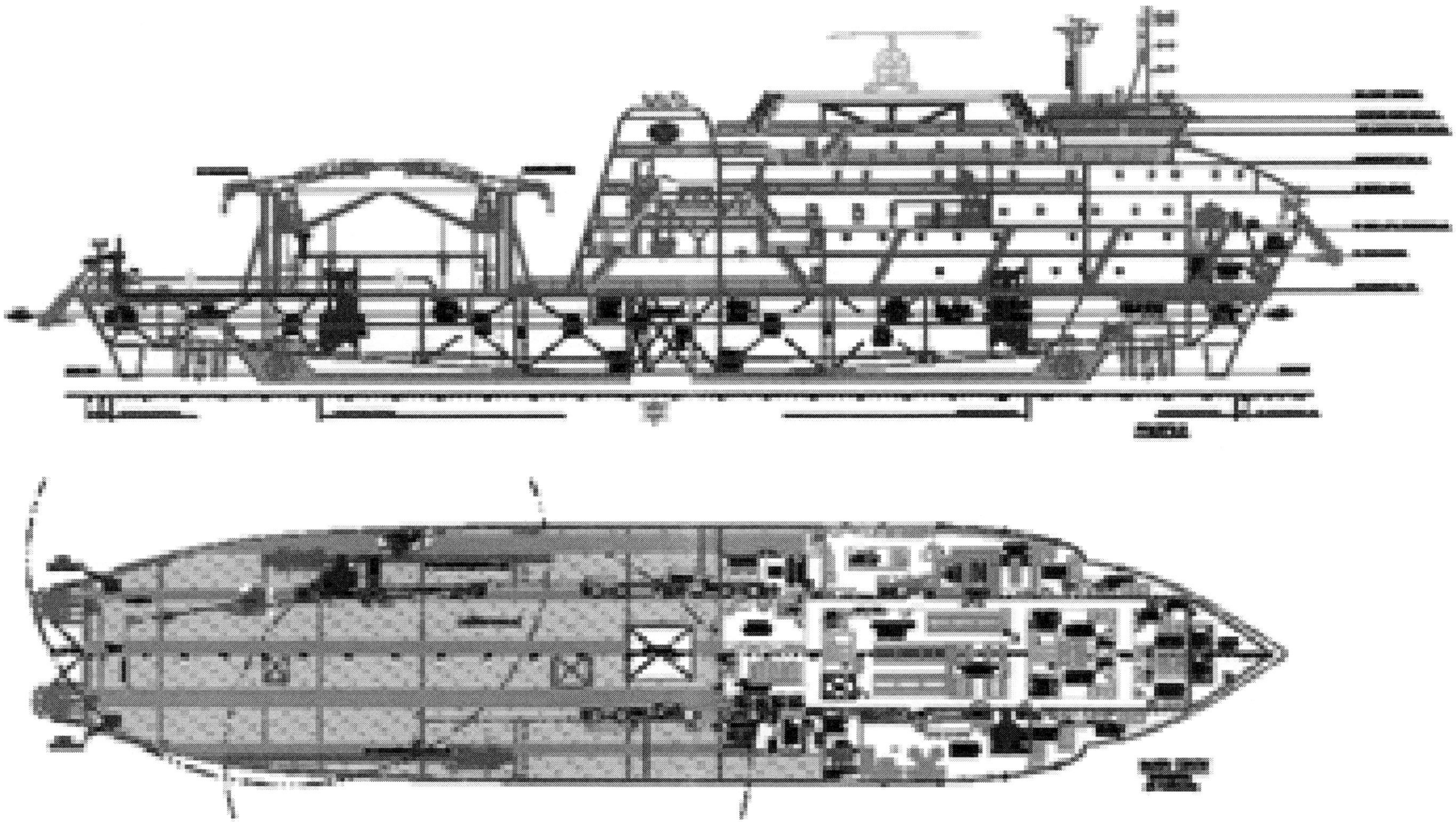

Main Features of converted MSV

- Diving support
- Accommodations for 250 persons
- Helicopter landing facilities
- Clear deck of strength 10 t/ m^2 for deck cargo

Main aspects of conversion

- Stability analysis
- Out fitting and engineering systems
- Powering analysis
- Structural analysis
  - Longitudinal strength
  - Increase in deck strength to 10 t/m^2
  - Structure iwo Moon pool
  - Helideck and supporting structure

Why Reliability based design methods ?

- Increase in complexity
- Advancement in science and technology
  - Better understanding wave
  - Improved methods of production
  - Improved materials
  - Large amount of available data
- Stringent safety norms
- Limitations of existing design codes
  - Inadequate account of uncertainties
  - Inconsistent level of safety

Why Reliability based design methods ?

- Limitations of existing design codes
  - Inadequate account of uncertainties
  - Inconsistent level of safety

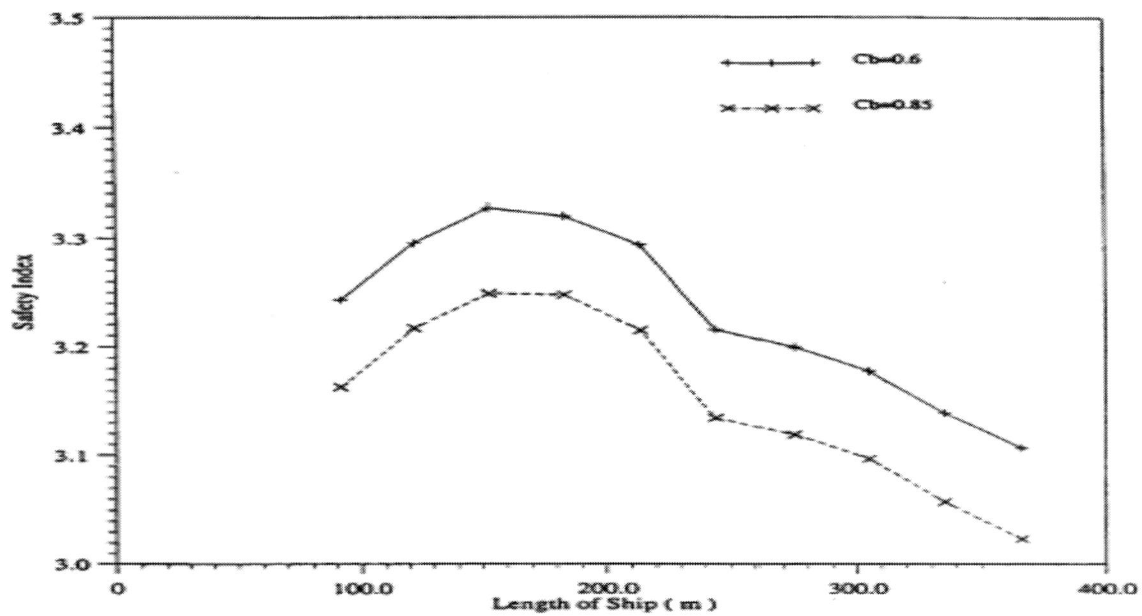

(Note :- Taken from SSC-368 Probability based ship design procedures: A demonstration)

Types of Reliability Methods

Level - I

- Simple
- Use Partial Safety Factor (PSF).
- Partial Safety Factor Format and LRFD

Level - II

- Use for calculation of Partial Safety Factor (PSF)
- Evaluation of reliability index (β) is based on the moments (mean , standard deviation) & distribution of random variables
- FOSM, FORM, and SORM

Level - III

- complex
- use complete probabilistic characteristic of random variables
- Space research

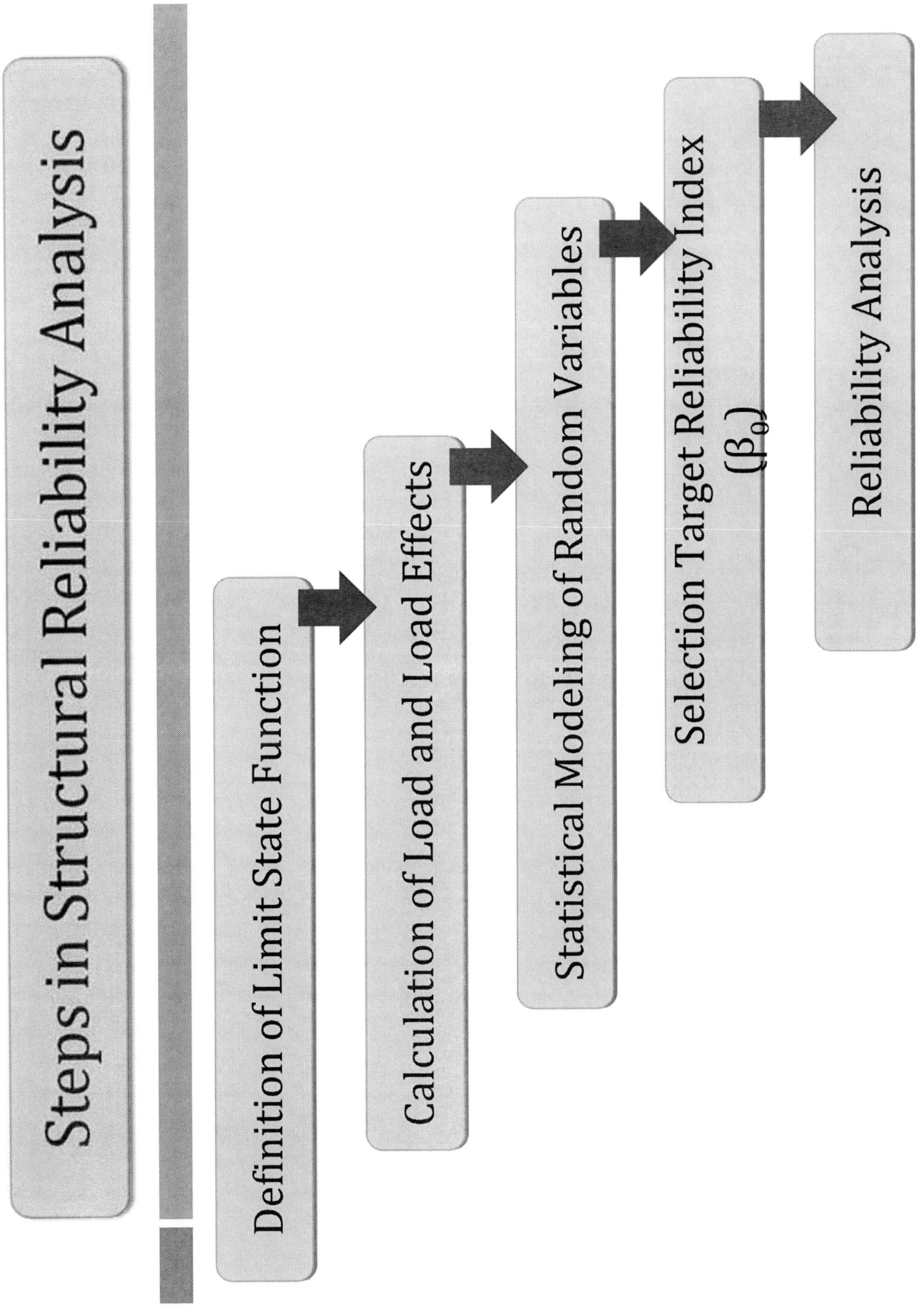

Steps in Structural Reliability Analysis
Definition of Limit State Function
Calculation of Load and Load Effects
Statistical Modeling of Random Variables
Selection Target Reliability Index (β_0)
Reliability Analysis

Limit State Function

- Identification of all the possible modes of failure
- Selection of a failure criterion for each failure mode
- Definition of the Limit state function

A limit state is defined as a condition for which a particular structural member or a complete structure is unable to perform the function for which it has been designed.

Types of Limit states

- Serviceability limit states
- Ultimate strength limit states
- Fatigue limit state
- Accidental limit states

Limit States for Hull Girder Failures

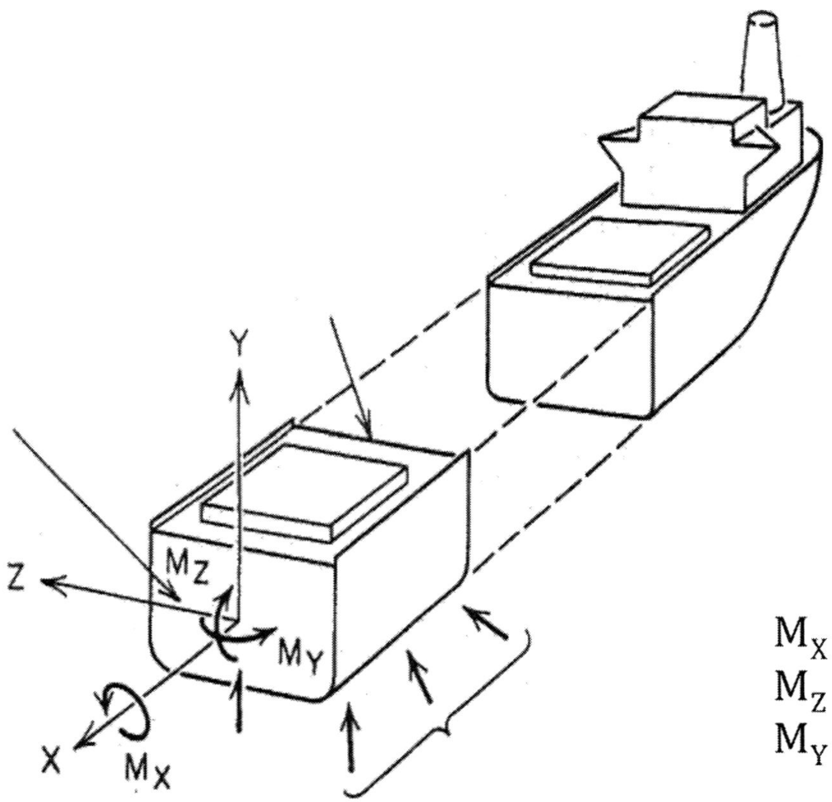

M_X = Torsional moment
M_Z = Vertical bending moment
M_Y = Horizontal bending moment

Hull Girder Model of a Ship (Hughes 1988)

Limit States for Hull Girder Failures

Initial yield failure of Hull girder	
Description	**Limit state**
Hull girder subjected to still-water bending moment M_{sw} and vertical wave bending moment M_{vw}.	$Z \geq \frac{M_{sw} + K_w M_{vw}}{\sigma_Y}$
Hull girder subjected to still-water bending moment M_{sw}, vertical wave bending moment M_{vw} and horizontal wave bending moment M_{hw}.	$Z \geq \frac{M_{sw} + K_w \sqrt{M_{vw}^2 + (Z/Z_H)^2 M_{hw}^2}}{\sigma_Y}$
Ultimate failure of hull girder	
Hull girder subjected to still-water bending moment M_{sw} and vertical wave bending moment M_{vw}.	$M_{vu,0} \geq M_{sw} + K_w \sqrt{M_{vw}^2 + M_{dw}^2}$
Hull girder subjected to still-water bending moment M_{sw}, vertical wave bending moment M_{vw} and horizontal wave bending moment M_{hw}.	$\left(\frac{M_{sw} + M_{vw}}{M_{vu}}\right)^{1.85} + \left(\frac{M_{hw}}{M_{hu}}\right) \leq 1$

Limit States for Stiffened Panel Failures

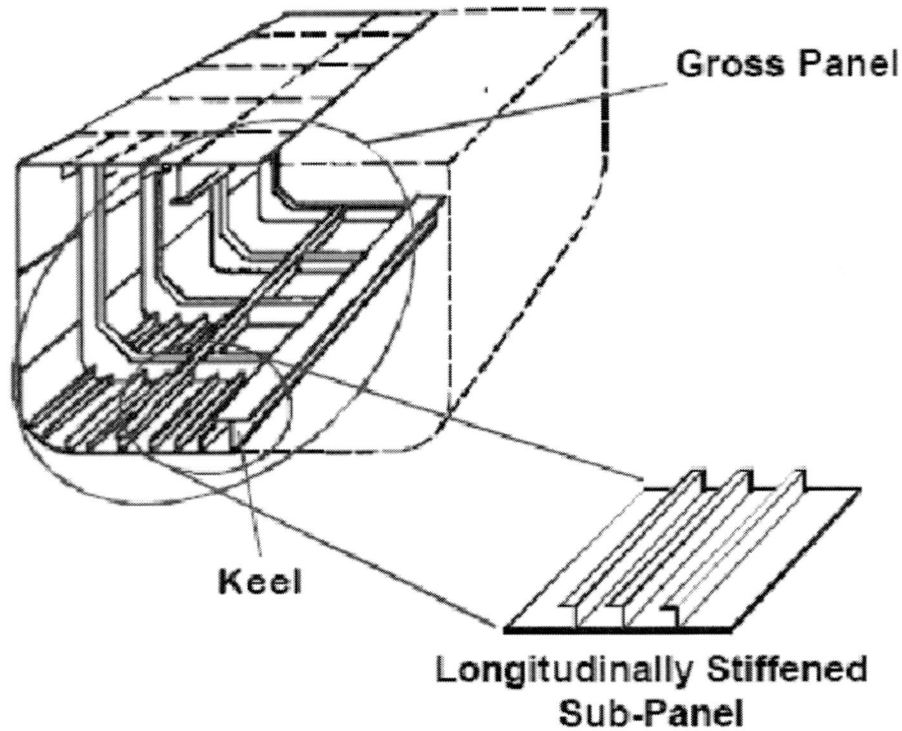

Portion of the Hull Girder Showing the Gross Panel and a Longitudinally Stiffened Sub-Panel (Hughes, 1988)

- Load effecting the strength of Stiffened panels
  - **Negative bending moment** – Load effect due to uniform lateral pressure which causes the plate to be in tension and the stiffener flanges to be in compression.
  - **Positive bending moment** - Load effect due to uniform lateral pressure which put the plating in compression and the stiffener flange in tension.
  - **In-plane Loads** – Axial loads, Shear loads

- Under these loads total 6 possible modes of failures are there, 3 for gross panel and 3 for longitudinal stiffened sub-panel

- **Three modes of failure for Gross panels are;**
  - Buckling of Gross panel
  - Biaxial compression under action of in- plane loads
  - Excessive plastic deformation under action of lateral pressure

- **Three modes of failure for Longitudinal Stiffened sub-panel;**
  - Mode I - Compressive collapse of the stiffener flange
  - Mode II - Compressive collapse of the plate flange
  - Mode III- Combined compressive collapse of the plate flange and tensile failure of the stiffener flange

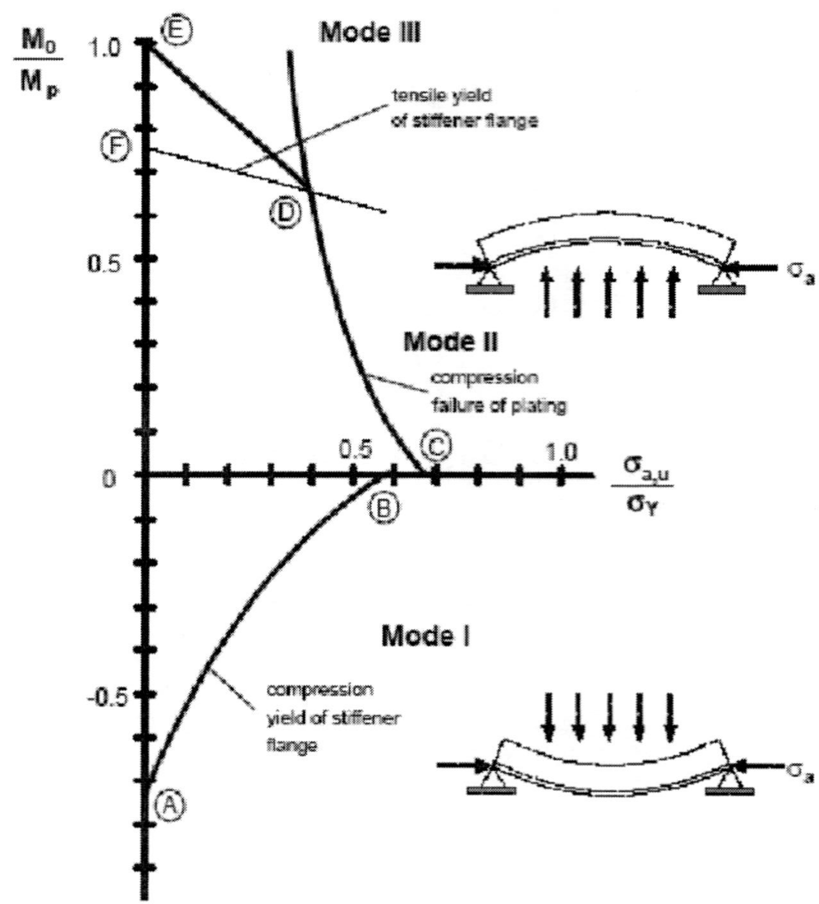

Diagram for Longitudinally Stiffened Sub-Panel Ultimate Strength (Hughes, 1998)

Limit States for Stiffened Panel Failures

Description	Limit state
Limit state for collapse of stiffened panel (gross panel), used for checking the minimum required rigidity of the transverse stiffeners.	$\frac{I_y a}{I_x b} - \frac{B_p^4}{\pi^2 C a^4}\left(1+\frac{1}{N}\right) \geq 0$
Limit state for excessive plastic deformation of stiffened panel (gross panel) under action of lateral pressure – Ultimate limit state	$\frac{P_c\left(n+1\right)}{a} - P_u \geq 0$
Limit state for initial yielding of the flange of laterally loaded stiffener subjected axial stresses and lateral pressure (Mode-I)	$\sigma_Y^{'} - \frac{M_{sw}+K_w M_{vw}}{Z} - \frac{M_{max}}{Z_s} \geq 0$
Limit state for buckling of stiffener subjected uni-axial	$\sigma_{cr} - \frac{M_{sw}+K_w M_{vw}}{Z} \geq 0$

Limit States for Plate Panel Failures

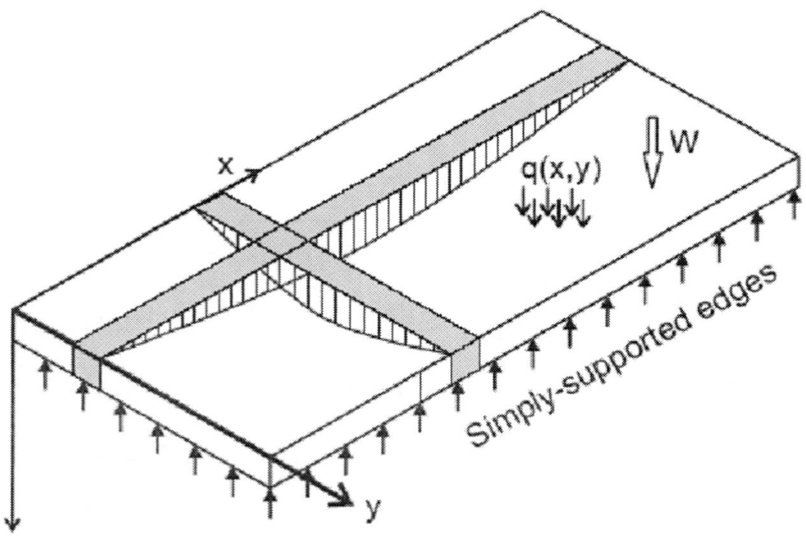

Plate Subjected to Lateral Loads

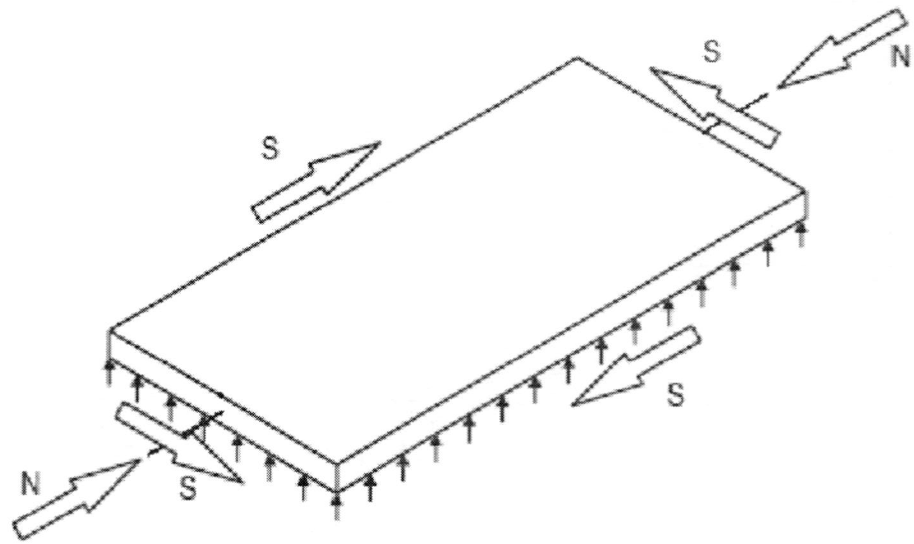

Plate Subjected to in - plane Loads

Limit States for Plate Panel Failures

Description	Limit state
Limit states for initial yielding of plate subjected to lateral pressure – serviceability limit state	$\sigma_Y - \sqrt{k_1^2 P^2 \left(\frac{b}{t}\right)^4 + k_2^2 P^2 \left(\frac{b}{t}\right)^4 - k_1 k_2 P^2 \left(\frac{b}{t}\right)^4} \geq 0$
Limit states for ultimate collapse failure of plate subjected to lateral pressure –ultimate strength limit state	$\sigma_u - (P_{st} + P_w) \geq 0$
Limit state for edge shear failure of plate panel	$\tau_{cr} - \tau_{sw} - K_w \tau_{vw} \geq 0$
Limit state for buckling failure of plate subjected uni-axial compressive loads	$\sigma_{cr}^* - \frac{M_{sw} + K_w M_{vw}}{Z} \geq 0$
Limit state for buckling failure of plate subjected bi-axial compressive loads	$\left(\frac{\sigma_x}{\sigma_{xcr}^*}\right)^n + \left(\frac{\sigma_y}{\sigma_{ycr}^*}\right)^n \leq 1$
Limit state for buckling failure of plate subjected bi-axial compressive loads and shear loads	$\left(\frac{\sigma_x}{R_{sx}\sigma_{xcr}^*}\right)^n + \left(\frac{\sigma_y}{R_{sy}\sigma_{ycr}^*}\right)^n \leq 1$

Statistical Modeling of Random Variables

Recommended Probabilistic characteristic of strength basic random variables (Assakkaf 1998 and Atua 1998) (From [12])

Variable	Nominal Value	Statistical Information		
		Mean	Standard Deviation	Distribution Type
t	t	t	0.02	normal
a	a	a	0.11	normal
b	b	b	0.09	normal
d_w	d_w	d_w	0.12	normal
f_w	f_w	f_w	0.07	normal
t_w	t_w	t_w	0.02	normal
t_f	t_f	t_f	0.02	normal
L	L	L	0.08	normal
B	B	B	0.01	normal
D	D	D	0.01	normal

Recommended Probabilistic characteristic of strength basic random variables (Assakkaf 1998 and Atua 1998) (From [12])

Variable	Nominal Value	Statistical Information		
		Mean	COV	Distribution Type
OS F_y	F_y	1.11 F_y	0.07	lognormal
HS F_y	F_y	1.22 F_y	0.09	lognormal
F_u	F_u	1.05 F_u	0.05	normal
E	E	1.04 E	0.02	normal
Z	Z_r	1.04 Z_r	0.05	lognormal
υ	0.3	0.3	0	
M_y	F_yZ	$\overline{F}_y\overline{Z}$	0.15	lognormal
M_P	F_yZ_P	$\overline{F}_y\overline{Z}_P$ or $c\overline{F}_y\overline{Z}$	0.18	lognormal

Recommended Probabilistic characteristic of Load random variables (Atua 1998) (From [12])

Variable	Distribution type	Mean to Nominal Ratio	COV
Still-water bending moment M_{sw}	Normal	0.4 to 0.6 for commercial ships, and 0.7 for naval vessels	0.3 to 0.9 for commercial ships, and 0.15 for naval vessels
Life-time Extreme wave- induced bending moment M_W	Largest extreme value (type I)	1.0	0.1 to 0.2
Whipping Bending Moment M_D	Extreme value (type I) Exponential	Mean value can be determined using formulae based on spectral analysis	0.2 to 0.3
Springing bending moment M_{SP}	Extreme value (type I)	1.0	0.3

Hydrostatic pressure due to Stillwater P_{sw}	Normal	0.4 to 0.6 for commercial ships, and 0.7 for naval vessels	0.15
Hydrostatic pressure due to waves P_W	Largest extreme value (type I)	1.0	0.15
Hydrostatic pressure due to dynamic effects P_D	Largest extreme value (type I)	1.0	0.25
Hydrostatic pressure due to combined wave and dynamic loads P_{WD}	Weiibull	1.0	0.25

Selection Target Reliability Index (β_o)

- **Agreeing upon a "reasonable" value.** This method is generally used for novel structures where there is no prior history.

- **Code calibration.** (calibrated reliability level that are implied in currently used codes) The level of risk is estimated for each provision of a successful code. Safety margins are adjusted to eliminate inconsistencies in the requirements. This method has been commonly used for code revisions.

- **Economic value analysis.** (cost benefit analysis). Target reliabilities are chosen to minimize the total expected costs over the service life of the structure. This approach is effectively used in design structures, for which failures results in only economical losses and consequences and not in loss of human life, as it is very difficult to assign a monetary value to human life.

Target Safety Indices – Based on Seriousness of Failure (From [15])

Seriousness of failure	Safety		Serviceability	
	P_f	β_0	P_f	β_0
Extreme	1.0×10^{-3}	3.1	5.1×10^{-6}	4.5
Severe	6.2×10^{-3}	2.5	3.1×10^{-5}	4.1
Moderate	2.3×10^{-2}	2.0	1.2×10^{-4}	3.7

Target Safety Indices – Mansour's Proposal (From [15])

Failure Mode	Commercial Ships		Naval Ships	
	P_f	β_0	P_f	β_0
Primary (Initial yield)	2.9×10^{-7}	5.0	1.0×10^{-9}	6.0
Primary (Ultimate strength)	2.3×10^{-4}	3.5	3.2×10^{-5}	4.0
Secondary (Ultimate strength)	6.2×10^{-3}	2.5	1.4×10^{-3}	3.0
Tertiary (Ultimate strength)	2.3×10^{-2}	2.0	6.2×10^{-3}	2.5
Fatigue Very serious Serious Not Serious	 1.4×10^{-3} 6.2×10^{-3} 1.6×10^{-1}	 3.1 2.5 1.0	 2.3×10^{-4} 1.4×10^{-3} 6.7×10^{-2}	 3.5 3.0 1.5

Target Safety Indices for Ship Structural Components

Recommendations for range target reliability index for Hull girder failure (Atua 1998) (From [12])

Range	Reference
4.0 to 6.0 (Sagging)	Mansour et al (1995)
5.0 to 6.0 (Hogging)	Mansour et al (1995)

Recommendations for range target reliability index for Stiffened panel Failures

Range	Reference
3.5 to 4.5	Mansour et al (1995)

Reliability Analysis

- Reliability analysis can be carried out with help Type-II and Type-III methods
- Reliability Analysis involves evaluation of Reliability Index (β) and compare it with Target Reliability index (β_o).
- The Partial Safety Factors (PSF) are calculated using following formula;

$$\gamma_i^* = \frac{X_i^*}{X_{in}}$$

- Here the Reliability software package- *Strurel* is used for carrying out reliability analysis.

Results and Discussions

- Ship Type – *Diving Support vessel* (DSV)

- Principle particulars

  LOA = 91.00 M

  LBP = 87.34 M

  B = 22.00 M

  D = 6.95 M

  T = 5.00 M

  Disp = 5780 T

  Cb = 0.61

Design Loads as per *rules for the classification of ships-2008.*

- <u>Still-water bending moment</u>
  - Hogging $M_{sw,0}$ = *166770 kN-m*
  - Sagging , $M_{sw,0}$ = *-24525 kN-m*
- <u>Vertical wave-induced bending moment</u>
  - Hogging $M_{vw,0}$ = *134577 kN-m*
  - Sagging $M_{vw,0}$ = *-168045 kN-m*
- <u>Local Pressure loads</u>
  - Still-water pressure $P_{st,0}$ = *50.30* kN/m^2
  - Wave pressure $P_{w,0}$ = *17.00* kN/m^2

Mid-ship Section

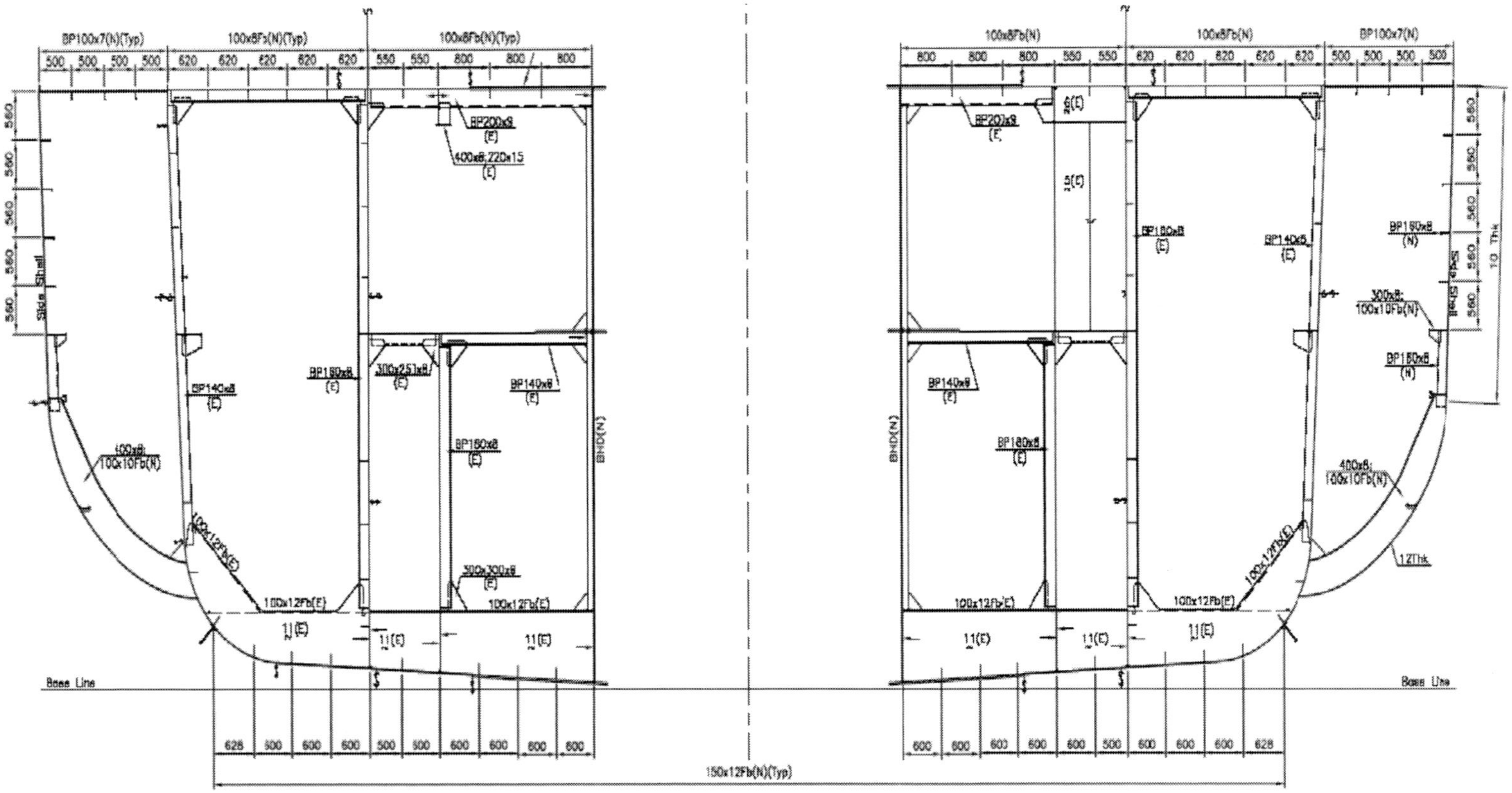

TYP ORDINARY FRAME SECTION IWO MOON POOL
SECTION AT FR.64

Available Section Modulus = 1.85 m^3

Initial Yielding Failure of Hull Girder

Limit state function for Initial Yielding Failure of Hull Girder is given as;

$$\sigma_a \geq \frac{M_{sw} + K_w M_{vw}}{Z}$$

$$g(x) = Z\sigma_Y - M_{sw} - K_w M_{vw}$$

or

$$M = g(X) = ZX_1 - M_{sw,0} X_2 - M_{vw,0} X_3$$

Range of Target Reliability Index (β_o) = (4 to 6)

Statistical information of random variables associated with hull girder failure

Variable	Description	Type of distribution	Mean	Standard deviation
X_1	Random variable corresponding to the yield strength of material	Normal	278735	22298.8
X_2	Uncertainties in the still-water bending moment	Normal	0.67	0.17
X_3	Uncertainties in the vertical wave bending moment	Gumbel (Max)	1.00	0.17
$M_{sw,0}$	Design still-water bending moment	Constant	166770	---
$M_{vw,0}$	Design vertical wave-induced bending moment	Constant	134577	---
Z	Mid-ship section modulus	Constant	1.85	---

Plot of Reliability Index (β) Vs Mid-ship Section Modulus (Z)

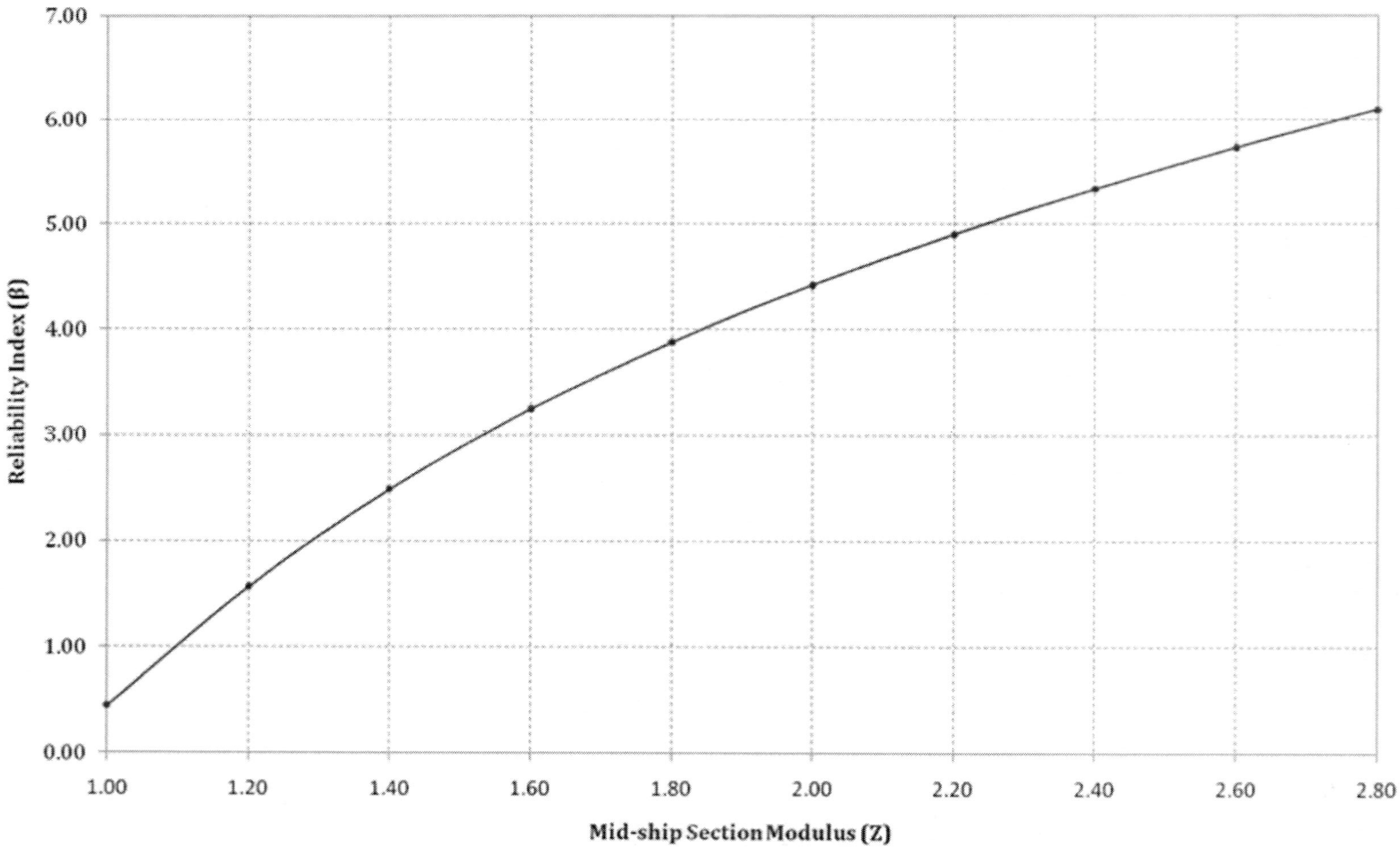

Plot of PSF's Vs Reliability Index (β)

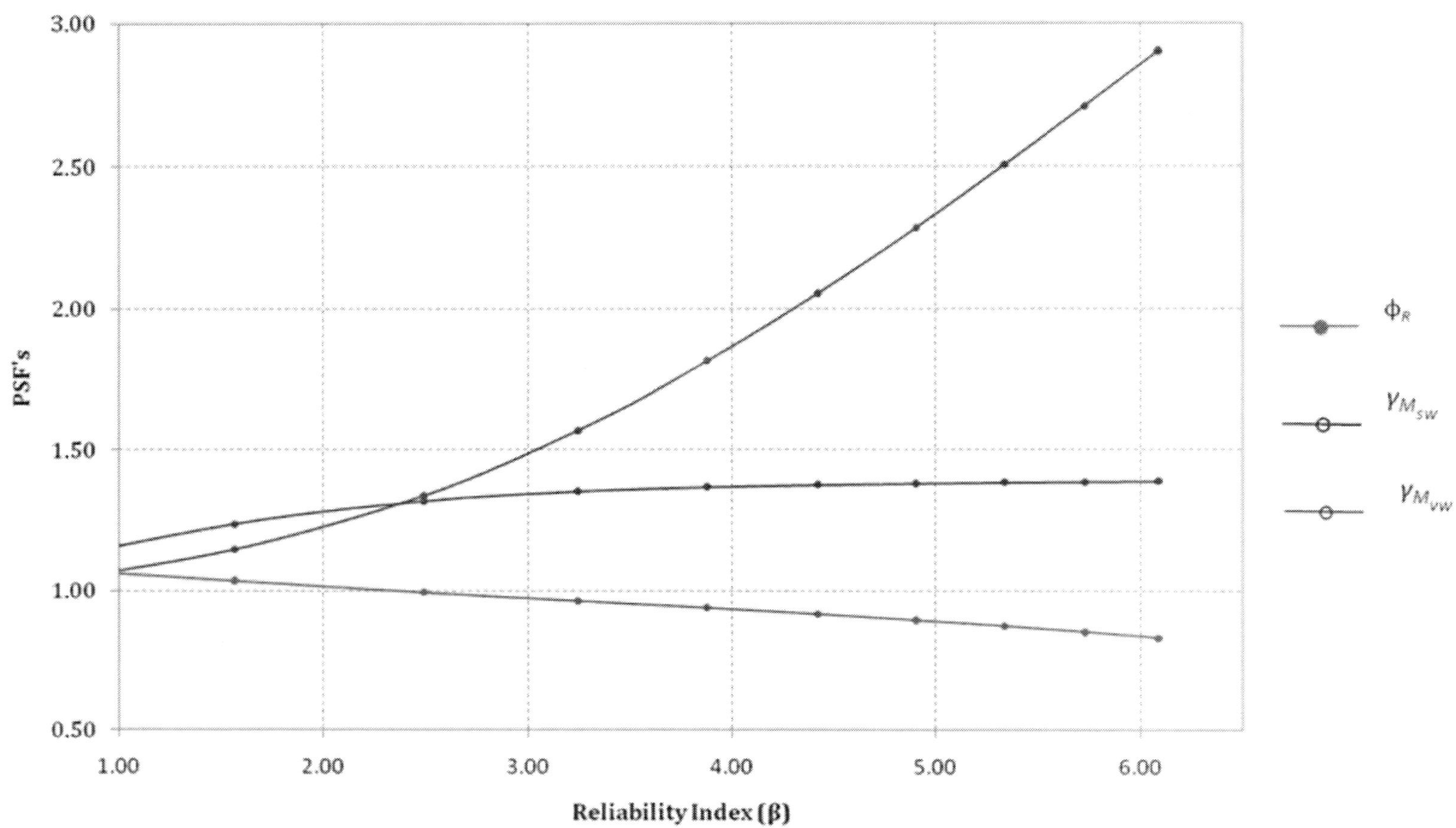

Partial Safety Factors for Hull girder failure

$$Z \geq \frac{\gamma_{M_{sw}} M_{sw,0} + \gamma_{M_{vw}} M_{vw,0}}{\Phi_R \left(\sigma_Y\right)_{\min}}$$

Partial safety Factors (PSF's)	Target reliability index (β_0)				
	4.0	4.5	5.0	5.5	6.0
Strength factor Φ_R	0.93	0.91	0.89	0.86	0.83
Load factor for still-water bending moment $\gamma_{M_{sw}}$	1.37	1.38	1.38	1.39	1.39
Load factor for vertical wave-induced bending moment $\gamma_{M_{vw}}$	1.87	2.09	2.33	2.59	2.85

Initial Yielding Failure of stiffener Flange (Mode – I)

Limit state function for Initial Yielding Failure of stiffener flange is given as;

$$g(x) = \sigma_Y' - \sigma_n - \frac{M_{max}}{Z_s}$$

$$M = g(X) = Z_s X_1' - \frac{Z_s}{Z}\left(M_{sw,0}X_2 + M_{vw,0}X_3\right) - X_2' \frac{\left(P_{st,0}X_4 + P_{w,0}X_5\right) s l^2}{m}$$

Where

$$X_1' = X_1\sqrt{1-3\left(\tau/X_1\right)^2} = X_1\sqrt{1-3\left(\frac{A_f}{Z_s t_w}\frac{\left(P_{st,0}X_4 + P_{w,0}X_5\right) s l}{2X_1}\right)^2}$$

$$X_2' = \frac{1}{1-0.18\frac{M_{sw,0}X_2 + M_{vw,0}X_3}{Z\sigma_E}}$$

Range of Target Reliability Index (β_o) = (3 to 4.5)

Statistical information of random variables associated initial yielding failure of stiffener flange

Variable	Description	Type of distribution	Mean	Standard deviation
X_1	Random variable corresponding to the yield strength of material	Normal	278735	22298.8
X_2	Uncertainties in the still-water bending moment	Normal	0.67	0.17
X_3	Uncertainties in the vertical wave bending moment	Gumbel (Max)	1.00	0.17
X_4	Uncertainties in the still-water pressure	Normal	0.67	0.17
X_5	Uncertainties in the wave-induced pressure	Gumbel (Max)	1.00	0.17
$M_{sw,0}$	Design still-water bending moment	Constant	166770	---
$M_{vw,0}$	Design vertical wave-induced bending moment	Constant	134577	---
$P_{st,0}$	Design still-water pressure	Constant	50.30	---
$P_{w,0}$	Design wave-induced pressure	Constant	17	---

Z	Mid-ship section modulus	Constant	1.85	---
k_w	Load combination factor	Constant	0.70	---
s	Stiffener spacing	Constant	600	---
l	Span of stiffener	Constant	2100	---
A_s	C/s area of stiffener without attached plating	Constant	3400	---
A_f	C/s area of face plate of stiffener	Constant	1000	---
Z_s	Section modulus of stiffener with attached plating	Constant	460000	---
I	Moment of inertia of stiffener with attached plating	Constant	1.05E8	---
t_p	Thickness of attached plating	Constant	10	---
t_w	Web thickness of stiffener	Constant	8	---

Plot of Reliability Index (β) Vs Section Modulus of Stiffener (Zs)

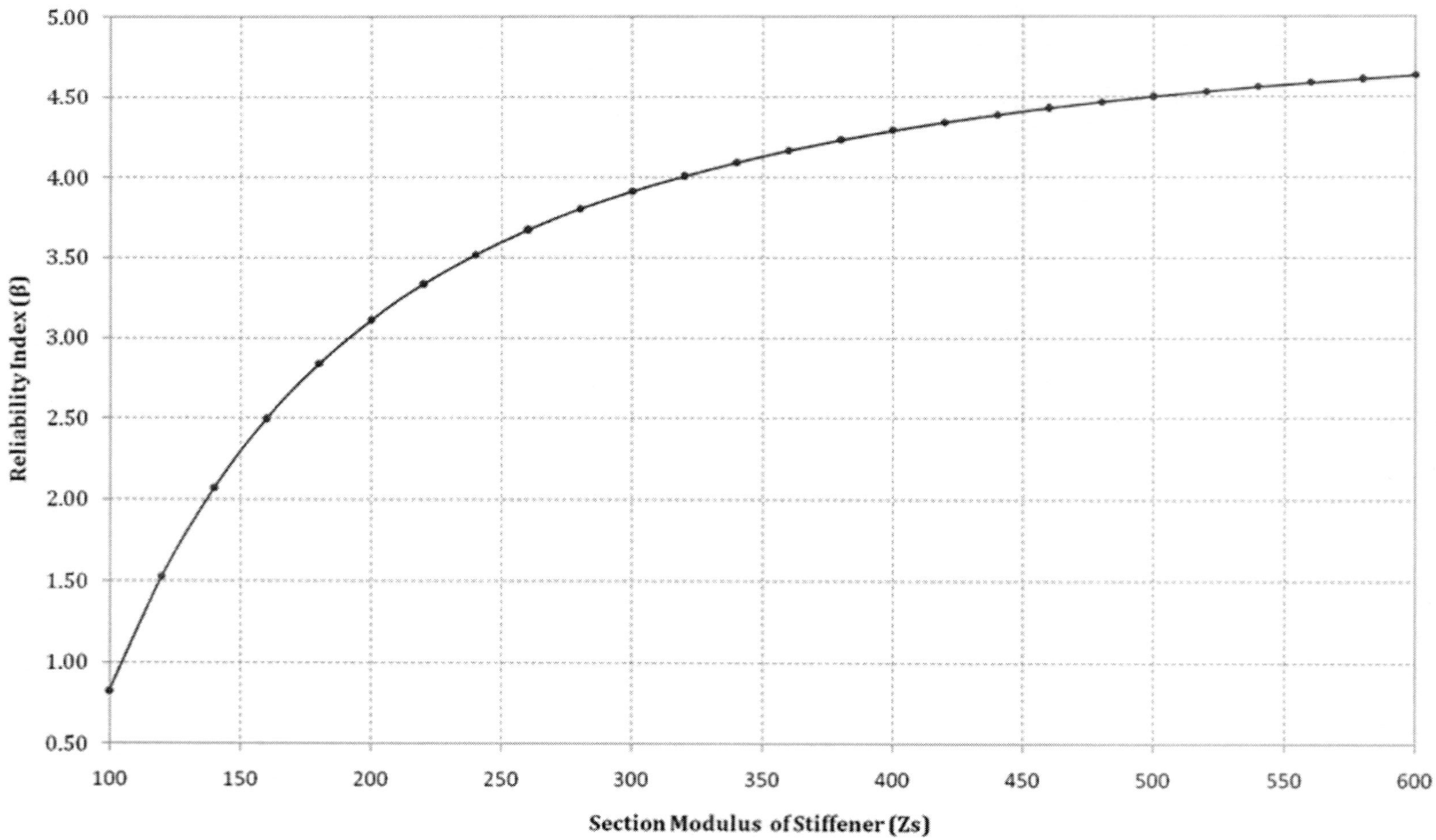

Plot of PSF's Vs Reliability Index (β)

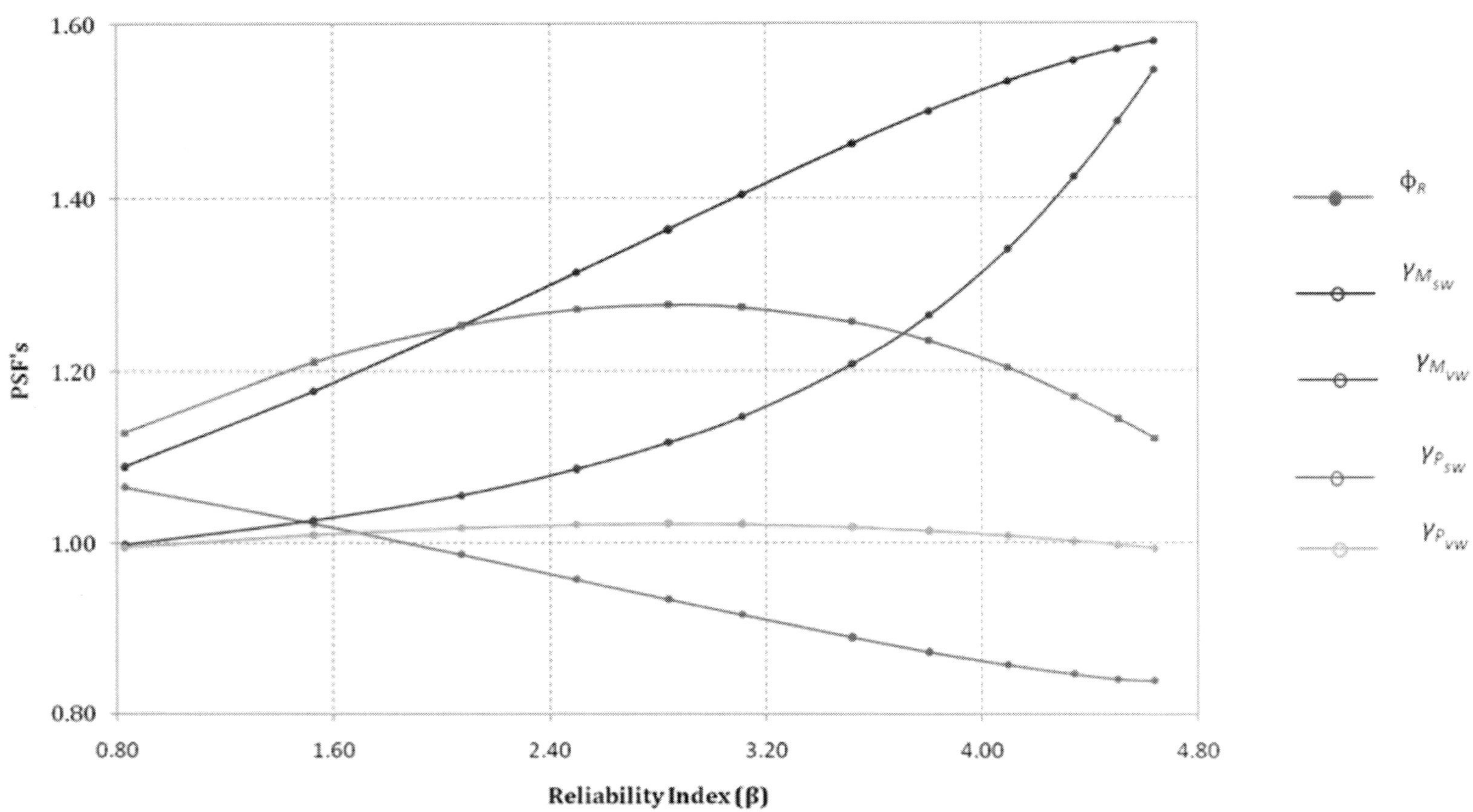

Partial Safety Factors for Stiffener failure

$$Z_s \geq \frac{\frac{Z_s}{Z}\left(\gamma_{M_{sw}} M_{sw,0} + \gamma_{M_{vw}} M_{vw,0}\right) + \left(\frac{1}{1 - 0.18\frac{\gamma_{M_{sw}} M_{sw,0} + \gamma_{M_{vw}} M_{vw,0}}{Z\sigma_E}}\right)\frac{\left(\gamma_{P_{sw}} P_{st,0} + \gamma_{P_{vw}} P_{w,0}\right) s l^2}{m}}{\Phi_R \left(\sigma_Y\right)_{\min}}$$

Partial safety Factors (PSF's)	Target reliability index (β_0)			
	3.0	3.5	4.0	4.5
Strength factor (Φ_S)	0.92	0.89	0.86	0.85
Load factor for still-water bending moment $\gamma_{M_{sw}}$	1.39	1.46	1.53	1.57
Load factor for vertical wave-induced bending moment $\gamma_{M_{vw}}$	1.14	1.21	1.34	1.49
Load factor for still-water pressure $\gamma_{P_{sw}}$	1.28	1.26	1.20	1.15
Load factor for wave-induced water pressure $\gamma_{P_{vw}}$	1.02	1.02	1.01	1.00

Initial Yielding Failure of Plate Panel Under Action of Lateral Pressure

Limit state function for Initial Yielding Failure of Plate Panel Under Action of Lateral Pressure is given as;

$$g(x) = \sigma_Y - (P_{st} + P_w)\left(\frac{b}{t}\right)^2 \sqrt{k_1^2 + k_2^2 - k_1 k_2}$$

or

$$g(x) = X_1 - (P_{st,0} X_4 + P_{w,0} X_5)\left(\frac{b}{t}\right)^2 \sqrt{k_1^2 + k_2^2 - k_1 k_2}$$

Range of Target Reliability Index (β_o) = (3 to 4.5)

Statistical information of random variables associated with initial yielding failure of plate under action of lateral pressure

Variable	Description	Type of distribution	Mean	Standard deviation
X_1	Random variable corresponding to the yield strength of material	Normal	278735	22298.8
X_4	Uncertainties in the still-water pressure	Normal	0.67	0.17
X_5	Uncertainties in the wave-induced pressure	Gumbel (Max)	1.00	0.17
$P_{st,0}$	Design still-water pressure	Constant	50.30	---
$P_{w,0}$	Design wave-induced pressure	Constant	17	---
a	Length of plate panel	Constant	2100	---
b	Width of plate panel	Constant	600	---
t	Plate thickness	Constant	12	---
k_1	Buckling factor	Constant	0.2378	---
k_2	Buckling factor	Constant	0.6272	---

Plot of Reliability Index (β) Vs Plate Thickness (t)

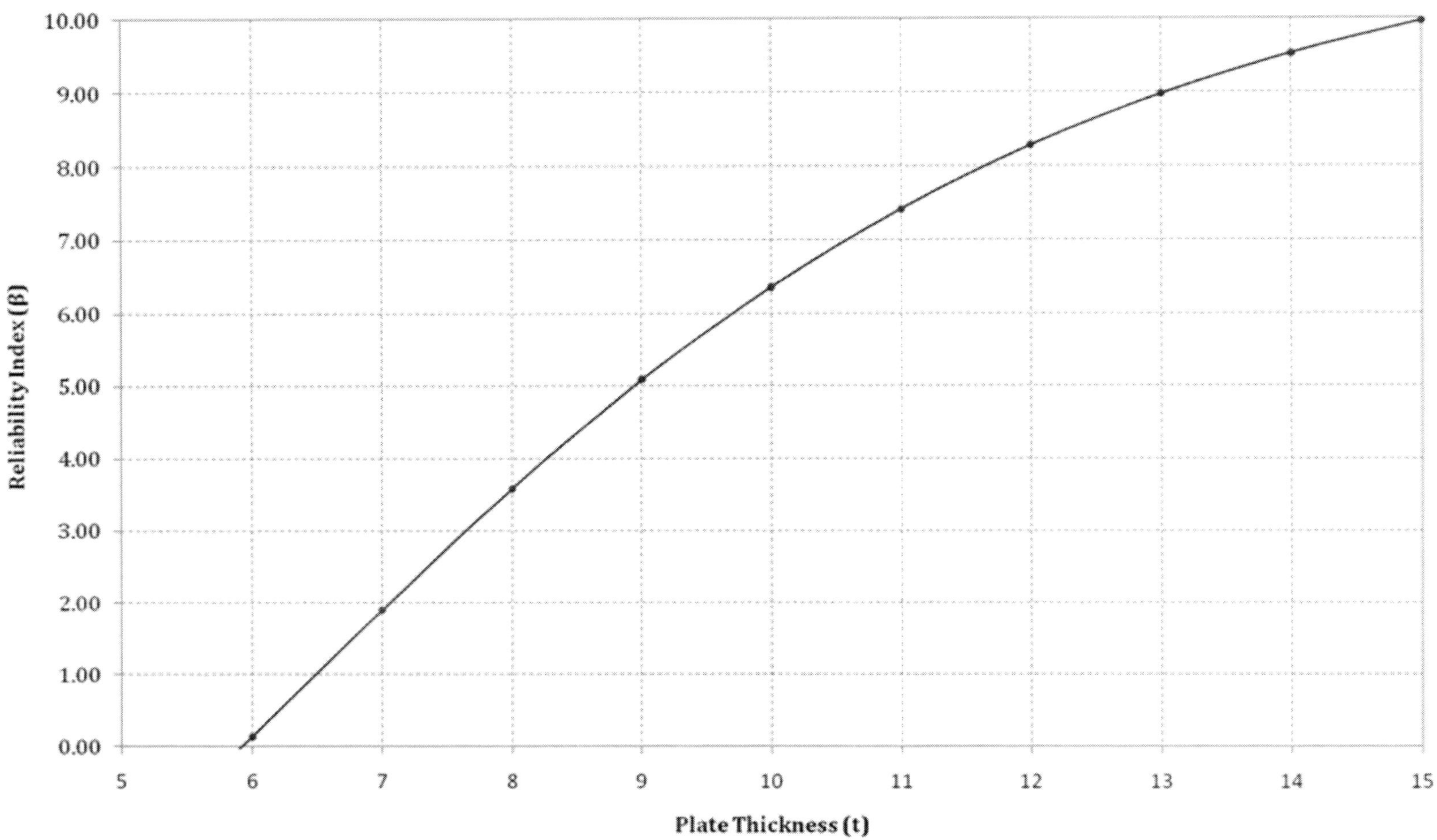

Plot of PSF's Vs Reliability Index (β)

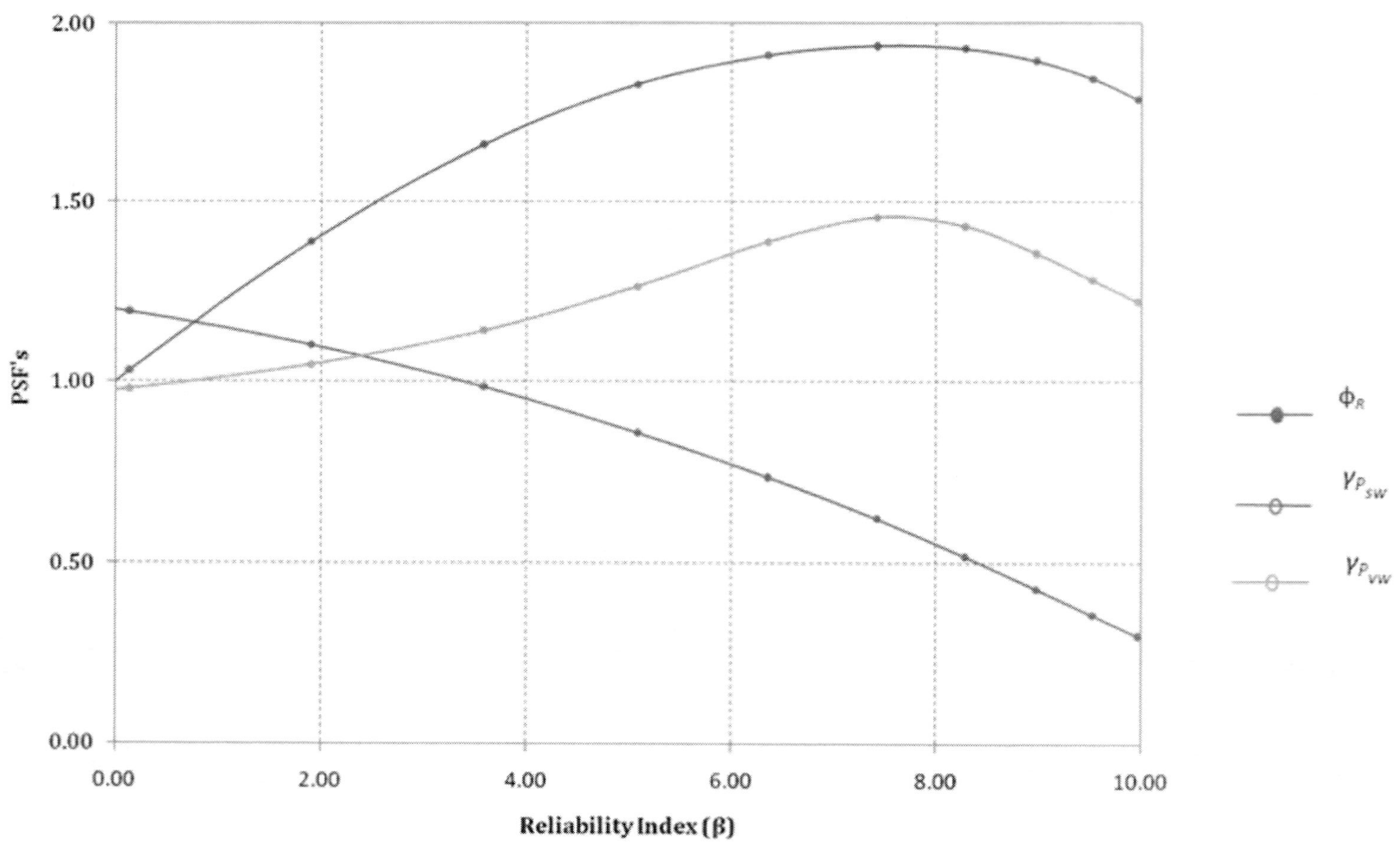

Partial Safety Factors for initial yielding failure of plate under action of Lateral Pressure

$$\left(\frac{t}{b}\right)^2 \geq \frac{\left(\gamma_{P_{sw}} P_{st,0} + \gamma_{P_{vw}} P_{w,0}\right)\sqrt{k_1^2 + k_2^2 - k_1 k_2}}{\Phi_R \left(\sigma_Y\right)_{min}}$$

Partial safety Factors (PSF's)	Target reliability index (β_0)			
	3.0	3.5	4.0	4.5
Strength factor (Φ_s)	1.03	0.99	0.95	0.91
Load factor for still-water pressure $\gamma_{P_{sw}}$	1.57	1.65	1.71	1.76
Load factor for wave-induced water pressure $\gamma_{P_{vw}}$	1.11	1.14	1.18	1.22

Buckling Failure of Plate Panel Under Uni-axial Stresses

Limit state function buckling failure of unstiffened plate panel under uni-axial stress is given as;

$$g(x) = \sigma_{cr} - \sigma_n$$

or

$$M = g(x) = X_1 \left(1 - X_1 / 4\sigma_E\right) - \frac{M_{sw,0} X_2 + M_{vw,0} X_3}{Z}$$

Range of Target Reliability Index (β_o) = (3 to 4.5)

Statistical information of random variables associated with buckling failure of plate under uni-axial stresses

Variable	Description	Type of distribution	Mean	Standard deviation
X_1	Random variable corresponding to the yield strength of material	Normal	278735	22298.8
X_2	Uncertainties in the still-water bending moment	Normal	0.67	0.17
X_3	Uncertainties in the vertical wave bending moment	Gumbel (Max)	1.00	0.17
$M_{sw,0}$	Design still-water bending moment	Constant	166770	---
$M_{vw,0}$	Design vertical wave-induced bending moment	Constant	134577	---
Z	Mid-ship section modulus	Constant	1.85	---
k_w	Load combination factor	Constant	0.70	---
b	Plate width	Constant	600	---
t	Plate thickness	Constant	12	---
k_1	Buckling factor	Constant	4	---

Plot of Reliability Index (β) Vs Plate Thickness (t)

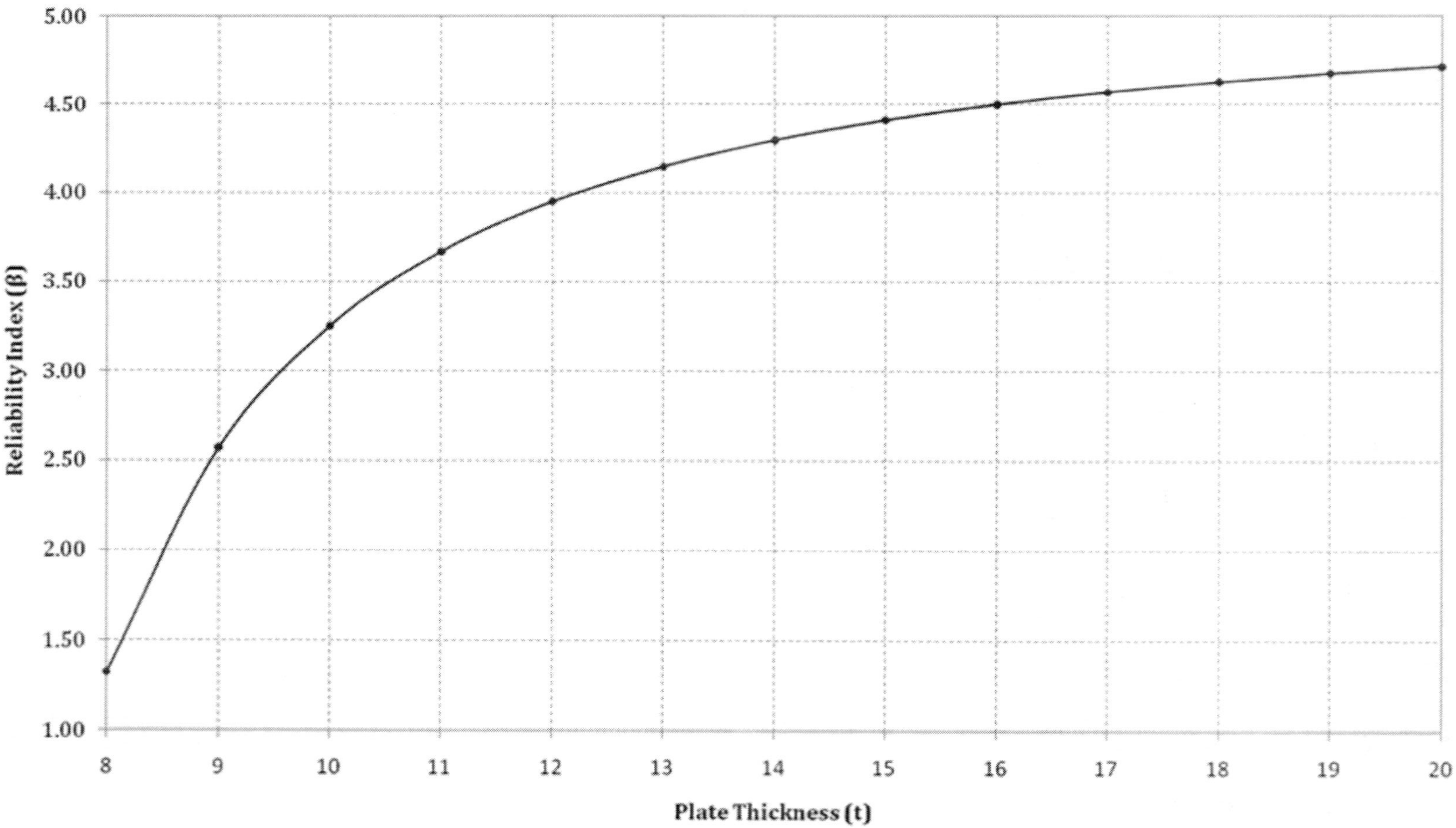

Plot of PSF's Vs Reliability Index (β)

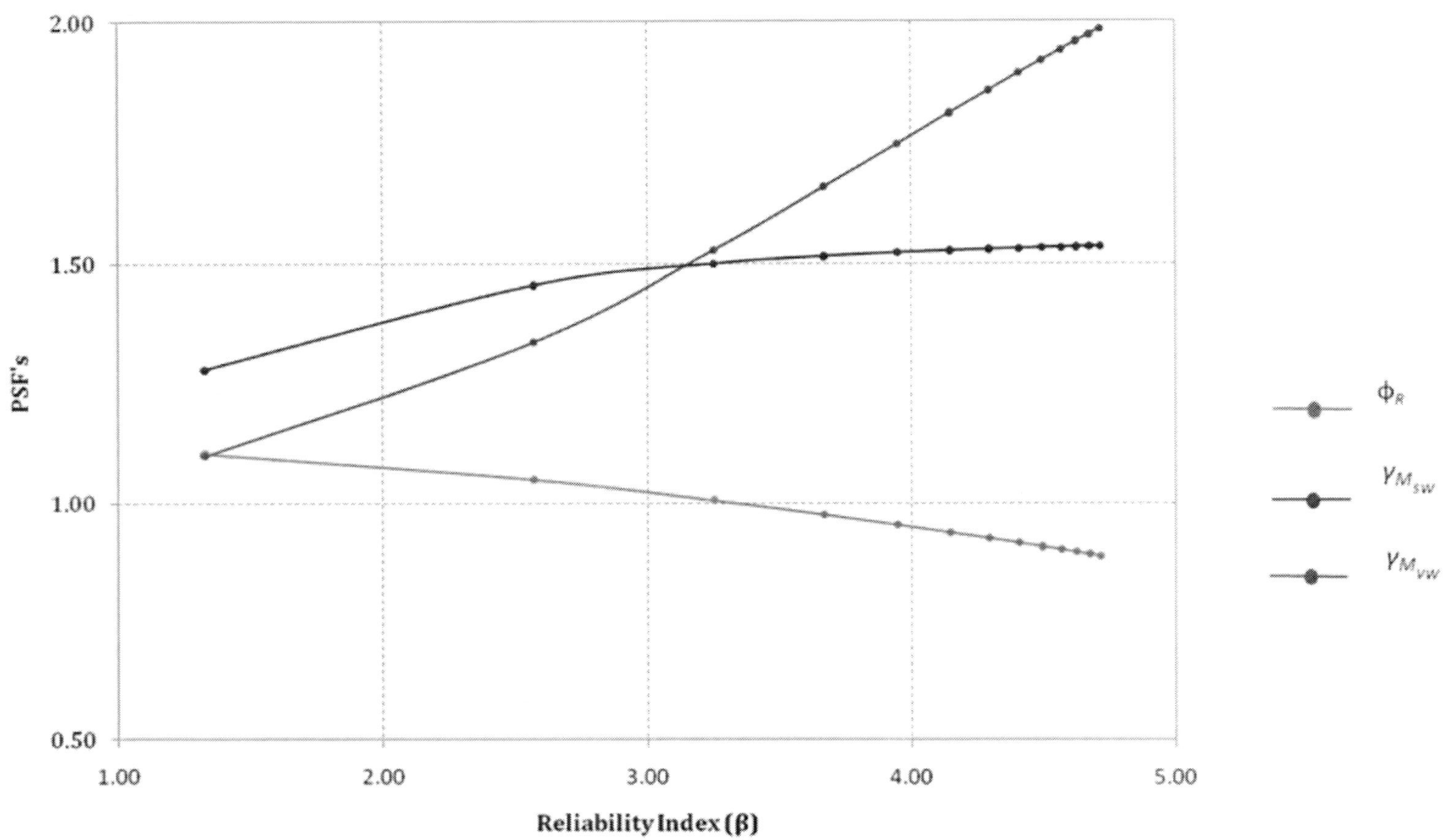

Partial Safety Factors for buckling failure of plate under uni-axial stresses

$$\Phi_R (\sigma_Y)_{\min} \left(1 - \Phi_R \frac{(\sigma_Y)_{\min}}{4\sigma_E} \right) \geq \frac{\gamma_{M_{sw}} M_{sw,0} + \gamma_{M_{vw}} M_{vw,0}}{Z}$$

Partial safety Factors (PSF's)	Target reliability index (β_0)			
	3.0	3.5	4.0	4.5
Strength factor (Φ_s)	1.02	0.99	0.95	0.91
Load factor for still-water bending moment $\gamma_{M_{sw}}$	1.48	1.51	1.52	1.53
Load factor for vertical wave-induced bending moment $\gamma_{M_{vw}}$	1.46	1.61	1.76	1.92

Conclusions

The reliability based design methods are excellent means for assessing the safety of ship structures under the certain condition of uncertainty. The Reliability methods used probabilistic means, for describing the random loads acting ship structure and the uncertainties associated with strength of ship structure. This accounts the uncertainties rationally and also provide the more freedom to designer.

Advantages of Reliability based deign methods:

1. Explicit consideration and evaluation of uncertainties associated with the design variables.
2. Provides means of achieving uniformity of safety within a given class of structures.
3. Gives an indication of level of safety / probability of failure.
4. Provides means for decomposition of global safety of a structure into partial safety factors associated with the individual design variables.
5. Provides the means to weight variables in terms of their significance.
6. Provide the potential to reduce weight without loss of reliability, or improve reliability without increasing weight.
7. The partial safety factor format (LRFD) used herein also provides a framework for extrapolating existing design practice to new ships where experience is limited.

Scope of Future Work

- Collection of data or carrying out experiments for accurate description of random loads (still-water loads and wave loads) acting on the different types of ships (container vessel, cargo, bulk carrier etc)
- Determining the limit state equations for less known modes of failure like Torsional Failure of container vessel.

References

1. Ranganathan, R. (1999) "*Structural reliability Analysis and Design*", Jaico Publication House, Mumbai.
2. Hughes, O.F., (1988) "*Ship Structural Design, A Rationally-Based, Computer-Aided Optimization Approach*" The Society of Naval Architects and Marine Engineers, Jersey City, New Jersey.
3. Timoshenko, S.P., and Goodier, J.N., (1951) *"Theory of Elasticity"*, McGraw- Hill.
4. Timoshenko, S.P., and krieger, S.W., (1951) *"Theory of Plates and Shells"*, McGraw- Hill.
5. Mansor, A. (1990) "*An Introduction to Structural Reliability Theory*" SSC-351, Ship Structure Committee, NTIS, Springfield, Virginia.
6. Mansor, A., Lin, M., Hovem, L., and Thayamballi, A., (1993) "Probability- *Based ship Design Procedures: A Demonstration*" SSC-368, Ship Structure Committee, NTIS, Springfield, Virginia.
7. Mansor, A., and Thayamballi, A., (1994) "Probability *Based ship Design: Loads and Load Combinations*" SSC-373, Ship Structure Committee, NTIS, Springfield, Virginia.
8. Mansor, A., Wirsching, P.H., White, G.J., and Ayyub, B.M., (1995) "Probability *Based ship Design: Implementation of Design Guidelines*" SSC-392, Ship Structure Committee, NTIS, Springfield, Virginia.

9. Ayyub, B.M., Beach, J., and Packard, T., (1995) "*Methodology for the Development of Reliability-Based Design Criteria for Surface Ship Structures,*" Naval Engineers Journal, ASNE, 107(1), Jan. 1995, 45-61.

10. Ayyub, B.M., Assakkaf., I.A., and Auta, K.I.,(2000) "*Reliability- Based Load and Resistance Factor Design (LRFD) of Hull Girders for Surface Ships* " Journal of Naval Engineers, Vol. 112, No.4, July.

11. Ayyub, B.M., Assakkaf., I.A., Beach, J., Metlton, W., Natale, N., and Colnley, J.A., (2002) "*Methodology for the Development of Reliability- Based Load and Resistance Factor Design (LRFD) guidelines for Ship Structures,*" Naval Engineers Journal, ASNE, 114(2).

12. Ayyub, B.M., Assakkaf., I.A., Sikora, J., Adamchack, J., and Auta, K., (2002) "*Reliability- Based Load and Resistance Factor Design (LRFD) guidelines for Hull Girder Bending,*" Naval Engineers Journal, ASNE, 114(2).

13. Assakkaf., I.A., Ayyub, B.M., Hess, P.E., and Atua, K.A., (2004) "*Reliability- Based Load and Resistance Factor Design (LRFD) guidelines for Stiffened Panels and Grillages of Ship Structure,*"

14. Assakkaf., I.A., Ayyub, B.M., Hess, P.E., and Knight, D.E., (2004) "*Reliability- Based Load and Resistance Factor Design (LRFD) guidelines for Unstiffened Panels of Ship Structure,*"

15. Dr. Dominique Beghin,(2007) " *Reliability- Based Structural Design*", Technical and Research Bulletin 2-32, SNAME.

16. *RINA rules for the classification of ships-2008.*

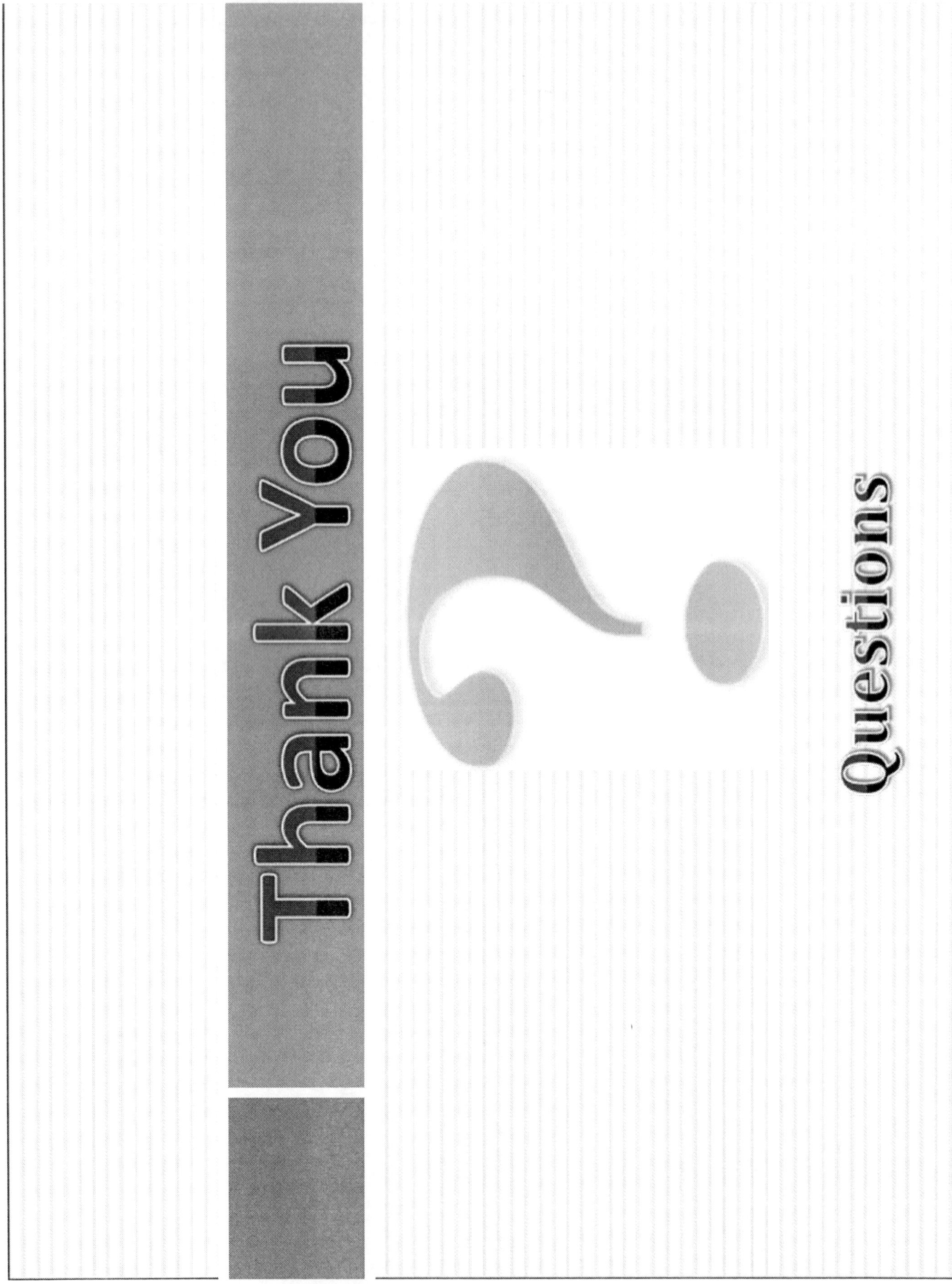
Thank You
Questions

ANALYSIS OF WELD INDUCED DISTORTION OF LARGE CURVED STIFFENED PANELS

P Biswas and **N R Mandal,** Department of Ocean Engineering and Naval Architecture, Indian Institute of Technology, Kharagpur, Indian

SUMMERY

Large curved and flat stiffened panels are used in ships and offshore structures. The joining process of stiffeners is mainly fusion welding. It leads to heat induced distortion. Prediction of weld induced distortion of such large curved stiffened panels is rarely found in published literature. These deformations adversely affect the subsequent fitup and alignment of the adjacent panels. It may also result in loss of structural integrity. These panels primarily suffer from angular and buckling distortions. The extent of distortion depends on several parameters like welding speed, plate thickness, welding current, voltage, restraints applied to the job while welding. Welding being a transient phenomenon, conventional finite element thermo mechanical analysis of large stiffened panels is virtually not possible. In the present work an analysis methodology based on the Finite Element inherent strain method for studying the distortion pattern of large curved stiffened plate panels was worked out.

Key words: Welding distortion, 3-D finite element analysis, Inherent strain, Curved Stiffened panel.

1. INTRODUCTION

Welding is one of the main joining methods and widely used in industries to assemble various products such as ships, trains, automobiles and bridges. The distortion in a welded structure results from the nonuniform expansion and contraction of the weld and surrounding base material, due to the heating and cooling cycle during welding process. It has undesirable effects on accuracy of assembly, external appearance and various strengths of the welded structures. For instance, the assembly process in shipbuilding essentially involves the joining of large blocks. Typically, these blocks are all-welded structures made of curved and flat stiffened plates. Distortions occur during the fabrication of these blocks due to a variety of causes, such as cutting and welding. Welding induced distortion not only degrades the performance but also increases the cost of a fabricated structure. In many cases, additional costs and schedule delays are incurred from straightening welding distortion. Therefore, the prediction of welding deformation beforehand has assumed critical importance. If the welding deformation can be predicted beforehand, appropriate distortion mitigation measures can be taken to maintain sufficient geometrical accuracy.

It is very important for large welded structures to quantitatively predict welding distortion during the assembly process. Generally, the major cause of geometrical error in the welded structure is local shrinkage [1] due to rapid heating and cooling in the weld zone. Basically, local shrinkage can be divided into three categories: longitudinal shrinkage, transverse shrinkage and angular distortion. These three kinds of local deformations can be taken as inherent deformations. They are strongly influenced by heat input, plate thickness and joint type.

At present, though thermal elastic–plastic finite element method can be effectively used to predict welding residual stress and distortion for small or medium welded structures [2–6], this method is inapplicable to simulate the welding distortion for large welded structures because of the large amount of computational time.

For seeking practical methods to predict welding residual stress and residual distortion, many numerical simulation procedures [7–11] have been developed in the past decades. Ueda and his co-workers [7, 8] studied the characteristic distributions of inherent strains in butt, T and I joints, and employed them to predict welding residual stress.

Michaleris and DeBiccari [9] developed a two-step numerical analysis procedure to predict welding induced deformation. In their method, a two-dimensional thermomechanical welding simulation was performed to determine the welding residual stresses. Then, the residual stresses were introduced to a three-dimensional structure for elastic analysis.

Jung and Tsai [10] developed a method named plasticity-based distortion analysis (PDA) and applied it to investigate the relationship between cumulative plastic strains and angular distortion in fillet welded T joint. In their research, the relationship between six cumulative plastic strain components and angular distortion was examined. They also studied the effect of external restraints and thermal management technique on the relationship between cumulative plastic strains and angular distortion using PDA procedure [11].

Experimental studies were carried out to investigate the thermal character and the effects of arc distance and welding parameters on the angular distortion in asymmetrical double-sided double arc welding [12] (ADSAW). Meanwhile, a 3-D thermal elastic plastic finite element model is developed to simulate transient temperature and welding deformation with different arc distance and heat input. In addition, the distortion mechanism of angular distortion were clarified through numerical simulation.

An experimental study was carried out to investigate the characteristics of welding deformation of fillet joints [13]. Meanwhile, a 3-D thermal elastic plastic finite element model was developed to simulate welding deformation. The influence of the flange thickness on welding deformation was investigated.

To ascertain the extent of deformation due to the thermal cycles caused by welding it calls for solving a complex thermal elasto-plastic problem, which is non-linear and involves plastic deformation of the medium at high temperature varying in both time and space. Analytical solutions turned out to be inadequate [14]. At the same time conventional numerical techniques proved to be highly time consuming and thereby prohibitively expensive in real life situations.

The increasing power and capability of computer systems mean that direct, detailed computational simulations are feasible to some extent. However, simplifications seem to be necessary in practice [15], as noted by Josefson [16] and Lindgren [17], it is important to keep computational times within a reasonable industrial time frame.

In the present work a simplified FE model based on inherent strain generated due to welding was developed to predict the welding distortion of large curved stiffened panels. The computational time required was about 30 to 40 minutes only in a Core (TM) 2 Duo, 2.66GHz computer with 2 GB RAM, which otherwise would have been not possible with other conventional computation techniques.

2. INHERENT STRAIN METHOD

Inherent strains are the inelastic permanent strains or residual plastic strains caused in welding process [1 and 9]. They exist in the welds and in its near vicinity which has undergone a large thermal cycle. These strains are considered responsible for causing welding deformations. Once the relation between the welding parameters and the inherent strain distribution is established, the residual stress and the deformation can be predicted by elastic analysis using inherent strain as initial strain.

Inherent strain resulting from welding is a combination of thermal and plastic strain. Thermal strains result from heating and cooling of the weld metal whereas plastic strains result when thermal straining is restrained by the cold metal surrounding the weld zone. In case of a long weld, the plate being restrained equally along its length, the welding gives rise to inherent strains which is independent of position along the weld line. Taking the inherent strain as the initial strain, welding deformation can easily be calculated by elastic analysis, thereby avoiding the complicated thermal elastic-plastic analysis.

In most of the fusion welding processes a structure undergoes locally a large thermal cycle. The thermal expansion of the weld metal and nearby areas is restricted by the constraint of surrounding metal where the temperature is much lower compared to the weld zone. This leads to formation of compressive plastic strains in the weld metal and nearby area.

The inherent strain caused due to welding can be divided as longitudinal inherent strain and transverse inherent strain. Let the total volume of longitudinal and transverse inherent strains in unit length be expressed as V_x and V_y respectively. The total volume of inherent strains in longitudinal and transverse directions is related to heat input q [18]. So, V_x and V_y can be expressed as given in equations 2.1 and 2.2 respectively.

$$V_x = Kq \tag{2.1}$$

$$V_x = \beta q \tag{2.2}$$

The range of coefficients K and β are taken as shown in equations 2.3 and 2.4 respectively:

$$K = (0.255 \; to \; 0.335)\frac{\alpha}{c\rho} \tag{2.3}$$

$$\beta = (0.255 \; to \; 1.0)\frac{\alpha}{c\rho} \tag{2.4}$$

where, α the thermal expansion coefficient; c the thermal capacity; ρ the density. Here the coefficients for welded structure made of mild steel are taken as given below [18]: $K = 8.6 \text{ X } 10^{-7}$ cm^3/J, $\beta = 2.8 \text{ X } 10^{-6}$ cm^3/J.

Longitudinal inherent strain ε_L and transverse inherent strain ε_T are defined as follows:

$$\varepsilon_L = \frac{V_x}{A_i} \text{ and } \varepsilon_T = \frac{V_y}{A_i} \tag{2.5}$$

A_i is the zone area where V_x and V_y are distributed. In the present analysis the calculated cross-section area A_i is 0.00027 m^2 as shown in Figure 1. For predicting welding deformation, using different area in a certain region has no effect on deformation analysis if the total volume of inherent strains and the location of its center do not change [18].

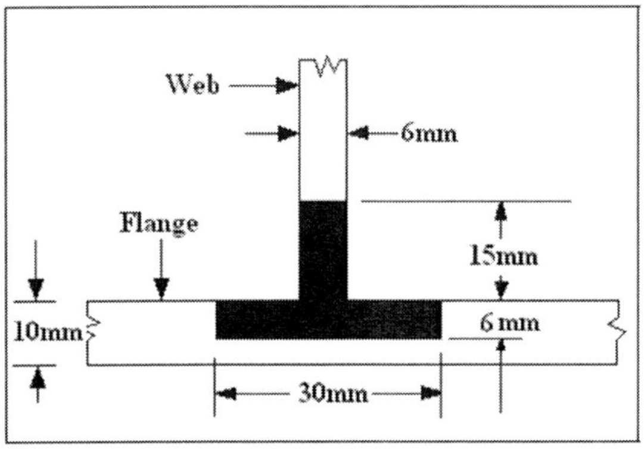

Figure 1: Cross section for calculation of A_i

3. FINITE ELEMENT MODEL

In the present study a large curved stiffened panel comprising of a 10mm thick, 9600mm × 3460mm C-Mn steel plate with 9 longitudinal and 9 transverse stiffeners was modeled as shown in Figure 2. The longitudinal frame spacing was 400mm and the spacing of the transverses was 1200mm. Eight nodded brick elements were used in the present analysis. The total number of elements and total number of nodes were 337898 and 681864 respectively.

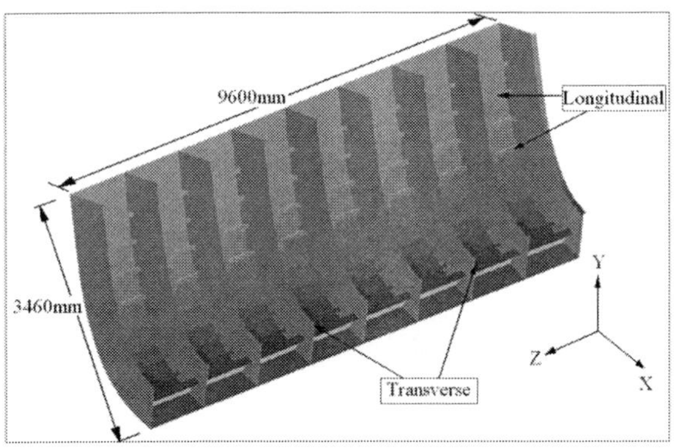

Figure 2: Co-ordinate system and dimensions of the curved stiffened plate panel

4. RESULTS AND DISCUSSION

The welding parameters used to weld the longitudinal and transverse stiffeners are shown in table 1. The deformation shape of UX and UY components are shown in Figures 3 and 4 respectively. The overall deformation shape i.e. the vector sum of all components of deformation is shown in Figure 5. It can be observed from Figure 5 that the maximum deflection of 34.24 mm occurred somewhere at the middle region of the stiffened panel.

Table 1: Plate thickness and welding parameters used in the study

Plate No.	Thickness (mm)	Current (A)	Voltage (V)	W.S (mm/s)
1	10	210	28.0	2.0

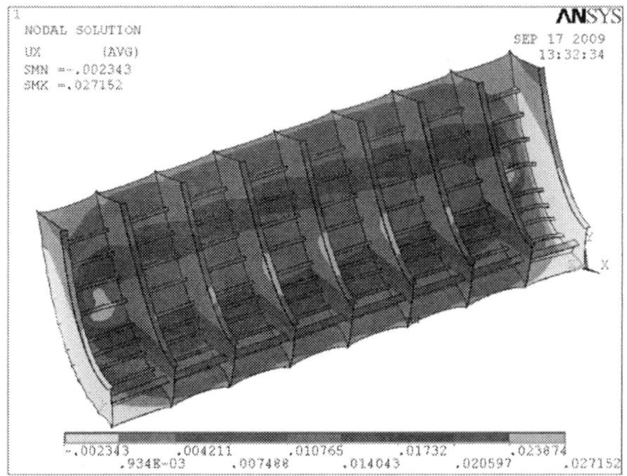

Figure 3: Deformation shape of UX component

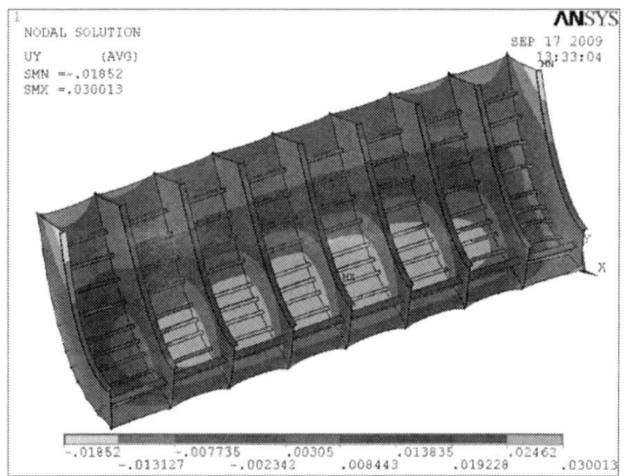

Figure 4: Deformation shape of UY component

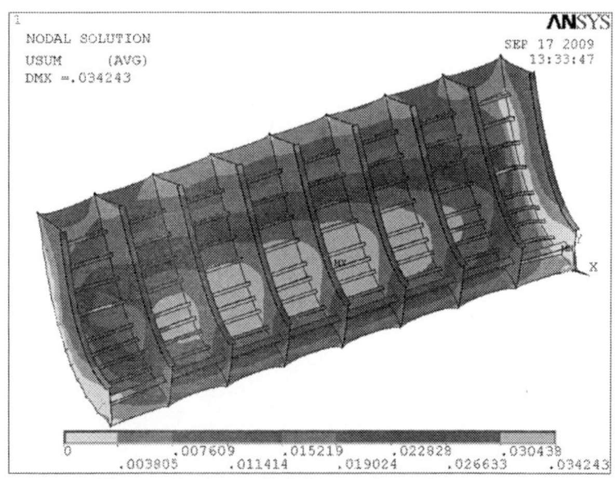

Figure 5: Overall deformation shape

The extent of distortion at the middle of the panel, the upper edge and the bottom edge of the stiffened panel along longitudinal direction are shown Figures 6 and 7 respectively.

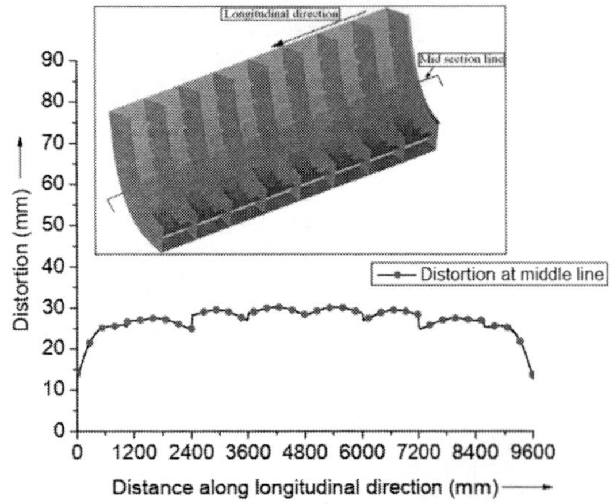

Figure 6: Distortions at middle line of the stiffened plate panel

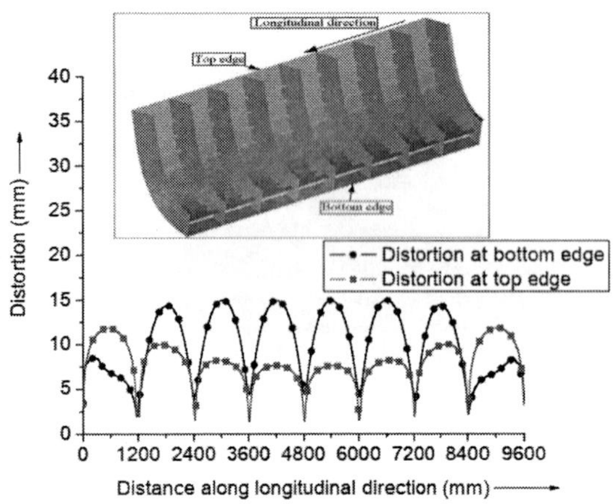

Figure 7: Distortions at top and bottom edge of the stiffened plate panel

From Figure 6 it can be observed that along the middle region the plate along with the transverse stiffeners underwent deformation. Whereas the extent of deformation of the transverse stiffeners along the panel edges were found to be much smaller compared to the deformation at the middle region as well as at the plate edges in between the transverse stiffeners. Also one can observe from Figure 7 that the deflection at the upper edge is generally less than that at the bottom edge.

5. CONCLUSIONS

The following conclusions can be derived from the present investigation:

i) A three-dimensional finite element model for predicting of distortions of large curved stiffened plate panels have been developed utilising the inherent strain method.

ii) The methodology developed here makes it possible to analyse and predict the distortion pattern and magnitude of the distortion which was otherwise not feasible with the existing FE transient nonlinear elasto-plastic thermo-mechanical modeling approach.

iii) It was observed that deformation was more near the middle region of the stiffened panel. The thermal stresses generated were strong enough to deform the transverses at these regions.

iv) The deflection at upper edge was less compared to that of at the bottom edge. However both were less than that observed at the middle region of the stiffened panel.

v) The deformation pattern was found to be convex in nature looking from the stiffener side.

vi) The computational methodology developed here was found to be very promising in analyzing large complex welded structures for overall distortion prediction.

vii) The computational time required was about 30 to 40 minutes only in a Core (TM) 2 Duo, 2.66GHz computer with 2 GB RAM, which otherwise would have been not possible with other conventional computation techniques.

6. REFERENCES

1. DENG, D., MURAKAWA, H., UEDA, Y., 'Theoretical prediction of welding distortion considering positioning and gap between parts', *Int. Offshore Polar Engrg.*, 14 (2), 138–144, 2004.

2. DENG, D., LUO, Y., SERIZAWA, H., SHIBAHARA, M. and MURAKAWA, H., 'Numerical simulation of residual stress and deformation considering phase transformation effect', *Trans. JWRI*, 32 (2), 325–333, 2003.

3. DENG, D. and MURAKAWA, H., 'Numerical simulation of temperature field and residual stress in multi-pass welds in stainless steel pipe and comparison with experimental measurements', *Comput. Mater. Sci.*, 37 (3), 269–277, 2006.

4. TENG, T. L. *et al.*, 'Analysis of residual stresses and distortions in T-Joint fillet welds', *Press. Vess. Piping*, 78, 523–538, 2001.

5. DENG, D., LIANG, W. and MURAKAWA, H., 'Determination of welding deformation in fillet-welded joint by means of numerical simulation and comparison with experimental measurements', *J. Mater. Process. Technol.*, 183, 219–225, 2007a.

6. NISHIKAWA, H., SERIZAWA, H., and MURAKAWA, H., 'Actual application of large-scaled FEM for analysis of mechanical problems in welding', Quart. *J. Jpn. Weld. Soc.*, 24 (2), 168–173, 2006.

7. UEDA, Y. and YUAN, M. G., 'Prediction of residual stresses in butt welded plates using inherent strains', *J. Engrg. Mater. Technol.*, 115 (10), 417–423, 1993.

8. YUAN, M. G. and UEDA, Y., 'Prediction of residual stresses in welded T and I-Joints using inherent strains', *J. Engrg. Mater. Techno.*, 118, 229–234, 1996.

9. MICHALERIS, P. and DEBICCARI, A., 'Prediction of welding distortion', *Welding Journal,* 76 (6), 172–181, 1997.

10. JUNG, H. and TSAI, C.L., 'Plasticity-based distortion analysis for fillet welded thin-plate T-joints', *Welding Journal,* 83 (6), 177–187, 2004.

11. JUNG, H. and TSAI, C.L., 'Fundamental studies on the effect of distortion control planes on angular distortion in fillet welded T-joints', *Welding Journal,* 83 (7), 213–223, 2004.

12. HUAJUN, Z. *et al.*, 'Fundamental studies on in-process controlling angular distortion in asymmetrical double-sided double arc welding', *Journal of materials processing technology*, 205, 214–223, 2008.

13. DEAN, DENG *et al.*, 'Determination of welding deformation in fillet-welded joint by means of numerical simulation and comparison with experimental measurements', *Journal of Materials Processing Technology*, 183, 219–225, 2007.

14. ADAK, M. and MANDAL, N. R., 'Pseudolinear equivalent constant rigidity concept for analyzing welding residual deformation', *Applied Mathematical Modelling*, 33, 2096–2108, 2009.

15. MOLLICONE, P. *et al.*, 'Simple thermo-elastic–plastic models for welding distortion simulation', Journal of Materials Processing Technology, 176, 77–86, 2006.

16. JOSEFSON, B. L. *et al.*, 'Welding residual distortions in ring-stiffened pipes', *J. Offshore Mech. Arctic Eng.* 118, 1996.

17. LINDGREN, L. E., 'Finite element modelling and simulation of welding. Part 1. Increased complexity', *Journal Thermal Stress.*, 24, 141–192, 2001.

18. JIAN-HUA, W. and HAO, L U., 'Prediction of Welding Deformations by FEM Based on Inherent Strains', *Journal of Shanghai Jiaotong University*, 12, 83s - 87s, 2000.

19. UEDA Y, KIM Y C. and YUAN M G., 'A predicting method of welding residual stress using source of residual stress (Report I) – Characteristics of inherent strain (source of residual stress)', *J. Trans of JWRI,* 18(1), 135s – 141s, 1989.

20. CHANG, D. J., HYUN S. R. and CHANG H. L., 'Prediction and Control of Welding Deformations in Stiffened Hull Blocks using Inherent Strain Approach', *Proceedings of the 14th Internaltional offshore and Polar Engineering Conference*, Toulon, France. May 23-28, 2004.

7. AUTHORS BIOGRAPHY

Pankaj Biswas holds the current position of Research Associate in the department of Ocean Engineering and Naval Architecture, Indian Institute of Technology, Kharagpur, India. **Nisith Ranjan Mandal** holds the current position of Professor, Head in the department of Ocean Engineering and Naval Architecture, Indian Institute of Technology, Kharagpur, India.

Analysis of Weld Induced Distortion of Large Curved Stiffened Panels

by

Dr. Pankaj Biswas and Prof. N.R. Mandal

DEPARTMENT OF OCEAN ENGINEERING AND NAVAL ARCHITECTURE
INDIAN INSTITUTE OF TECHNOLOGY

11TH DECEMBER 2009

INTRODUCTION:

Basic structural component of ships & offshore platforms

- ➢ Orthogonally stiffened plate panels

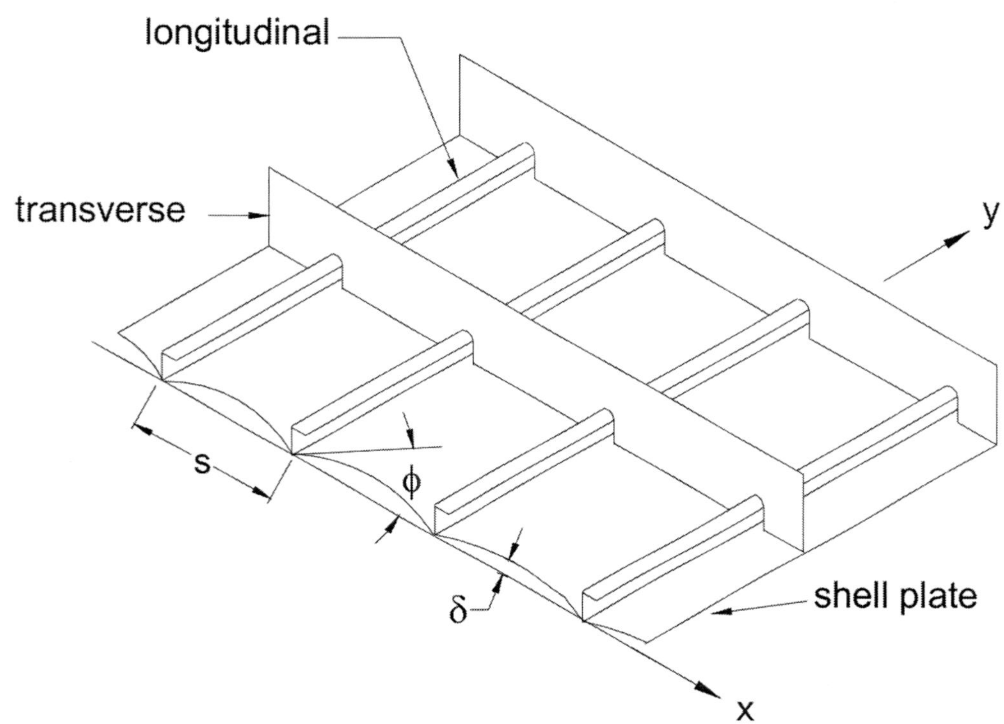

- The joining process of stiffeners is mainly fusion welding. It leads to heat induced distortion.

- These deformations adversely affect the subsequent fitup and alignment of the adjacent panels.

- It may also result in loss of structural integrity. These panels primarily suffer from angular and buckling distortions.

❑ The extent of distortion depends on several parameters

- ✓ Welding speed,
- ✓ Plate thickness,
- ✓ Welding current,
- ✓ Arc voltage,
- ✓ Restraints applied to the job while welding.

Finite Element Inherent strain method

❖Introduction:

- ✓ Generally FE transient nonlinear thermo-mechanical analysis is highly time consumable. It was seen for solving a 300x200x10mm size welding plate the computational time required is around 3 &1/2 hour. So for solving big stiffened panel used in ship, FE transient nonlinear thermo-mechanical is not feasible. For that reason the simplified technique is required to analyze welding distortion.

- ✓ Inherent strain method is simplifying Finite Element method for welding distortion prediction which is computationally feasible and efficient method.

- ✓ Inherent strains are the residual plastic strains caused in welding process. They exist in the welds and in its near vicinity which has undergone a large thermal cycle. These strains are considered as a source causing the welding deformations.

Concept of Inherent strain:

- The total strain incase of welding can be divided into the sum of the elastic strain, the Plastic strain and the thermal strain, respectively,

$$\varepsilon_{total} = \varepsilon^{e} + \varepsilon^{p} + \varepsilon^{th}$$

- Since the thermal strain disappears when the temperature returns to room temperature after the thermal cycle, so the total strain becomes

$$\varepsilon_{total} = \varepsilon^{e} + \varepsilon^{p}$$

- Where elastic strain is produced by residual stress and the total strain corresponds to the residual deformation.

$$\varepsilon_{inherent\ strain} = \varepsilon^{p} = \varepsilon_{residual\ deformation} - \varepsilon_{residual\ stress}$$

❖ So, it is seen, conceptually, that residual stress and residual deformation are produced by inherent strain.

Classification and calculation of inherent strain:

- The inherent strain caused due to the welding can be divided as

  i. Longitudinal inherent strain & **ii.** Transverse inherent strain.

- Calculation of inherent strains:

❖ The total volume of inherent strains in longitudinal and transverse directions is related to rate of heat input q as given below.

$$V_x = Kq \quad , \quad V_y = \beta q \quad \text{and} \quad q = \sum_{i=1}^{n} \frac{\eta V_i I_i}{(ws)_i}$$

Where, V= Welding voltage
I = Welding current
ws= Welding speed
i = No. of welding pass.

Calculation of inherent strains continued:

❖From thick plates to thin plates, coefficients K and β can be introduced as:

$$K = (0.255\ to\ 0.335)\frac{\alpha}{c\rho} \qquad \text{and} \qquad \beta = (0.255\ to\ 1.0)\frac{\alpha}{c\rho}$$

➢ In this study the value of β and K are:

β =2.8 X 10^{-6} cm^3/J & K = 8.6 X 10^{-7} cm^3/J

✓Longitudinal inherent strain = $\varepsilon_L = \frac{V_x}{A_i}$

✓Transverse inherent strain = $\varepsilon_T = \frac{V_y}{A_i}$

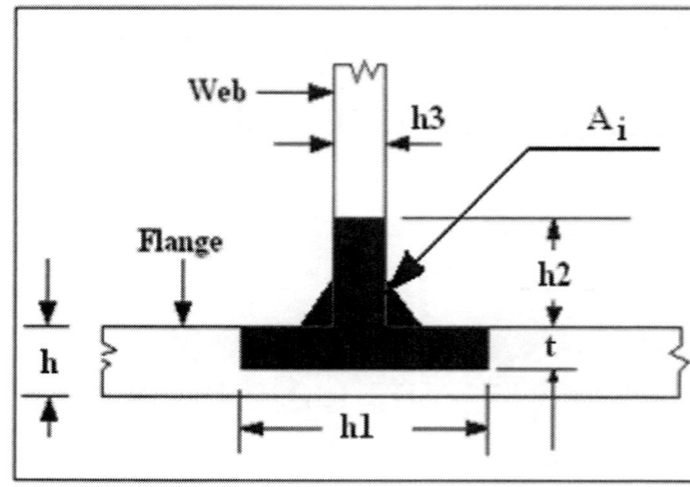

Where Ai is the zone area where longitudinal and transverse inherent strains are distributed. In this study the cross-section area Ai is 0.00027 m2 .

FE transient nonlinear thermo-mechanical analysis

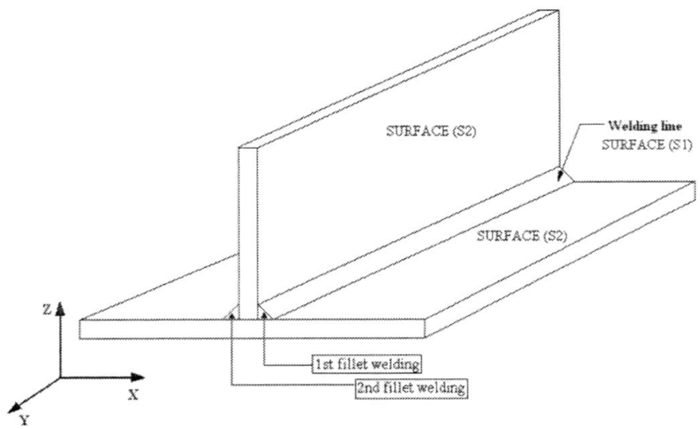

Fig. Axis system and model

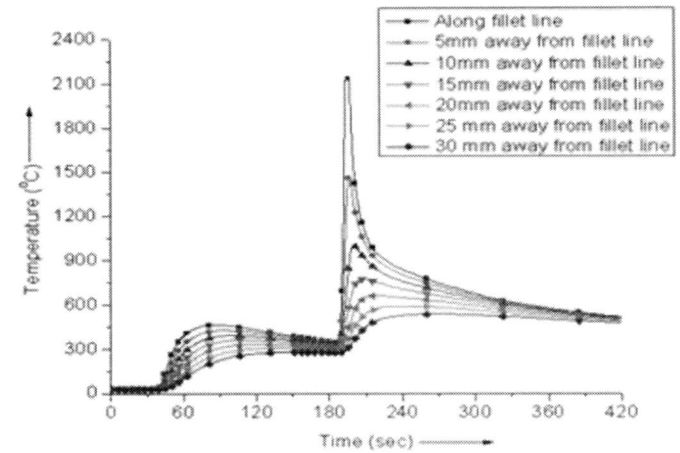

Fig.Temperature distribution on plate top surface in 2nd side

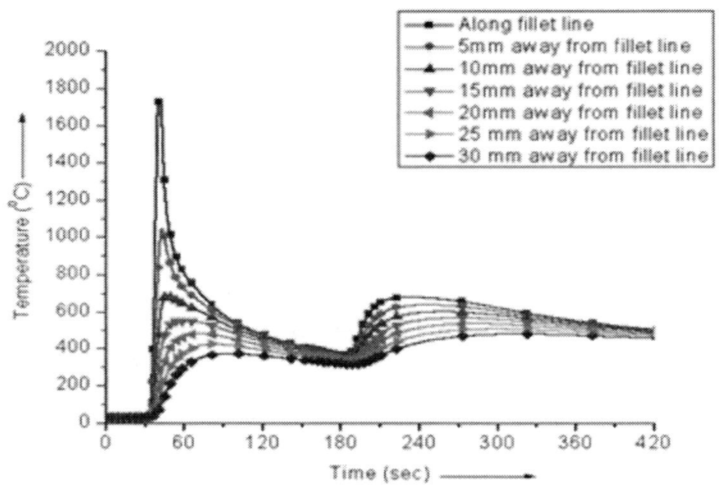

Fig.Temperature distribution on plate top surface in 1st side

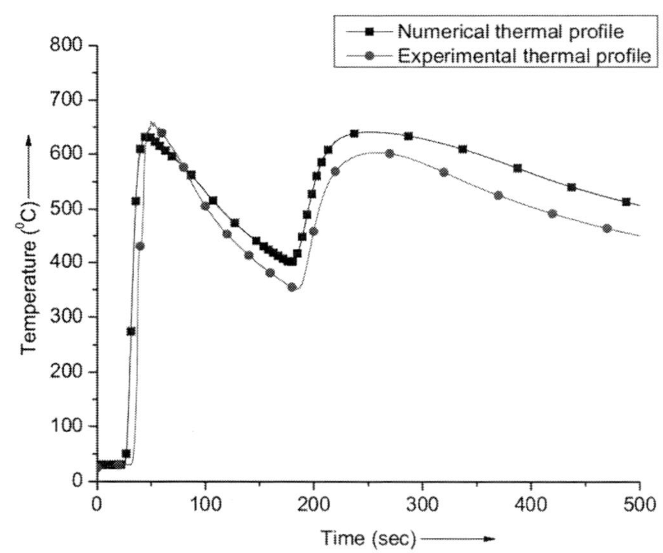

Verification of experimental and numerical distortion:

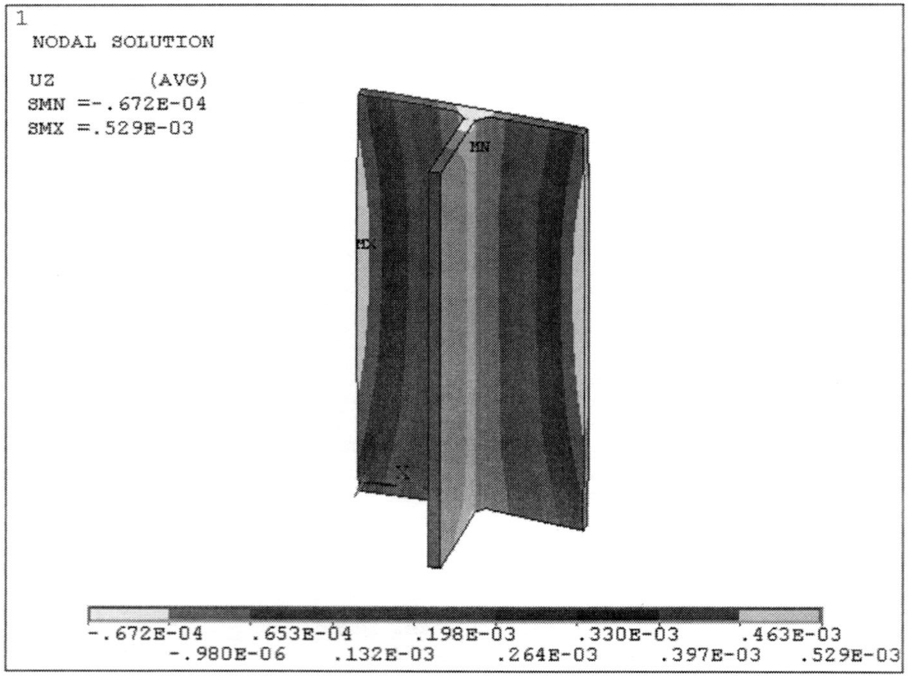

Fig. Distortion pattern for 10mm thick plate

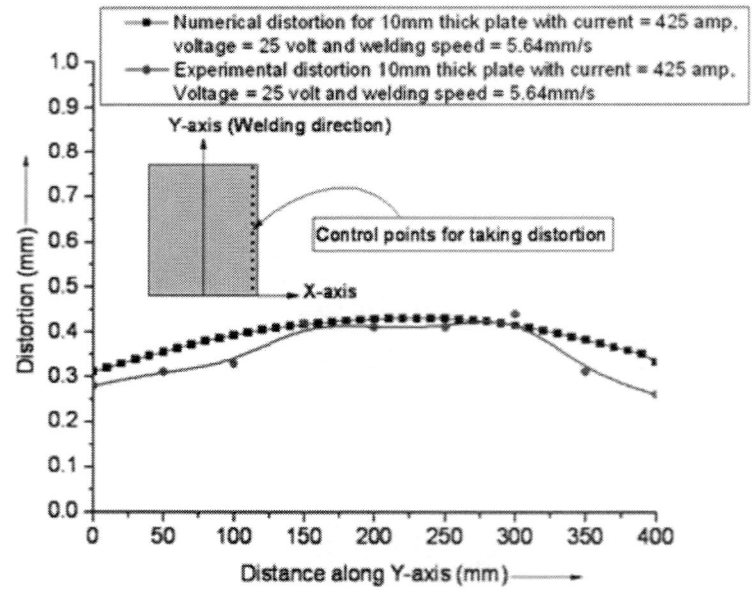

Fig. Experimental and numerical comparison of welding distortion for 10mm thick plate

A COMPARATIVE STUDY BETWEEN INHERENT STRAIN METHOD AND TRANSIENT ELASTO-PLASTIC THERMO-MECHANICAL ANALYSIS

- An orthogonally stiffened panel as used in ship structures was created. However, the dimensions of the panel and the scantling of the stiffeners were kept very small. The panels stiffened by 3 longitudinal members and 2 transverse members.

Table : Plate thickness and welding parameters used in the study

Plate No.	Thickness (mm)	Current (amp)	Voltage (volt)	W.S (m/min)
1	6	165	24.5	0.175

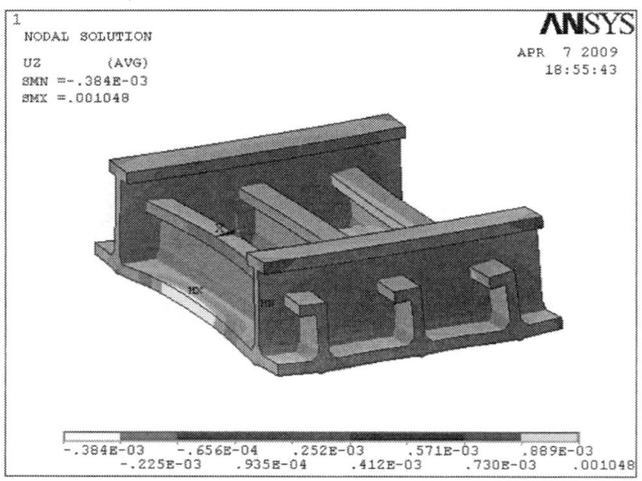

Fig. Deformed shape obtained by transient elasto -plastic thermo-mechanical analysis

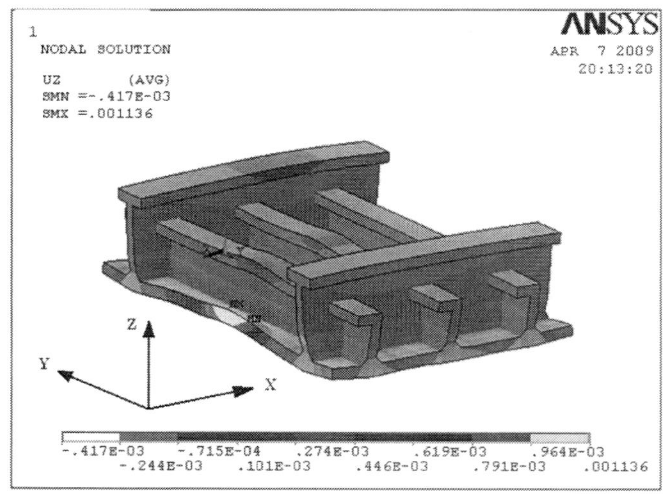

Fig. Deformed shape obtained by inherent strain method

COMPARISON BETWEEN INHERENT STRAIN METHOD AND TRANSIENT ELASTO-PLASTIC THERMO-MECHANICAL ANALYSIS RESULTS:

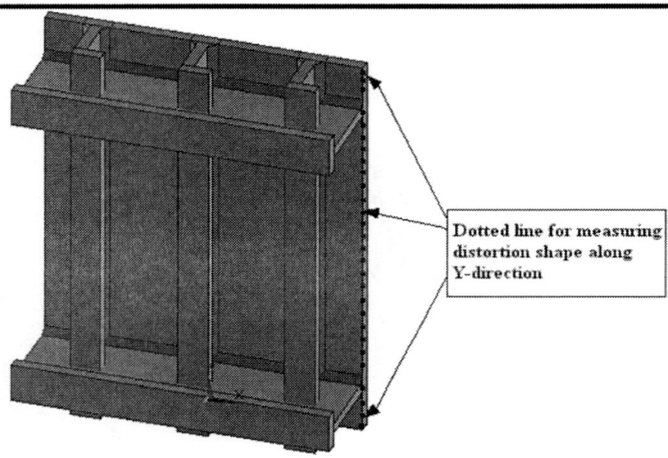

Fig. Control points along the plate edge for measuring welding distortion of stiffened plate panel

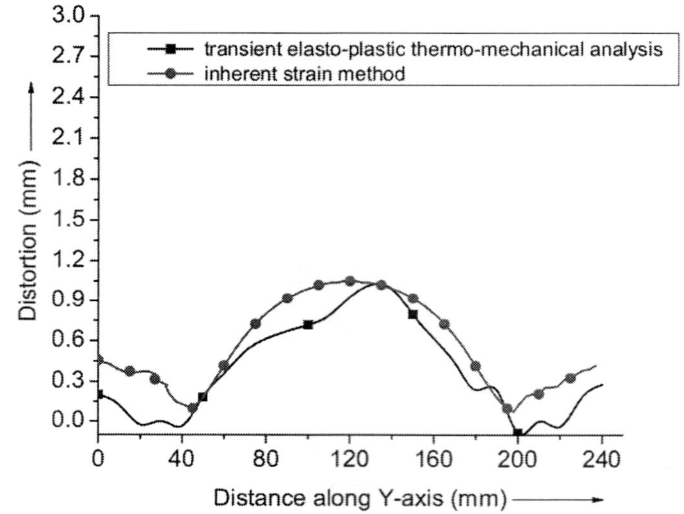

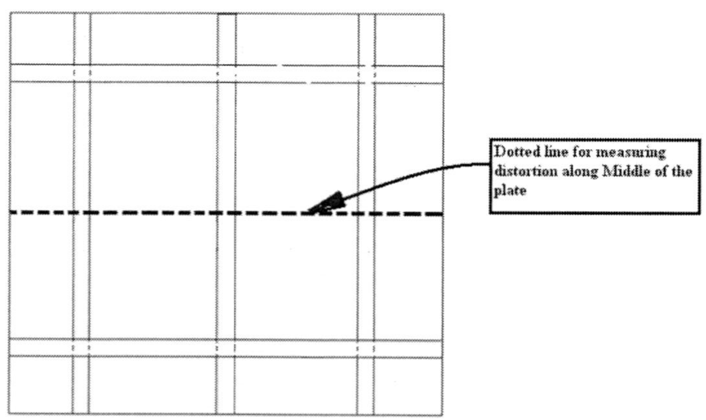

Fig. Control points along the middle of the panel for measuring welding distortion of stiffened plate panel

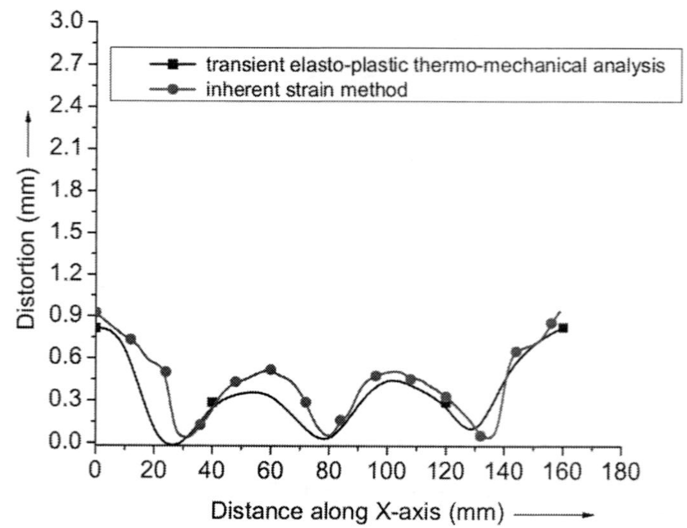

DISCUSSION BETWEEN INHERENT STRAIN METHOD AND TRANSIENT ELASTO-PLASTIC THERMO-MECHANICAL ANALYSIS :

- Of these two methods of analysis, the transient thermo-mechanical analysis can be considered as the one which attempts to consider all aspects of actual process of welding and there gives near exact solution of resulting deformation.

- Whereas in Inherent strain method simplifying assumptions were made to achieve computationally feasible and efficient method of solving this highly nonlinear transient thermo-mechanical problem.

- Of these two methods, the inherent strain method was found to be highly computationally efficient leading to almost 70 times less CPU time compared to that of transient elasto-plastic analysis. At the same time the deviation in the magnitude of distortion was found to vary within 7 to 10% only.

- Hence the inherent strain method enables us to predict weld induced distortions of full size complex structures as used in ship construction.

EXPERIMENTAL VERIFICATION:

- A trapezoidal box was fabricated using 10 mm thick mild steel plates.

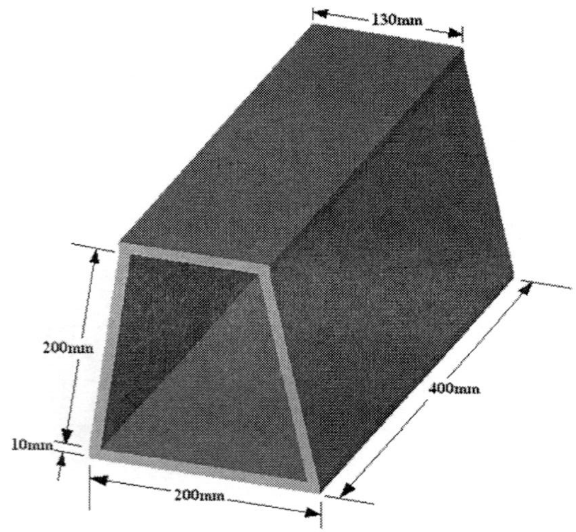

Fig. Dimensions of Trapezoidal Box

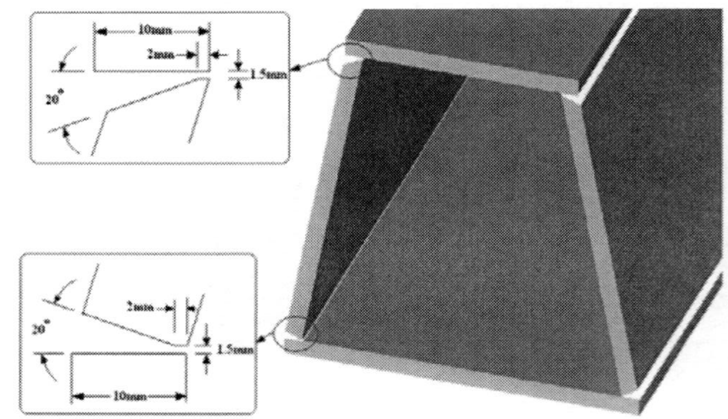

Fig. Details of weld groove for 10mm thick steel plate

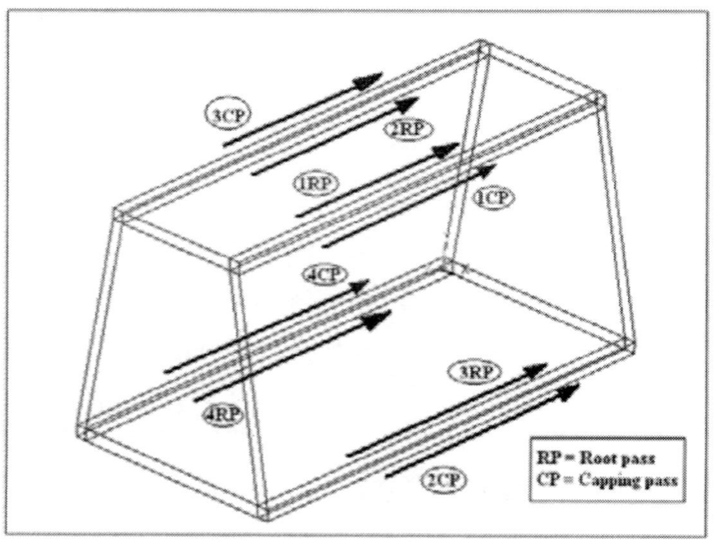

Table : Welding parameters of root and filler pass

Pass type	Current (amp)	Voltage (volt)	Welding speed (mm/min)
Root pass 1	165-172	10.1-10.6	116
Root pass 2	168-174	9.7-10.1	124
Root pass 3	162-170	10.5-10.7	141
Root pass 4	165-174	9.8-10.2	136
Filler pass 1	162-174	10.7-10.9	75.3
Filler pass 2	163-174	11.8-12.4	63.5
Filler pass 3	168-179	11.5-12.1	74.7
Filler pass 4	165-178	12.3-12.5	62.5

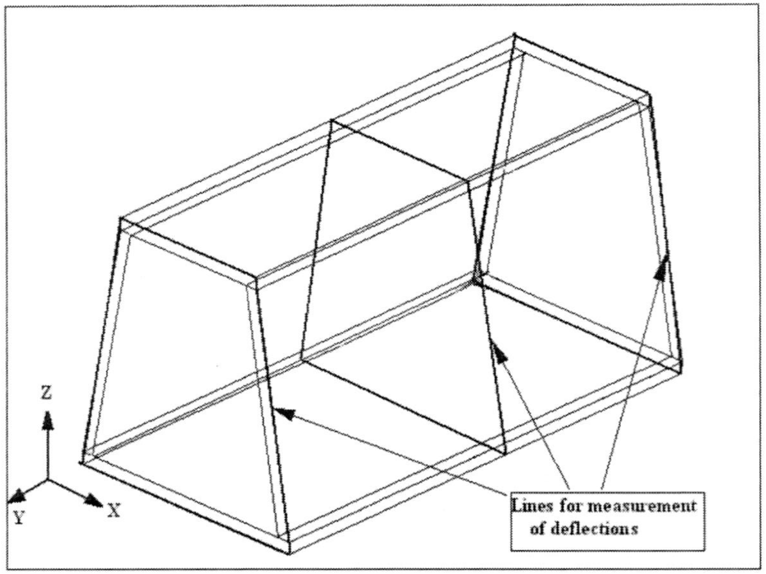

Fig. Lines for measurement of deflections of box

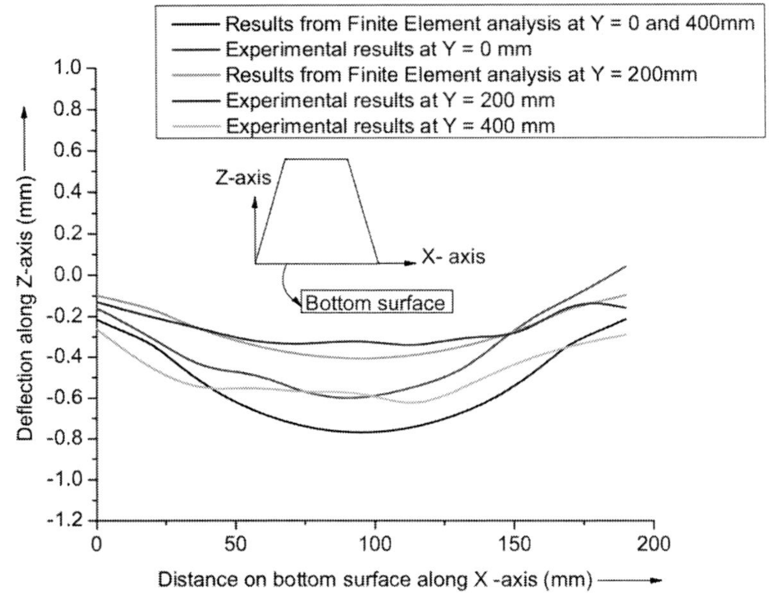

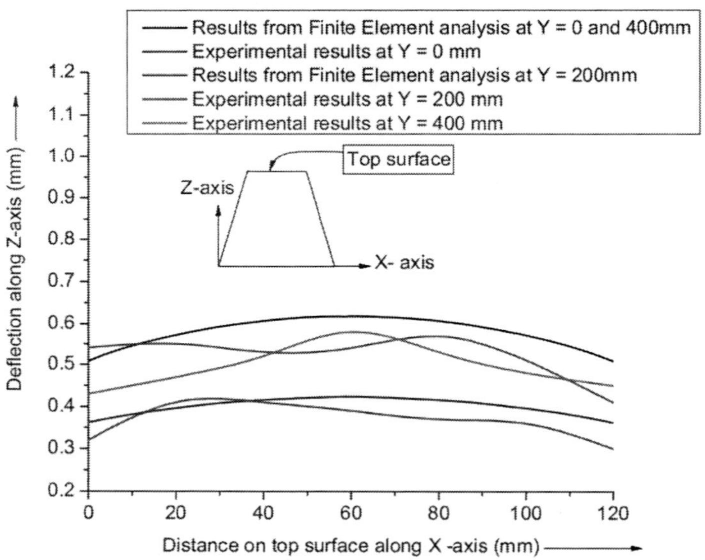

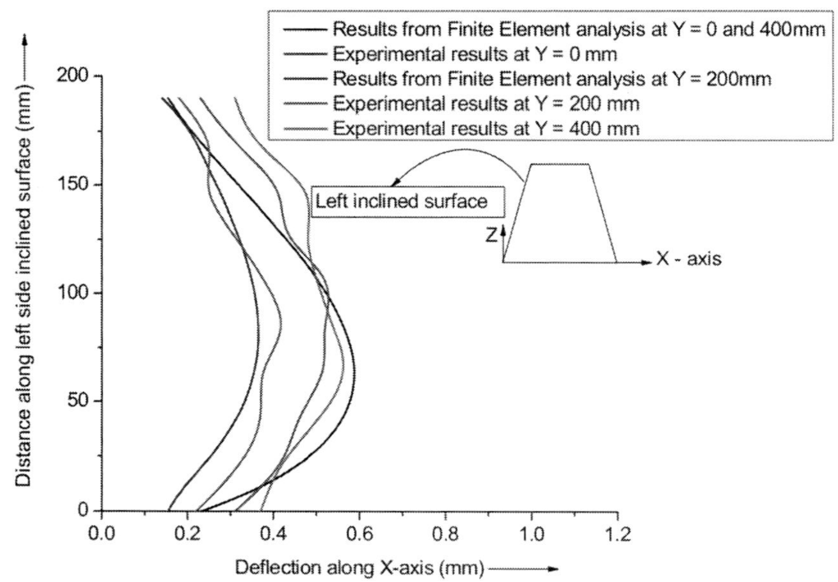

The deviation of magnitude of the deformations varied between about 7 to 15%.

3-D large curved stiffened panels distortion analysis:

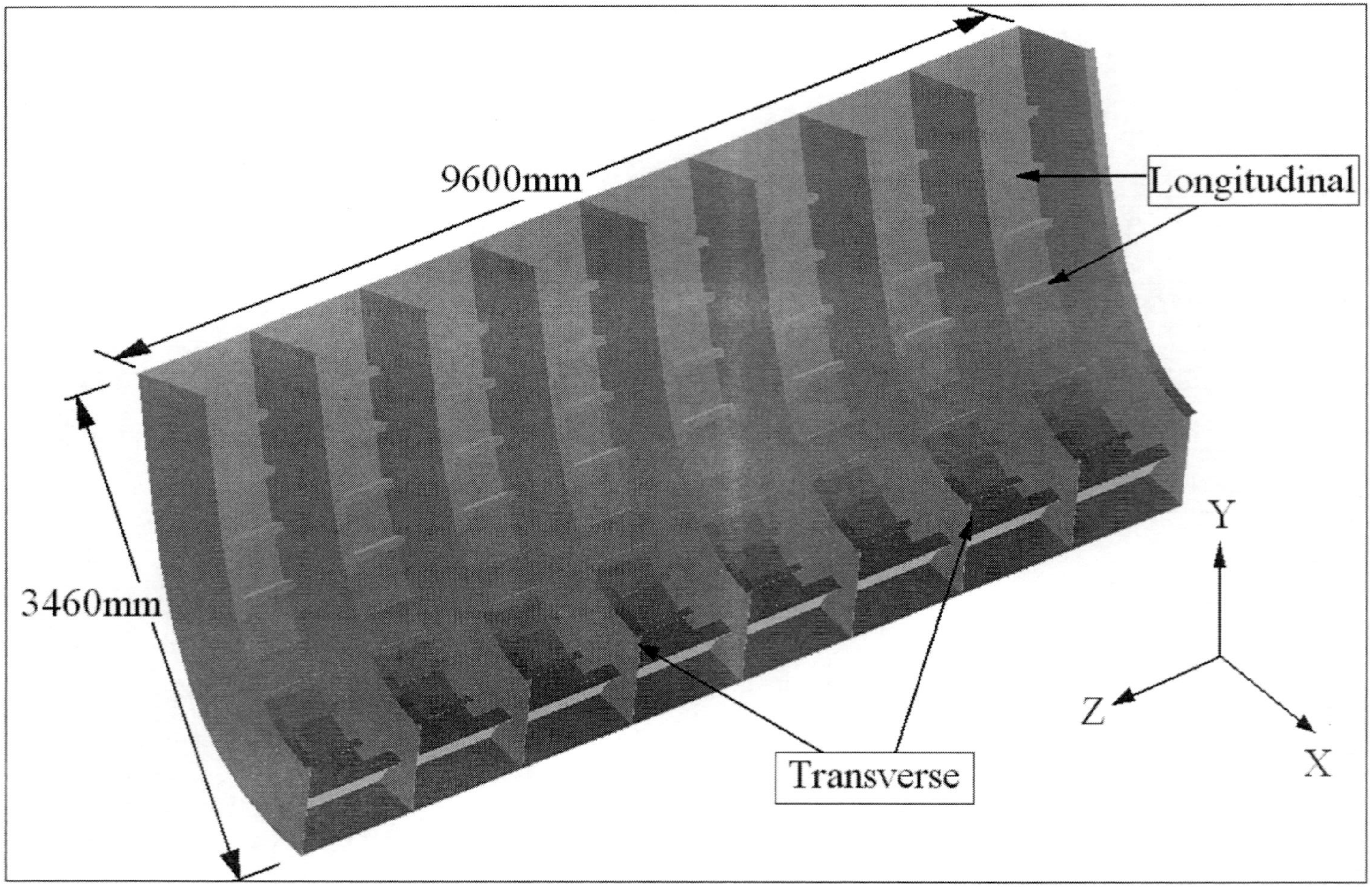

Fig. Detail dimensions of 3-D large curved stiffened panel

Results of 3-D large curved stiffened panels :

Table : Welding parameters used in the study

Plate No.	Thickness (mm)	Current (A)	Voltage (V)	W.S (mm/s)
1	10	210	28.0	2.0

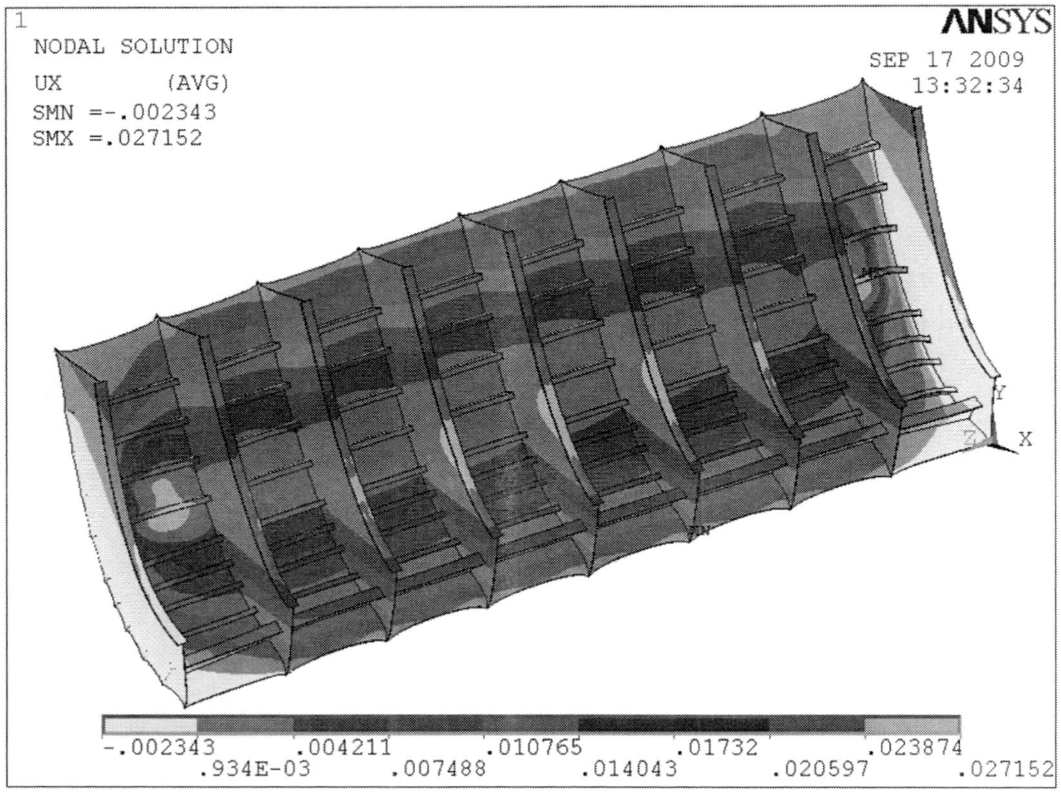

Fig. Deformation shape of UX component

Deform shape 3-D large curved stiffened panels :

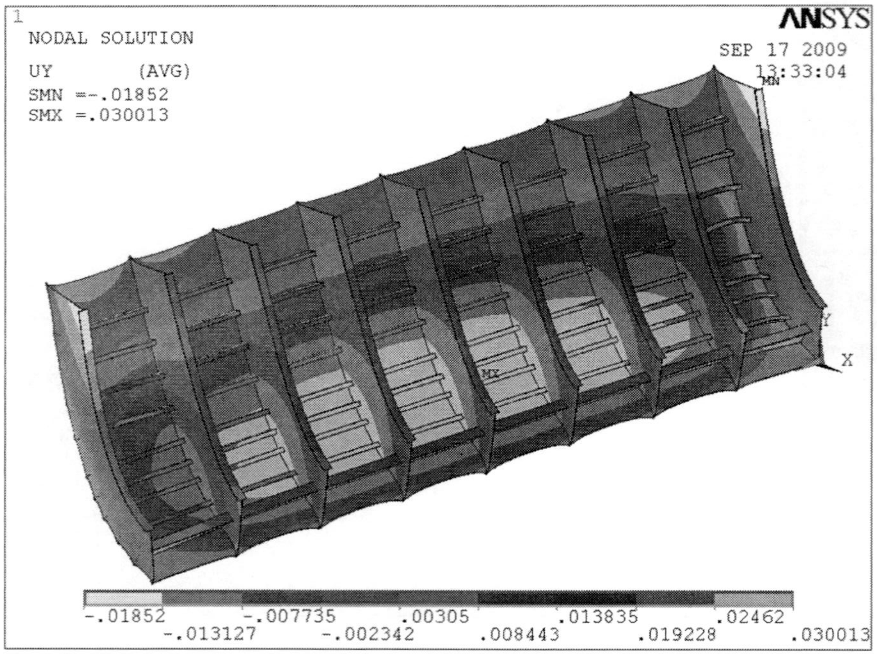

Fig. Deformation shape of UY component

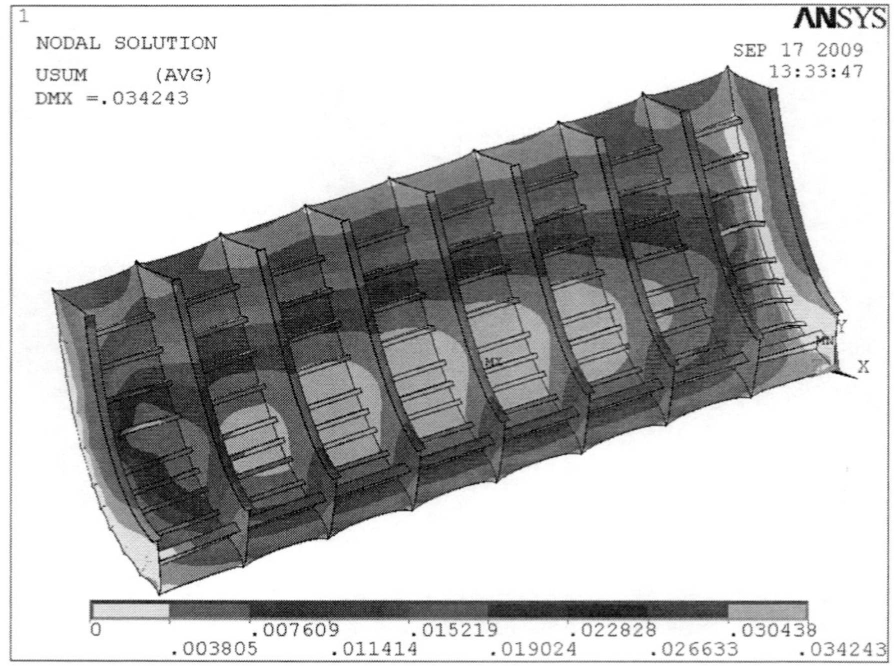

Fig. Overall deformation shape

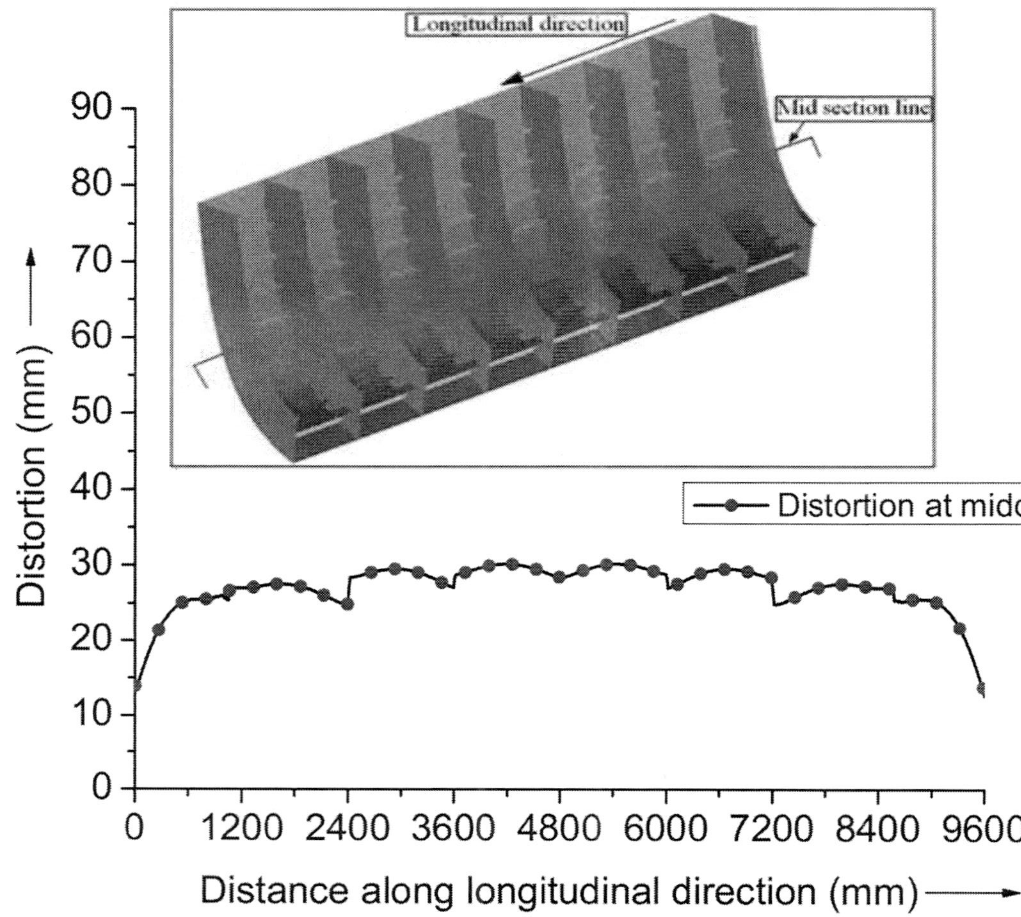

Fig. Distortions at middle line of the stiffened plate panel

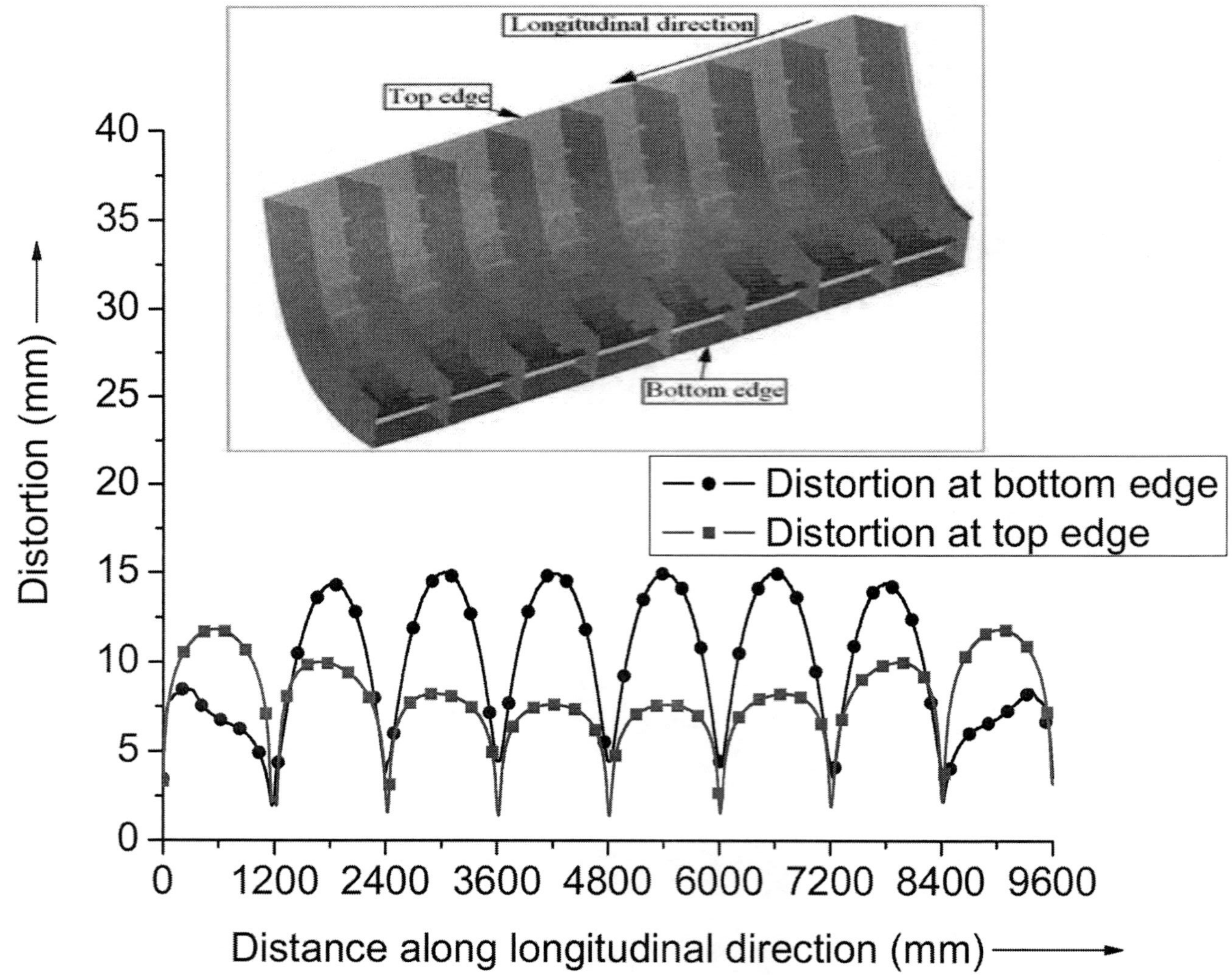

Fig. Distortions at top and bottom edge of the stiffened plate panel

CONCLUSIONS

The following conclusions can be derived from the present investigation:

- ❑ The methodology developed here makes it possible to analyse and predict the distortion pattern and magnitude of the distortion which was otherwise not feasible with the existing FE transient nonlinear elasto-plastic thermo-mechanical modeling approach.
- ❑ It was observed that deformation was more near the middle region of the stiffened panel. The thermal stresses generated were strong enough to deform the transverses at these regions.
- ❑ The deflection at upper edge was less compared to that of at the bottom edge. However both were less than that observed at the middle region of the stiffened panel.
- ❑ The deformation pattern was found to be convex in nature looking from the stiffener side.
- ❑ The computational methodology developed here was found to be very promising in analyzing large complex welded structures for overall distortion prediction.
- ❑ The computational time required was about 30 to 40 minutes only in a Core (TM) 2 Duo, 2.66GHz computer with 2 GB RAM.

NAMASKAR

Strength of Stiffened Composite Plates with Opening

Authors
P. Alagusundaramoorthy
S. Rahima Shabeen

Introduction

- Stiffened plates form the basic components of ships, off-shore structures and are subjected to out-of-plane load due to cargo and water pressure.
- Openings are provided in the plates for access and service.
- Advanced composites offer advantages of high strength to weight ratio, high stiffness to weight ratio, corrosion resistance and resistance to many degrading chemicals.
- The plates with bonded stiffeners fail by flange plate-stiffener debonding mode of failure under out-of-plane load before reaching their full capacity. This mode of failure can be avoided by having a few layers of fabric continuous between the flange plate and the stiffeners.

Objective

- The ultimate strength of stiffened composite plates with integral stiffeners with and without opening under out-of-plane load is studied in this paper.

Test Specimen

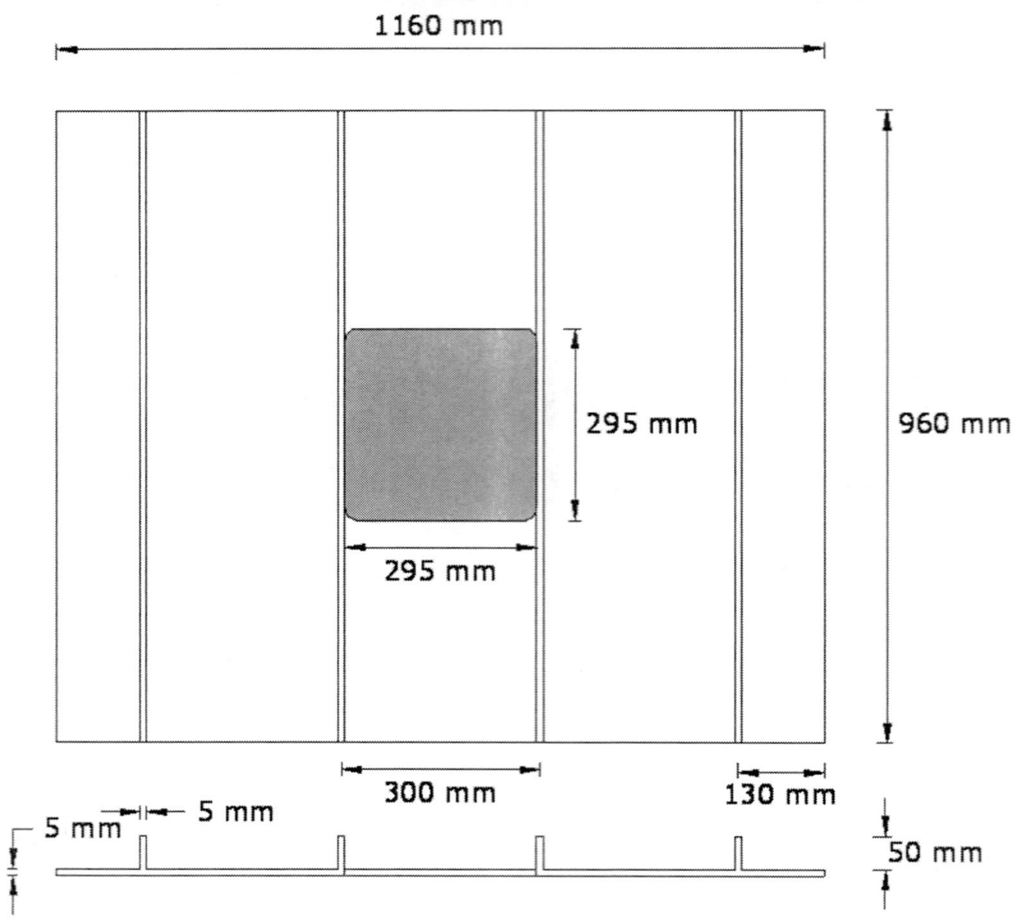

Properties of GFRP

S. No	Property	Value
1	Tensile strength in warp direction	250 MPa
2	Tensile modulus in warp direction	15800 MPa
3	Tensile strength in weft direction	211 MPa
4	Tensile modulus in weft direction	15333 MPa
5	Compressive strength	135 MPa
6	Major Poisson's ratio	0.139
7	Minor Poisson's ratio	0.125
8	Flexural strength	384 MPa
9	Flexural modulus	15388 MPa

Fabrication of Specimens

1. Application of Releasing Agent

2. Application of Resin

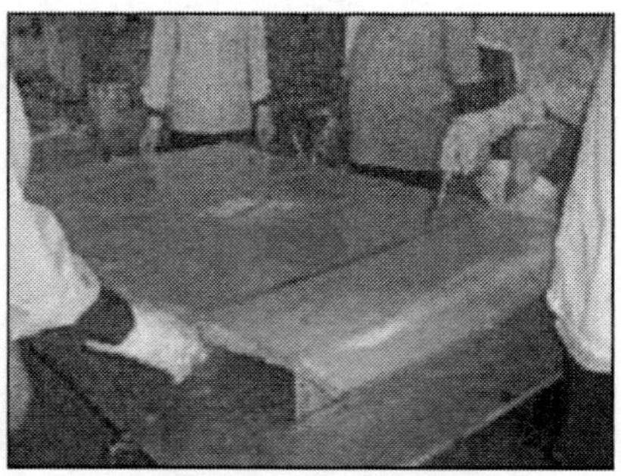

3. Laying of WRM Fabric

4. Removal of Air Voids using Rollers

Fabrication by Hand Lay-up Process

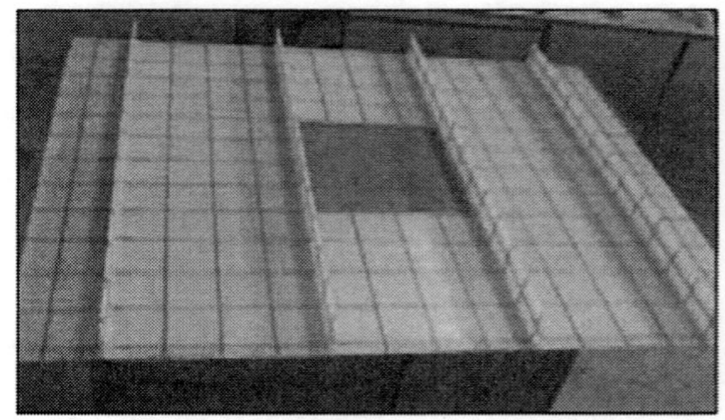

Stiffened Composite Plate

Test Set-up

A- Self-straining test rig, B- Test specimen C- Imperfection measurement set-up

Loading Set-up

Inflatable Air Balloon

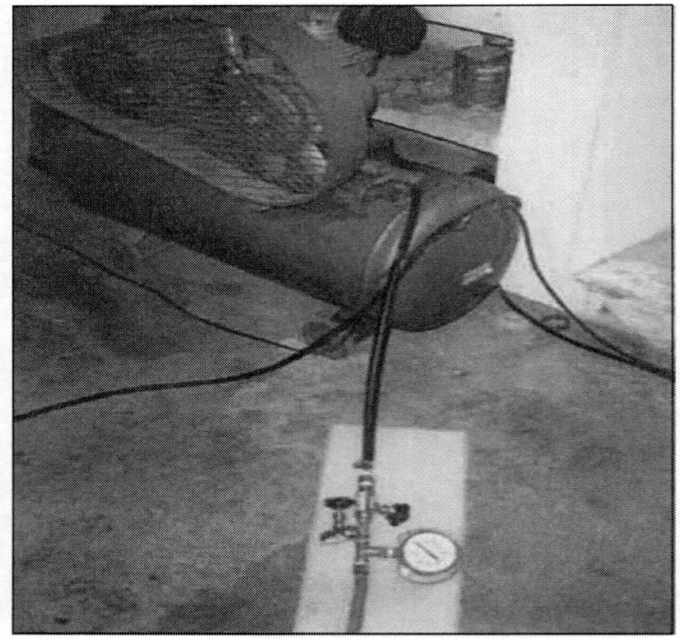

Compressor and Pressure Gauge

Initial Imperfection Profile

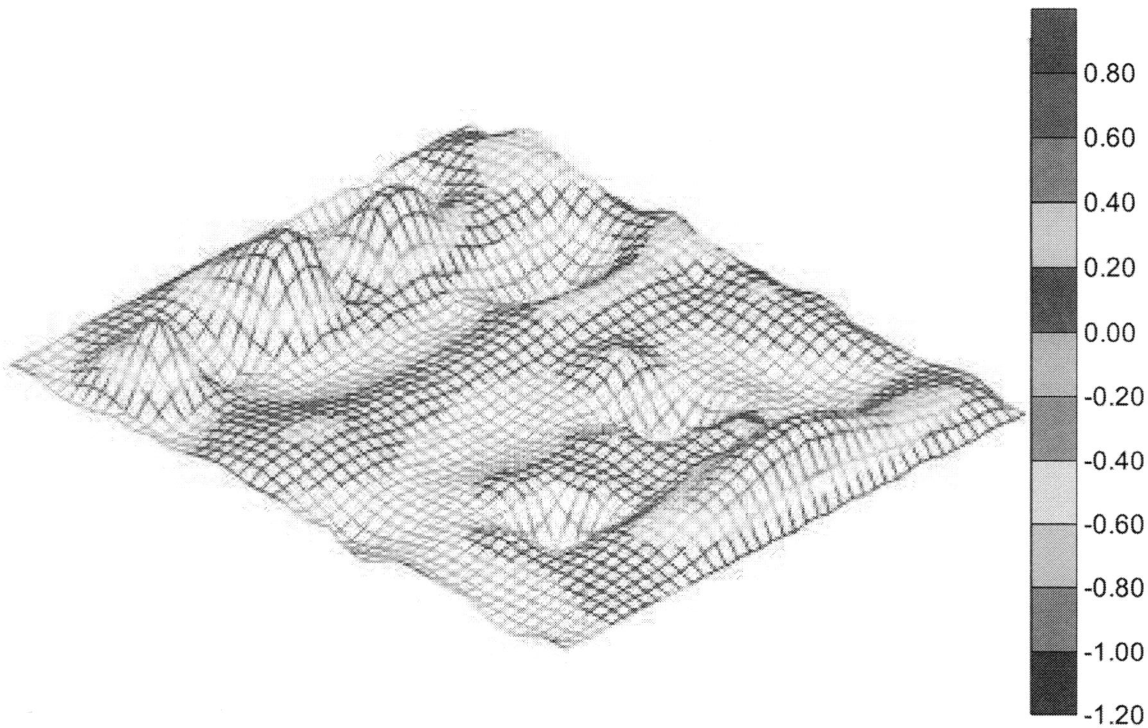

Finite Element Model of Stiffened Composite Plate

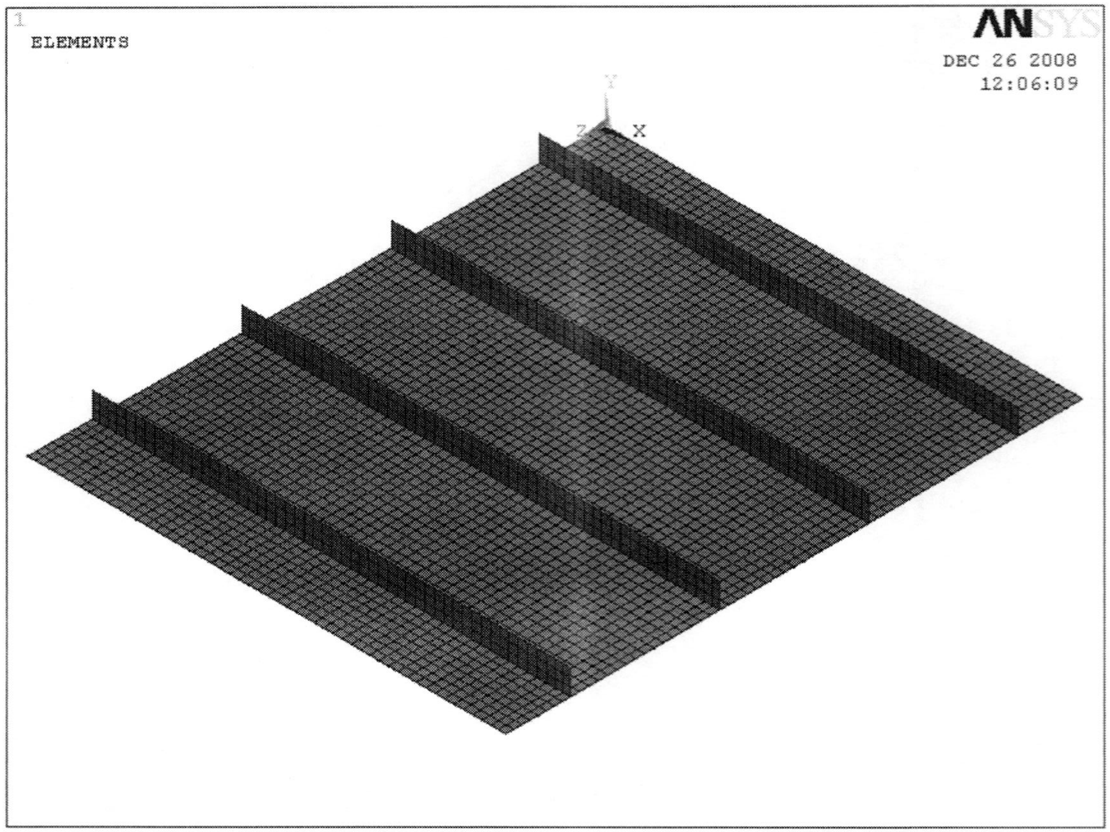

Ultimate Load and Mode of Failure

Specimen	Ultimate Load (kN)		P_{FEA}/P_E	Mode of Failure
	Expt. P_E	FEA P_{FEA}		
SCP	61	56	0.914	Stiffener Tension
SCPO	56	52	0.929	Stiffener Tension

Failure of SCP by Rupture of fabric

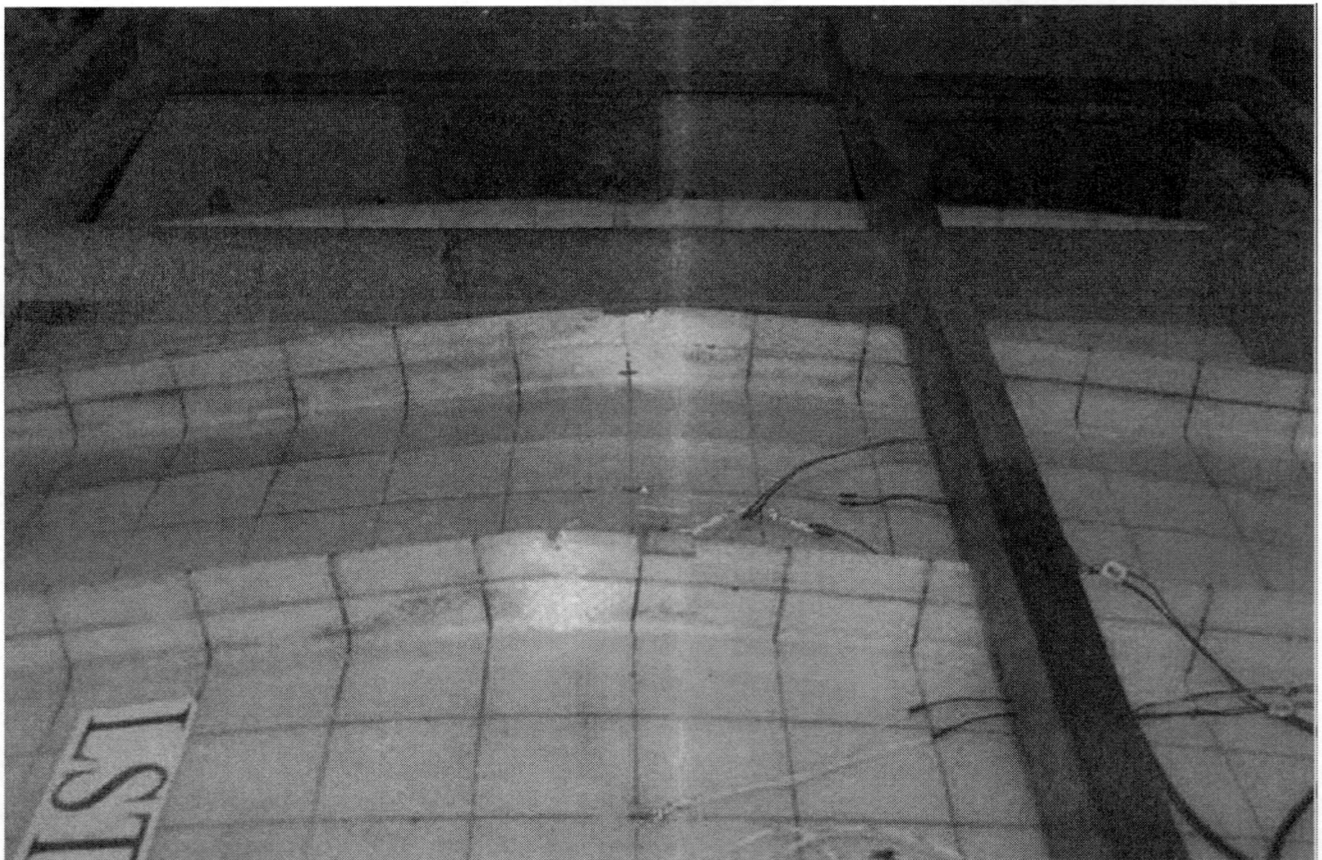

Failure of SCPO by Rupture of Fabric

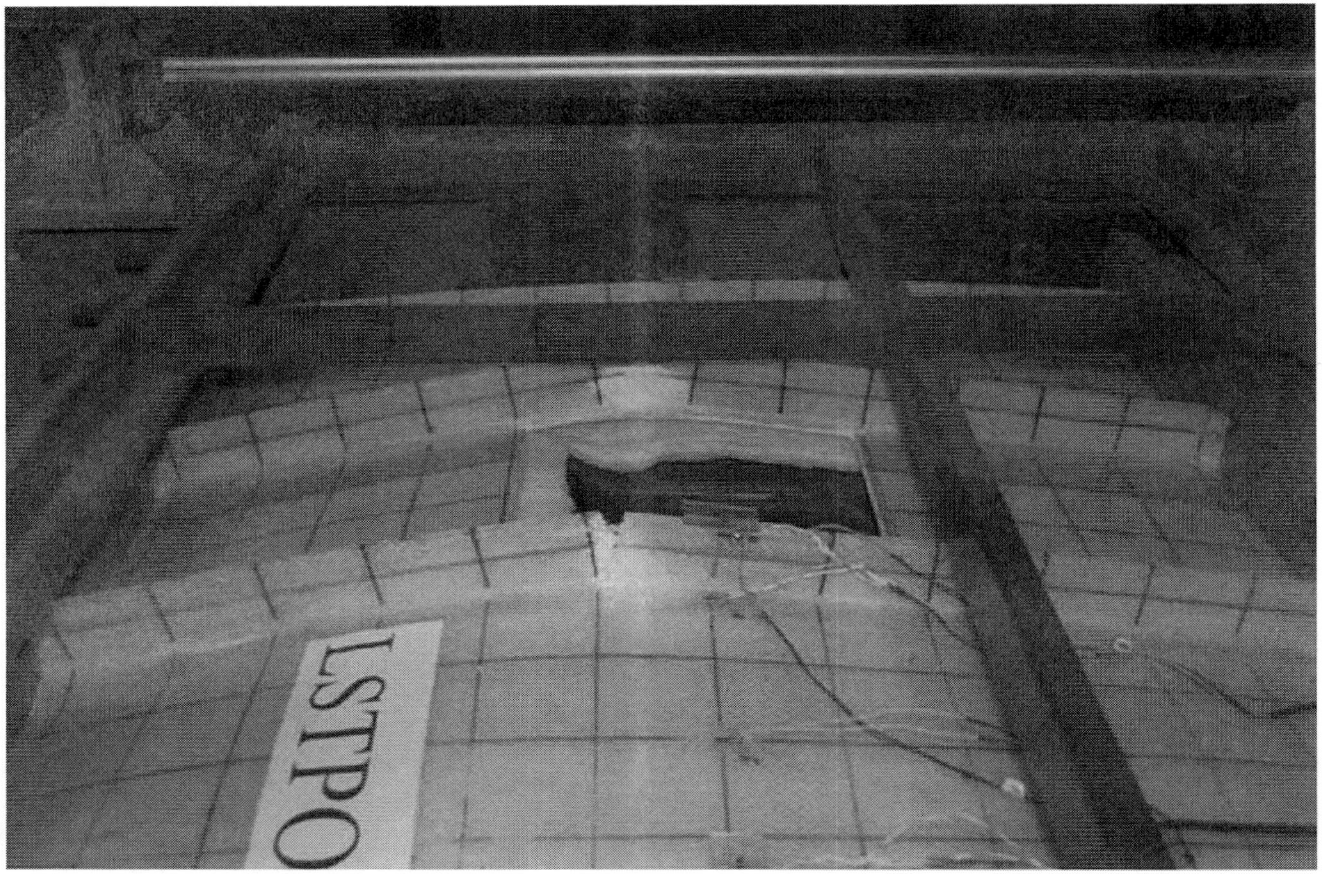

11/12/2009 ICSOT 2009: Development in Ship Design and Construction

Stress Contours of SCP at Failure

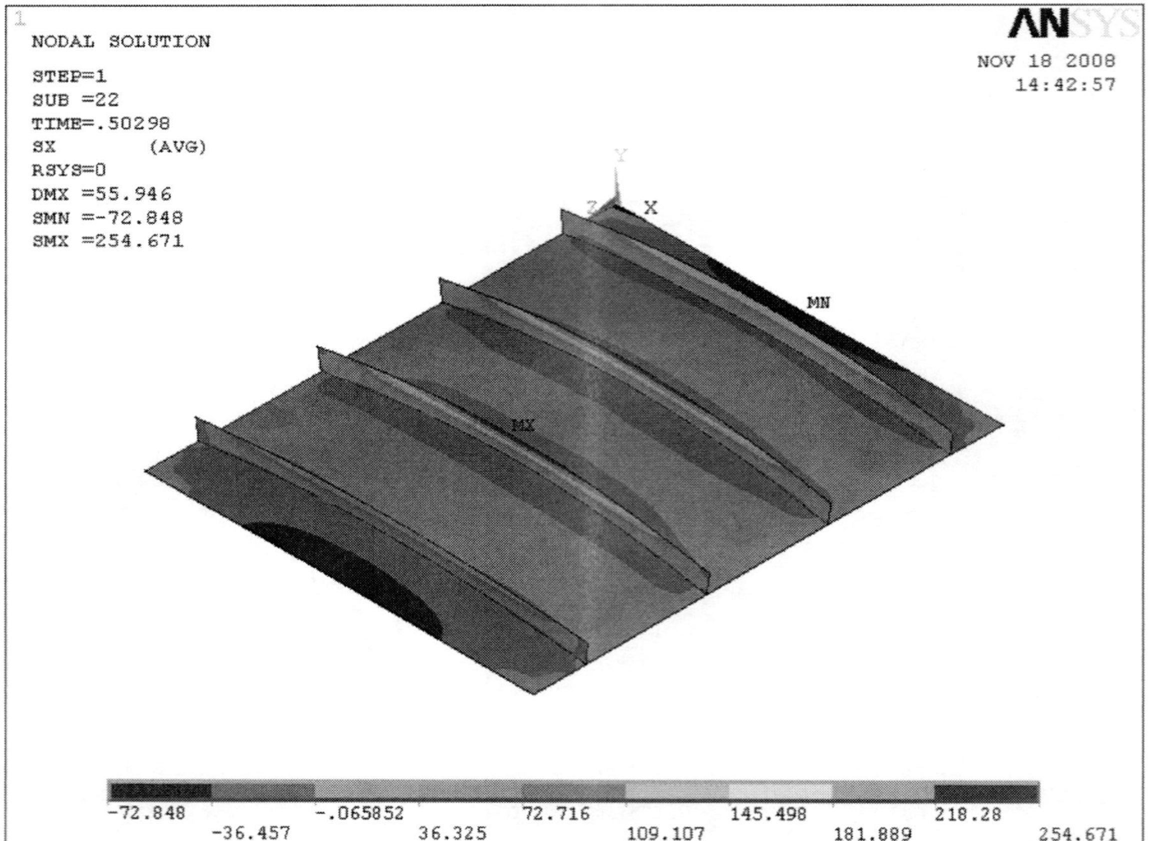

Stress Contours of SCPO at Failure

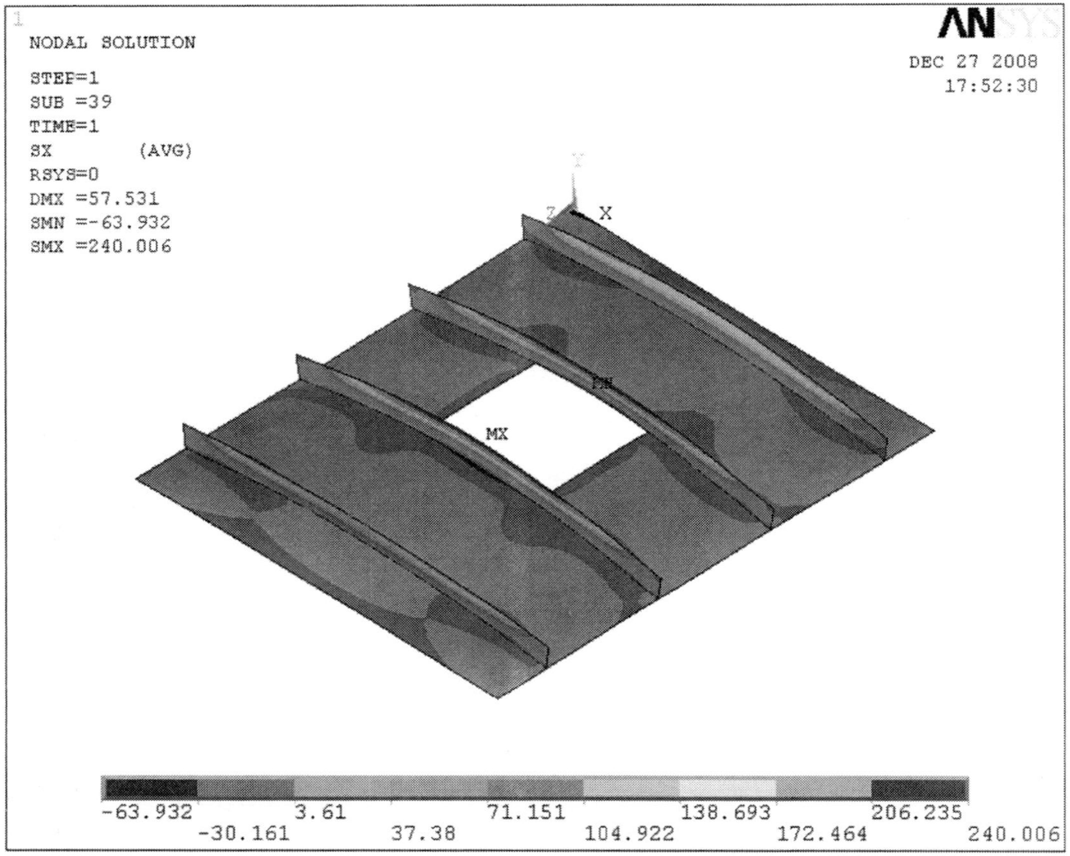

Effect of reinforcement around Opening

Specimen	Ultimate Load (kN)	Reduction in Strength	Regain in Strength
SCP	61	-	-
SCPO	56	8 %	-
SCPRO	60	-	7 %

Conclusions

- The presence of square opening reduces the out-of-plane load carrying capacity of stiffened composite plates by about 8 %.
- Flange plate-stiffener debonding mode of failure can be avoided by adopting the special lay-up structure.
- The developed FE model can be used for the analysis of stiffened composite plates with and without opening.
- The loss in strength of stiffened composite plate due to the presence of opening can be regained by slightly increasing the size of the longitudinal stiffeners adjacent to the opening.

References

- "Experimental and theoretical investigations on stiffened and unstiffened panels under uniform transverse loading." Composite Structures, 63, 293-304.
- Kong, C. W., Lee. I. C., Kim. C. G. and Hong. C. S. (1998). "Postbuckling and failure of stiffened composite panels under axial compression." Composite Structures, 42, 13-21.
- Rahima Shabeen, S. (2009). "Ultimate strength of stiffened composite plates." PhD thesis, Department of Civil Engineering, IIT Madras, India.

Thank you

AUTHOR INDEX